THEORY AND CALCULATION OF
TRANSIENT ELECTRIC PHENOMENA
AND OSCILLATIONS

Theory And Calculation

Of

Transient Electric Phenomena
And Oscillations

By

Charles Proteus Steinmetz

Third Edition
Reviewed and Enlarged
Sixth Impression

McGraw-Hill Book Company, Inc.
1920
Wexford College Press
2003

THE MAPLE PRESS · YORK PA

DEDICATED

TO THE

MEMORY OF MY FRIEND AND TEACHER

RUDOLF EICKEMEYER

PREFACE TO THE THIRD EDITION

SINCE the appearance of the first edition, ten years ago, the study of transients has been greatly extended and the term "transient" has become fully established in electrical literature. As the result of the increasing importance of the subject and our increasing knowledge, a large part of this book had practically to be rewritten, with the addition of much new material, especially in Sections III and IV.

In Section III, the chapters on "Final Velocity of the Electric Field" and on "High-frequency Conductors" have been rewritten and extended.

As Section V, an entirely new section has been added, comprising six new chapters.

The effect of the finite velocity of the electric field, that is, the electric radiation in creating energy components of inductance and of capacity and thereby effective series and shunt resistances is more fully discussed. These components may assume formidable values at such high frequencies as are not infrequent in transmission circuits, and thereby dominate the phenomena. These energy components and the equations of the unequal current distribution in the conductor are then applied to a fuller discussion of high-frequency conduction.

In Section IV, a chapter has been added discussing the relation of the common types of currents: direct current, alternating current, etc., to the general equations of the electric circuit. A discussion is also given of the interesting case of a direct current with distributed leakage, as such gives phenomena analogous to wave propagation, such as reflection, etc., which are usually familiar only with alternating or oscillating currents.

A new chapter is devoted to impulse currents, as a class of non-periodic but transient currents reciprocal to the periodic but permanent alternating currents.

Hitherto in theoretical investigations of transients, the circuit constants r L C and g have been assumed as constant. This, however, disagrees with experience at very high frequencies or steep wave fronts, thereby limiting the usefulness of the theoretical investigation, and makes the calculation of many important phenomena, such as the determination of the danger zone of steep wave fronts, the conditions of circuit design limiting the danger zone, etc., impossible. The study of these phenomena has been undertaken and four additional chapters devoted to the change of circuit constants with the frequency, the increase of attenuation constant resulting therefrom, and the degeneration, that is rounding off of complex waves, the flattening of wave fronts with the time and distance of travel, etc., added.

The method of symbolic representation has been changed from the time diagram to the crank diagram, in accordance with the international convention, and in conformity with the other books; numerous errors of the previous edition corrected, etc.

CHARLES P. STEINMETZ.

Jan., 1920.

PREFACE TO THE FIRST EDITION

THE following work owes its origin to a course of instruction given during the last few years to the senior class in electrical engineering at Union University and represents the work of a number of years. It comprises the investigation of phenomena which heretofore have rarely been dealt with in text-books but have now become of such importance that a knowledge of them is essential for every electrical engineer, as they include some of the most important problems which electrical engineering will have to solve in the near future to maintain its thus far unbroken progress.

A few of these transient phenomena were observed and experimentally investigated in the early days of electrical engineering, for instance, the building up of the voltage of direct-current generators from the remanent magnetism. Others, such as the investigation of the rapidity of the response of a compound generator or a booster to a change of load, have become of importance with the stricter requirements now made on electric systems. Transient phenomena which were of such short duration and small magnitude as to be negligible with the small apparatus of former days have become of serious importance in the huge generators and high power systems of to-day, as the discharge of generator fields, the starting currents of transformers, the short-circuit currents of alternators, etc. Especially is this the case with two classes of phenomena closely related to each other: the phenomena of distributed capacity and those of high frequency currents. Formerly high frequency currents were only a subject for brilliant lecture experiments; now, however, in the wireless telegraphy they have found an important industrial use. Telephony has advanced from the art of designing elaborate switchboards to an engineering science, due to the work of M. I. Pupin and others, dealing with the fairly high frequency of sound waves. Especially lightning and all the kindred high voltage and high frequency phenomena in electric systems have become of great and still rapidly increasing importance, due to the great increase in extent and in power of the modern electric systems, to the interdependence of all the electric power users in a large territory, and to the destructive capabilities resulting from such disturbances. Where hundreds of miles of high and medium potential circuits, overhead lines and underground cables, are interconnected, the phenomena of distributed capacity, the effects of charging currents of lines and cables, have become such as to require careful study. Thus phenomena which once were of scientific interest only, as the unequal current distribution in conductors carrying alternating currents, the finite velocity of propagation of the electric field, etc., now require careful study by the electrical engineer, who meets them in the rail return of the single-phase railway, in the effective impedance interposed to the lightning discharge on which the safety of the entire system depends, etc.

The characteristic of all these phenomena is that they are transient functions of the independent variable, time or distance, that is, decrease with increasing value of the independent variable, gradually or in an oscillatory manner, to zero at infinity, while the functions representing the steady flow of electric energy are constants or periodic functions.

While thus the phenomena of alternating currents are represented by the periodic function, the sine wave and its higher harmonics or overtones, most of the transient phenomena lead to a function which is the product of exponential and trigonometric terms, and may be called an oscillating function, and its overtones or higher harmonics.

A second variable, distance, also enters into many of these phenomena; and while the theory of alternating-current apparatus and phenomena usually has to deal only with functions of one independent variable, time, which variable is eliminated by the introduction of the complex quantity, in this volume we have frequently to deal with functions of time and of distance.

We thus have to consider alternating functions and transient functions of time and of distance.

The theory of alternating functions of time is given in "Theory and Calculation of Alternating Current Phenomena." Transient functions of time are studied in the first section of the present work, and in the second section are given periodic transient phenomena, which have become of industrial importance, for instance, in rectifiers, for circuit control, etc. The third section gives the theory of phenomena which are alternating in time and transient in distance, and the fourth and last section gives phenomena transient in time and in distance.

To some extent this volume can thus be considered as a continuation of "Theory and Calculation of Alternating Current Phenomena."

In editing this work, I have been greatly assisted by Prof. O. Ferguson, of Union University, who has carefully revised the manuscript, the equations and the numerical examples and checked the proofs, so that it is hoped that the errors in the work are reduced to a minimum.

Great credit is due to the publishers and their technical staff for their valuable assistance in editing the manuscript and for the representative form of the publication they have produced.

CHARLES P. STEINMETZ.

SCHENECTADY, December, 1908.

PREFACE TO THE SECOND EDITION

Due to the relatively short time which has elapsed since the appearance of the first edition, no material changes or additions were needed in the preparation of the second edition. The work has been carefully perused and typographical and other errors, which had passed into the first edition, were eliminated. In this, thanks are due to those readers who have drawn my attention to errors.

Since the appearance of the first edition, the industrial importance of transients has materially increased, and considerable attention has thus been devoted to them by engineers. The term "transient" has thereby found an introduction, as noun, into the technical language, instead of the more cumbersome expression "transient phenomenon," and the former term is therefore used to some extent in the revised edition.

As appendix have been added tables of the velocity functions of the electric field, sil x and col x, and similar functions, together with explanation of their mathematical relations, as tables of these functions are necessary in calculations of wave propagation, but are otherwise difficult to get. These tables were derived from tables of related functions published by J. W. L. Glaisher, Philosophical Transactions of the Royal Society of London, 1870, Vol. 160.

CONTENTS

SECTION I. TRANSIENTS IN TIME.

SECTION II. PERIODIC TRANSIENTS.

SECTION IV. TRANSIENTS IN TIME AND SPACE.

SECTION I

TRANSIENTS IN TIME

TRANSIENTS IN TIME

CHAPTER I.

THE CONSTANTS OF THE ELECTRIC CIRCUIT.

1. To transmit electric energy from one place where it is
generated to another place where it is used, an electric circuit is required, consisting of conductors which connect the
point of generation with the point of utilization.

When electric energy flows through a circuit, phenomena
take place inside of the conductor as well as in the space outside of the conductor.

In the conductor, during the flow of electric energy through
the circuit, electric energy is consumed continuously by being
converted into heat. Along the circuit, from the generator
to the receiver circuit, the flow of energy steadily decreases
by the amount consumed in the conductor, and a power gradient exists in the circuit along or parallel with the conductor.

(Thus, while the voltage may decrease from generator to
receiver circuit, as is usually the case, or may increase, as in
an alternating-current circuit with leading current, and while
the current may remain constant throughout the circuit, or
decrease, as in a transmission line of considerable capacity
with a leading or non-inductive receiver circuit, the flow of
energy always decreases from generating to receiving circuit,
and the power gradient therefore is characteristic of the direction of the flow of energy.)

In the space outside of the conductor, during the flow of
energy through the circuit, a condition of stress exists which
is called the *electric field* of the conductor. That is, the
surrounding space is not uniform, but has different electric
and magnetic properties in different directions.

No power is required to maintain the electric field, but energy is required to produce the electric field, and this energy is returned, more or less completely, when the electric field disappears by the stoppage of the flow of energy.

Thus, in starting the flow of electric energy, before a permanent condition is reached, a finite time must elapse during which the energy of the electric field is stored, and the generator therefore gives more power than consumed in the conductor and delivered at the receiving end; again, the flow of electric energy cannot be stopped instantly, but first the energy stored in the electric field has to be expended. As result hereof, where the flow of electric energy pulsates, as in an alternating-current circuit, continuously electric energy is stored in the field during a rise of the power, and returned to the circuit again during a decrease of the power.

The electric field of the conductor exerts magnetic and electrostatic actions.

The magnetic action is a maximum in the direction concentric, or approximately so, to the conductor. That is, a needle-shaped magnetizable body, as an iron needle, tends to set itself in a direction concentric to the conductor.

The electrostatic action has a maximum in a direction radial, or approximately so, to the conductor. That is, a light needle-shaped conducting body, if the electrostatic component of the field is powerful enough, tends to set itself in a direction radial to the conductor, and light bodies are attracted or repelled radially to the conductor.

Thus, the electric field of a circuit over which energy flows has three main axes which are at right angles with each other:

The electromagnetic axis, concentric with the conductor.

The electrostatic axis, radial to the conductor.

The power gradient, parallel to the conductor.

This is frequently expressed pictorially by saying that the lines of magnetic force of the circuit are concentric, the lines of electrostatic force radial to the conductor.

Where, as is usually the case, the electric circuit consists of several conductors, the electric fields of the conductors superimpose upon each other, and the resultant lines of magnetic and of electrostatic forces are not concentric and radial respectively, except approximately in the immediate neighborhood of the conductor.

In the electric field between parallel conductors the magnetic and the electrostatic lines of force are conjugate pencils of circles.

2. Neither the power consumption in the conductor, nor the electromagnetic field, nor the electrostatic field, are proportional to the flow of energy through the circuit.

The product, however, of the intensity of the magnetic field, Φ, and the intensity of the electrostatic field, Ψ, is proportional to the flow of energy or the power, P, and the power P is therefore resolved into a product of two components, i and e, which are chosen proportional respectively to the intensity of the magnetic field Φ and of the electrostatic field Ψ.

That is, putting

$$P = ie \tag{1}$$

we have

$$\Phi = Li = \text{the intensity of the electromagnetic field.} \tag{2}$$

$$\Psi = Ce = \text{the intensity of the electrostatic field.} \tag{3}$$

The component i, called the *current*, is defined as that factor of the electric power P which is proportional to the magnetic field, and the other component e, called the voltage, is defined as that factor of the electric power P which is proportional to the electrostatic field.

Current i and voltage e, therefore, are mathematical fictions, factors of the power P, introduced to represent respectively the magnetic and the electrostatic or " dielectric " phenomena.

The current i is measured by the magnetic action of a circuit, as in the ammeter; the voltage e, by the electrostatic action of a circuit, as in the electrostatic voltmeter, or by producing a current i by the voltage e and measuring this current i by its magnetic action, in the usual voltmeter.

The coefficients L and C, which are the proportionality factors of the magnetic and of the dielectric component of the electric field, are called the *inductance* and the *capacity* of the circuit, respectively.

As electric power P is resolved into the product of current i and voltage e, the power loss in the conductor, P_l, therefore can also be resolved into a product of current i and voltage e_l which is consumed in the conductor. That is,

$$P_l = ie_l.$$

It is found that the voltage consumed in the conductor, e_l, is proportional to the factor i of the power P, that is,

$$e_l = ri, \tag{4}$$

where r is the proportionality factor of the voltage consumed by the loss of power in the conductor, or by the power gradient, and is called the *resistance* of the circuit.

Any electric circuit therefore must have three constants, r, L, and C, where

r = circuit constant representing the power gradient, or the loss of power in the conductor, called *resistance.*

L = circuit constant representing the intensity of the electro-magnetic component of the electric field of the circuit, called *inductance.*

C = circuit constant representing the intensity of the electro-static component of the electric field of the circuit, called *capacity.*

In most circuits, there is no current consumed in the conductor, i_l, and proportional to the voltage factor e of the power P, that is:

$$i_l = ge$$

where g is the proportionality factor of the current consumed by the loss of power in the conductor, which depends on the voltage, such as dielectric losses, etc. Where such exist, a fourth circuit constant appears, the *conductance* g, regarding which see sections III and IV.

3. A change of the magnetic field of the conductor, that is, if the number of lines of magnetic force Φ surrounding the conductor, generates an e.m.f.

$$e' = \frac{d\Phi}{dt} \tag{5}$$

in the conductor and thus absorbs a power

$$P' = ie' = i\frac{d\Phi}{dt} \tag{6}$$

or, by equation (2): $\Phi = Li$ by definition, thus:

$$\frac{d\Phi}{dt} = L\frac{di}{dt}, \text{ and: } P' = Li\frac{di}{dt}, \tag{7}$$

and the total energy absorbed by the magnetic field during the rise of current from zero to i is

$$W_M = \int P' dt \tag{8}$$

$$= L \int i\, di,$$

that is,

$$W_M = \frac{i^2 L}{2}. \tag{9}$$

A change of the dielectric field of the conductor, Ψ, absorbs a current proportional to the change of the dielectric field:

$$i' = \frac{d\Psi}{dt}, \tag{10}$$

and absorbs the power

$$P'' = ei' = e\frac{d\Psi}{dt}, \tag{11}$$

or, by equation (3),

$$P'' = Ce\frac{de}{dt}, \tag{12}$$

and the total energy absorbed by the dielectric field during a rise of voltage from 0 to e is

$$W_K = \int P'' dt \tag{13}$$

$$= C \int e\,de,$$

that is

$$W_K = \frac{e^2 C}{2}. \tag{14}$$

The power consumed in the conductor by its resistance r is

$$P_r = ie_l, \tag{15}$$

and thus, by equation (4),

$$P_r = i^2 r. \tag{16}$$

That is, when the electric power

$$P = ei \tag{1}$$

exists in a circuit, it is

$P_r = i^2 r$ = power lost in the conductor,　(16)

$W_M = \dfrac{i^2 L}{2}$ = energy stored in the magnetic field of the circuit, (9)

$W_K = \dfrac{e^2 C}{2}$ = energy stored in the dielectric field of the circuit, (14) and the three circuit constants r, L, C therefore appear as the components of the energy conversion into heat, magnetism, and electric stress, respectively, in the circuit.

4. The circuit constant, resistance r, depends only on the size and material of the conductor, but not on the position of the conductor in space, nor on the material filling the space surrounding the conductor, nor on the shape of the conductor section.

The circuit constants, inductance L and capacity C, almost entirely depend on the position of the conductor in space, on the material filling the space surrounding the conductor, and on the shape of the conductor section, but do not depend on the material of the conductor, except to that small extent as represented by the electric field inside of the conductor section.

5. The resistance r is proportional to the length and inversely proportional to the section of the conductor,

$$r = \rho \frac{l}{A}, \tag{17}$$

where ρ is a constant of the material, called the *resistivity* or *specific resistance*.

For different materials, ρ varies probably over a far greater range than almost any other physical quantity. Given in ohms per centimeter cube,* it is, approximately, at ordinary temperatures:

Metals: Cu...........................1.6×10^{-6}
Al............................2.8×10^{-6}
Fe............................10×10^{-6}
Hg............................94×10^{-6}
Gray cast iron..............up to 100×10^{-6}
High-resistance alloys.......up to 150×10^{-6}

Electrolytes: NO_3H.............down to 1.3 at 30 per cent
KOH.............down to 1.9 at 25 per cent
NaCl..............down to 4.7 at 25 per cent
up to
Pure river water ....10^4

and over alcohols, oils, etc., to practically infinity.

* Meaning a conductor of one centimeter length and one square centimeter section.

So-called *"insulators"*:

Fiber..................................about 10^{12}
Paraffin oil............................about 10^{13}
Paraffin..........................about 10^{14} to 10^{16}
Mica..................................about 10^{14}
Glass..........................about 10^{14} to 10^{16}
Rubber...............................about 10^{16}
Air................................practically ∞

In the wide gap between the highest resistivity of metal alloys, about $\rho = 150 \times 10^{-6}$, and the lowest resistivity of electrolytes, about $\rho = 1$, are

Carbon: metallic................down to 100×10^{-6}
amorphous (dense)..........0.04 and higher
anthracite.......................very high

Silicon and *Silicon Alloys:*
Cast silicon..........................1 down to 0.04
Ferro silicon...............0.04 down to 50×10^{-6}

The resistivity of *arcs* and of *Geissler tube discharges* is of about the same magnitude as electrolytic resistivity.

The resistivity, ρ, is usually a function of the temperature, rising slightly with increase of temperature in metallic conductors and decreasing in electrolytic conductors. Only with few materials, as silicon, the temperature variation of ρ is so enormous that ρ can no longer be considered as even approximately constant for all currents i which give a considerable temperature rise in the conductor. Such materials are commonly called pyroelectrolytes.

6. The inductance L is proportional to the section and inversely proportional to the length of the magnetic circuit surrounding the conductor, and so can be represented by

$$L = \frac{\mu A}{l},$$ (18)

where μ is a constant of the material filling the space surrounding the conductor, which is called the magnetic *permeability*.

As in general neither section nor length is constant in different parts of the magnetic circuit surrounding an electric con-

* See "Theory and Calculation of Electric Circuits."

ductor, the magnetic circuit has as a rule to be calculated piecemeal, or by integration over the space occupied by it.

The permeability, μ, is constant and equals unity or very closely $\mu = 1$ for all substances, with the exception of a few materials which are called the magnetic materials, as iron, cobalt, nickel, etc., in which it is very much higher, reaching sometimes and under certain conditions in iron values as high as $\mu = 6000$ and even as high as $\mu = 30,000$.

In these magnetic materials the permeability μ is not constant but varies with the magnetic flux density, or number of lines of magnetic force per unit section, $\mathfrak{B}$, decreasing rapidly for high values of $\mathfrak{B}$.

In such materials the use of the term μ is therefore inconvenient, and the inductance, L, is calculated by the relation between the magnetizing force as given in ampere-turns per unit length of magnetic circuit, or by "field intensity," and magnetic induction $\mathfrak{B}$.

The magnetic induction $\mathfrak{B}$ in magnetic materials is the sum of the "space induction" $\mathfrak{IC}$, corresponding to unit permeability, plus the "metallic induction" $\mathfrak{B}'$, which latter reaches a finite limiting value. That is,

$$\mathfrak{B} = \mathfrak{IC} + \mathfrak{B}'. \tag{19}$$

The limiting values, or so-called "saturation values," of $\mathfrak{B}'$ are approximately, in lines of magnetic force per square centimeter:

Iron...21,000
Cobalt.......................................12,000
Nickel....................................... 6,000
Magnetite.................................... 5,000
Manganese alloys .... .up to 5,000

The inductance, L, therefore is a constant of the circuit if the space surrounding the conductor contains no magnetic material, and is more or less variable with the current, i, if magnetic material exists in the space surrounding the conductor. In the latter case, with increasing current, i, the inductance, L, first slightly increases, reaches a maximum, and then decreases, approaching as limiting value the value which it would have in the absence of the magnetic material.

7. The capacity, C, is proportional to the section and inversely proportional to the length of the electrostatic field of the conductor:

$$C = \frac{\kappa A}{l}, \qquad (20)$$

where κ is a constant of the material filling the space surrounding the conductor, which is called the "dielectric constant," or the "specific capacity," or " permittivity."

Usually the section and the length of the different parts of the electrostatic circuit are different, and the capacity therefore has to be calculated piecemeal, or by integration.

The dielectric constant κ of different materials varies over a relative narrow range only. It is approximately:

$\kappa = 1$ in the vacuum, in air and in other gases,

$\kappa = 2$ to 3 in oils, paraffins, fiber, etc.,

$\kappa = 3$ to 4 in rubber and gutta-percha,

$\kappa = 3$ to 5 in glass, mica, etc.,

reaching values as high as 7 to 8 in organic compounds of heavy metals, as lead stearate, and about 12 in sulphur.

The dielectric constant, κ, is practically constant for all voltages e, up to that voltage at which the electrostatic field intensity, or the electrostatic gradient, that is, the "volts per centimeter," exceeds a certain value δ, which depends upon the material and which is called the "dielectric strength" or "disruptive strength" of the material. At this potential gradient the medium breaks down mechanically, by puncture, and ceases to insulate, but electricity passes and so equalizes the potential gradient.

The disruptive strength, δ, given in volts per centimeter is approximately:

Air: 30,000.

Oils: 250,000 to 1,000,000.

Mica: up to 4,000,000.

The capacity, C, of a circuit therefore is constant up to the voltage e, at which at some place of the electrostatic field the dielectric strength is exceeded, disruption takes place, and a part of the surrounding space therefore is made conducting, and by this increase of the effective size of the conductor the capacity C is increased.

8. Of the amount of energy consumed in creating the electric field of the circuit not all is returned at the disappearance of the electric field, but a part is consumed by conversion into heat in producing or in any other way changing the electric field. That is, the conversion of electric energy into and from the electromagnetic and electrostatic stress is not complete, but a loss of energy occurs, especially with the magnetic field in the so-called magnetic materials, and with the electrostatic field in unhomogeneous dielectrics.

The energy loss in the production and reconversion of the magnetic component of the field can be represented by an effective resistance r' which adds itself to the resistance r_0 of the conductor and more or less increases it.

The energy loss in the electrostatic field can be represented by an effective resistance r'', shunting across the circuit, and consuming an energy current i'', in addition to the current i in the conductor. Usually, instead of an effective resistance r'', its reciprocal is used, that is, the energy loss in the electrostatic field represented by a shunted conductance g.

In its most general form the electric circuit therefore contains the constants:

1. Inductance L, storing the energy, $\dfrac{i^2 L}{2}$,

2. Capacity C, storing the energy, $\dfrac{e^2 C}{2}$,

3. Resistance $r = r_0 + r'$, consuming the power, $i^2 r = i^2 r_0 + i^2 r'$,
4. Conductance g, consuming the power, $e^2 g$,

where r_0 is the resistance of the conductor, r' the effective resistance representing the power loss in the magnetic field L, and g represents the power loss in the electrostatic field C.

9. If of the three components of the electric field, the electromagnetic stress, electrostatic stress, and the power gradient, one equals zero, a second one must equal zero also. That is, either all of the three components exist or only one exists.

Electric systems in which the magnetic component of the field is absent, while the electrostatic component may be considerable, are represented for instance by an electric generator or a battery on open circuit, or by the electrostatic machine. In such systems the disruptive effects due to high voltage, therefore, are most pronounced, while the power is negligible, and phenomena of this character are usually called "static."

Electric systems in which the electrostatic component of the field is absent, while the electromagnetic component is considerable, are represented for instance by the short-circuited secondary coil of a transformer, in which no potential difference and, therefore, no electrostatic field exists, since the generated e.m.f. is consumed at the place of generation. Practically negligible also is the electrostatic component in all low-voltage circuits.

The effect of the resistance on the flow of electric energy in industrial applications is restricted to fairly narrow limits: as the resistance of the circuit consumes power and thus lowers the efficiency of the electric transmission, it is uneconomical to permit too high a resistance. As lower resistance requires a larger expenditure of conductor material, it is usually uneconomical to lower the resistance of the circuit below that which gives a reasonable efficiency.

As result hereof, practically always the relative resistance, that is, the ratio of the power lost in the resistance to the total power, lies between 2 per cent and 20 per cent.

It is different with the inductance L and the capacity C. Of the two forms of stored energy, the magnetic $\dfrac{i^2 L}{2}$ and electrostatic $\dfrac{e^2 C}{2}$, usually one is so small that it can be neglected compared with the other, and the electric circuit with sufficient approximation treated as containing resistance and inductance, or resistance and capacity only.

In the so-called electrostatic machine and its applications, frequently only capacity and resistance come into consideration.

In all lighting and power distribution circuits, direct current or alternating current, as the 110- and 220-volt lighting circuits, the 500-volt railway circuits, the 2000-volt primary distribution circuits, due to the relatively low voltage, the electrostatic energy $\dfrac{e^2 C}{2}$ is still so very small compared with the electromagnetic energy, that the capacity C can for most purposes be neglected and the circuit treated as containing resistance and inductance only.

Of approximately equal magnitude is the electromagnetic energy $\frac{i^2L}{2}$ and the electrostatic energy $\frac{e^2C}{2}$ in the high-potential long-distance transmission circuit, in the telephone circuit, and in the condenser discharge, and so in most of the phenomena resulting from lightning or other disturbances. In these cases all three circuit constants, r, L, and C, are of essential importance.

10. In an electric circuit of negligible inductance L and negligible capacity C, no energy is stored, and a change in the circuit thus can be brought about instantly without any disturbance or intermediary transient condition.

In a circuit containing only resistance and capacity, as a static machine, or only resistance and inductance, as a low or medium voltage power circuit, electric energy is stored essentially in one form only, and a change of the circuit, as an opening of the circuit, thus cannot be brought about instantly, but occurs more or less gradually, as the energy first has to be stored or discharged.

In a circuit containing resistance, inductance, and capacity, and therefore capable of storing energy in two different forms, the mechanical change of circuit conditions, as the opening of a circuit, can be brought about instantly, the internal energy of the circuit adjusting itself to the changed circuit conditions by a transfer of energy between static and magnetic and inversely, that is, after the circuit conditions have been changed, a transient phenomenon, usually of oscillatory nature, occurs in the circuit by the readjustment of the stored energy.

These transient phenomena of the readjustment of stored electric energy with a change of circuit conditions require careful study wherever the amount of stored energy is sufficiently large to cause serious damage. This is analogous to the phenomena of the readjustment of the stored energy of mechanical motion: while it may be harmless to instantly stop a slowly moving light carriage, the instant stoppage, as by collision, of a fast railway train leads to the usual disastrous result. So also, in electric systems of small stored energy, a sudden change of circuit conditions may be safe, while in a high-potential power system of very great stored electric energy any change of circuit conditions requiring a sudden change of energy is liable to be destructive.

Where electric energy is stored in one form only, usually little danger exists, since the circuit protects itself against sudden change by the energy adjustment retarding the change, and only where energy is stored electrostatically and magnetically, the mechanical change of the circuit conditions, as the opening of the circuit, can be brought about instantly, and the stored energy then surges between electrostatic and magnetic energy.

In the following, first the phenomena will be considered which result from the stored energy and its readjustment in circuits storing energy in one form only, which usually is as electromagnetic energy, and then the general problem of a circuit storing energy electromagnetically and electrostatically will be considered.

CHAPTER II.

INTRODUCTION.

11. In the investigation of electrical phenomena, currents and potential differences, whether continuous or alternating, are usually treated as stationary phenomena. That is, the assumption is made that after establishing the circuit a sufficient time has elapsed for the currents and potential differences to reach their final or permanent values, that is, become constant, with continuous current, or constant periodic functions of time, with alternating current. In the first moment, however, after establishing the circuit, the currents and potential differences in the circuit have not yet reached their permanent values, that is, the electrical conditions of the circuit are not yet the normal or permanent ones, but a certain time elapses while the electrical conditions adjust themselves.

12. For instance, a continuous e.m.f., e_0, impressed upon a circuit of resistance r, produces and maintains in the circuit a current,

$$i_0 = \frac{e_0}{r}.$$

In the moment of closing the circuit of e.m.f. e_0 on resistance r, the current in the circuit is zero. Hence, after closing the circuit the current i has to rise from zero to its final value i_0. If the circuit contained only resistance but no inductance, this would take place instantly, that is, there would be no transition period. Every circuit, however, contains some inductance. The inductance L of the circuit means L interlinkages of the circuit with lines of magnetic force produced by unit current in the circuit, or iL interlinkages by current i. That is, in establishing current i_0 in the circuit, the magnetic flux i_0L must be produced. A change of the magnetic flux iL surrounding a circuit generates in the circuit an e.m.f.,

$$e = \frac{d}{dt}(iL).$$

16

This opposes the impressed e.m.f. e_0, and therefore lowers the e.m.f. available to produce the current, and thereby the current, which then cannot instantly assume its final value, but rises thereto gradually, and so between the starting of the circuit and the establishment of permanent condition a transition period appears. In the same manner and for the same reasons, if the impressed e.m.f. e_0 is withdrawn, but the circuit left closed, the current i does not instantly disappear but gradually dies out, as shown in Fig. 1, which gives the rise and the decay of a

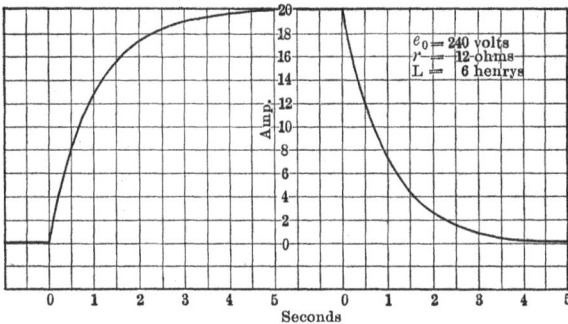

Fig. 1. Rise and decay of continuous current in an inductive circuit.

continuous current in an inductive circuit: the exciting current of an alternator field, or a circuit having the constants $r = 12$ ohms; $L = 6$ henrys, and $e_0 = 240$ volts; the abscissas being seconds of time.

13. If an electrostatic condenser of capacity C is connected to a continuous e.m.f. e_0, no current exists, in stationary condition, in this direct-current circuit (except that a very small current may leak through the insulation or the dielectric of the condenser), but the condenser is charged to the potential difference e_0, or contains the electrostatic charge

$$Q = Ce_0.$$

In the moment of closing the circuit of e.m.f. e_0 upon the capacity C, the condenser contains no charge, that is, zero potential difference exists at the condenser terminals. If there were no resistance and no inductance in the circuit in the moment of closing the circuit, an infinite current would exist charging the condenser instantly to the potential difference e_0. If r is the resistance of the direct-current circuit containing the condenser, and this circuit contains no inductance, the current

starts at the value $i = \dfrac{e_0}{r}$, that is, in the first moment after

closing the circuit all the impressed e.m.f. is consumed by the current in the resistance, since no charge and therefore no potential difference exists at the condenser. With increasing charge of the condenser, and therefore increasing potential difference at the condenser terminals, less and less e.m.f. is available for the resistance, and the current decreases, and ultimately becomes zero, when the condenser is fully charged.

If the circuit also contains inductance L, then the current cannot rise instantly but only gradually: in the moment after closing the circuit the potential difference at the condenser is still zero, and rises at such a rate that the increase of magnetic flux iL in the inductance produces an e.m.f. Ldi/dt, which consumes the impressed e.m.f. Gradually the potential difference at the condenser increases with its increasing charge, and the current and thereby the e.m.f. consumed by the resistance increases, and so less e.m.f. being available for consumption by the inductance, the current increases more slowly, until ultimately it ceases to rise, has reached a maximum, the inductance consumes no e.m.f., but all the impressed e.m.f. is consumed by the current in the resistance and by the potential difference at the condenser. The potential difference at the condenser continues to rise with its increasing charge; hence less e.m.f. is available for the resistance, that is, the current decreases again, and ultimately becomes zero, when the condenser is fully charged. During the decrease of current the decreasing magnetic flux iL in the inductance produces an e.m.f., which assists the impressed e.m.f., and so retards somewhat the decrease of current.

Fig. 2 shows the charging current of a condenser through an inductive circuit, as i, and the potential difference at the condenser terminals, as e, with a continuous impressed e.m.f. e_0, for the circuit constants $r = 250$ ohms; $L = 100$ mh.; $C = 10$ mf., and $e_0 = 1000$ volts.

If the resistance is very small, the current immediately after closing the circuit rises very rapidly, quickly charges the condenser, but at the moment where the condenser is fully charged to the impressed e.m.f. e_0, current still exists. This current cannot instantly stop, since the decrease of current and therewith the decrease of its magnetic flux iL generates an e.m.f.,

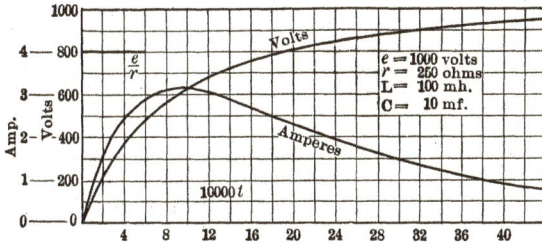

Fig. 2. Charging a condenser through a circuit having resistance and inductance. Constant potential. Logarithmic charge: high resistance.

which maintains the current, or retards its decrease. Hence electricity still continues to flow into the condenser for some time after it is fully charged, and when the current ultimately stops, the condenser is overcharged, that is, the potential difference at the condenser terminals is higher than the impressed e.m.f. e_0, and as result the condenser has partly to discharge again, that is, electricity begins to flow in the opposite direction, or out of the condenser. In the same manner this reverse current, due to the inductance of the circuit, overreaches and discharges the condenser farther than down to the impressed e.m.f. e_0, so that after the discharge current stops again a charging current — now less than the initial charging current — starts, and so by a series of oscillations, overcharges and undercharges, the condenser gradually charges itself, and ultimately the current dies out.

Fig. 3 shows the oscillating charge of a condenser through an inductive circuit, by a continuous impressed e.m.f. e_0. The current is represented by i, the potential difference at the condenser terminals by e, with the time as abscissas. The constants of the circuit are: $r = 40$ ohms; $L = 100$ mh.; $C = 10$ mf., and $e_0 = 1000$ volts.

In such a continuous-current circuit, containing resistance, inductance, and capacity in series to each other, the current at the moment of closing the circuit as well as the final current is zero, but a current exists immediately after closing the circuit, as a transient phenomenon; a temporary current, steadily increasing and then decreasing again to zero, or consisting of a number of alternations of successively decreasing amplitude: an oscillating current.

If the circuit contains no resistance and inductance, the current into the condenser would theoretically be infinite. That

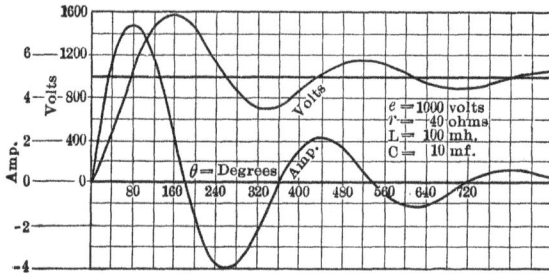

Fig. 3. Charging a condenser through a circuit having resistance and inductance. Constant potential. Oscillating charge: low resistance.

is, with low resistance and low inductance, the charging current of a condenser may be enormous, and therefore, although only transient, requires very serious consideration and investigation. If the resistance is very low and the inductance appreciable, the overcharge of the condenser may raise its voltage above the impressed e.m.f., e_0 sufficiently to cause disruptive effects.

14. If an alternating e.m.f.,

$$e = E \cos \theta,$$

is impressed upon a circuit of such constants that the current lags 45°, that is, the current is

$$i = I \cos (\theta - 45°),$$

and the circuit is closed at the moment $\theta = 45°$, at this moment the current should be at its maximum value. It is, however, zero, and since in a circuit containing inductance (that is, in practically any circuit) the current cannot change instantly, it follows that in this case the current gradually rises from zero as initial value to the permanent value of the sine wave i.

This approach of the current from the initial value, in the

present case zero, to the final value of the curve i, can either be gradual, as shown by the curve i_1 of Fig. 4, or by a series of oscillations of gradually decreasing amplitude, as shown by curve i_2 of Fig. 4.

15. The general solution of an electric current problem therefore includes besides the permanent term, constant or periodic,

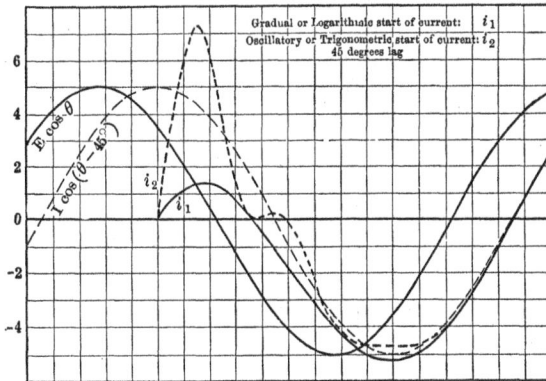

Fig. 4. Starting of an alternating-current circuit having inductance.

a transient term, which disappears after a time depending upon the circuit conditions, from an extremely small fraction of a second to a number of seconds.

These transient terms appear in closing the circuit, opening the circuit, or in any other way changing the circuit conditions, as by a change of load, a change of impedance, etc.

In general, in a circuit containing resistance and inductance only, but no capacity, the transient terms of current and voltage are not sufficiently large and of long duration to cause harmful nor even appreciable effects, and it is mainly in circuits containing capacity that excessive values of current and potential difference may be reached by the transient term, and therewith serious results occur. The investigation of transient terms therefore is largely an investigation of the effects of electrostatic capacity.

16. No transient terms result from the resistance, but only those circuit constants which represent storage of energy, magnetically by the inductance L, electrostatically by the capacity C, give rise to transient phenomena, and the more the resistance predominates, the less is therefore the severity and duration of the transient term.

When closing a circuit containing inductance or capacity or both, the energy stored in the inductance and the capacity has first to be supplied by the impressed e.m.f. before the circuit conditions can become stationary. That is, in the first moment after closing an electric circuit, or in general changing the circuit conditions, tne impressed e.m.f., or rather the source producing the impressed e.m.f., has, in addition to the power consumed in maintaining the circuit, to supply the power which stores energy in inductance and capacity, and so a transient term appears immediately after any change of circuit condition. If the circuit contains only one energy-storing constant, as either inductance or capacity, the transient term, which connects the initial with the stationary condition of the circuit, necessarily can be a steady logarithmic term only, or a gradual approach. An oscillation can occur only with the existence of two energy-storing constants, as capacity and inductance, which permit a surge of energy from the one to the other, and therewith an overreaching.

17. Transient terms may occur periodically and in rapid succession, as when rectifying an alternating current by synchronously reversing the connections of the alternating impressed e.m.f. with the receiver circuit (as can be done mechanically or without moving apparatus by unidirectional conductors, as arcs). At every half wave the circuit reversal starts a transient term, and usually this transient term has not yet disappeared, frequently not even greatly decreased, when the next reversal again starts a transient term. These transient terms may predominate to such an extent that the current essentially consists of a series of successive transient terms.

18. If a condenser is charged through an inductance, and the condenser shunted by a spark gap set for a lower voltage than the impressed, then the spark gap discharges as soon as the condenser charge has reached a certain value, and so starts a transient term; the condenser charges again, and discharges, and so by the successive charges and discharges of the condenser a series of transient terms is produced, recurring at a frequency depending upon the circuit constants and upon the ratio of the disruptive voltage of the spark gap to the impressed e.m.f.

Such a phenomenon for instance occurs when on a high-potential alternating-current system a weak spot appears in the cable insulation and permits a spark discharge to pass to the ground, that is, in shunt to the condenser formed by the cable conductor and the cable armor or ground.

19. In most cases the transient phenomena occurring in electric circuits immediately after a change of circuit conditions are of no importance, due to their short duration. They require serious consideration, however,—

(a) In those cases where they reach excessive values. Thus in connecting a large transformer to an alternator the large initial value of current may do damage. In short-circuiting a large alternator, while the permanent or stationary short-circuit current is not excessive and represents little power, the very much larger momentary short-circuit current may be beyond the capacity of automatic circuit-opening devices and cause damage by its high power. In high-potential transmissions the potential differences produced by these transient terms may reach values so high above the normal voltage as to cause disruptive effects. Or the frequency or steepness of wave front of these transients may be so great as to cause destructive voltages across inductive parts of the circuits, as reactors, end turns of transformers and generators, etc.

(b) Lightning, high-potential surges, etc., are in their nature essentially transient phenomena, usually of oscillating character.

(c) The periodical production of transient terms of oscillating character is one of the foremost means of generating electric currents of very high frequency as used in wireless telegraphy, etc.

(d) In alternating-current rectifying apparatus, by which the direction of current in a part of the circuit is reversed every half wave, and the current so made unidirectional, the stationary condition of the current in the alternating part of the circuit is usually never reached, and the transient term is frequently of primary importance.

(e) In telegraphy the current in the receiving apparatus essentially depends on the transient terms, and in long-distance cable telegraphy the stationary condition of current is never approached, and the speed of telegraphy depends on the duration of the transient terms.

(f) Phenomena of the same character, but with space instead

of time as independent variable, are the distribution of voltage and current in a long-distance transmission line; the phenomena occurring in multigap lightning arresters; the transmission of current impulses in telephony; the distribution of alternating current in a conductor, as the rail return of a single-phase railway; the distribution of alternating magnetic flux in solid magnetic material, etc.

Some of the simpler forms of transient terms are investigated and discussed in the following pages.

CHAPTER III.

INDUCTANCE AND RESISTANCE IN CONTINUOUS-CURRENT CIRCUITS.

20. In continuous-current circuits the inductance does not enter the equations of stationary condition, but, if e_0 = impressed e.m.f., r = resistance, L = inductance, the permanent value of current is $i_0 = \dfrac{e_0}{r}$.

Therefore less care is taken in direct-current circuits to reduce the inductance than in alternating-current circuits, where the inductance usually causes a drop of voltage, and direct-current circuits as a rule have higher inductance, especially if the circuit is used for producing magnetic flux, as in solenoids, electro-magnets, machine-fields.

Any change of the condition of a continuous-current circuit, as a change of e.m.f., of resistance, etc., which leads to a change of current from one value i_0 to another value i_1, results in the appearance of a transient term connecting the current values i_0 and i_1, and into the equation of the transient term enters the inductance.

Count the time t from the moment when the change in the continuous-current circuit starts, and denote the impressed e.m.f. by e_0, the resistance by r, and the inductance by L.

$i_1 = \dfrac{e_0}{r}$ = current in permanent or stationary condition after the change of circuit condition.

Denoting by i_0 the current in circuit before the change, and therefore at the moment $t = 0$, by i the current during the change, the e.m.f. consumed by resistance r is

$$ir,$$

and the e.m.f. consumed by inductance L is

$$L \frac{di}{dt},$$

where i = current in the circuit.

Hence,
$$e_0 = ir + L\frac{di}{dt}, \tag{1}$$

or, substituting $e_0 = i_1 r$, and transposing,

$$-\frac{r}{L}dt = \frac{di}{i - i_1}. \tag{2}$$

This equation is integrated by

$$-\frac{r}{L}t = \log(i - i_1) - \log c,$$

where $-\log c$ is the integration constant, or,

$$i - i_1 = c\varepsilon^{-\frac{r}{L}t}.$$

However, for $t = 0$, $i = i_0$.

Substituting this, gives

$$i_0 - i_1 = c,$$

hence,
$$i = i_1 + (i_0 - i_1)\varepsilon^{-\frac{r}{L}t}, \tag{3}$$

the equation of current in the circuit.

The counter e.m.f. of self-inductance is

$$e_1 = -L\frac{di}{dt} = r(i_0 - i_1)\varepsilon^{-\frac{r}{L}t}, \tag{4}$$

hence a maximum for $t = 0$, thus:

$$e_1^0 = r(i_0 - i_1). \tag{5}$$

The e.m.f. of self-inductance e_1 is proportional to the change of current $(i_0 - i_1)$, and to the resistance r of the circuit after the change, hence would be ∞ for $r = \infty$, or when opening the circuit. That is, an inductive circuit cannot be opened instantly, but the arc following the break maintains the circuit for some time, and the voltage generated in opening an inductive circuit is the higher the quicker the break. Hence in a highly inductive circuit, as an electromagnet or a machine field, the insulation may be punctured by excessive generated e.m.f. when quickly opening the circuit.

As example, some typical circuits may be considered.

21. *Starting of a continuous-current lighting circuit, or non-inductive load.*

Let $e_0 = 125$ volts = impressed e.m.f. of the circuit, and $i_1 = 1000$ amperes = current in the circuit under stationary condition; then the effective resistance of the circuit is

$$r = \frac{e_0}{i_1} = 0.125 \text{ ohm.}$$

Assuming 10 per cent drop in feeders and mains, or 12.5 volts, gives a resistance, $r_0 = 0.0125$ ohm of the supply conductors. In such large conductor the inductance may be estimated as 10 mh. per ohm; hence, $L = 0.125$ mh. $= 0.000125$ henry.

The current at the moment of starting is $i_0 = 0$, and the general equation of the current in the circuit therefore is, by substitution in (3),

$$i = 1000 \ (1 - \varepsilon^{-1000\,t}). \tag{6}$$

The time during which this current reaches half value, or $i = 500$ amperes, is given by substitution in (6)

$$500 = 1000 \ (1 - \varepsilon^{-1000\,t}),$$

hence $\quad\quad\quad \varepsilon^{-1000\,t} = 0.5,$

$$t = 0.00069 \text{ seconds.}$$

The time during which the current reaches 90 per cent of its full value, or $i = 900$ amperes, is $t = 0.0023$ seconds, that is, the current is established in the circuit in a practically inappreciable time, a fraction of a hundredth of a second.

22. *Excitation of a motor field.*

Let, in a continuous-current shunt motor, $e_0 = 250$ volts = impressed e.m.f., and the number of poles $= 8$.

Assuming the magnetic flux per pole, $\Phi_0 = 12.5$ megalines, and the ampere-turns per pole required to produce this magnetic flux as $\mathfrak{F} = 9000$.

Assuming 1000 watts used for the excitation of the motor field gives an exciting current

$$i_1 = \frac{1000}{250} = 4 \text{ amperes,}$$

and herefrom the resistance of the total motor field circuit as

$$r = \frac{e_0}{i_1} = 62.5 \text{ ohms.}$$

To produce $\mathfrak{F} = 9000$ ampere-turns, with $i_1 = 4$ amperes, requires $\dfrac{\mathfrak{F}}{i_1} = 2250$ turns per field spool, or a total of $n = 18,000$ turns.

$n = 18,000$ turns interlinked with $\Phi_0 = 12.5$ megalines gives a total number of interlinkages for $i_1 = 4$ amperes of $n\Phi_0 = 225 \times 10^9$, or 562.5×10^9 interlinkages per unit current, or 10 amperes, that is, an inductance of the motor field circuit $L = 562.5$ henrys.

The constants of the circuit thus are $e_0 = 250$ volts; $r = 62.5$ ohms; $L = 562.5$ henrys, and $i_0 = 0 =$ current at time $t = 0$.

Hence, substituting in (3) gives the equation of the exciting current of the motor field as

$$i = 4 \left(1 - \varepsilon^{-0.1111 t}\right) \tag{7}$$

Half excitation of the field is reached after the time $t = 6.23$ seconds;

90 per cent of full excitation, or $i = 3.6$ amperes, after the time $t = 20.8$ seconds.

That is, such a motor field takes a very appreciable time after closing the circuit before it has reached approximately full value and the armature circuit may safely be closed.

Assume now the motor field redesigned, or reconnected so as to consume only a part, for instance half, of the impressed e.m.f., the rest being consumed in non-inductive resistance. This may be done by connecting the field spools by two in multiple.

In this case the resistance and the inductance of the motor field are reduced to one-quarter, but the same amount of external resistance has to be added to consume the impressed e.m.f., and the constants of the circuit then are: $e_0 = 250$ volts; $r = 31.25$ ohms; $L = 140.6$ henrys, and $i_0 = 0$.

The equation of the exciting current (3) then is

$$i = 8 \left(1 - \varepsilon^{-0.2222 t}\right), \tag{8}$$

that is, the current rises far more rapidly. It reaches 0.5 value after $t = 3.11$ seconds, 0.9 value after $t = 10.4$ seconds.

An inductive circuit, as a motor field circuit, may be made to respond to circuit changes more rapidly by inserting non-inductive resistance in series with it and increasing the impressed e.m.f., that is, the larger the part of the impressed e.m.f. consumed by non-inductive resistance, the quicker is the change.

Disconnecting the motor field winding from the impressed e.m.f. and short-circuiting it upon itself, as by leaving it connected in shunt with the armature (the armature winding resistance and inductance being negligible compared with that of the field winding), causes the field current and thereby the field magnetism to decrease at the same rate as it increased in (7) and (8), provided the armature instantly comes to a standstill, that is, its e.m.f. of rotation disappears. This, however, is usually not the case, but the motor armature slows down gradually, its momentum being consumed by friction and other losses, and while still revolving an e.m.f. of gradually decreasing intensity is generated in the armature winding; this e.m.f. is impressed upon the field.

The discharge of a motor field winding through the armature winding, after shutting off the power, therefore leads to the case of an inductive circuit with a varying impressed e.m.f.

23. *Discharge of a motor field winding.*

Assume that in the continuous-current shunt motor discussed under 22, the armature comes to rest $t_1 = 40$ seconds after the energy supply has been shut off by disconnecting the motor from the source of impressed e.m.f., while leaving the motor field winding still in shunt with the motor armature winding.

The resisting torque, which brings the motor to rest, may be assumed as approximately constant, and therefore the deceleration of the motor armature as constant, that is, the motor speed decreasing proportionally to the time.

If then S = full motor speed, $S\left(1 - \dfrac{t}{t_1}\right)$ is the speed of the motor at the time t after disconnecting the motor from the source of energy.

Assume the magnetic flux Φ of the motor as approximately proportional to the exciting current, at exciting current i the magnetic flux of the motor is $\Phi = \dfrac{i}{i_1}\,\Phi_0$, where $\Phi_0 = 12.5$ megalines is the flux corresponding to full excitation $i_1 = 4$ amperes.

The e.m.f. generated in the motor armature winding and thereby impressed upon the field winding is proportional to the magnetic flux of the field, Φ, and to the speed $S\left(1 - \dfrac{t}{t_1}\right)$, and since full speed S and full flux Φ_0 generate an e.m.f. $e_0 = 250$ volts, the e.m.f. generated by the flux Φ and speed $S\left(1 - \dfrac{t}{t_1}\right)$, that is, at time t is

$$e = e_0 \frac{i}{i_1}\left(1 - \frac{t}{t_1}\right), \tag{9}$$

and since

$$\frac{e_0}{i_1} = r,$$

we have

$$e = ir\left(1 - \frac{t}{t_1}\right); \tag{10}$$

or for $r = 62.5$ ohms, and $t_1 = 40$ seconds, we have

$$e = 62.5\, i\,(1 - 0.025\, t). \tag{11}$$

Substituting this equation (10) of the impressed e.m.f. into the differential equation (1) gives the equation of current i during the field discharge,

$$ir\left(1 - \frac{t}{t_1}\right) = ir + L\,\frac{di}{dt}, \tag{12}$$

hence,

$$-\frac{rt\,dt}{t_1 L} = \frac{di}{i}, \tag{13}$$

integrated by

$$-\frac{rt^2}{2\,t_1 L} = \log ci,$$

where the integration constant c is found by

$$t = 0, \quad i = i_1, \quad \log ci_1 = 0, \quad c = \frac{1}{i_1},$$

hence,

$$-\frac{rt^2}{2\,t_1 L} = \log \frac{i}{i_1}, \tag{14}$$

or,

$$i = i_1 \varepsilon^{-\frac{rt^2}{2\,t_1 L}}, \tag{15}$$

This is the equation of the field current during the time in which the motor armature gradually comes to rest.

At the moment when the motor armature stops, or for

$$t = t_1,$$

it is

$$i_2 = i_1 \varepsilon^{-\frac{r t_1}{2 L}}. \tag{16}$$

This is the same value which the current would have with the armature permanently at rest, that is, without the assistance of the e.m.f. generated by rotation, at the time $t = \dfrac{t_1}{2}$.

The rotation of the motor armature therefore reduces the decrease of field current so as to require twice the time to reach value i_2, that it would without rotation.

These equations cease to apply for $t > t_1$, that is, after the armature has come to rest, since they are based on the speed equation $S\left(1 - \dfrac{t}{t_1}\right)$, and this equation applies only up to $t = t_1$, but for $t > t_1$ the speed is zero, and not negative, as given by $S\left(1 - \dfrac{t}{t_1}\right)$.

That is, at the moment $t = t_1$ a break occurs in the field discharge curve, and after this time the current i decreases in accordance with equation (3), that is,

$$i = i_2 \varepsilon^{-\frac{r}{L}\left(t - t_1\right)}, \tag{17}$$

or, substituting (16),

$$i = i_1 \varepsilon^{-\frac{r}{L}\left(t - \frac{t_1}{2}\right)}. \tag{18}$$

Substituting numerical values in these equations gives:
for $t < t_1$,

$$i = 4 \, \varepsilon^{-0.001388 \, t^2}; \tag{19}$$

for $t = t_1 = 40$,

$$i = 0.436; \tag{20}$$

for $t > t_1$,

$$i = 4 \, \varepsilon^{-0.1111 \, (t - 20)}. \tag{21}$$

Hence, the field has decreased to half its initial value after the time $t = 22.15$ seconds, and to one tenth of its initial value after $t = 40.73$ seconds.

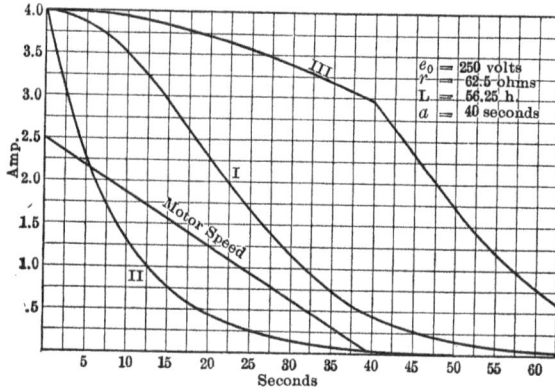

Fig. 5.   Field discharge current.

Fig. 5 shows as curve I the field discharge current, by equations (19), (20), (21), and as curve II the current calculated by the equation

$$i = 4 \, \varepsilon^{-0.1111t},$$

that is, the discharge of the field with the armature at rest, or when short-circuited upon itself and so not assisted by the e.m.f. of rotation of the armature.

The same Fig. 5 shows as curve III the beginning of the field discharge current for $L = 4200$, that is, the case that the field circuit has a much higher inductance, as given by the equation

$$i = 4 \, \varepsilon^{-0.000185 t^2}.$$

As seen in the last case, the decrease of field current is very slow, the field decreasing to half value in 47.5 seconds.

24. *Self-excitation of direct-current generator.*

In the preceding, the inductance L of the machine has been assumed as constant, that is, the magnetic flux Φ as proportional to the exciting current i. For higher values of Φ, this is not even approximately the case. The self-excitation of the direct-current generator, shunt or series wound, that is, the feature that the voltage of the machine after the start gradually builds up from the value given by the residual magnetism to its full value, depends upon the disproportionality of the magnetic flux with the magnetizing current. When considering this phenomenon, the inductance cannot therefore be assumed as constant.

When investigating circuits in which the inductance L is not constant but varies with the current, it is preferable not to use the term "inductance" at all, but to introduce the magnetic flux Φ.

The magnetic flux Φ varies with the magnetizing current i by an empirical curve, the magnetic characteristic or saturation curve of the machine. This can approximately, within the range considered here, be represented by a hyperbolic curve, as was first shown by Fröhlich in 1882:

$$\Phi = \frac{\phi i}{1 + bi}, \tag{22}$$

where ϕ = magnetic flux per ampere, in megalines, at low density.

$\dfrac{\phi}{b}$ = magnetic saturation value, or maximum magnetic flux, in megalines, and

$$\frac{i}{\Phi} = \frac{1 + bi}{\phi} \tag{23}$$

can be considered as the magnetic exciting reluctance of the machine field circuit, which here appears as linear function of the exciting current i.

Considering the same shunt-wound commutating machine as in (12) and (13), having the constants $r = 62.5$ ohms = field resistance; $\Phi_0 = 12.5$ megalines = magnetic flux per pole at normal m.m.f.; $\mathcal{F} = 9000$ ampere-turns = normal m.m.f. per pole; $n = 18{,}000$ turns = total field turns (field turns per pole $= \dfrac{18{,}000}{8} = 2250$), and $i_1 = 4$ amperes = current for full excitation, or flux, $\Phi_0 = 12.5$ megalines.

Assuming that at full excitation, Φ_0, the magnetic reluctance has already increased by 50 per cent above its initial value, that is, that the ratio $\dfrac{\text{ampere-turns}}{\text{magnetic flux}}$, or $\dfrac{i}{\Phi}$, at $\Phi = \Phi_0 = 12.5$ megalines and $i = i_1 = 4$ amperes, is 50 per cent higher than at low excitation, it follows that

$$1 + bi_1 = 1.5, \left.\begin{array}{c} \\ \\ \end{array}\right\} \tag{24}$$
$$b = 0.125.$$

or

Since $i = i_1 = 4$ produces $\Phi = \Phi_0 = 12.5$, it follows, from (22) and (24)

$$\phi = 4.69.$$

That is, the magnetic characteristic (22) of the machine is approximated by

$$\Phi = \frac{4.69\, i}{1 + .125\, i}. \tag{25}$$

Let now e_c = e.m.f. generated by the rotation of the armature per megaline of field flux.

This e.m.f. e_c is proportional to the speed, and depends upon the constants of the machine. At the speed assumed in (12) and (13), $\Phi_0 = 12.5$ megalines, $e_0 = 250$ volts, that is,

$$e_c = \frac{e_0}{\Phi_0} = 20 \text{ volts.}$$

Then, in the field circuit of the machine, the impressed e.m.f., or e.m.f. generated in the armature by its rotation through the magnetic field is,

$$e = e_c \Phi = 20\Phi;$$

the e.m.f. consumed by the field resistance r is

$$ir = 62.5\, i;$$

the e.m.f. consumed by the field inductance, that is, generated in the field coils by the rise of magnetic flux Φ, is

$$n\frac{d\Phi}{dt}10^{-2} = 180\frac{d\Phi}{dt}.$$

(Φ being given in megalines, e_0 in volts.)

The differential equation of the field circuit therefore is (1)

$$e_c\Phi = ir + \frac{n}{100}\frac{d\Phi}{dt}.$$ (26)

Since this equation contains the differential quotient of Φ, it is more convenient to make Φ and not i the dependent variable; then substitute for i from equation (22),

$$i = \frac{\Phi}{\phi - b\Phi},$$ (27)

which gives

$$e_.\Phi = \frac{\Phi r}{\phi - b\Phi} + \frac{n}{100}\frac{d\Phi}{dt},$$ (28)

or, transposed,

$$\frac{100\,dt}{n} = \frac{(\phi - b\Phi)\,d\Phi}{\Phi\{(\phi e_. - r) - be_.\Phi\}}.$$ (29)

This equation is integrated by resolving into partial fraction by the identity

$$\frac{\phi - b\Phi}{\Phi\{(\phi e_. - r) - be_c\Phi\}} = \frac{A}{\Phi} + \frac{B}{\phi e_c - r - be_c\Phi};$$ (30)

resolved, this gives

$$\phi - b\Phi = A(\phi e_c - r) - (Abe_c\Phi - B\Phi);$$

hence,

$$\left.\begin{aligned} A &= \frac{\phi}{\phi e_c - r}, \\[1em] B &= \frac{br}{\phi e_c - r}, \end{aligned}\right\}$$ (31)

and

$$\frac{100\,dt}{n} = \frac{\phi d\Phi}{(\phi e_c - r)\Phi} + \frac{brd\Phi}{(\phi e_c - r)(\phi e_c - r - be_.\Phi)}.$$ (32)

This integrates by the logarithmic functions

$$\frac{100\,t}{n} = \frac{\phi}{\phi e_c - r}\log\Phi - \frac{r}{e_c(\phi e_c - r)}\log(\phi e_c - r - be_c\Phi) + C.$$ (33)

The integration constant C is calculated from the residual magnetic flux of the machine, that is, the remanent magnetism of the field poles at the moment of start.

Assume, at the time, $t = 0$, $\Phi = \Phi_r = 0.5$ megalines $=$ residual magnetism and substituting in (33),

$$0 = \frac{\phi}{\phi e_c - r} \log \Phi_r - \frac{r}{e_c\,(\phi e_c - r)} \log (\phi e_c - r - be_c\Phi_r) + C,$$

and herefrom calculate C.

C substituted in (33) gives

$$\frac{100\,t}{n} = \frac{\phi}{\phi e_c - r} \log \frac{\Phi}{\Phi_r} - \frac{r}{e_c\,(\phi e_c - r)} \log \frac{\phi e_c - r - be_c\Phi}{\phi e_c - r - be_c\Phi_r}, \quad (34)$$

or,

$$t = \frac{n}{100\,e_c\,(\phi e_c - r)} \left\{ \phi e_c \log \frac{\Phi}{\Phi_r} - r \log \frac{\phi e_c - r - be_c\Phi}{\phi e_c - r - be_c\Phi_r} \right\}. \quad (35)$$

substituting

$$e = e_c\Phi$$

and

$$e_m = e_c\Phi_r,$$

where $e_m =$ e.m.f. generated in the armature by the rotation in the residual magnetic field,

$$t = \frac{n}{100\,e_c\,(\phi e_c - r)} \left\{ \phi e_c \log \frac{e}{e_m} - r \log \frac{\phi e_c - r - be}{\phi e_c - r - be_m} \right\}. \quad (36)$$

This, then, is the relation between e and t, or the equation of the building up of a continuous-current generator from its residual magnetism, its speed being constant.

Substituting the numerical values $n = 18{,}000$ turns; $\phi = 4.69$ megalines; $b = 0.125$; $e_c = 20$ volts; $r = 62.5$ ohms; $\Phi_r = 0.5$ megaline, and $e_m = 10$ volts, we have

$$t = 26.8 \log \Phi - 17.9 \log (31.25 - 2.5\,\Phi) + 79.6 \quad (37)$$

and

$$t = 26.8 \log e - 17.9 \log (31.25 - 0.125\,e) - 0.98. \quad (38)$$

Fig. 6 shows the e.m.f. *e* as function of the time *t*. As seen,
under the conditions assumed here, it takes several minutes
before the e.m.f. of the machine builds up to approximately
full value.

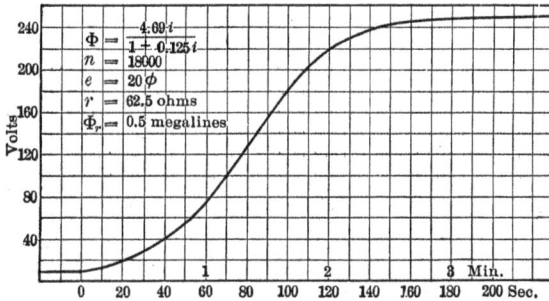

Fig. 6.   Building-up curve of a shunt generator.

The phenomenon of self-excitation of shunt generators therefore is a transient phenomenon which may be of very long
duration.

From equations (35) and (36) it follows that

$$e = \frac{\phi e_c - r}{b} = 250 \text{ volts} \tag{39}$$

is the e.m.f. to which the machine builds up at $t = \infty$, that is,
in stationary condition.

To make the machine self-exciting, the condition

$$\phi e_c - r > 0 \tag{40}$$

must obtain, that is, the field winding resistance must be

$$\left.\begin{array}{c} r < \phi e_c, \\[2mm] r < 93.8 \text{ ohms}, \end{array}\right\} \tag{41}$$

or,

or, inversely, e_c, which is proportional to the speed, must be

$$\left.\begin{array}{c} e_c > \dfrac{r}{\phi}, \\[3mm] e_c > 13.3 \text{ volts}. \end{array}\right\} \tag{42}$$

or,

The time required by the machine to build up decreases with increasing e_c, that is, increasing speed; and increases with increasing r, that is, increasing field resistance.

25. *Self-excitation of direct-current series machine.*

Of interest is the phenomenon of self-excitation in a series machine, as a railway motor, since when using the railway motor as brake, by closing its circuit upon a resistance, its usefulness depends upon the rapidity of building up as generator.

Assuming a 4-polar railway motor, designed for $e_0 = 600$ volts and $i_1 = 200$ amperes, let, at current $i = i_1 = 200$ amperes, the magnetic flux per pole of the motor be $\Phi_0 = 10$ megalines, and 8000 ampere-turns per field pole be required to produce this flux. This gives 40 exciting turns per pole, or a total of $n = 160$ turns.

Estimating 8 per cent loss in the conductors of field and armature at 200 amperes, this gives a resistance of the motor circuit $r_0 = 0.24$ ohms.

To limit the current to the full load value of $i_1 = 200$ amperes, with the machine generating $e_0 = 600$ volts, requires a total resistance of the circuit, internal plus external, of

$$r = 3 \text{ ohms,}$$

or an external resistance of 2.76 ohms.

600 volts generated by 10 megalines gives

$$e_c = 60 \text{ volts per megaline per field pole.}$$

Since in railway motors at heavy load the magnetic flux is carried up to high values of saturation, at $i_1 = 200$ amperes the magnetic reluctance of the motor field may be assumed as three times the value which it has at low density, that is, in equation (22),

$$1 + bi_1 = 3,$$

or,

$$b = 0.01,$$

and since for $i = 200$, $\Phi = 10$, we have in (22)

$$\phi = 0.15,$$

hence,

$$\Phi = \frac{0.15\, i}{1 + 0.01\, i} \tag{43}$$

represents the magnetic characteristic of the machine.

Assuming a residual magnetism of 10 per cent, or $\Phi_r = 1$ megaline, hence $e_m = e_c \Phi_r = 60$ volts, and substituting in equation (36) gives $n = 160$ turns; $\phi = 0.15$ megaline; $b = 0.01$; $e_c = 60$ volts; $r = 3$ ohms; $\Phi_r = 1$ megaline, and $e_m = 60$ volts,

$$t = 0.04 \log e - 0.01333 \log (600 - e) - 0.08. \qquad (44)$$

This gives for $e = 300$, or 0.5 excitation, $t = 0.072$ seconds; and for $e = 540$, or 0.9 excitation, $t = 0.117$ seconds; that is, such a motor excites itself as series generator practically instantly, or in a small fraction of a second.

The lowest value of e at which self-excitation still takes place is given by equation (42) as

$$e_c = \frac{r}{\phi} = 20,$$

that is, at one-third of full speed.

If this series motor, with field and armature windings connected in generator position,—that is, reverse position,—short-circuits upon itself,

$$r = 0.24 \text{ ohms},$$

we have

$$t = 0.0274 \log e - 0.00073 \log (876 - e) - 0.1075, \qquad (45)$$

that is, self-excitation is practically instantaneous:

$e = 300$ volts is reached after $t = 0.044$ seconds.

Since for $e = 300$ volts, the current $i = \frac{e}{r} = 1250$ amperes, the power is $p = ei = 375$ kw., that is, a series motor short-circuited in generator position instantly stops.

Short-circuited upon itself, $r = 0.24$, this series motor still builds up at $e_c = \frac{r}{\phi} = 1.6$, and since at full load speed $e_c = 60$, $e_c = 1.6$ is 2.67 per cent of full load speed, that is, the motor acts as brake down to 2.67 per cent of full speed.

It must be considered, however, that the parabolic equation (22) is only an approximation of the magnetic characteristic, and the results based on this equation therefore are approximate only.

One of the most important transient phenomena of direct-current circuits is the reversal of current in the armature coil short-circuited by the commutator brush in the commutating machine. Regarding this, see " Theoretical Elements of Electrical Engineering," Part II, Section B.

CHAPTER IV.

INDUCTANCE AND RESISTANCE IN ALTERNATING-CURRENT CIRCUITS.

26. In alternating-current circuits, the inductance L, or, as it is usually employed, the reactance $x = 2\pi fL$, where f = frequency, enters the expression of the transient as well as the permanent term.

At the moment $\theta = 0$, let the e.m.f. $e = E \cos(\theta - \theta_0)$ be impressed upon a circuit of resistance r and inductance L, thus inductive reactance $x = 2\pi fL$; let the time $\theta = 2\pi ft$ be counted from the moment of closing the circuit, and θ_0 be the phase of the impressed e.m.f. at this moment.

In this case the e.m.f. consumed by the resistance $= ir$, where i = instantaneous value of current.

The e.m.f. consumed by the inductance L is proportional to L and to the rate of change of the current, $\dfrac{di}{dt}$, thus, is $L\dfrac{di}{dt}$, or, by substituting $\theta = 2\pi ft$, $x = 2\pi fL$, the e.m.f. consumed by inductance is $x\dfrac{di}{d\theta}$.

Since $e = E\cos(\theta - \theta_0)$ = impressed e.m.f.,

$$E \cos(\theta - \theta_0) = ir + x\frac{di}{d\theta} \qquad (1)$$

is the differential equation of the problem.

This equation is integrated by the function

$$i = I\cos(\theta - \delta) + A\varepsilon^{-a\theta}, \qquad (2)$$

where ε = basis of natural logarithms = 2.7183.

Substituting (2) in (1),

$$E\cos(\theta - \theta_0) = Ir\cos(\theta - \delta) + Ar\varepsilon^{-a\theta} - Ix\sin(\theta - \delta) - Aax\varepsilon^{-a\theta},$$

or, rearranged:

$$(E\cos\theta_0 - Ir\cos\delta - Ix\sin\delta)\cos\theta + (E\sin\theta_0 - Ir\sin\delta$$
$$+ Ix\cos\delta)\sin\theta - A\varepsilon^{-a\theta}(ax - r) = 0.$$

Since this equation must be fulfilled for any value of θ, if (2) is the integral of (1), the coefficients of $\cos \theta$, $\sin \theta$, $\varepsilon^{-a\theta}$ must vanish separately.

That is,

$$\left.\begin{aligned} E \cos \theta_0 - Ir \cos \delta - Ix \sin \delta = 0, \\ E \sin \theta_0 - Ir \sin \delta + Ix \cos \delta = 0, \end{aligned}\right\} \tag{3}$$

and
$$ax - r = 0.$$

Herefrom it follows that

$$a = \frac{r}{x}. \tag{4}$$

Substituting in (3),

$$\left.\begin{aligned} \tan \theta_1 = \frac{x}{r} \\ z = \sqrt{r^2 + x^2}, \end{aligned}\right\} \tag{5}$$

and

where $\theta_1 = $ lag angle and $z = $ impedance of circuit, we have

$$\left.\begin{aligned} E \cos \theta_0 - Iz \cos (\delta - \theta_1) = 0 \\ E \sin \theta_0 - Iz \sin (\delta - \theta_1) = 0, \end{aligned}\right\}$$

and

and herefrom

$$\left.\begin{aligned} I = \frac{E}{z} \\ \delta = \theta_0 + \theta_1. \end{aligned}\right\} \tag{6}$$

and

Thus, by substituting (4) and (6) in (2), the integral equation becomes

$$i = \frac{E}{z} \cos (\theta - \theta_0 - \theta_1) + A\varepsilon^{-\frac{r}{x}\theta}, \tag{7}$$

where A is still indefinite, and is determined by the initial conditions of the circuit, as follows:

for $\qquad \theta = 0, \qquad i = 0;$

hence, substituting in (7).

$$0 = \frac{E}{z} \cos (\theta_0 + \theta_1) + A,$$

or,

$$A = -\frac{E}{z} \cos (\theta_0 + \theta_1),\qquad (8)$$

and, substituted in (7),

$$i = \frac{E}{z} \left\{ \cos (\theta - \theta_0 - \theta_1) - \varepsilon^{-\frac{r}{x}\theta} \cos (\theta_0 + \theta_1) \right\}\qquad (9)$$

is the general expression of the current in the circuit.

If at the starting moment $\theta = 0$ the current is not zero but $= i_0$, we have, substituted in (7),

$$i_0 = \frac{E}{z} \cos (\theta_0 + \theta_1) + A,$$

$$A = i_0 - \frac{E}{z} \cos (\theta_0 + \theta_1),$$

$$i = \frac{E}{z} \left\{ \cos (\theta - \theta_0 - \theta_1) - \left(\cos (\theta_0 + \theta_1) - \frac{i_0 z}{E} \right) \varepsilon^{-\frac{r}{x}\theta} \right\}.\qquad (10)$$

27. The equation of current (9) contains a permanent term $\frac{E}{z} \cos (\theta - \theta_0 - \theta_1)$, which usually is the only term considered, and a transient term $\frac{E}{z} \varepsilon^{-\frac{r}{x}\theta} \cos (\theta_0 + \theta_1)$.

The greater the resistance r and smaller the reactance x, the more rapidly the term $\frac{E}{z} \varepsilon^{-\frac{r}{x}\theta} \cos (\theta_0 + \theta_1)$ disappears.

This transient term is a maximum if the circuit is closed at the moment $\theta_0 = -\theta_1$, that is, at the moment when the permanent value of current, $\frac{E}{z} \cos (\theta - \theta_0 - \theta_1)$, should be a maximum, and is then

$$\frac{E}{z} \varepsilon^{-\frac{r}{x}\theta}.$$

The transient term disappears if the circuit is closed at the moment $\theta_0 = 90° - \theta_1$, or when the stationary term of current passes the zero value.

As example is shown, in Fig. 7, the starting of the current under the conditions of maximum transient term, or $\theta_0 = -\theta_1$, in a circuit of the following constants: $\dfrac{x}{r} = 0.1$, corresponding approximately to a lighting circuit, where the permanent value

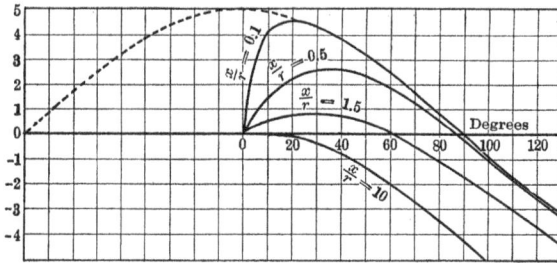

Fig. 7.   Starting current of an inductive circuit.

of current is reached in a small fraction of a half wave; $\dfrac{x}{r} = 0.5$, corresponding to the starting of an induction motor with rheostat in the secondary circuit; $\dfrac{x}{r} = 1.5$, corresponding to an unloaded transformer, or to the starting of an induction motor with short-circuited secondary, and $\dfrac{x}{r} = 10$, corresponding to a reactive coil.

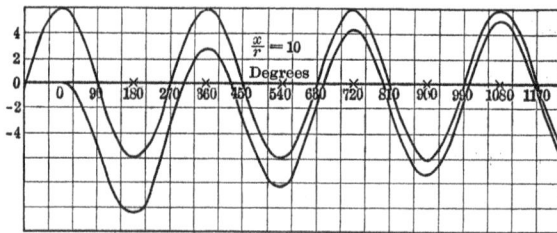

Fig. 8.   Starting current of an inductive circuit.

Of the last case, $\dfrac{x}{r} = 10$, a series of successive waves are plotted in Fig. 8, showing the very gradual approach to permanent condition.

Fig. 9 shows, for the circuit $\frac{x}{r} = 1.5$, the current when closing the circuit 0°, 30°, 60°, 90°, 120°, 150° respectively behind the zero value of permanent current.

The permanent value of current is shown in Fig. 7 in dotted line.

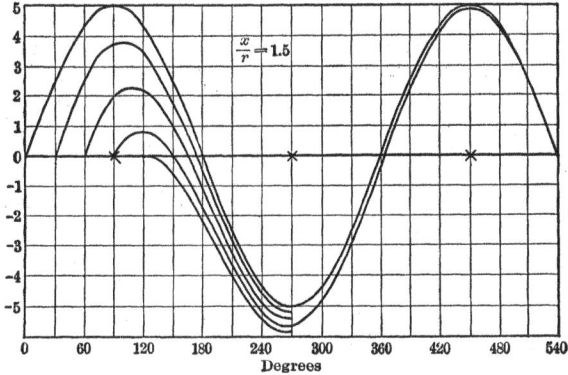

Fig. 9. Starting current of an inductive circuit.

28. Instead of considering, in Fig. 9, the current wave as consisting of the superposition of the permanent term $I \cos(\theta - \theta_0)$ and the transient term $- I\varepsilon^{-\frac{r}{x}\theta} \cos \theta_0$ the current wave can directly be represented by the permanent term

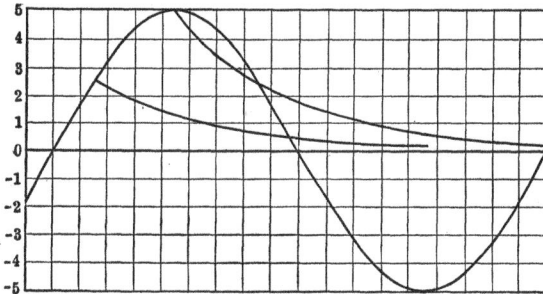

Fig. 10. Current wave represented directly.

$I \cos(\theta - \theta_0)$ by considering the zero line of the diagram as deflected exponentially to the curve $I\varepsilon^{-\frac{r}{x}\theta} \cos \theta_0$ in Fig. 10. That is, the instantaneous values of current are the vertical distances of the sine wave $I \cos (\theta - \theta_0)$ from the exponential curve $I\varepsilon^{-\frac{r}{x}\theta} \cos \theta_0$, starting at the initial value of permanent current.

In polar coordinates, in this case $I \cos (\theta - \theta_0)$ is the circle, $I\varepsilon^{-\frac{r}{x}\theta} \cos \theta_0$ the exponential or loxodromic spiral.

As a rule, the transient term in alternating-current circuits containing resistance and inductance is of importance only in circuits containing iron, where hysteresis and magnetic saturation complicate the phenomenon, or in circuits where unidirectional or periodically recurring changes take place, as in rectifiers, and some such cases are considered in the following chapters.

CHAPTER V.

29. If a continuous e.m.f. e is impressed upon a circuit containing resistance, inductance, and capacity in series, the stationary condition of the circuit is zero current, $i = o$, and the potential difference at the condenser equals the impressed e.m.f., $e_1 = e$, no permanent current exists, but only the transient current of charge or discharge of the condenser.

The capacity C of a condenser is defined by the equation

$$i = C \frac{de}{dt},$$

that is, the current into a condenser is proportional to the rate of increase of its e.m.f. and to the capacity.

It is therefore

$$de = \frac{1}{C} i dt,$$

and

$$e = \frac{1}{C} \int i dt \qquad (1)$$

is the potential difference at the terminals of a condenser of capacity C with current i in the circuit to the condenser.

Let then, in a circuit containing resistance, inductance, and capacity in series, $e =$ impressed e.m.f., whether continuous, alternating, pulsating, etc.; $i =$ current in the circuit at time t; $r =$ resistance; $L =$ inductance, and $C =$ capacity; then the e.m.f. consumed by resistance r is

$$ri;$$

the e.m.f. consumed by inductance L is

$$L \frac{di}{dt},$$

and the e.m.f. consumed by capacity C is

$$e_1 = \frac{1}{C} \int idt;$$

hence, the impressed e.m.f. is

$$e = ri + L\frac{di}{dt} + \frac{1}{C}\int idt, \qquad (2)$$

and herefrom the potential difference at the condenser terminals is

$$e_1 = \frac{1}{C}\int idt = e - ri - L\frac{di}{dt}. \qquad (3)$$

Equation (2) differentiated and rearranged gives

$$L\frac{d^2i}{dt^2} + r\frac{di}{dt} + \frac{1}{C}i = \frac{de}{dt} \qquad (4)$$

as the general differential equation of a circuit containing resistance, inductance, and capacity in series.

30. If the impressed e.m.f. is constant,

$$e = \text{constant},$$

then

$$\frac{de}{dt} = 0,$$

and equation (4) assumes the form, for continuous-current circuits,

$$L\frac{d^2i}{dt^2} + r\frac{di}{dt} + \frac{1}{C}i = 0. \qquad (5)$$

This equation is a linear relation between the dependent variable, i, and its differential quotients, and as such is integrated by an exponential function of the general form

$$i = A\varepsilon^{-at}. \qquad (6)$$

(This exponential function also includes the trigonometric functions sine and cosine, which are exponential functions with imaginary exponent a.)

Substituting (6) in (5) gives

$$\left(a^2L - ar + \frac{1}{C}\right)A\varepsilon^{-at} = 0;$$

this must be an identity, irrespective of the value of t, to make (6) the integral of (5). That is,

$$a^2L - ar + \frac{1}{C} = 0. \tag{7}$$

A is still indefinite, and therefore determined by the terminal conditions of the problem.

From (7) follows

$$a = \frac{r \pm \sqrt{r^2 - \frac{4L}{C}}}{2L}, \tag{8}$$

hence the two roots,

$$\left. \begin{array}{l} a_1 = \dfrac{r - s}{2L} \\[3mm] \text{and} \\[3mm] a_2 = \dfrac{r + s}{2L}, \end{array} \right\} \tag{9}$$

where

$$s = \sqrt{r^2 - \frac{4L}{C}}. \tag{10}$$

Since there are two roots, a_1 and a_2, either of the two expressions (6), ε^{-a_1t} and ε^{-a_2t}, and therefore also any combination of these two expressions, satisfies the differential equation (5).

That is, the general integral equation, or solution of differential equation (5), is

$$i = A_1\varepsilon^{-\frac{r-s}{2L}t} + A_2\varepsilon^{-\frac{r+s}{2L}t}. \tag{11}$$

Substituting (11) and (9) in equation (3) gives the potential difference at the condenser terminals as

$$e_1 = e - \left\{ \frac{r+s}{2}A_1\varepsilon^{-\frac{r-s}{2L}t} + \frac{r-s}{2}A_2\varepsilon^{-\frac{r+s}{2L}t} \right\} \tag{12}$$

31. Equations (11) and (12) contain two indeterminate constants, A_1 and A_2, which are the integration constants of the differential equation of second order, (5), and determined by the terminal conditions, the current and the potential difference at the condenser at the moment $t = 0$.

Inversely, since in a circuit containing inductance and capacity two electric quantities must be given at the moment of start of the phenomenon, the current and the condenser potential — representing the values of energy stored at the moment $t = 0$ as electromagnetic and as electrostatic energy, respectively — the equations must lead to two integration constants, that is, to a differential equation of second order.

Let $i = i_0$ = current and $e_1 = e_0$ = potential difference at condenser terminals at the moment $t = 0$; substituting in (11) and (12),

$$i_0 = A_1 + A_2$$

and

$$e_0 = e - \frac{r+s}{2} A_1 - \frac{r-s}{2} A_2;$$

hence,

$$A_1 = - \frac{e_0 - e + \dfrac{r-s}{2} i_0}{s}$$

and

$$A_2 = + \frac{e_0 - e + \dfrac{r+s}{2} i_0}{s},$$ $$(13)$$

and therefore, substituting in (11) and (12), the *current* is

$$i = \frac{e_0 - e + \dfrac{r+s}{2} i_0}{s} \varepsilon^{-\frac{r+s}{2L}t} - \frac{e_0 - e + \dfrac{r-s}{2} i_0}{s} \varepsilon^{-\frac{r-s}{2L}t}, \qquad (14)$$

the *condenser potential* is

$$e_1 = e - \frac{1}{2} \left\{ (r-s) \frac{e_0 - e + \dfrac{r+s}{2} i_0}{s} \varepsilon^{-\frac{r+s}{2L}t} - (r+s) \frac{e_0 - e + \dfrac{r-s}{2} i_0}{s} \varepsilon^{-\frac{r-s}{2L}t} \right\}$$ $$(15)$$

For *no condenser charge*, or $i_0 = 0$, $e_0 = 0$, we have

$$A_1 = \frac{e}{s}$$

and

$$A_2 = -\frac{e}{s} = -A_1;$$

substituting in (11) and (12), we get the *charging current* as

$$i = \frac{e}{s}\left\{ \varepsilon^{-\frac{r-s}{2L}t} - \varepsilon^{-\frac{r+s}{2L}t} \right\}. \tag{16}$$

The *condenser potential as*

$$e_1 = e\left\{ 1 - \frac{1}{2s}\left[(r+s)\,\varepsilon^{-\frac{r-s}{2L}t} - (r-s)\,\varepsilon^{-\frac{r+s}{2L}t} \right] \right\}. \tag{17}$$

For a *condenser discharge* or $i_0 = 0$, $e = e_0$, we have

$$A_1 = -\frac{e_0}{s}$$

and

$$A_2 = +\frac{e_0}{s} = -A_1;$$

hence, the *discharging current* is

$$i = -\frac{e_0}{s}\left\{ \varepsilon^{-\frac{r-s}{2L}t} - \varepsilon^{-\frac{r+s}{2L}t} \right\}. \tag{18}$$

The *condenser potential* is

$$e_1 = \frac{e_0}{2s}\left\{ (r+s)\,\varepsilon^{-\frac{r-s}{2L}t} - (r-s)\,\varepsilon^{-\frac{r+s}{2L}t} \right\}, \tag{19}$$

that is, in condenser discharge and in condenser charge the currents are the same, but opposite in direction, and the condenser potential rises in one case in the same way as it falls in the other.

32. As example is shown, in Fig. 11, the charge of a condenser of $C = 10$ mf. capacity by an impressed e.m.f. of

$e = 1000$ volts through a circuit of $r = 250$ ohms resistance and $L = 100$ mh. inductance; hence, $s = 150$ ohms, and the charging current is

$$i = 6.667 \left\{ \varepsilon^{-500t} - \varepsilon^{-2000t} \right\} \text{ amperes.}$$

The condenser potential is

$$e_1 = 1000 \left\{ 1 - 1.333 \, \varepsilon^{-500t} + 0.333 \, \varepsilon^{-2000t} \right\} \text{ volts.}$$

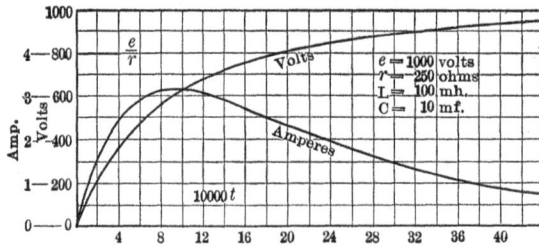

Fig. 11.   Charging a condenser through a circuit having resistance and inductance.   Constant potential.   Logarithmic charge.

33. The equations (14) to (19) contain the square root,

$$s = \sqrt{r^2 - \frac{4L}{C}},$$

hence, they apply in their present form only when

$$r^2 > \frac{4L}{C}.$$

If $r^2 = \dfrac{4L}{C}$, these equations become indeterminate, or $= \dfrac{0}{0}$, and if $r^2 < \dfrac{4L}{C}$, s is imaginary, and the equations assume a complex imaginary form.   In either case they have to be rearranged to assume a form suitable for application.

Three cases have thus to be distinguished:

(a) $r^2 > \dfrac{4L}{C}$, in which the equations of the circuit can be used in their present form.   Since the functions are exponential or logarithmic, this is called the *logarithmic* case.

(b) $r^2 = \dfrac{4L}{C}$ is called the *critical* case, marking the transition between (a) and (c), but belonging to neither.

(c) $r^2 < \dfrac{4L}{C}$. In this case trigonometric functions appear; it is called the *trigonometric* case, or *oscillation*.

34. In the *logarithmic* case,

$$r^2 > \frac{4L}{C}$$

or,
$$4L < Cr^2,$$

that is, with high resistance, or high capacity, or low inductance, equations (14) to (19) apply.

The term $\varepsilon^{-\frac{r-s}{2L}t}$ is always greater than $\varepsilon^{-\frac{r+s}{2L}t}$, since the former has a lower coefficient in the exponent, and the difference of these terms, in the equations of condenser charge and discharge, is always positive. That is, the current rises from zero at $t = 0$, reaches a maximum and then falls again to zero at $t = \infty$, but it never reverses. The maximum of the current is less than $i = \dfrac{e}{s}$.

The exponential term in equations (17) and (19) also never reverses. That is, the condenser potential gradually changes, without ever reversing or exceeding the impressed e.m.f. in the charge or the starting potential in the discharge.

Hence, in the case $r^2 > \dfrac{4L}{C}$, no abnormal voltage is produced in the circuit, and the transient term is of short duration, so that a condenser charge or discharge under these conditions is relatively harmless.

In charging or discharging a condenser, or in general a circuit containing capacity, the insertion of a resistance in series in the circuit of such value that $r^2 > \dfrac{4L}{C}$ therefore eliminates the danger from abnormal electrostatic or electromagnetic stresses.

In general, the higher the resistance of a circuit, compared with inductance and capacity, the more the transient term is suppressed.

35. In a circuit containing resistance and capacity but no inductance, $L = 0$, we have, substituting in (5),

$$r \frac{di}{dt} + \frac{1}{C} i = 0, \tag{20}$$

or, transposing,

$$\frac{di}{i} = - \frac{dt}{rC},$$

which is integrated by

$$i = c\varepsilon^{-\frac{t}{rC}}, \tag{21}$$

where c = integration constant.

Equation (21) gives for $t = 0$, $i = c$; that is, the current at the moment of closing the circuit must have a finite value, or must jump instantly from zero to c. This is not possible, but so also it is not possible to produce a circuit without any inductance whatever.

Therefore equation (21) does not apply for very small values of time, t, but for very small t the inductance, L, of the circuit, however small, determines the current.

The potential difference at the condenser terminals from (3) is

$$e_1 = e - ri$$

hence

$$e_1 = e - rc\varepsilon^{-\frac{t}{rC}} \tag{22}$$

The integration constant c cannot be determined from equation (21) at $t = 0$, since the current i makes a jump at this moment.

But from (22) it follows that if at the moment $t = 0$, $e_1 = e_0$, $e_0 = e - rc$,

hence,

$$c = \frac{e - e_0}{r},$$

and herefrom the equations of the non-inductive condenser circuit,

$$i = \frac{(e - e_0)\varepsilon^{-\frac{t}{rC}}}{r} \tag{23}$$

and

$$e_1 = e - (e - e_0)\varepsilon^{-\frac{t}{rC}}. \tag{24}$$

As seen, these equations do not depend upon the current i_0 in the circuit at the moment before $t = 0$.

36. These equations do not apply for very small values of t, but in this case the inductance, L, has to be considered, that is, equations (14) to (19) used.

For $L = 0$ the second term in (14) becomes indefinite, as it contains $\varepsilon^{\frac{0}{0}t}$, and therefore has to be evaluated as follows:

For $L = 0$, we have

$$s = r,$$

$$\frac{r + s}{2} = r,$$

and

$$\frac{r - s}{2} = 0$$

and, developed by the binomial theorem, dropping all but the first term,

$$r - s = r \left\{ 1 - \sqrt{1 - \frac{4\,L}{r^2 C}} \right\}$$

$$= \frac{2\,L}{rC},$$

and

$$\frac{r - s}{2\,L} = \frac{1}{rC},$$

$$\frac{r + s}{2\,L} = \frac{r}{L}.$$

Substituting these values in equations (14) and (15) gives the current as

$$i = \frac{e - e_0}{r} \varepsilon^{-\frac{t}{rC}} - \frac{e - e_0 - r\dot{i}_0}{r} \varepsilon^{-\frac{r}{L}t} \tag{25}$$

and the potential difference at the condenser as

$$e_1 = e - (e - e_0)\,\varepsilon^{-\frac{t}{rC}}; \tag{26}$$

that is, in the equation of the current, the term

$$- \frac{e - e_0 - r\dot{i}_0}{r} \varepsilon^{-\frac{r}{L}t}$$

has to be added to equation (23). This term makes the transition from the circuit conditions before $t = 0$ to those after $t = 0$, and is of extremely short duration.

For instance, choosing the same constants as in § 32, namely: $e = 1000$ volts; $r = 250$ ohms; $C = 10$ mf., but choosing the inductance as low as possible, $L = 5$ mh., gives the equations of condenser charge, i.e., for $i_0 = 0$ and $e_0 = 0$,

$$i = 4 \left\{ \varepsilon^{-400t} - \varepsilon^{-50,000t} \right\}$$

and

$$e_1 = 1000 \left\{ 1 - \varepsilon^{-400t} \right\}.$$

The second term in the equation of the current, $\varepsilon^{-50,000t}$, has decreased already to 1 per cent after $t = 17.3 \times 10^{-6}$ seconds, while the first term, ε^{-400t}, has during this time decreased only by 0.7 per cent, that is, it has not yet appreciably decreased.

37. In the *critical* case,

$$r^2 = \frac{4 L}{C}$$

and

$$s = 0,$$

$$a_1 = a_2 = \frac{r}{2 L},$$

$$A_1 = - A_2 = \frac{e - e_0 - \dfrac{r}{2} i_0}{s}.$$

Hence, substituting in equation (14) and rearranging,

$$i = \varepsilon^{-\frac{r}{2L}t} \left\{ i_0 \left(\varepsilon^{\frac{s}{2L}t} + \varepsilon^{-\frac{s}{2L}t} \right) + \left(e - e_0 - \frac{r}{2} i_0 \right) \frac{\varepsilon^{\frac{s}{2L}t} - \varepsilon^{-\frac{s}{2L}t}}{s} \right\} \cdot \quad (27)$$

The last term of this equation,

$$F = \frac{N}{D} = \frac{\varepsilon^{\frac{s}{2L}t} - \varepsilon^{-\frac{s}{2L}t}}{s} = \frac{0}{0},$$

that is, becomes indeterminate for $s = 0$, and therefore is evaluated by differentiation,

$$F = \frac{\dfrac{dN}{ds}}{\dfrac{dD}{ds}} = \frac{t}{L}. \tag{28}$$

Substituting (28) in (27) gives the equation of *current*,

$$i = \left[i_0 + \frac{t}{L}\left(e - e_0 - \frac{r}{2} i_0 \right) \right] \epsilon^{-\frac{r}{2L}t}. \tag{29}$$

The condenser potential is found, by substituting in (15), to be

$$e_1 = e - \frac{1}{2}\epsilon^{-\frac{r}{2L}t} \left\{ (e - e_0)(\epsilon^{\frac{s}{2L}t} + \epsilon^{-\frac{s}{2L}t}) - \frac{r}{s}\left(e - e_0 - \frac{r i_0}{2} \right)\left(\epsilon^{\frac{s}{2L}t} - \epsilon^{-\frac{s}{2L}t} \right) \right\} \tag{30}$$

The last term of this equation is, for $s = 0$:

$$\frac{r}{s}\left(e - e_0 - \frac{r i_0}{2} \right)\left(\epsilon^{\frac{s}{2L}t} - \epsilon^{-\frac{s}{2L}t} \right) = \frac{rt}{L}\left(e - e_0 - \frac{r i_0}{2} \right) \tag{31}$$

This gives the *condenser potential* as:

$$e_1 = e - \epsilon^{-\frac{r}{2L}t}\left\{ (e - e_0) + \frac{rt}{2L}\left(e - e_0 - \frac{r i_0}{2} \right) \right\} \tag{32}$$

Herefrom it follows that for the condenser charge, $i_0 = 0$ and $e_0 = 0$,

$$i = \frac{t}{L}e\,\varepsilon^{-\frac{r}{2L}t}$$

and

$$e_1 = e\left\{ 1 - \left(1 + \frac{rt}{2L} \right)\varepsilon^{-\frac{r}{2L}t} \right\};$$

for the condenser discharge, $i_0 = 0$ and $e = 0$,

$$i = -\frac{t}{L} e_0 \varepsilon^{-\frac{r}{2L}t}$$

and

$$e_1 = \left(1 + \frac{rt}{2L}\right) e_0 \varepsilon^{-\frac{r}{2L}t}.$$

38. As an example are shown, in Fig. 12, the charging current and the potential difference at the terminals of the condenser,

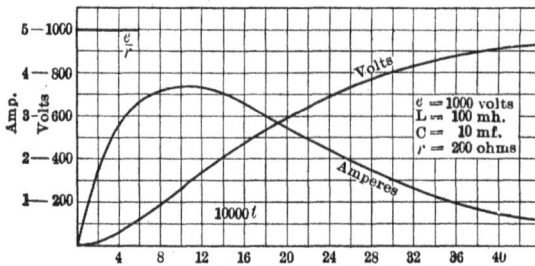

Fig. 12.   Charging a condenser through a circuit having resistance and inductance.   Constant potential.   Critical charge.

in a circuit having the constants, $e = 1000$ volts; $C = 10$ mf.; $L = 100$ mh., and such resistance as to give the critical start, that is,

$$r = \sqrt{\frac{4L}{C}} = 200 \text{ ohms.}$$

In this case,

$$i = 10,000 \, t\varepsilon^{-1000t}$$

and

$$e_1 = 1000 \left\{1 - (1 + 1000 \, t) \, \varepsilon^{-1000t}\right\}.$$

39. In the *trigonometric* or *oscillating* case,

$$r^2 < \frac{4L}{C}.$$

The term under the square root (10) is negative, that is, the square root, s, is imaginary, and a_1 and a_2 are complex imaginary quantities, so that the equations (11) and (12) appear in imaginary form.   They obviously can be reduced to real terms, since the phenomenon is real. Since an exponential function with imaginary exponents is a trigonometric function, and inversely, the solution of the equation thus leads to trigonometric functions, that is, the phenomenon is periodic or oscillating.

Substituting $s = jq$, we have

$$q = \sqrt{\frac{4L}{C} - r^2} \tag{33}$$

and

$$\left. \begin{aligned} a_1 &= \frac{r - jq}{2}, \\ a_2 &= \frac{r + jq}{2}. \end{aligned} \right\} \tag{34}$$

Substituting (34) in (11) and (12), and rearranging,

$$i = \varepsilon^{-\frac{r}{2L}t} \left\{ A_1 \varepsilon^{+\frac{jq}{2L}t} + A_2 \varepsilon^{-\frac{jq}{2L}t} \right\}, \tag{35}$$

$$e_1 = e - \varepsilon^{-\frac{r}{2L}t} \left\{ \frac{r + jq}{2} A_1 \varepsilon^{+\frac{jq}{2L}t} + \frac{r - jq}{2} A_2 \varepsilon^{-\frac{jq}{2L}t} \right\}. \tag{36}$$

Between the exponential function and the trigonometric functions exist the relations

$$\left. \begin{aligned} \varepsilon^{+jv} &= \cos v + j \sin v \\ \varepsilon^{-jv} &= \cos v - j \sin v. \end{aligned} \right\} \tag{37}$$

and

Substituting (37) in (35), and rearranging, gives

$$i = \varepsilon^{-\frac{r}{2L}t} \left\{ (A_1 + A_2) \cos \frac{q}{2L} t + j (A_1 - A_2) \sin \frac{q}{2L} t \right\}.$$

Substituting the two new integration constants,

$$\left. \begin{aligned} B_1 &= A_1 + A_2 \\ B_2 &= j (A_1 - A_2), \end{aligned} \right\} \tag{38}$$

and

gives

$$i = \varepsilon^{-\frac{r}{2L}t} \left\{ B_1 \cos \frac{q}{2L} t + B_2 \sin \frac{q}{2L} t \right\}. \tag{39}$$

In the same manner, substituting (37) in (36), rearranging, and substituting (38), gives

$$e_1 = e - \varepsilon^{-\frac{r}{2L}t} \left\{ \frac{rB_1 + qB_2}{2} \cos \frac{q}{2L} t + \frac{rB_2 - qB_1}{2} \sin \frac{q}{2L} t \right\}. \quad (40)$$

B_1 and B_2 are now the two integration constants, determined by the terminal conditions. That is, for $t = 0$, let $i = i_0 =$ current and $e_1 = e_0 =$ potential difference at condenser terminals, and substituting these values in (39) and (40) gives

$$i_0 = B_1$$

and

$$e_0 = e - \frac{rB_1 + qB_2}{2},$$

hence,

and

$$\left. \begin{aligned} B_1 &= i_0 \\[2mm] B_2 &= \frac{2(e - e_0) - ri_0}{q} \end{aligned} \right\} \quad (41)$$

Substituting (41) in (39) and (40) gives the general *equations of condenser oscillation:*
the current is

$$i = \varepsilon^{-\frac{r}{2L}t} \left\{ i_0 \cos \frac{q}{2L} t + \frac{2(e - e_0) - ri_0}{q} \sin \frac{q}{2L} t \right\}, \quad (42)$$

and the potential difference at condenser terminals is

$$e_1 = e - \varepsilon^{-\frac{r}{2L}t} \left\{ (e - e_0) \cos \frac{q}{2L} t + \frac{r(e - e_0) - \frac{r^2 + q^2}{2} i_0}{q} \sin \frac{q}{2L} t \right\}. \quad (43)$$

Herefrom follow the equations of condenser charge and discharge, as special case:
For *condenser charge,* $i_0 = 0$; $e_0 = 0$, we have

$$i = \frac{2e}{q} \varepsilon^{-\frac{r}{2L}t} \sin \frac{q}{2L} t \quad (44)$$

and
$$e_1 = e\left\{1 - \varepsilon^{-\frac{r}{2L}t}\left(\cos\frac{q}{2L}t + \frac{r}{q}\sin\frac{q}{2L}t\right)\right\}, \tag{45}$$

and for *condenser discharge*, $i_0 = 0$, $e = 0$, we have

$$i = -\frac{2e_0}{q}\varepsilon^{-\frac{r}{2L}t}\sin\frac{q}{2L}t \tag{46}$$

and

$$e_1 = e_0\varepsilon^{-\frac{r}{2L}t}\left\{\cos\frac{q}{2L}t + \frac{r}{q}\sin\frac{q}{2L}t\right\}. \tag{47}$$

40. As an example is shown the oscillation of condenser charge in a circuit having the constants, $e = 1000$ volts; $L = 100$ mh., and $C = 10$ mf.

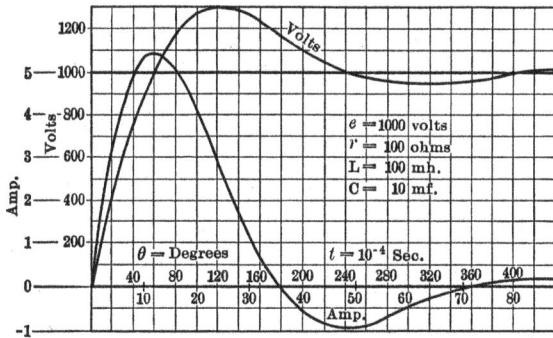

Fig. 13. Charging a condenser through a circuit having resistance and inductance. Constant potential. Oscillating charge.

(*a*) In Fig. 13, $r = 100$ ohms, hence, $q = 173$ and the current is
$$i = 11.55\,\varepsilon^{-500\,t}\sin 866\,t;$$
the condenser potential is
$$e_1 = 1000\left\{1 - \varepsilon^{-500\,t}\left(\cos 866\,t + 0.577\sin 866\,t\right)\right\}.$$

(*b*) In Fig. 14, $r = 40$ ohms, hence, $q = 196$ and the current is
$$i = 10.2\,\varepsilon^{-200\,t}\sin 980\,t;$$
the condenser potential is
$$e_1 = 1000\left\{1 - \varepsilon^{-200\,t}\left(\cos 980\,t + 0.21\sin 980\,t\right)\right\}.$$

41. Since the equations of current and potential difference (42) to (47) contain trigonometric functions, the phenomena are periodic or waves, similar to alternating currents. They differ from the latter by containing an exponential factor $\varepsilon^{-\frac{r}{2L}t}$, which steadily decreases with increase of t. That is, the suc-

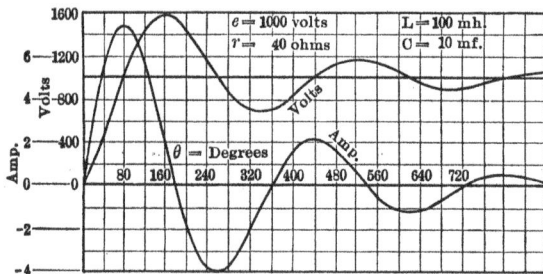

Fig. 14. Charging a condenser through a circuit having resistance and inductance. Constant potential. Oscillating charge.

cessive half waves of current and of condenser potential progressively decrease in amplitude. Such alternating waves of progressively decreasing amplitude are called oscillating waves.

Since equations (42) to (47) are periodic, the time t can be represented by an angle θ, so that one complete period is denoted by 2π or one complete revolution,

$$\theta = \frac{q}{2L}t = 2\pi ft. \tag{48}$$

$$2\pi f = \frac{q}{2L},$$

hence, the frequency of oscillation is

$$f = \frac{q}{4\pi L}, \tag{49}$$

or, substituting

$$q = \sqrt{\frac{4L}{C} - r^2}$$

gives the frequency of oscillation as

$$f = \frac{1}{2\pi}\sqrt{\frac{1}{LC} - \left(\frac{r}{2L}\right)^2}. \tag{50}$$

This frequency decreases with increasing resistance r, and becomes zero for $\left(\dfrac{r}{2\,L}\right)^2 = \dfrac{1}{LC}$, that is, $r^2 = \dfrac{4\,L}{C}$, or the critical case, where the phenomenon ceases to be oscillating.

If the resistance is small, so that the second term in equation (50) can be neglected, the frequency of oscillation is

$$f = \frac{1}{2\,\pi\sqrt{LC}}\,. \tag{51}$$

Substituting θ for t by equation (48)

$$t = \frac{2\,L}{q}\,\theta$$

in equations (42) and (43) gives the *general equations*,

$$i = \varepsilon^{-\frac{r}{q}\theta}\left\{i_0 \cos \theta + \frac{(e - e_0) - \dfrac{r}{2}\,i_0}{2\,q}\sin \theta\right\}, \tag{52}$$

$$e_1 = e - \varepsilon^{-\frac{r}{q}\theta}\left\{(e - e_0)\cos \theta + \frac{r\,(e - e_0) - \dfrac{r^2 + q^2}{2}\,i_0}{q}\sin \theta\right\}, \tag{53}$$

$$\theta = 2\,\pi ft \tag{48}$$

and

$$f = \frac{q}{4\,\pi L} = \frac{1}{2\,\pi}\sqrt{\frac{1}{LC} - \left(\frac{r}{2\,L}\right)^2}\,. \tag{50}$$

42. If the resistance r can be neglected, that is, if r^2 is small compared with $\dfrac{4\,L}{C}$, the following equations are approximately exact:

$$q = 2\sqrt{\frac{L}{C}} \tag{54}$$

and

$$f = \frac{1}{2\,\pi\sqrt{LC}},$$

or,

$$2\,\pi f = \frac{1}{\sqrt{LC}}\,. \tag{55}$$

Introducing now $x = 2 \pi f L =$ inductive reactance and $x' = \dfrac{1}{2 \pi f C} =$ capacity reactance, and substituting (55), we have

$$x = \sqrt{\frac{L}{C}}$$

and

$$x' = \sqrt{\frac{L}{C}},$$

hence, $x' = x,$

that is, the frequency of oscillation of a circuit containing inductance and capacity, but negligible resistance, is that frequency f which makes the condensive reactance $x' = \dfrac{1}{2 \pi f C}$ equal the inductive reactance $x = 2 \pi f L$:

$$x' = x = \sqrt{\frac{L}{C}}. \tag{56}$$

Then (54),

$$q = 2 x, \tag{57}$$

and the general equations (52) and (53) are

$$i = \varepsilon^{-\frac{r}{2x}\theta} \left\{ i_0 \cos \theta + \frac{(e - e_0) - \dfrac{r}{2} i_0}{4 x} \sin \theta \right\}; \tag{58}$$

$$e_1 = e - \varepsilon^{-\frac{r}{2x}\theta} \left\{ (e - e_0) \cos \theta + \frac{r (e - e_0) - 2 x^2 i_0}{2 x} \sin \theta \right\}; \tag{59}$$

$$x = \sqrt{\frac{L}{C}} \tag{56}$$

and by (48) and (55):

$$\theta = \frac{t}{\sqrt{LC}}.$$

43. Due to the factor $\varepsilon^{-\frac{r}{2L}t}$, successive half waves of oscillation decrease the more in amplitude, the greater the resistance r.

The ratio of the amplitude of successive half waves, or the decrement of the oscillation, is $\Delta = \varepsilon^{-\frac{r}{2L}t_1}$. where $t_1 =$ duration of one half wave or one half cycle, $= \dfrac{1}{2f}$.

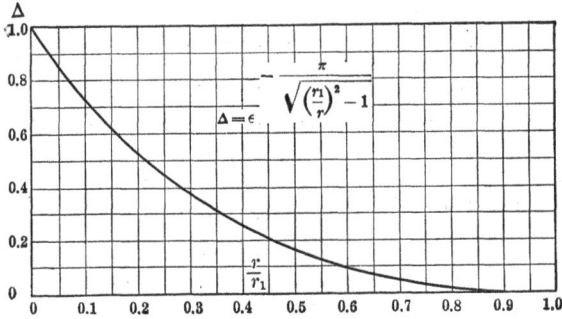

Fig. 15.   Decrement of Oscillation.

Hence, from (50),

$$t_1 = \frac{\pi}{\sqrt{\dfrac{1}{LC} - \left(\dfrac{r}{2L}\right)^2}},$$

and

$$\Delta = \varepsilon^{-\delta} = \varepsilon^{-\frac{\pi}{\sqrt{\frac{4L}{r^2C}-1}}}. \tag{60}$$

Denoting the critical resistance as

$$r_1^2 = \frac{4L}{C}, \tag{61}$$

we have

$$\left. \begin{aligned} \Delta &= \varepsilon^{-\delta} = \varepsilon^{-\frac{\pi}{\sqrt{\left(\frac{r_1}{r}\right)^2-1}}}, \\ \text{or,} \qquad \delta &= \frac{\pi}{\sqrt{\left(\frac{r_1}{r}\right)^2-1}}; \end{aligned} \right\} \tag{62}$$

that is, the decrement of the oscillating wave, or the decay of the oscillation, is a function only of the ratio of the resistance of the circuit to its critical resistance, that is, the minimum resistance which makes the phenomenon non-oscillatory.

In Fig. 15 are shown the numerical values of the decrement Δ, for different ratios of actual to critical resistance $\dfrac{r}{r_1}$.

As seen, for $r > 0.21\ r_1$, or a resistance of the circuit of more than 21 per cent of its critical resistance, the decrement Δ is below 50 per cent, or the second half wave less than half the first one, etc.; that is, very little oscillation is left.

Where resistance is inserted into a circuit to eliminate the danger from oscillations, one-fifth of the critical resistance, or $r = 0.4\ \sqrt{\dfrac{L}{C}}$, seems sufficient to practically dampen out the oscillation.

CHAPTER VI.

OSCILLATING CURRENTS.

44. The charge and discharge of a condenser through an inductive circuit produces periodic currents of a frequency depending upon the circuit constants.

The range of frequencies which can be produced by electrodynamic machinery is rather limited: synchronous machines or ordinary alternators can give economically and in units of larger size frequencies from 10 to 125 cycles. Frequencies below 10 cycles are available by commutating machines with low frequency excitation. Above 125 cycles the difficulties rapidly increase, due to the great number of poles, high peripheral speed, high power required for field excitation, poor regulation due to the massing of the conductors, which is required because of the small pitch per pole of the machine, etc., so that 1000 cycles probably is the limit of generation of constant potential alternating currents of appreciable power and at fair efficiency. For smaller powers, by using capacity for excitation, inductor alternators have been built and are in commercial service for wireless telegraphy and telephony, for frequencies up to 100,000 and even 200,000 cycles per second.

Still, even going to the limits of peripheral speed, and sacrificing everything for high frequency, a limit is reached in the frequency available by electrodynamic generation.

It becomes of importance, therefore, to investigate whether by the use of the condenser discharge the range of frequencies can be extended.

Since the oscillating current approaches the effect of an alternating current only if the damping is small, that is, the resistance low, the condenser discharge can be used as high frequency generator only by making the circuit of as low resistance as possible.

This, however, means limited power. When generating oscillating currents by condenser discharge, the load put on the circuit, that is, the power consumed in the oscillating-current circuit, represents an effective resistance, which increases the rapidity of the decay of the oscillation, and thus limits the power, and, when approaching the critical value, also lowers the frequency. This is obvious, since the oscillating current is the dissipation of the energy stored electrostatically in the condenser, and the higher the resistance of the circuit, the more rapidly is this energy dissipated, that is, the faster the oscillation dies out.

With a resistance of the circuit sufficiently low to give a fairly well sustained oscillation, the frequency is, with sufficient approximation,

$$f = \frac{1}{2\pi\sqrt{LC}}.$$

45. The constants, capacity, C, inductance, L, and resistance, r, have no relation to the size or bulk of the apparatus. For instance, a condenser of 1 mf., built to stand continuously a potential of 10,000 volts, is far larger than a 200-volt condenser of 100 mf. capacity. The energy which the former is able to store is $\dfrac{Ce^2}{2} = 50$ joules, while the latter stores only 2 joules, and therefore the former is 25 times as large.

A reactive coil of 0.1 henry inductance, designed to carry continuously 100 amperes, stores $\dfrac{Li^2}{2} = 500$ joules; a reactive coil of 1000 times the inductance, 100 henrys, but of a current-carrying capacity of 1 ampere, stores 5 joules only, therefore is only about one-hundredth the size of the former.

A resistor of 1 ohm, carrying continuously 1000 amperes, is a ponderous mass, dissipating 1000 kw.; a resistor having a resistance a million times as large, of one megohm, may be a lead pencil scratch on a piece of porcelain.

Therefore the size or bulk of condensers and reactors depends not only on C and L but also on the voltage and current which can be applied continuously, that is, it is approximately proportional to the energy stored, $\dfrac{Ce^2}{2}$ and $\dfrac{Li^2}{2}$, or since in electrical engineering energy is a quantity less frequently used than power, condensers and reactors are usually characterized by the power or rather apparent power which can be impressed upon them continuously by referring to a standard frequency, for which 60 cycles is generally used.

That means that reactors, condensers, and resistors are rated in kilowatts or kilovolt-amperes, just as other electrical apparatus, and this rating characterizes their size within the limits of design, while a statement like "a condenser of 10 mf." or "a reactor of 100 mh." no more characterizes the size than a statement like "an alternator of 100 amperes capacity" or "a transformer of 1000 volts."

A bulk of 1 cu. ft. in condenser can give about 5 to 10 kv-amp. at 60 cycles. Hence, 100 kv-amp. constitutes a very large size of condenser.

In the oscillating condenser discharge, the frequency of oscillation is such that the inductive reactance equals the condensive reactance. The same current is in both at the same terminal voltage. That means that the volt-amperes consumed by the inductance equal the volt-amperes consumed by the capacity.

The kilovolt-amperes of a condenser as well as of a reactor are proportional to the frequency. With increasing frequency, at constant voltage impressed upon the condenser, the current varies proportionally with the frequency; at constant alternating current through the reactor, the voltage varies proportionally with the frequency.

If then at the frequency of oscillation, reactor and condenser have the same kv-amp., they also have the same at 60 cycles.

A 100-kv-amp. condenser requires a 100-kv-amp. reactive coil for generating oscillating currents. A 100-kv-amp. reactive coil has approximately the same size as a 50-kw. transformer and can indeed be made from such a transformer, of ratio 1 : 1, by connecting the two coils in series and inserting into the magnetic circuit an air gap of such length as to give the rated magnetic density at the rated current.

A very large oscillating-current generator, therefore, would consist of 100-kv-amp. condenser and 100-kv-amp. reactor.

46. Assuming the condenser to be designed for 10,000 volts alternating impressed e.m.f. at 60 cycles, the 100 kv-amp. condenser consumes 10 amperes: its condensive reactance is $x_c = \dfrac{E}{I} = 1000$ ohms, and the capacity $C = \dfrac{1}{2\,\pi f_0 x_c} = 2.65$ mf.

Designing the reactor for different currents, and therewith different voltages, gives different values of inductance L, and therefore of frequency of oscillation f.

From the equations of the instantaneous values of the condenser discharge, (46) and (47), follow their effective values, or $\sqrt{\text{mean square}}$,

$$e_1 = \frac{e_0}{\sqrt{2}}\,\varepsilon^{-\frac{r}{2L}t}$$

and

$$i = \frac{e_0\sqrt{2}}{q}\,\varepsilon^{-\frac{r}{2L}t} = \frac{e_0}{\sqrt{2}}\sqrt{\frac{C}{L}}\,\varepsilon^{-\frac{r}{2L}t},$$

$$(63)$$

and thus the power,

$$p_1 = e_1 i = \frac{e_0^2}{2}\sqrt{\frac{C}{L}}\,\varepsilon^{-\frac{r}{L}t}, \qquad (64)$$

since for small values of r

$$q = 2\sqrt{\frac{L}{C}}.$$

Herefrom would follow that the energy of each discharge is

$$W = \int_0^\infty p_1 dt = \frac{e_0^2}{2\,r}\sqrt{CL}. \qquad (65)$$

Therefore, for 10,000 volts effective at 60 cycles at the condenser terminals, the e.m.f. is

$$e_0 = 10,000\sqrt{2},$$

and the condenser voltage is

$$e_1 = 10,000\,\varepsilon^{-\frac{r}{2L}t}.$$

Designing now the 100-kv-amp. reactive coil for different voltages and currents gives for an oscillation of 10,000 volts:

Reactive Coil.		React- ance.	Inductance.	Frequency of Oscillation.	Oscillating Current.	Oscillating Power.
Amp. i_0.	Volts, e_0.	$\frac{e_0}{i_0} = x_0$.	$\frac{x_0}{2\pi f_0} = L$.	$f = \dfrac{1}{2\pi\sqrt{LC}}$.	Amp., i.	Kv-amp., p_1.
1	100,000	10^5	265	6	1	10
10	10,000	10^3	2.65	60	10	100
100	1,000	10	2.65×10^{-2}	600	100	1,000
1,000	100	10^{-1}	2.65×10^{-4}	6,000	1,000	10,000
10,000	10	10^{-3}	2.65×10^{-6}	60,000	10,000	100,000
100,000	1	10^{-5}	2.65×10^{-8}	600,000	100,000	1,000,000
					$\times \varepsilon^{-\frac{r}{2L}t}$	$\times \varepsilon^{-\frac{r}{L}t}$

As seen, with the same kilovolt-ampere capacity of condenser and of reactive coil, practically any frequency of oscillation can be produced, from low commercial frequencies up to hundred thousands of cycles.

At frequencies between 500 and 2000 cycles, the use of iron in the reactive coil has to be restricted to an inner core, and at frequencies above this iron cannot be used, since hysteresis and eddy currents would cause excessive damping of the oscillation. The reactive coil then becomes larger in size.

47. Assuming 96 per cent efficiency of the reactive coil and 99 per cent of the condenser,

$$r = 0.05\,x,$$

gives

$$r = 0.05\sqrt{\frac{L}{C}},$$

since

$$x = 2\pi f L,$$

$$f = \frac{1}{2\pi\sqrt{LC}},$$

and the energy of the discharge, by (65), is

$$W = \frac{e_0^{\,2}}{2\,r}\sqrt{LC} = 10\,e_0^{\,2}\,C \text{ volt-ampere-seconds;}$$

thus the power factor is

$$\cos\theta_0 = 0.05.$$

Since the energy stored in the capacity is

$$W_0 = \frac{e_0^2 C}{2} \text{ joules,}$$

the critical resistance is

$$r_1 = 2\sqrt{\frac{L}{C}};$$

hence,

$$\frac{r}{r_1} = 0.025,$$

and the decrement of the oscillation is

$$\Delta = 0.92,$$

that is, the decay of the wave is very slow at no load.

Assuming, however, as load an external effective resistance equal to three times the internal resistance, that is, an electrical efficiency of 75 per cent, gives the total resistance as

$$r + r' = 0.2\,x;$$

hence,

$$\frac{r + r'}{r_1} = 0.1,$$

and the decrement is

$$\Delta = 0.73;$$

hence a fairly rapid decay of the wave.

At high frequencies, electrostatic, inductive, and radiation losses greatly increase the resistance, thus giving lower efficiency and more rapid decay of the wave.

48. The frequency of oscillation does not directly depend upon the size of apparatus, that is, the kilovolt-ampere capacity of condenser and reactor. Assuming, for instance, the size, in kilovolt-amperes, reduced to $\frac{1}{n}$, then, if designed for the same voltage, condenser and reactor, each takes $\frac{1}{n}$ the current, that is, the condensive reactance is n times as great, and therefore the capacity of the condenser, C, reduced to $\frac{1}{n}$, the inductance, L, is increased n-fold, so that the product CL, and thereby the frequency, remains the same; the power output, however, of the oscillating currents is reduced to $\dfrac{1}{n}$.

The limit of frequency is given by the mechanical dimensions.

With a bulk of condenser of 10 to 20 cu. ft., the minimum length of the discharge circuit cannot well be less than 10 ft.; 10 ft. of conductor of large size have an inductance of at least 0.002 mh. $= 2 \times 10^{-6}$, and the frequency of oscillation would therefore be limited to about 60,000 cycles per second, even without any reactive coil, in a straight discharge path.

The highest frequency which can be reached may be estimated about as follows:

The minimum length of discharge circuit is the gap between the condenser plates.

The minimum condenser capacity is given by two spheres, since small plates give a larger capacity, due to the edges.

The minimum diameter of the spheres is 1.5 times their distance, since a smaller sphere diameter does not give a clean spark discharge, but a brush discharge precedes the spark.

With $e_0 = 10{,}000\,\sqrt{2}$, the spark gap length between spheres is $e = 0.3$ in., and the diameter of the spheres therefore 0.45 in. The oscillating circuit then consists of two spheres of 0.45 in., separated by a gap of 0.3 in.

This gives an approximate length of oscillating circuit of 0.5 in., or an inductance $L = \dfrac{0.3 \times 10^{-3}}{24{,}000} = 0.125 \times 10^{-7}$ henry.

The capacity of the spheres against each other may be estimated as $C = 50 \times 10^{-8}$ mf.; this gives the frequency of oscillation as

$$f = \frac{1}{2\,\pi\sqrt{LC}} = 2 \times 10^{9},$$

or, 2 billion cycles.

At $\qquad\qquad e_0 = 10{,}000\,\sqrt{2}$ volts,

$$e_1 = 10{,}000\,\varepsilon^{-\frac{r}{2L}t} \text{ volts,}$$

$$i = 2.83\,\varepsilon^{-\frac{r}{2L}t} \text{ amp.,}$$

and $\qquad\qquad p_1 = 28.3\,\varepsilon^{-\frac{r}{L}t}$ kv-amp.

Reducing the size and spacing of the spheres proportionally, and proportionally lowering the voltage, or increasing the dielectric strength of the gap by increasing the air pressure, gives still higher frequencies.

As seen, however, the power of the oscillation decreases with increasing frequency, due to the decrease of size and therewith of storage ability, of capacity, and of inductance.

With a frequency of billions of cycles per second, the effective resistance must be very large, and therefore the damping rapid.

Such an oscillating system of two spheres separated by a gap would have to be charged by induction, or the spheres charged separately and then brought near each other, or the spheres may be made a part of a series of spheres separated by gaps and connected across a high potential circuit, as in some forms of lightning arresters.

Herefrom it appears that the highest frequency of oscillation of appreciable power which can be produced by a condenser discharge reaches billions of cycles per second, thus is enormously higher than the highest frequencies which can be produced by electrodynamic machinery.

At five billion cycles per second, the wave length is about 6 cm., that is, the frequency only a few octaves lower than the lowest frequencies observed as heat radiation or ultra red light.

The average wave length of visible light, 55×10^{-6} cm., corresponding to a frequency of 5.5×10^{14} cycles, would require spheres 10^{-5} cm. in diameter, that is, approaching molecular dimensions.

OSCILLATING-CURRENT GENERATOR.

49. A system of constant impressed e.m.f., e, charging a condenser C through a circuit of inductance L and resistance r, with a discharge circuit of the condenser, C, comprising an air gap in series with a reactor of inductance L_0 and a resistor of resistance r_0, is a generator of oscillating current if the air gap is set for such a voltage e_0 that it discharges before the voltage of the condenser C has reached the maximum, and if the resistance r_0 is such as to make the condenser discharge oscillatory, that is,

$$r_0^2 < \frac{4 L_0}{C}.$$

In such a system, as shown diagrammatically in Fig. 16, as soon, during the charge of the condenser, as the terminal voltage at C and thereby at the spark gap has reached the value e_0, the condenser C discharges over this spark gap, its potential difference falls to zero, then it charges again up to potential difference e_0, discharges, etc. Thus a series of oscillating discharges

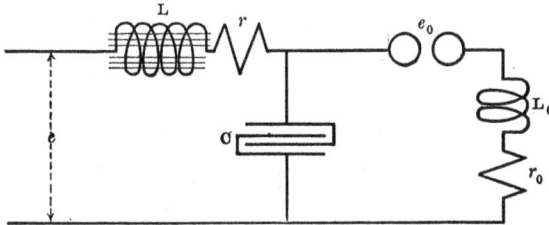

Fig. 16.   Oscillating-current generator.

occur in the circuit, L_0, r_0, at intervals equal to the time required to charge condenser C over reactor L and resistor r, up to the potential difference e_0, with an impressed e.m.f. e.

The resistance, r, obviously should be as low as possible, to get good efficiency of transformation; the inductance, L, must be so large that the time required to charge condenser C to potential e_0 is sufficient for the discharge over L_0, r_0 to die out and also the spark gap e_0 to open, that is, the conducting products of the spark in the gap e_0 to dissipate. This latter takes a considerable time, and an air blast directed against the spark gap e_0, by carrying away the products of the discharge, permits a more rapid recurrence of the discharge. The velocity of the air blast (and therefore the pressure of the air) must be such as to carry the ionized air or the metal vapors which the discharge forms in the gap e_0 out of the discharge path faster than the condenser recharges.

Assuming, for instance, the spark gap, e_0, set for 20,000 volts, or about 0.75 in., the motion of the air blast during successive discharges then should be large compared with 0.75 in., hence at least 3 to 6 in. With 1000 discharges per second, this would require an air velocity of $v = 250$ to 500 feet per second, with 5000 discharges per second an air velocity of $v = 1250$ to 2500 feet per second, corresponding to an air pressure of approximately $p = 14.7 \{(1 + 2 v^2 10^{-7})^{3.5} - 1\}$ lb. per sq. in., or 0.66 to 2.75 lb. in the first, 23 to 230 lb. in the second case.

While the condenser charge may be oscillatory or logarithmic, efficiency requires a low value of r, that is, an oscillatory charge.

With a frequency of discharge in L_0, r_0 very high compared with the frequency of charge, the duration of the discharge is short compared with the duration of the charge, that is, the oscillating currents consist of a series of oscillations separated by relatively long periods of rest. Thus the current in L does not appreciably change during the time of the discharge, and at the end of the condenser charge the current in the reactor, L, is the same as the current in L, with which the next condenser charge starts. The charging current of the condenser, C, in L thus changes from i_0 at the beginning of the charge, or condenser e.m.f., $e_0 = 0$, to the same value i_0 at the end of the charge, or condenser e.m.f., $e_1 = e_0$.

50. Counting, therefore, the time, t, from the moment when the condenser charge begins, we have the terminal conditions:

$t = 0$, $i = i_0$, $e_1 = 0$ at the beginning of the condenser charge.
$t = t_0$, $i = i_0$, $e_1 = e_0$ at the end of the condenser charge.

In the condenser discharge, through circuit L_0, r_0, counting the time t' from the moment when the condenser discharge begins, that is, $t' = t - t_0$, we have

$t' = 0$, $i = 0$, $e_1 = e_0$ the terminal condition.

e_0, thus, is that value of the voltage e_1 at which discharge takes place across the spark gap, and t_0 is the time elapsing between $e_1 = 0$ and $e_1 = e_0$, or the time required to build up the voltage e_1 sufficiently to break down the spark gap.

Under the assumption that the period of oscillation of the condenser charge through L, r, is large compared with the period of oscillation of the condenser discharge through L_0, r_0, the equations are:

(*A*) Condenser discharge:

$$i = \frac{2 e_0}{q_0} \varepsilon^{-\frac{r_0}{2 L_0} t'} \sin \frac{q_0}{2 L_0} t', \tag{66}$$

$$e_1 = e_0 \varepsilon^{-\frac{r_0}{2 L_0} t'} \left\{ \cos \frac{q_0}{2 L_0} t' + \frac{r_0}{q_0} \sin \frac{q_0}{2 L_0} t' \right\}, \tag{67}$$

where

$$q_0 = \sqrt{\frac{4 L_0}{C} - r_0^2}. \tag{68}$$

(B) Condenser charge:

$$i = \varepsilon^{-\frac{r}{2L}t} \left\{ i_0 \cos\frac{q}{2L}t + \frac{2e - ri_0}{q} \sin\frac{q}{2L}t \right\}, \quad (69)$$

$$e_1 = e - \varepsilon^{-\frac{r}{2L}t} \left\{ e \cos\frac{q}{2L}t + \frac{re - \frac{r^2+q^2}{2}i_0}{q} \sin\frac{q}{2L}t \right\}, \quad (70)$$

where

$$q = \sqrt{\frac{4L}{C} - r^2}. \quad (71)$$

Substituting in (69) and (70) the above discussed terminal conditions,

$$t = t_0, \quad i = i_0, \quad e_1 = e_0,$$

gives

$$i_0 = \varepsilon^{-\frac{r}{2L}t_0} \left\{ i_0 \cos\frac{q}{2L}t_0 + \frac{2e - ri_0}{q} \sin\frac{q}{2L}t_0 \right\} \quad (72)$$

and

$$e_0 = e - \varepsilon^{-\frac{r}{2L}t_0} \left\{ e \cos\frac{q}{2L}t_0 + \frac{re - \frac{r^2+q^2}{2}i_0}{q} \sin\frac{q}{2L}t_0 \right\}. \quad (73)$$

Denoting, for convenience,

$$\left. \begin{aligned} \frac{r}{2L}t_0 &= s, \\[2mm] \frac{q}{2L}t_0 &= \phi, \\[2mm] \frac{r}{q} &= a, \end{aligned} \right\} \quad (74)$$

and

and resolving (72) for i_0, gives

$$i_0 = \frac{2e}{q} \frac{\varepsilon^{-s}\sin\phi}{1 - \varepsilon^{-s}\cos\phi + a\varepsilon^{-s}\sin\phi}, \quad (75)$$

and substituting (75) in (73) and rearranging,

$$e_0 = e \frac{1 - 2\varepsilon^{-s}\cos\phi + \varepsilon^{-2s}}{1 - \varepsilon^{-s}\cos\phi + a\varepsilon^{-s}\sin\phi}. \quad (76)$$

The two equations (75), (76) permit the calculation of two of the three quantities i_0, e_0, t_0: the time, t_0, of condenser charge appears in the exponential function, in s, and in the trigonometric function, in ϕ.

Since in an oscillating-current generator of fair efficiency, that is, when r is as small as possible, s is a small quantity, ε^{-s} can be resolved into the series

$$\varepsilon^{-s} = 1 - s + \frac{s^2}{2} - + \cdots . \tag{77}$$

Substituting (77) in (75), and dropping all terms higher than s^2, gives

$$i_0 = \frac{2\,e}{q} \frac{\left(1 - s + \dfrac{s^2}{2}\right)\sin\phi}{1 - \cos\phi + s\cos\phi - \dfrac{s^2}{2}\cos\phi + a\sin\phi - as\sin\phi} .$$

Multiplying numerator and denominator by $\left(1 + \dfrac{s}{2}\right)$, and rearranging, gives

$$\left. \begin{aligned}
i_0 &= \frac{2\,e}{q} \frac{\sin\phi}{\dfrac{2+s}{2-s} - \cos\phi + a\sin\phi} \\[2mm]
&= \frac{2\,e}{q} \frac{\sin\phi}{\dfrac{2\,s}{2-s} + 2\sin^2\dfrac{\phi}{2} + a\sin\phi} .
\end{aligned} \right\} \tag{78}$$

Substituting (77) in (76), dropping terms higher than s^2 and as, multiplying numerator and denominator by $\left(1 + \dfrac{s}{2}\right)$, and rearranging, gives

$$e_0 = 2\,e\; \frac{2\sin^2\dfrac{\phi}{2} + \dfrac{s^2}{2}}{\dfrac{2\,s}{2-s} + 2\sin^2\dfrac{\phi}{2} + a\sin\phi} . \tag{79}$$

Substituting t_0 in (78) and (79) gives

$$i_0 = \frac{2\,e}{q} \cdot \frac{\sin \dfrac{q}{2\,L}\,t_0}{\dfrac{2\,r t_0}{4\,L - r t_0} + 2\sin^2 \dfrac{q}{4\,L}\,t_0 + \dfrac{r}{q}\sin \dfrac{q}{2\,L}\,t_0} \qquad (80)$$

and

$$e_0 = 2\,e \cdot \frac{2\sin^2 \dfrac{q}{4\,L}\,t_0 + \dfrac{r^2 t_0^{\,2}}{8\,L^2}}{\dfrac{2\,r t_0}{4\,L - r t_0} + 2\sin^2 \dfrac{q}{4\,L}\,t_0 + \dfrac{r}{q}\sin \dfrac{q}{2\,L}\,t_0} \qquad (81)$$

as approximate equations giving i_0 and e_0 as functions of t_0, or the time of condenser charge.

51. The time, t_0, during which the condenser charges, increases with increasing e_0, that is, increasing length of the spark gap in the discharge circuit, at first almost proportionally, then, as e_0 approaches $2\,e$, more slowly.

As long as e_0 is appreciably below $2\,e$, that is, about $e_0 < 1.75\,e$, t_0 is relatively short, and the charging current i, which increases from i_0 to a maximum, and then decreases again to i_0, does not vary much, but is approximately constant, with an average value very little above i_0, so that the power supplied by the impressed e.m.f., e, to the charging circuit can approximately be assumed as

$$p_0 = e i_0. \qquad (82)$$

The condenser discharge is intermittent, consisting of a series of oscillations, with a period of rest between the oscillations, which is long compared with the duration of the oscillation, and during which the condenser charges again.

The discharge current of the condenser is, (66),

$$i = \frac{2\,e_0}{q_0}\,\varepsilon^{-\frac{r_0}{2\,L_0}t}\,\sin \frac{q_0}{2\,L_0}\,t, \text{ in amp.,}$$

and since such an oscillation recurs at intervals of t_0 seconds, the effective value, or square root of mean square of the discharge current, is

$$i_1 = \sqrt{\frac{1}{t_0}\int_0^{t_0} i^2\,dt.} \qquad (83)$$

Long before $t = t_0$, i is practically zero, and as upper limit of the integral can therefore be chosen ∞ instead of t_0.

Substituting (66) in (83), and taking the constant terms out of the square root, gives the effective value of discharge current as

$$i_1 = \frac{2\,e_0}{q_0}\sqrt{\frac{1}{t_0}\int_0^\infty \varepsilon^{-\frac{r_0}{L_0}t}\sin^2\frac{q_0}{2\,L_0}t\,dt}$$

$$= \frac{2\,e_0}{q_0}\sqrt{\frac{1}{2\,t_0}\left\{\int_0^\infty \varepsilon^{-\frac{r_0}{L_0}t}\,dt - \int_0^\infty \varepsilon^{-\frac{r_0}{L_0}t}\cos\frac{q_0}{L_0}t\,dt\right\}}\;; \qquad (84)$$

however,

$$\int_0^\infty \varepsilon^{-\frac{r_0}{L_0}t}\,dt = -\frac{L_0}{r_0}\left[\varepsilon^{-\frac{r_0}{L_0}t}\right]_0^\infty = \frac{L_0}{r_0},$$

and by fractional integration,

$$\int_0^\infty \varepsilon^{-\frac{r_0}{L_0}t}\cos\frac{q_0}{L_0}t\,dt$$

$$= \frac{\dfrac{L_0}{q_0}}{1+\left(\dfrac{r_0}{q_0}\right)^2}\left[\varepsilon^{-\frac{r_0}{L_0}t}\left\{\sin\frac{q_0}{L_0}t - \frac{r_0}{q_0}\cos\frac{q_0}{L_0}t\right\}\right]_0^\infty$$

$$= \frac{L_0 r_0}{r_0{}^2 + q_0{}^2}\;;$$

hence, substituting in (84),

$$i_1 = e_0\sqrt{\frac{2\,L_0}{t_0 r_0\,(r_0{}^2 + q_0{}^2)}}. \qquad (85)$$

Since

$$q^2 = \frac{4\,L}{C} - r^2,$$

we have, substituting in (85),

$$i_1 = e_0\sqrt{\frac{C}{2\,t_0 r_0}}, \qquad (86)$$

and, denoting by

$$f_1 = \frac{1}{t_0},$$

the frequency of condenser charge, or the number of complete trains of discharge oscillations per second,

$$i_1 = e_0 \sqrt{\frac{Cf_1}{2\,r_0}}, \tag{87}$$

that is, the effective value of the discharge current is proportional to the condenser potential, e_0, proportional to the square root of the capacity, C, and the frequency of charge, f_1, and inversely proportional to the square root of the resistance, r_0, of the discharge circuit; but it does not depend upon the inductance L_0 of the discharge circuit, and therefore does not depend on the frequency of the discharge oscillation.

The power of the discharge is

$$p_1 = i_1^2 r_0 = f_1 \frac{e_0^2 C}{2}. \tag{88}$$

Since $\dfrac{e_0^2 C}{2}$ is the energy stored in the condenser of capacity C at potential e_0, and f_1 the frequency or number of discharges of this energy per second, equation (88) is obvious.

Inversely therefore, from equation (88), that is, the total energy stored in the condenser and discharging per second, the effective value of discharge current can be directly calculated as

$$i_1 = \sqrt{\frac{p_1}{r_0}} = e_0 \sqrt{\frac{Cf_1}{2\,r_0}}.$$

The ratio of effective discharge current, i_1, to mean charging current, i_0, is

$$\frac{i_1}{i_0} = \frac{e_0}{i_0} \sqrt{\frac{Cf_1}{2\,r_0}}, \tag{89}$$

and substituting (80) and (81) in (89),

$$\frac{i_1}{i_0} = q \sqrt{\frac{Cf_1}{2r_0}} \; \frac{2 \sin^2 \dfrac{q}{4L} t_0 + \dfrac{r^2 t_0^2}{8L}}{\sin \dfrac{q}{2L} t_0}. \tag{90}$$

The magnitude of this quantity can be approximated by neglecting r compared with $\dfrac{4\,L}{C}$, that is, substituting $q = \sqrt{\dfrac{4\,L}{C}}$ and replacing the sine-function by the arcs. This gives

$$\frac{i_1}{i_0} = \frac{1}{\sqrt{2\,r_0 C f_1}}, \tag{91}$$

that is, the ratio of currents is inversely proportional to the square root of the resistance of the discharge circuit, of the capacity, and of the frequency of charge.

52. Example: Assume an oscillating-current generator, feeding a Tesla transformer for operating X-ray tubes, or directly supplying an iron arc (that is, a condenser discharge between iron electrodes) for the production of ultraviolet light.

The constants of the charging circuit are: the impressed e.m.f., $e = 15,000$ volts; the resistance, $r = 10,000$ ohms; the inductance, $L = 250$ henrys, and the capacity, $C = 2 \times 10^{-8}$ farads $= 0.02$ mf.

The constants of the discharge circuit are: (a) operating Tesla transformer, the estimated resistance, $r_0 = 20$ ohms (effective) and the estimated inductance, $L_0 = 60 \times 10^{-6}$ henry $= 0.06$ mh.; (b) operating ultraviolet arc, the estimated resistance, $r_0 = 5$ ohms (effective) and the estimated inductance, $L_0 = 4 \times 10^{-6}$ henry $= 0.004$ mh.

Therefore in the charging circuit,

$$q = 223,400 \text{ ohms}, \qquad \frac{r}{q} = 0.0448,$$

$$\frac{q}{2\,L} = 446.8, \qquad \frac{r}{2\,L} = 20,$$

$$\frac{L}{r} = 0.025;$$

then

$$i_0 = 0.1344 \frac{\sin 446.8\,t_0}{\dfrac{2\,t_0}{0.1 - t_0} + 2\sin^2 223.4\,t_0 + 0.0448 \sin 446.8\,t_0},$$

and

$$e_0 = 30,000 \frac{2\sin^2 223.4\,t_0 + 200\,t_0^2}{\dfrac{2\,t_0}{0.1 - t_0} + 2\sin^2 223.4\,t_0 + 0.0448 \sin 446.8\,t_0}. \tag{92}$$

Fig. 17 shows i_0 and e_0 as ordinates, with the time of charge t_0 as abscissas.

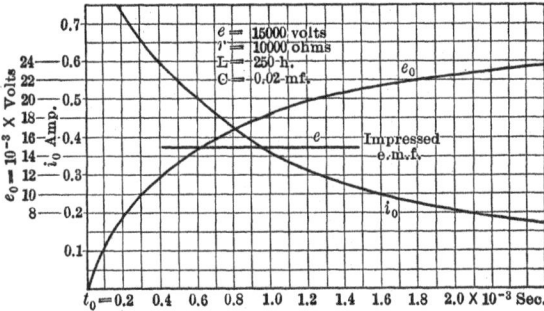

Fig. 17. Oscillating-current generator charge.

The frequency of the charging oscillation is

$$f = \frac{q}{4\,\pi L} = 71.2 \text{ cycles per sec.;}$$

for

$$i_0 = 0.365 \text{ amp.,}$$

substituting in equations (69) and (70) we have

$$\left. \begin{aligned} &i = \varepsilon^{-20t}\left\{0.365 \cos 446.8\,t + 0.118 \sin 446.8\,t\right\}, \text{ in amp.,}\\ &\text{and}\\ &e_1 = 15{,}000\left\{1 - \varepsilon^{-20t}\left[\cos 446.8\,t - 2.67 \sin 446.8\,t\right]\right\}, \text{ in volts,} \end{aligned} \right\} \quad (93)$$

the equations of condenser charge.

From these equations the values of i and e_1 are plotted in Fig. 18, with the time t as abscissas.

As seen, the value $i = i_0 = 0.365$ amp., is reached again at the time $t_0 = 0.0012$, that is, after 30.6 time-degrees or about $\frac{1}{12}$ of a period. At this moment the condenser e.m.f. is $e_1 = e_0 = 22{,}300$ volts; that is, by setting the spark gap for 22,300 volts the duration of the condenser charge is 0.0012 second, or in other words, every 0.0012 second, or 833 times per second, discharge oscillations are produced.

With this spark gap, the charging current at the beginning and at the end of the condenser charge is 0.365 amp., and the average charging current is 0.3735 amp. at 15,000 volts, consuming 5.6 kva.

Assume that the e.m.f. at the condenser terminals at the end of the charge is $e_0 = 22,300$ volts; then consider two cases, namely: (a) the condenser discharges into a Tesla transformer, and (b) the condenser discharges into an iron arc.

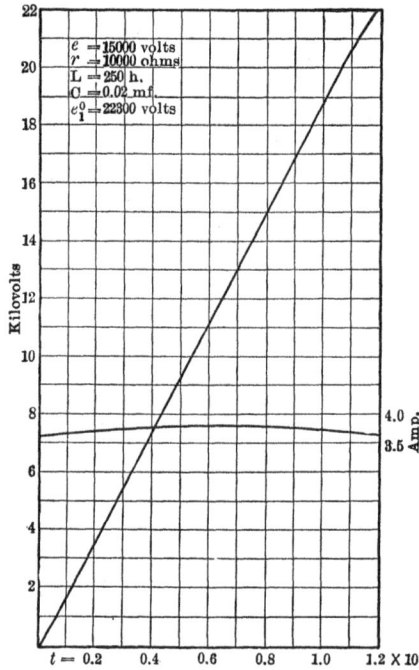

Fig. 18. Oscillating-current generator condenser charge.

(a) The Tesla transformer, that is, an oscillating-current transformer, has no iron, but a primary coil of very few turns (20) and a secondary coil of a larger number of turns (360), both immersed in oil.

While the actual ohmic resistance of the discharge circuit is only 0.1 ohm, the load on the secondary of the Tesla transformer, the dissipation of energy into space by brush discharge, etc., and the increase of resistance by unequal current distribution in the conductor, increase the effective resistance to many times the ohmic resistance. We can, therefore, assign the following estimated values: $r_0 = 20$ ohms; $L_0 = 60 \times 10^{-6}$ henry, and $C = 2 \times 10^{-8}$ farad.

Then

$$q_0 = 108 \text{ ohms}, \qquad\qquad \frac{r_0}{q_0} = 0.186,$$

$$\frac{q_0}{2 L_0} = 0.898 \times 10^6, \qquad\qquad \frac{r_0}{2 L_0} = 0.1667 \times 10^6,$$

which give

$$\left.\begin{array}{l} i = 415 \, \varepsilon^{-0.1667 \times 10^6 t} \sin 0.898 \times 10^6 \, t, \text{ amp.} \\ \text{and} \\ e_1 = 22{,}300 \, \varepsilon^{-0.1667 \times 10^6 t} \{\cos 0.898 \times 10^6 \, t + 0.186 \sin 0.898 \times 10^6 \, t\}, \\ \qquad \text{volts.} \end{array}\right\} \tag{94}$$

The frequency of oscillation is

$$f_0 = \frac{0.898 \times 10^6}{2 \pi} = 143{,}000 \text{ cycles per sec.} \tag{95}$$

Fig. 19 shows the current i and the condenser potential e_1 during the discharge, with the time t as abscissas. As seen, the discharge frequency is very high compared with the fre-

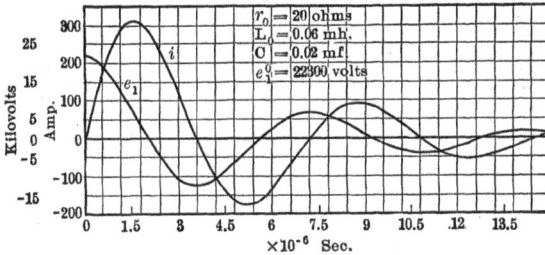

Fig. 19.   Oscillating-current generator condenser discharge.

quency of charge, the duration of discharge very short, and the damping very great; a decrement of 0.55, so that the oscillation dies out very rapidly. The oscillating current, however, is enormous compared with the charging current; with a mean charging current of 0.3735 amp., and a maximum charging current of 0.378 amp. the maximum discharge current is 315 amp.,or 813 times as large as the charging current.

The effective value of the discharge current, from equation (87), is $i_1 = 14.4$ amp., or nearly 40 times the charging current.

53. (b) When discharging the condenser directly, through an ultraviolet or iron arc, in a straight path, and estimating $r_0 = 5$ ohms and $L_0 = 4 \times 10^{-6}$ henry, we have

$$q_0 = 27.84 \text{ ohms}, \qquad \frac{r_0}{q_0} = 0.1795,$$

$$\frac{q_0}{2 L_0} = 3.48 \times 10^6, \qquad \frac{r_0}{2 L_0} = 0.625 \times 10^6;$$

then,

$$i = 1600\, \varepsilon^{-0.625 \times 10^6 t} \sin 3.48 \times 10^6\, t, \text{ in amp.,}$$

and

$$e_1 = 22{,}300\, \varepsilon^{-0.625 \times 10^6 t} \left\{ \cos 3.48 \times 10^6\, t + 0.1795 \sin 3.48 \times 10^6\, t \right\},$$
$$\text{in volts,}$$

$$(96)$$

and the frequency of oscillation is

$$f_0 = 562{,}000 \text{ cycles per sec.;} \qquad (97)$$

that is, the frequency is still higher, over half a million cycles; the maximum discharge current over 1000 amperes; however, the duration of the discharge is still shorter, the oscillations dying out more rapidly.

The effective value of the discharge current, from (87), is $i_1 = 28.88$ amp., or 77 times the charging current. A hot wire ammeter in the discharge circuit in this case showed 29 amp.

As seen, with a very small current supply, of 0.3735 amp., at $e = 15{,}000$ volts, in the discharge circuit a maximum voltage of 22,300, or nearly 50 per cent higher than the impressed voltage, is found, and a very large current, of an effective value very many times larger than the supply current.

As a rule, instead of a constant impressed e.m.f., e, a low frequency alternating e.m.f. is used, since it is more conveniently generated by a step-up transformer. In this case the condenser discharges occur not at constant intervals of t_0 seconds, but only during that part of each half wave when the e.m.f. is sufficient to jump the gap e_0, and at intervals which are shorter at the maximum of the e.m.f. wave than at its beginning and end.

For instance, using a step-up transformer giving 17,400 volts effective (by the ratio of turns 1 : 150, with 118 volts impressed at 60 cycles), or a maximum of 24,700 volts, then during each half wave the first discharge occurs as soon as the voltage has reached 22,300, sufficient to jump the spark gap, and then a series of discharges occurs, at intervals decreasing with the increase of the impressed e.m.f., up to its maximum, and then with increasing intervals, until on the decreasing wave the e.m.f. has fallen below that which, during the charging oscillation, can jump the gap e_0, that is, about 13,000 volts. Then the oscillating discharges stop, and start again during the next half wave.

Hence the phenomenon is of the same character as investigated above for constant impressed e.m.f., except that it is intermittent, with gaps during the zero period of impressed voltage and unequal time intervals t_0 between the successive discharges.

54. An underground cable system can act as an oscillating-current generator, with the capacity of the cables as condenser, the internal inductance of the generators as reactor, and a short-circuiting arc as discharge circuit.

In a cable system where this phenomenon was observed the constants were approximately as follows: capacity of the cable system, $C = 102$ mf.; inductance of 30,000-kw. in generators, $L = 6.4$ mh.; resistance of generators and circuit up to the short-circuiting arc, $r = 0.1$ ohm and $r = 1.0$ ohm respectively; impressed e.m.f., 11,000 volts effective, and the frequency 25 cycles per second.

The frequency of charging oscillation in this case is

$$f = \frac{q}{4\,\pi L} = 197 \text{ cycles per sec.}$$

since

$$q = \sqrt{\frac{4\,L}{C} - r^2} = 15.8 \text{ ohms.}$$

Substituting these values in the preceding equations, and estimating the constants of the discharge circuit, gives enormous values of discharge current and e.m.f.

CHAPTER VII.

RESISTANCE, INDUCTANCE, AND CAPACITY IN SERIES IN ALTERNATING-CURRENT CIRCUIT.

55. Let, at time $t = 0$ or $\theta = 0$, the e.m.f.,

$$e = E \cos (\theta - \theta_0), \tag{1}$$

be impressed upon a circuit containing in series the resistance, r, the inductance, L, and the capacity, C.

The inductive reactance is $\quad x = 2 \pi f L$

and the condensive reactance is $x_c = \dfrac{1}{2 \pi f C}$, $\left.\vphantom{\dfrac{1}{2}}\right\}$ $\quad$ (2)

where f = frequency and $\theta = 2 \pi f t$. $\tag{3}$

Then the e.m.f. consumed by resistance is ri; the e.m.f. consumed by inductance is

$$L \frac{di}{dt} = x \frac{di}{d\theta},$$

and the e.m.f. consumed by capacity is

$$e_1 = \frac{1}{C} \int i \, dt = x_c \int i \, d\theta, \tag{4}$$

where i = instantaneous value of the current.

Hence, $\qquad e = ri + x \dfrac{di}{d\theta} + x_c \int i \, d\theta, \tag{5}$

or, $\qquad E \cos (\theta - \theta_0) = ri + x \dfrac{di}{d\theta} + x_c \int i \, d\theta, \tag{6}$

and hence, the difference of potential at the condenser terminals is

$$e_1 = x_c \int i \, d\theta = E \cos (\theta - \theta_0) - ri - x \frac{di}{d\theta}. \tag{7}$$

88

Equation (6) differentiated gives

$$E \sin (\theta - \theta_0) + x \frac{d^2 i}{d\theta^2} + r \frac{di}{d\theta} + x_c i = 0. \tag{8}$$

The integral of this equation (8) is of the general form

$$i = A \varepsilon^{-a\theta} + B \cos (\theta - \sigma). \tag{9}$$

Substituting (9) in (8), and rearranging, gives

$$A \varepsilon^{-a\theta} \left\{ a^2 x - ar + x_c \right\} + \sin \theta \left\{ E \cos \theta_0 - rB \cos \sigma - B (x - x_c) \sin \sigma \right\}$$
$$- \cos \theta \left\{ E \sin \theta_0 - rB \sin \sigma + B (x - x_c) \cos \sigma \right\} = 0,$$

and, since this must be an identity,

$$\left. \begin{array}{c} a^2 x - ar + x_c = 0, \\ E \cos \theta_0 - rB \cos \sigma - B (x - x_c) \sin \sigma = 0, \\ E \sin \theta_0 - rB \sin \sigma + B (x - x_c) \cos \sigma = 0. \end{array} \right\} \tag{10}$$

Substituting

$$\left. \begin{array}{c} s = \sqrt{r^2 - 4 x x_c}, \\ z_0 = \sqrt{r^2 + (x - x_c)^2}, \\ \tan \gamma = \dfrac{x - x_c}{r}, \end{array} \right\} \tag{11}$$

in equations (10) gives

$$\left. \begin{array}{c} a = \dfrac{r \pm s}{2 x}, \\[2mm] B = \dfrac{E}{z_0}, \\[2mm] \sigma = \theta_0 + \gamma \end{array} \right\} \tag{12}$$

and $A = $ indefinite,

and the equation of current, (9), thus is

$$i = \frac{E}{z_0} \cos (\theta - \theta_0 - \gamma) + A_1 \varepsilon^{-\frac{r-s}{2x}\theta} + A_2 \varepsilon^{-\frac{r+s}{2x}\theta}, \tag{13}$$

and, substituting (12) in (7), and rearranging, the potential difference at the condenser terminals is

$$e_1 = \frac{Ex_c}{z_0} \sin (\theta - \theta_0 - \gamma) - \frac{r+s}{2} A_1 \varepsilon^{-\frac{r-s}{2x}\theta} - \frac{r-s}{2} A_2 \varepsilon^{-\frac{r+s}{2x}\theta}. \quad (14)$$

The two integration constants A_1 and A_2 are given by the terminal conditions of the problem.

Let, at the moment of start,

$$\theta = 0,$$

$i = i_0 =$ instantaneous value of current and

$e_1 = e_0 =$ instantaneous value of condenser potential difference. $\qquad\qquad (15)$

Substituting in (13) and (14),

$$i_0 = \frac{E}{z_0} \cos (\theta_0 + \gamma) + A_1 + A_2$$

and

$$e_0 = -\frac{Ex_c}{z_0} \sin (\theta_0 + \gamma) - \frac{r+s}{2} A_1 - \frac{r-s}{2} A_2.$$

Therefore

$$A_1 + A_2 = i_0 - \frac{E}{z_0} \cos (\theta_0 + \gamma)$$

and

$$A_1 - A_2 = -\frac{ri_0 + 2e_0}{s} + \frac{E}{sz_0} \{r \cos (\theta_0 + \gamma) - 2 x_c \sin (\theta_0 + \gamma)\}, \qquad (16)$$

or,

$$A_1 = -\frac{\frac{r-s}{2} i_0 + e_0}{s} + \frac{E}{sz_0} \left\{\frac{r-s}{2} \cos (\theta_0 + \gamma) - x_c \sin (\theta_0 + \gamma)\right\}$$

and

$$A_2 = +\frac{\frac{r+s}{2} i_0 + e_0}{s} - \frac{E}{sz_0} \left\{\frac{r+s}{2} \cos (\theta_0 + \gamma) - x_c \sin (\theta_0 + \gamma)\right\}. \qquad (17)$$

Substituting (17) in (13) and (14) gives the integral equations of the problem.

The *current* is

$$i = \frac{E}{z_0}\cos(\theta - \theta_0 - \gamma) + \frac{E}{sz_0}\left\{ \varepsilon^{-\frac{r-s}{2x}\theta}\left[\frac{r-s}{2}\cos(\theta_0 + \gamma) - x_c\sin(\theta_0 + \gamma)\right]\right.$$

$$\left. - \varepsilon^{-\frac{r+s}{2x}\theta}\left[\frac{r+s}{2}\cos(\theta_0 + \gamma) - x_c\sin(\theta_0 + \gamma)\right]\right\}$$

$$- \frac{1}{s}\left\{\varepsilon^{-\frac{r-s}{2x}\theta}\left[\frac{r-s}{2}i_0 + e_0\right] - \varepsilon^{-\frac{r+s}{2x}\theta}\left[\frac{r+s}{2}i_0 + e_0\right]\right\}, \tag{18}$$

and the *potential difference at the condenser terminals* is

$$e_1 = \frac{Ex_c}{z_0}\sin(\theta - \theta_0 - \gamma)$$

$$- \frac{E}{2\,sz_0}\left\{(r+s)\varepsilon^{-\frac{r-s}{2x}\theta}\left[\frac{r-s}{2}\cos(\theta_0 + \gamma) - x_c\sin(\theta_0 + \gamma)\right]\right.$$

$$\left. - (r-s)\varepsilon^{-\frac{r+s}{2x}\theta}\left[\frac{r+s}{2}\cos(\theta_0 + \gamma) - x_c\sin(\theta_0 + \gamma)\right]\right\}$$

$$+ \frac{1}{2\,s}\left\{(r+s)\varepsilon^{-\frac{r-s}{2x}\theta}\left[\frac{r-s}{2}i_0 + e_0\right] - (r-s)\varepsilon^{-\frac{r+s}{2x}\theta}\left[\frac{r+s}{2}i_0 + e_0\right]\right\}, \tag{19}$$

where

$$z_0 = \sqrt{r^2 + (x - x_c)^2},$$

$$\tan\gamma = \frac{x - x_c}{r}, \tag{11}$$

and

$$s = \sqrt{r^2 - 4\,x\,x_c}.$$

The expressions of i and e_1 consist of three terms each:

(1) The permanent term, which is the only one remaining after some time;

(2) A transient term depending upon the constants of the circuit, r, s, x_c, z_0, x, the impressed e.m.f., E, and its phase θ_0 at the moment of starting, but independent of the conditions existing in the circuit before the start; and

(3) A term depending, besides upon the constants of the circuit, upon the instantaneous values of current and potential difference, i_0 and e_0, at the moment of starting the circuit, and thereby upon the electrical conditions of the circuit before impressing the e.m.f., e. This term disappears if the circuit is dead before the start.

Equations (18) and (19) contain the term $s = \sqrt{r^2 - 4\, x\, x_c}$ $= \sqrt{r^2 - 4\,\dfrac{L}{C}}$; hence apply only when $r^2 > 4\, x\, x_c$, but become indeterminate if $r^2 = 4\, x\, x_c$, and imaginary if $r^2 < 4\, x\, x_c$; in the latter cases they have to be rearranged so as to appear in real form, in manner similar to that in Chapter V.

56. In the *critical case*, $r^2 = 4\, x x_c$ and $s = 0$, equation (18), rearranged, assumes the form

$$i = \frac{E}{z_0} \cos(\theta - \theta_0 - \gamma) + \frac{E}{z_0} \varepsilon^{-\frac{r}{2x}\theta}$$

$$\left\{ \left[\frac{r}{2}\cos(\theta_0 + \gamma) - x_c \sin(\theta_0 + \gamma) \right] \frac{\varepsilon^{+\frac{s}{2x}\theta} - \varepsilon^{-\frac{s}{2x}\theta}}{s} - \cos(\theta_0 + \gamma) \right\}$$

$$- \varepsilon^{-\frac{r}{2x}\theta} \left\{ \left[\frac{r}{2}i_0 + e_0 \right] \frac{\varepsilon^{+\frac{s}{2x}\theta} - \varepsilon^{-\frac{s}{2x}\theta}}{s} - i_0 \right\}.$$

However, developing in a series, and canceling all but the first term as infinitely small, we have

$$\frac{\varepsilon^{+\frac{s}{2x}\theta} - \varepsilon^{-\frac{s}{2x}\theta}}{s} = \frac{\theta}{x} ;$$

hence the *current* is

$$i = \frac{E}{z_0} \cos(\theta - \theta_0 - \gamma) + \frac{E}{z_0} \varepsilon^{-\frac{r}{2x}\theta}$$

$$\left\{ \left[\frac{r}{2}\cos(\theta_0 + \gamma) - x_c \sin(\theta_0 + \gamma) \right] \frac{\theta}{x} - \cos(\theta_0 + \gamma) \right\}$$

$$+ \varepsilon^{-\frac{r}{2x}\theta} \left\{ i_0 - \left[\frac{r}{2}i_0 + e_0 \right] \frac{\theta}{x} \right\}, \tag{20}$$

and in the same manner the *potential difference at condenser terminals* is

$$e_1 = \frac{E x_c}{z_0} \sin(\theta - \theta_0 - \gamma) - \frac{E}{2 z_0} \varepsilon^{-\frac{r}{2x}\theta}$$

$$\left\{ \left[\frac{r^2}{2} \cos(\theta_0 + \gamma) - x_c r \sin(\theta_0 + \gamma) \right] \frac{\theta}{x} - 2 x_c \sin(\theta + \gamma) \right\}$$

$$+ \frac{1}{2} \varepsilon^{-\frac{r}{2x}\theta} \left\{ \left[\frac{r^2}{2} i_0 + r e_0 \right] \frac{\theta}{x} + 2 e_0 \right\}. \tag{21}$$

Here again three terms exist, namely: a permanent term, a transient term depending only on E and θ_0, and a transient term depending on i_0 and e_0.

57. In the *trigonometric* or *oscillatory* case, $r^2 < 4 x x_c$, s becomes imaginary, and equations (18) and (19) therefore contain complex imaginary exponents, which have to be eliminated, since the complex imaginary form of the equation obviously is only apparent, the phenomenon being real.

Substituting

$$q = \sqrt{4 x x_c - r^2} = js \tag{22}$$

in equations (13) and (14), and also substituting the trigonometric expressions

and
$$\left. \begin{array}{l} \varepsilon^{+j\frac{q}{2x}\theta} = \cos \frac{q}{2x}\theta + j \sin \frac{q}{2x}\theta \\[3mm] \varepsilon^{-j\frac{q}{2x}\theta} = \cos \frac{q}{2x}\theta - j \sin \frac{q}{2x}\theta, \end{array} \right\} \tag{23}$$

and separating the imaginary and the real terms, gives

$$i = \frac{E}{z_0} \cos(\theta - \theta_0 - \gamma) + \varepsilon^{-\frac{r}{2x}\theta}$$

$$\left\{ (A_1 + A_2) \cos \frac{q}{2x}\theta - j (A_1 - A_2) \sin \frac{q}{2x}\theta \right\}$$

and

$$e_1 = \frac{Ex_c}{z_0} \sin (\theta - \theta_\bullet - \gamma) - \varepsilon^{-\frac{r}{2x}\theta}$$

$$\left\{ \frac{A_1 + A_2}{2} \left[r \cos \frac{q}{2x}\theta - q \sin \frac{q}{2x}\theta \right] + j \frac{A_1 - A_2}{2} \right.$$

$$\left. \left[q \cos \frac{q}{2x}\theta + r \sin \frac{q}{2x}\theta \right] \right\};$$

then substituting herein the equations (16) and (22) the imaginary disappears, and we have the *current*,

$$i = \frac{E}{z_0} \cos (\theta - \theta_0 - \gamma) - \frac{E}{z_0} \varepsilon^{-\frac{r}{2x}\theta}$$

$$\left\{ \cos (\theta_0 + \gamma) \cos \frac{q}{2x}\theta + \left[\frac{2x_c}{q} \sin (\theta_0 + \gamma) - \frac{r}{q} \cos (\theta_0 + \gamma) \right] \sin \frac{q}{2x}\theta \right\}$$

$$+ \varepsilon^{-\frac{r}{2x}\theta} \left\{ i_0 \cos \frac{q}{2x}\theta - \frac{2e_0 + ri_0}{q} \sin \frac{q}{2x}\theta \right\}, \qquad (24)$$

and the *potential difference at the condenser terminals*,

$$e_1 = \frac{Ex_c}{z_0} \sin (\theta - \theta_0 - \gamma) + \frac{Ex_c}{z_0} \varepsilon^{-\frac{r}{2x}\theta}$$

$$\left\{ \sin (\theta_0 + \gamma) \cos \frac{q}{2x}\theta + \left[\frac{r}{q} \sin (\theta_0 + \gamma) - \frac{2x}{q} \cos (\theta_0 + \gamma) \right] \sin \frac{q}{2x}\theta \right\}$$

$$+ \varepsilon^{-\frac{r}{2x}\theta} \left\{ e_0 \cos \frac{q}{2x}\theta + \frac{2re_0 + 4xx_ci_0}{2q} \sin \frac{q}{2x}\theta \right\}. \qquad (25)$$

Here the three component terms are seen also.

58. As examples are shown in Figs. 20 and 21, the starting of the current i, its permanent term i', and the two transient terms i_1 and i_2, and their difference, for the constants $E = 1000$ volts = maximum value of impressed e.m.f.; $r = 200$ ohms = resistance; $x = 75$ ohms = inductive reactance, and $x_c = 75$ ohms = condensive reactance. We have

$$4\,x\,x_c = 22,500$$

and

$$r^2 = 40,000;$$

therefore

$$r^2 > 4\,x\,x_c,$$

that is, the start is logarithmic, and $z_0 = 200$, $s = 132$, and $\gamma = 0$.

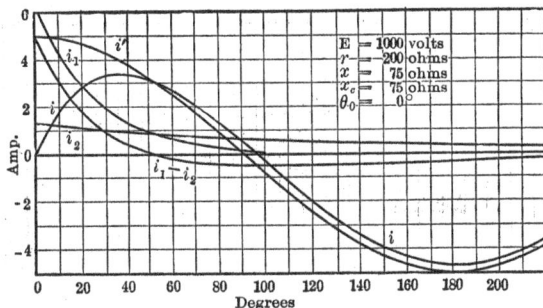

Fig. 20. Starting of an alternating-current circuit, having capacity, inductance and resistance in series. Logarithmic start.

In Fig. 20 the circuit is closed at the moment $\theta_0 = 0$, that is, at the maximum value of the impressed e.m.f., giving from the equations (18) and (19), since $i_0 = 0$, $e_0 = 0$,

$$i = 5 \left\{ \cos \theta - 1.26 \, \varepsilon^{-2.22 \theta} + 0.26 \, \varepsilon^{-0.452 \theta} \right\}$$

and

$$e_1 = 375 \left\{ \sin \theta + 0.57 \left(\varepsilon^{-2.22 \theta} - \varepsilon^{-0.452 \theta} \right) \right\}.$$

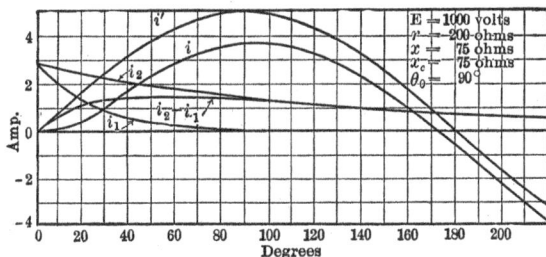

Fig. 21. Starting of an alternating-current circuit having capacity, inductance and resistance in series. Logarithmic start.

In Fig. 21 the circuit is closed at the moment $\theta_0 = 90°$, that is, at the zero value of the impressed e.m.f., giving the equations

$$i = 5 \left\{ \sin \theta + 0.57 \left(\varepsilon^{-2.22 \theta} - \varepsilon^{-0.452 \theta} \right) \right\}$$

and

$$e_1 = - 375 \left\{ \cos \theta + 0.26 \, \varepsilon^{-2.22 \theta} - 1.26 \, \varepsilon^{-0.452 \theta} \right\}.$$

There exists no value of θ_0 which does not give rise to a transient term.

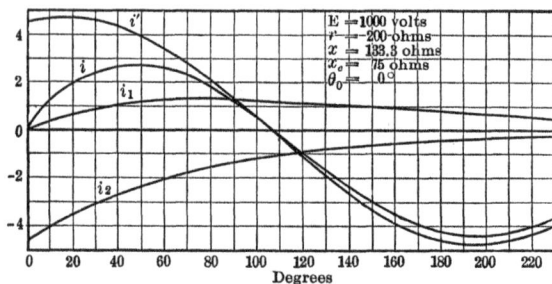

Fig. 22.　Starting of an alternating-current circuit having capacity, inductance and resistance in series.　Critical start.

In Fig. 22 the start of a circuit is shown, with the inductive reactance increased so as to give the critical condition,

$$r^2 = 4 \, x \, x_c,$$

but otherwise the constants are the same as in Figs. 20 and 21, that is, $E = 1000$ volts; $r = 200$ ohms; $x = 133.3$ ohms, and $x_c = 75$ ohms;

therefore　　$z_0 = 208.3,$

$$\tan \gamma = \frac{58.3}{200} = 0.2915, \quad \text{or} \quad \gamma = 16°,$$

assuming that the circuit is started at the moment $\theta_0 = 0$, or at the maximum value of impressed e.m.f.

Then (20) and (21) give

$$i = 4.78 \cos (\theta - 16°) + \varepsilon^{-0.75 \, \theta} (2.7 \, \theta - 4.6)$$

and

$$e_1 = 358 \sin (\theta - 16°) - \varepsilon^{-0.75 \, \theta} (410 \, \theta - 99).$$

Here also no value of θ_0 exists at which the transient term disappears. ·

59. The most important is the oscillating case, $r^2 < 4 \, x \, x_c$, since it is the most common in electrical circuits, as underground cable systems and overhead high potential circuits, and also is practically the only one in which excessive currents or excessive voltages, and thereby dangerous phenomena, may occur.

If the condensive reactance x_c is high compared with the resistance r and the inductive reactance x, the equations of start for the circuit from dead condition, that is, $i_0 = 0$ and $e_0 = 0$, are found by substitution into the general equations (24) and (25), which give the *current* as

$$i = -\frac{E}{x_c} \left\{ \sin(\theta - \theta_0) + \varepsilon^{-\frac{r}{2x}\theta} \left[\sin\theta_0 \cos\sqrt{\frac{x_c}{x}}\theta \right. \right.$$

$$\left. \left. - \sqrt{\frac{x_c}{x}}\cos\theta_0 \sin\sqrt{\frac{x_c}{x}}\theta \right] \right\} \tag{26}$$

and the *potential difference at the condenser terminals* as

$$e_1 = E \left\{ \cos(\theta - \theta_0) - \varepsilon^{-\frac{r}{2x}\theta} \right.$$

$$\left[\cos\theta_0 \cos\sqrt{\frac{x_c}{x}}\theta + \left(\frac{r}{2\sqrt{x\,x_c}}\cos\theta_0 + \sqrt{\frac{x}{x_c}}\sin\theta_0\right)\sin\sqrt{\frac{x_c}{x}}\theta \right] \right\}, \tag{27}$$

where
$$q = 2\sqrt{x\,x_c},\ z_0 = x_c,\ \text{and}\ \gamma = -90°. \tag{28}$$

In this case an oscillating term always exists whatever the value of θ_0, that is, the point of the wave, where the circuit is started.

The frequency of oscillation therefore is

$$f_0 = \frac{q}{2x}f = f\sqrt{\frac{x_c}{x} - \frac{r^2}{4x^2}}$$

or, approximately, $\qquad\qquad\qquad\qquad\qquad\qquad\qquad$ (29)

$$f_0 = \sqrt{\frac{x_c}{x}}f,$$

where f = fundamental frequency.

Substituting $x = 2\pi fL$ and $x_c = \dfrac{1}{2\pi fC}$, we have

$$f_0 = \frac{1}{2\pi}\sqrt{\frac{1}{CL} - \frac{r^2}{4L^2}},$$

or, approximately, $\qquad\qquad\qquad\qquad\qquad\qquad\qquad$ (30)

$$f_0 = \frac{1}{2\pi\sqrt{CL}}.$$

60. The oscillating start, or, in general, change of circuit conditions, is the most important, since in circuits containing capacity the transient effect is almost always oscillating.

The most common examples of capacity are distributed capacity in transmission lines, cables, etc., and capacity in the form of electrostatic condensers for neutralizing lagging currents, for constant potential-constant current transformation, etc.

(*a*) In transmission lines or cables the charging current is a fraction of full-load current i_0, and the e.m.f. of self-inductance consumed by the line reactance is a fraction of the impressed e.m.f. e_0. Since, however, the charging current is (approximately)

$= \dfrac{e_0}{x_c}$ and the e.m.f. of self-inductance $= xi_0$, we have

$$\frac{e_0}{x_c} < i_0, \quad xi_0 < e_0;$$

hence, multiplying,

$$\frac{x}{x_c} < 1 \text{ and } x < x_c.$$

The resistance r is of the same magnitude as x; thus

$$4\,x\,x_c > r^2.$$

For instance, with 10 per cent resistance drop, 30 per cent reactance voltage, and 20 per cent charging current in the line, assuming half the resistance and reactance as in series with the capacity (that is, representing the distributed capacity of the line by one condenser shunted across its center) and denoting

$$p = \frac{e_0}{i_0},$$

where e_0 = impressed voltage, i_0 = full-load current, we have

$$\left. \begin{array}{l} x_c = \dfrac{p}{0.2} = 5\,p, \\[2mm] x = 0.5 \times 0.3\,p = 0.15\,p, \\[2mm] r = 0.5 \times 0.1\,p = 0.05\,p, \end{array} \right\}$$

and

$$r \div x \div x_c = 1 \div 3 \div 100,$$

and

$$4\,x\,x_c \div r^2 = 1200 \div 1.$$

In this case, to make the start non-oscillating, we must have $x < \dfrac{1}{400}r$, or $x < 0.000125\,p$, which is not possible; or $r > p\sqrt{3}$, which can be done only by starting the circuit through a very large non-inductive resistance (of such size as to cut the starting current down to less than $\dfrac{1}{\sqrt{3}}$ of full-load current). Even in this case, however, oscillations would appear by a change of load, etc., after the start of the circuit.

(*b*) When using electrostatic condensers for producing watt-less leading currents, the resistance in series with the condensers is made as low as possible, for reasons of efficiency. Even with the extreme value of 10 per cent resistance, or $r \div x_c = 1 \div 10$, the non-oscillating condition is $x < \dfrac{1}{40}r$, or 0.25 per cent, which is not feasible.

In general, if

x consumes........ 1 2 4 9 16 per cent of the condenser potential difference,

r must consume > 20 28.3 40 60 80 per cent of the condenser potential difference.

That is, a very high non-inductive resistance is required to avoid oscillations.

The frequency of oscillation is approximately $f_0 = \sqrt{\dfrac{x_c}{x}}f$ that is, is lower than the impressed frequency if $x_c < x$ (or the permanent current lags), and higher than the impressed frequency if $x_c > x$ (or the permanent current leads). In transmission lines and cables the latter is always the case.

Since in a transmission line $\dfrac{p}{x_c}$ is approximately the charging current, as fraction of full-load current, and $\dfrac{x}{p}$ half the line e.m.f. of self-inductance, or reactance voltage, as fraction of impressed voltage, the following is approximately true:

The frequency of oscillation of a transmission line is the impressed frequency divided by the square root of the product of charging current and of half the reactance voltage of the line, given respectively as fractions of full-load current and of impressed voltage. For instance, 10 per cent charging current, 20 per cent reactance voltage, gives an oscillation frequency

$$f_0 = \frac{f}{\sqrt{0.1 \times 0.1}} = 10 f.$$

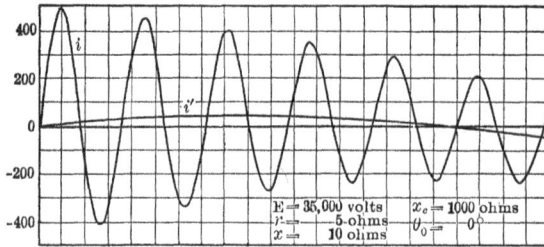

Fig. 23. Starting of an alternating-current circuit having capacity, inductance and resistance in series. Oscillating start of transmission line.

61. In Figs. 23 and 24 is given as example the start of current in a circuit having the constants, $E = 35,000$ cos $(\theta - \theta_0)$; $r = 5$ ohms; $x = 10$ ohms, and $x_c = 1000$ ohms.

In Fig. 23 for $\theta_0 = 0°$, or approximately maximum oscillation,

$$i = - 35 \left\{ \sin \theta - 10\, \varepsilon^{- 0.25\, \theta} \sin 10\, \theta \right\}$$

and

$$e_1 = 35,000 \left\{ \cos \theta - \varepsilon^{- 0.25\, \theta} [\cos 10\, \theta + 0.025 \sin 10\, \theta] \right\}.$$

In Fig. 24 for $\theta_0 = 90°$, or approximately minimum oscillation,

$$i = 35 \left\{ \cos \theta - \varepsilon^{- 0.25\, \theta} \cos 10\, \theta \right\}$$

and

$$e_1 = 35,000 \left\{ \sin \theta + 0.1\, \varepsilon^{- 0.25\, \theta} \sin 10\, \theta \right\}.$$

As seen, the frequency is 10 times the fundamental, and in starting the potential difference nearly doubles.

As further example, Fig. 25 shows the start of a circuit of a frequency of oscillation of the same magnitude as the fundamental, in resonance condition, $x = x_{c}$, and of high resistance.

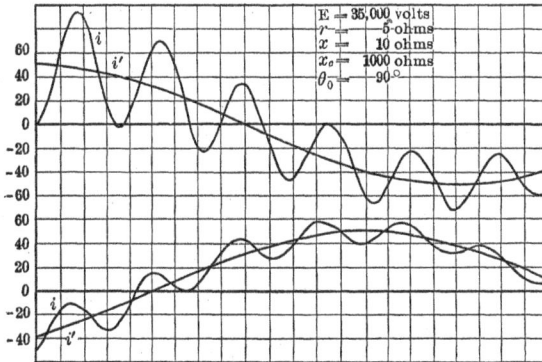

Fig. 24. Starting of an alternating-current circuit having capacity, inductance and resistance in series. Oscillating start of transmission line.

The circuit constants are $E = 1500$ volts; $r = 30$ ohms; $x = 20$ ohms; $x_c = 20$ ohms, and $\theta_0 = -\gamma$; which give $q = 26.46$; $z_0 = 30$; $\gamma = 0$, and $\theta_0 = 0$.

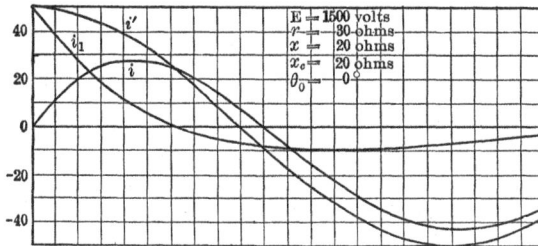

Fig. 25. Starting of an alternating-current circuit having capacity, inductance and resistance in series. Oscillating start. High resistance.

Substituting in equations (24) and (25) gives

$$i = 50\left\{\cos\theta - \varepsilon^{-0.75\,\theta}\left[\cos 0.661\,\theta - 1.14\sin 0.661\,\theta\right]\right\}$$
and
$$e_1 = 1000\left\{\sin\theta - 1.51\,\varepsilon^{-0.75\,\theta}\sin 0.661\,\theta\right\}.$$

As example of an oscillation of long wave, Fig. 26 represents the start of a circuit having the constants $E = 1500$ volts; $r = 10$ ohms; $x = 62.5$ ohms; $x_c = 10$ ohms, and $\theta_0 = -\gamma$; which give $q = 49$; $z_0 = 53.4$; $\gamma = 79°$, and $\theta_0 = -79°$.

Substituting in equations (24) and (25) gives

$$i = 28 \left\{ \cos\theta - \varepsilon^{-0.08\,\theta} \left[\cos 0.39\,\theta - 0.2 \sin 0.39\,\theta \right] \right\}$$

and

$$e_1 = 280 \left\{ \sin\theta - 2.55\,\varepsilon^{-0.08\,\theta} \sin 0.396\,\theta \right\}.$$

Such slow oscillations for instance occur in a transmission line connected to an open circuited transformer.

62. While in the preceding examples, Figs. 23 to 26, constants of transmission lines have been used, as will be shown in the following chapters, in the case of a transmission line

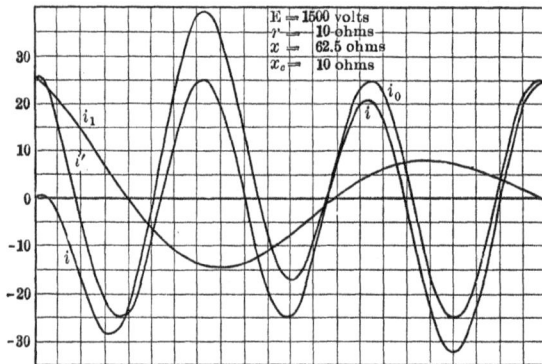

Fig. 26. Starting of an alternating-current circuit having capacity, inductance and resistance in series. Oscillating start of long period.

with distributed capacity and inductance, the oscillation does not consist of one definite frequency but an infinite series of frequencies, and the preceding discussion thus approximates only the fundamental frequency of the system. This, however, is the frequency which usually predominates in a high power low frequency surge of the system.

In an underground cable system the preceding discussion applies more closely, since in such a system capacity and inductance are more nearly localized: the capacity is in the underground cables, which are of low inductance, and the inductance is in the generating system, which has practically no capacity.

In an underground cable system the tendency therefore is either towards a local, very high frequency oscillation, or traveling wave, of very limited power, in a part of the cables, or a low frequency high power surge, frequently of destructive magnitude, of the joint capacity of the cables, against the inductance of the generating system.

63. The physical meaning of the transient terms can best be understood by reviewing their origin.

In a circuit containing resistance and inductance only, but a single transient term appears of exponential nature. In such a circuit at any moment, and thus at the moment of start, the current should have a certain definite value, depending on the constants of the circuit. In the moment of start, however, the current may have a different value, depending on the preceding condition, as for instance the value zero if the circuit has been open before. The current thus adjusts itself from the initial value to the permanent value on an exponential curve, which disappears if the initial value happens to coincide with the final value, as for instance if the circuit is closed at the moment of the e.m.f. wave, when the permanent current should be zero. The approach of current to the permanent value is retarded by the inductance, accelerated by the resistance of the circuit.

In a circuit containing inductance and capacity, at any moment the current has a certain value and the condenser a certain charge, that is, potential difference. In the moment of start, current intensity and condenser charge have definite values, depending on the previous condition, as zero, if the circuit was open, and thus two transient terms must appear, depending upon the adjustment of current and of condenser e.m.f. to their permanent values.

Since at the moment when the current is zero the condenser e.m.f. is maximum, and inversely, in a circuit containing inductance and capacity, the starting of a circuit always results in the appearance of a transient term.

If the circuit is closed at the moment when the condenser e.m.f. should be zero, that is, about the maximum value of current, the transient term of current cannot exceed in amplitude its final value, since its maximum or initial value equals the value which the current should have at this moment. If, however, the circuit is closed at the moment where the current should be zero and the condenser e.m.f. maximum, the condenser being without charge acts in the first moment like a short circuit, that is, the current begins at a value corresponding to the impressed e.m.f. divided by the line impedance. Thus if we neglect the resistance and if the condenser reactance equals n^2 times line reactance, the current starts at n^2 times its final rate; thus it would, in a half wave, give n^2 times the full charge of the condenser, or in other words, charge the condenser in $\dfrac{1}{n}$ of the time of a half wave. That is, the period of the starting current is $\dfrac{1}{n}$ and the amplitude n times that of the final current. However, as soon as the condenser is charged, in $\dfrac{1}{n}$ of a period of the impressed e.m.f., the magnetic field of the charging current produces a return current, discharging the condenser again at the same rate.

Thus the normal condition of start is an oscillation of such a frequency as to give the full condenser charge at a rate which when continued up to full frequency would give an amplitude equal to the impressed e.m.f. divided by the line reactance. The effect of the line resistance is to consume e.m.f. and thus dampen the oscillation, until the resistance consumes during the condenser charge as much energy as the magnetic field would store up, and then the oscillation disappears and the start becomes exponential.

Analytically the double transient term appears as the result of the two roots of a quadratic equation, as seen above.

CHAPTER VIII.

LOW FREQUENCY SURGES IN HIGH POTENTIAL SYSTEMS.

64. In electric circuits of considerable capacity, that is, in extended high potential systems, as long distance transmission lines and underground cable systems, occasionally destructive high potential low frequency surges occur; that is, oscillations of the whole system, of the same character as in the case of localized capacity and inductance discussed in the preceding chapter.

While a system of distributed capacity has an infinite number of frequencies, which usually are the odd multiples of a fundamental frequency of oscillation, in those cases where the fundamental frequency predominates and the effect of the higher frequencies is negligible, the oscillation can be approximated by the equations of oscillation given in Chapters V and VII, which are far simpler than the equations of an oscillation of a system of distributed capacity.

Such low frequency surges take in the total system, not only the transmission lines but also the step-up transformers, generators, etc., and in an underground cable system in such an oscillation the capacity and inductance are indeed localized to a certain extent, the one in the cables, the other in the generating system. In an underground cable system, therefore, of the infinite series of frequencies of oscillations which theoretically exist, only the fundamental frequency and those very high harmonics which represent local oscillations of sections of cables can be pronounced, and the first higher harmonics of the fundamental frequency must be practically absent. That is, oscillations of an underground cable system are either

(a) Low frequency high power surges of the whole system, of a frequency of a few hundred cycles, frequently of destructive character, or,

(b) Very high frequency low power oscillations, local in character, so called "static," probably of frequencies of hundred thousands of cycles, rarely directly destructive, but indirectly harmful in their weakening action on the insulation and the possibility of their starting a low frequency surge.

The former ones only are considered in the present chapter. Their causes may be manifold, — changes of circuit conditions, as starting, opening a short circuit, existence of a flaring arc on the system, etc.

In the circuit from the generating system to the capacity of the transmission line or the underground cables, we have always $r^2 < \dfrac{4\,L}{C}$; that is, the phenomenon is always oscillatory, and equations (24) and (25), Chapter VII, apply, and for the *current* we have

$$i = \frac{E}{z_0}\cos(\theta - \theta_0 - \gamma) + \varepsilon^{-\frac{r}{2\,x}\theta}\left\{\left[i_0 - \frac{E}{z_0}\cos(\theta_0 + \gamma)\right]\cos\frac{q}{2\,x}\theta\right.$$

$$\left. - \left[\frac{2\,e_0 + r i_0}{q} + \frac{E}{q z_0}\left(2\,x_c\sin(\theta_0 + \gamma) - r\cos(\theta_0 + \gamma)\right)\right]\sin\frac{q}{2\,x}\theta\right\}, \quad (1)$$

and for the *condenser potential* we have

$$e_1 = \frac{E x_c}{z_0}\sin(\theta - \theta_0 - \gamma) + \varepsilon^{-\frac{r}{2\,x}\theta}\left\{\left[e_0 + \frac{E x_c}{z_0}\sin(\theta_0 + \gamma)\right]\cos\frac{q}{2\,x}\theta\right.$$

$$\left. + \left[\frac{2\,r e_0 + 4\,x\,x_c i_0}{2\,q} + \frac{E x_c}{q z_0}\left(r\sin(\theta_0 + \gamma) - 2\,x\cos(\theta_0 + \gamma)\right)\right]\sin\frac{q}{2\,x}\theta\right\}$$

$$\tag{2}$$

65. These equations (1) and (2) can be essentially simplified by neglecting terms of secondary magnitude.

x_c is in high potential transmission lines or cables always very large compared with r and x.

The full-load resistance and reactance voltage may vary from less than 5 per cent to about 20 per cent of the impressed e.m.f., the charging current of the line from 5 per cent to about 20 per cent of full-load current, at normal voltage and frequency.

In this case, x_c is from 25 to more than 400 times as large as r or x, and r and x thus negligible compared with x_c.

It is then, in close approximation:

$$\left.\begin{aligned}
z_0 &= x_c, \\
q &= 2\sqrt{x\,x_c}, \\
\gamma &= -\frac{\pi}{2} = -90^\circ.
\end{aligned}\right\} \qquad (3)$$

Substituting these values in equations (1) and (2) gives the *current* as

$$i = -\frac{E}{x_c}\sin(\theta-\theta_0) + \varepsilon^{-\frac{r}{2x}\theta}\left\{\left[i_0 - \frac{E}{x_c}\sin\theta_0\right]\cos\sqrt{\frac{x_c}{x}}\,\theta\right.$$

$$\left. -\left[\frac{2e_0+ri_0}{2\sqrt{x\,x_c}} - \frac{E}{2\sqrt{x\,x_c}}\left(2\cos\theta_0 + \frac{r}{x_c}\sin\theta_0\right)\right]\sin\sqrt{\frac{x_c}{x}}\,\theta\right\}, \qquad (4)$$

and the *potential difference at the condenser* as

$$e_1 = E\cos(\theta-\theta_0) + \varepsilon^{-\frac{r}{2x}\theta}\left\{[e_0 - E\cos\theta_0]\cos\sqrt{\frac{x_c}{x}}\,\theta\right.$$

$$+\left[\frac{2re_0 + 4xx_ci_0}{4\sqrt{xx_c}} - \frac{E}{4\sqrt{xx_c}}\right.$$

$$\left.\left. (2r\cos\theta_0 + 4x\sin\theta_0)\right]\sin\sqrt{\frac{x_c}{x}}\,\theta\right\}.$$

These equations consist of three terms:

$$\left.\begin{aligned}
i &= i' + i'' + i''', \\
e_1 &= e_1' + e_1'' + e_1''';
\end{aligned}\right\} \qquad (5)$$

$$\left.\begin{aligned}
i' &= -\frac{E}{x_c}\sin(\theta-\theta_0), \\
e_1' &= E\cos(\theta-\theta_0);
\end{aligned}\right\} \qquad (6)$$

$$i'' = -\frac{E}{x_c}\varepsilon^{-\frac{r}{2x}\theta}\left\{\sin\theta_0\cos\sqrt{\frac{x_c}{x}}\theta - \left[\sqrt{\frac{x_c}{x}}\cos\theta_0\right.\right.$$

$$\left.\left. + \frac{r}{2\sqrt{x\,x_c}}\sin\theta_0\right]\sin\sqrt{\frac{x_c}{x}}\theta\right\},$$

$$e_1'' = -E\varepsilon^{-\frac{r}{2x}\theta}\left\{\cos\theta_0\cos\sqrt{\frac{x_c}{x}}\theta + \left(\frac{r}{2\sqrt{x\,x_c}}\cos\theta_0\right.\right.$$

$$\left.\left. + \sqrt{\frac{x}{x_c}}\sin\theta_0\right)\sin\sqrt{\frac{x_c}{x}}\theta\right\};$$

$$(7)$$

or, by dropping terms of secondary order,

$$i'' = \frac{E}{\sqrt{x\,x_c}}\varepsilon^{-\frac{r}{2x}\theta}\cos\theta_0\sin\sqrt{\frac{x}{x_c}}\theta,$$

$$e_1'' = -E\varepsilon^{-\frac{r}{2x}\theta}\cos\theta_0\cos\sqrt{\frac{x_c}{x}}\theta;$$

$$(8)$$

and:

$$i''' = \varepsilon^{-\frac{r}{2x}\theta}\left\{i_0\cos\sqrt{\frac{x_c}{x}}\theta - \frac{2\,e_0 + r i_0}{2\sqrt{x\,x_c}}\sin\sqrt{\frac{x_c}{x}}\theta\right\},$$

$$e_1''' = \varepsilon^{-\frac{r}{2x}\theta}\left\{e_0\cos\sqrt{\frac{x_c}{x}}\theta + \frac{2\,r e_0 + 4\,x x_c i_0}{4\sqrt{x x_c}}\sin\sqrt{\frac{x_c}{x}}\theta\right\};$$

$$(9)$$

or, by dropping terms of secondary order,

$$i''' = \varepsilon^{-\frac{r}{2x}\theta}\left\{i_0\cos\sqrt{\frac{x_c}{x}}\theta - \frac{e_0}{\sqrt{x\,x_c}}\sin\sqrt{\frac{x_c}{x}}\theta\right\},$$

$$e_1''' = \varepsilon^{-\frac{r}{2x}\theta}\left\{e_0\cos\sqrt{\frac{x_c}{x}}\theta + i_0\sqrt{x\,x_c}\sin\sqrt{\frac{x_c}{x}}\theta\right\}.$$

$$(10)$$

Thus the *total current* is approximately

$$i = -\frac{E}{x_c}\sin(\theta - \theta_0) + \varepsilon^{-\frac{r}{2x}\theta}\left\{i_0\cos\sqrt{\frac{x_c}{x}}\theta\right.$$

$$\left. - \frac{e_0 - E\cos\theta_0}{\sqrt{x\,x_c}}\sin\sqrt{\frac{x_c}{x}}\theta\right\},$$

and the *difference of potential at the condenser* is

$$e_1 = E\cos(\theta - \theta_0) + \varepsilon^{-\frac{r}{2x}\theta}\left\{(e_0 - E\cos\theta_0)\cos\sqrt{\frac{x_c}{x}}\theta\right.$$

$$\left. + i_0\sqrt{x\,x_c}\sin\sqrt{\frac{x_c}{x}}\theta\right\}.$$

$$(11)$$

Of the three terms: i', e_1'; i'', e_1''; i''', e_1''', the first obviously represents the stationary condition of charging current and condenser potential, since the two other terms disappear for $t = \infty$.

The second term, i'', e_1'', represents that component of oscillation which depends upon the phase of impressed e.m.f., or the point of the impressed e.m.f. wave, at which the oscillation begins, while the third term, i''', e_1''', represents the component of oscillation which depends upon the instantaneous values of current and e.m.f. respectively, at the moment at which the oscillation begins. $\varepsilon^{-\frac{r}{2x}\theta}$ is the decrement of the oscillation.

66. The frequency of oscillation is

$$f_0 = \sqrt{\frac{x_c}{x}}\, f,$$

where f is the impressed frequency. That is, the frequency of oscillation equals the impressed frequency times the square root of the ratio of condensive reactance and inductive reactance of the circuit, or is the impressed frequency divided by the square root of inductance voltage and capacity current, as fraction of impressed voltage and full-load current.

Since

$$x_c = \frac{1}{2\,\pi f C} \quad \text{and} \quad x = 2\,\pi f L,$$

the frequency of oscillation is

$$f_0 = \frac{1}{2\,\pi\sqrt{CL}};$$

that is, is independent of the frequency of the impressed e.m.f.

Substituting

$$\theta = 2\,\pi f t, \quad x_c = \frac{1}{2\,\pi f C} \quad \text{and} \quad x = 2\,\pi f L$$

in equations (8), (10), and (11), we have

$$\left.\begin{aligned}
i'' &= \sqrt{\frac{C}{L}}\, E\varepsilon^{-\frac{r}{2L}t} \cos\theta_0 \sin\frac{t}{\sqrt{CL}}, \\[2ex]
e_1'' &= -\, E\varepsilon^{-\frac{r}{2L}t} \cos\theta_0 \cos\frac{t}{\sqrt{CL}};
\end{aligned}\right\} \quad (12)$$

$$i''' = \varepsilon^{-\frac{r}{2L}t}\left\{i_0 \cos \frac{t}{\sqrt{CL}} - e_0 \sqrt{\frac{C}{L}} \sin \frac{t}{\sqrt{CL}}\right\},$$

$$e_1''' = \varepsilon^{-\frac{r}{2L}t}\left\{e_0 \cos \frac{t}{\sqrt{CL}} + i_0 \sqrt{\frac{L}{C}} \sin \frac{t}{\sqrt{CL}}\right\}; \qquad (13)$$

$$i = -2\pi f CE \sin(\theta - \theta_0) + \varepsilon^{-\frac{r}{2L}t}\left\{i_0 \cos \frac{t}{\sqrt{CL}}\right.$$

$$\left. - (e_0 - E \cos \theta_0)\sqrt{\frac{C}{L}} \sin \frac{t}{\sqrt{CL}}\right\},$$

$$e_1 = E \cos(\theta - \theta_0) + \varepsilon^{-\frac{r}{2L}t}\left\{(e_0 - E \cos \theta_0)\cos \frac{t}{\sqrt{CL}}\right.$$

$$\left. + i_0 \sqrt{\frac{L}{C}} \sin \frac{t}{\sqrt{CL}}\right\}. \qquad (14)$$

The oscillating terms of these equations are independent of the impressed frequency. That is, the oscillating currents and potential differences, caused by a change of circuit conditions (as starting, change of load, or opening circuit), are independent of the impressed frequency, and thus also of the wave shape of the impressed e.m.f., or its higher harmonics (except as regards terms of secondary order).

The first component of oscillation, equation (12), depends not only upon the line constants and the impressed e.m.f., but principally upon the phase, or the point of the impressed e.m.f. wave, at which the oscillation starts; however, it does not depend upon the previous condition of the circuit. Therefore this component of oscillation is the same as the oscillation produced in starting the transmission line, that is, connecting it, unexcited, to the generator terminals.

There exists no point of the impressed e.m.f. wave where no oscillation occurs (while, when starting a circuit containing resistance and inductance only, at the point of the impressed e.m.f. wave where the final current passes zero the stationary condition is instantly reached).

With capacity in circuit, any change of circuit conditions involves an electric oscillation.

The maximum intensities of the starting oscillation occur near the value $\theta_0 = 0$, and are

and

$$i'' = \frac{E}{\sqrt{x\,x_c}}\, \varepsilon^{-\frac{r}{2x}\theta} \sin \sqrt{\frac{x_c}{x}}\,\theta$$

$$e_1'' = -E\, \varepsilon^{-\frac{r}{2x}\theta} \cos \sqrt{\frac{x_c}{x}}\,\theta. \tag{15}$$

Since

$$i' = -\frac{E}{x_c} \sin(\theta - \theta_0)$$

is the stationary value of charging current, it follows that the maximum intensity which the oscillating current, produced in starting a transmission line, may reach is $\sqrt{\dfrac{x_c}{x}}$ times the stationary charging current, or the initial current bears to the stationary value the same ratio as the frequency of oscillation to the impressed frequency.

The maximum oscillating e.m.f. generated in starting a transmission line is of the same value as the impressed e.m.f. Thus the maximum value of potential difference occurring in a transmission line at starting is less than twice the impressed e.m.f. and no excessive voltages can be generated in starting a circuit.

The minimum values of the starting oscillation occur near $\theta_0 = 90°$, and are, from equations (7),

and

$$i'' = -\frac{E}{x_c}\, \varepsilon^{-\frac{r}{2x}\theta} \cos \sqrt{\frac{x_c}{x}}\,\theta$$

$$e_1'' = -\sqrt{\frac{x}{x_c}}\,E\, \varepsilon^{-\frac{r}{2x}\theta} \sin \sqrt{\frac{x_c}{x}}\,\theta; \tag{16}$$

that is, the oscillating current is of the same intensity as the charging current, and the maximum rush of current thus is less than twice the stationary value. The potential difference in the circuit rises only little above the impressed e.m.f.

The second component of the oscillation, equation (13), does not depend upon the point of the impressed e.m.f. wave at which the oscillation starts, θ_0, nor upon the impressed e.m.f. as a whole, E, but, besides upon the constants of the circuit, it depends only upon the instantaneous values of current and of potential difference in the circuit at the moment when the oscillation starts, i_0 and e_0.

Thus, if $i_0 = 0$, $e_0 = 0$, or in starting a transmission line, unexcited, by connecting it to the impressed e.m.f., this term disappears. It is this component which may cause excessive potential differences. Two cases shall more fully be discussed, namely:

(*a*) Opening the circuit of a transmission line under load, and

(*b*) rupturing a short-circuit on the transmission line.

67. (*a*) If i_0 is the instantaneous value of full-load current, e_0 the instantaneous value of difference of potential at the condenser, ri_0 is small compared with e_0, and $\sqrt{x\,x_c}\,i_0$ is of the same magnitude as e_0.

Writing

$$\tan\delta = \frac{e_0}{i_0\sqrt{x\,x_c}},$$

and substituting in equations (10), we have

$$
\left.
\begin{aligned}
i''' &= \sqrt{i_0{}^2 + \frac{e_0{}^2}{x\,x_c}}\ \varepsilon^{-\frac{r}{2x}\theta}\cos\left(\sqrt{\frac{x_c}{x}}\,\theta + \delta\right) \\[2mm]
e_1''' &= \sqrt{e_0{}^2 + i_0{}^2 x\,x_c}\ \varepsilon^{-\frac{r}{2x}\theta}\sin\left(\sqrt{\frac{x_c}{x}}\,\theta + \delta\right);
\end{aligned}
\right\}
\quad (17)
$$

that is, the amplitude of oscillation is $\sqrt{i_0{}^2 + \dfrac{e_0{}^2}{x\,x_c}}$ for the current, and $\sqrt{e_0{}^2 + i_0{}^2 x\,x_c}$ for the e.m.f. Thus the generated e.m.f. can be larger than the impressed e.m.f., but is, as a rule, still of the same magnitude, except when x_c is very large.

In the expressions of the total current and potential difference at condenser, in equations (11), $(e_0 - E\cos\theta_0)$ is the difference between the potential difference at the condenser and the impressed e.m.f., at the instant of starting of the oscillation, or the voltage consumed by the line impedance, and this is small if the current is not excessive. Thus, neglecting the terms with $(e_0 - E \cos \theta_0)$, equations (11) assume the form

$$
\left.\begin{aligned}
i &= -\frac{E}{x_c} \sin (\theta - \theta_0) + i_0 \varepsilon^{-\frac{r}{2x}\theta} \cos \sqrt{\frac{x_c}{x}}\, \theta \\[2mm]
e_1 &= E \cos (\theta - \theta_0) + i_0 \sqrt{x\, x_c}\, \varepsilon^{-\frac{r}{2x}\theta} \sin \sqrt{\frac{x_c}{x}}\, \theta ;
\end{aligned}\right\} \quad (18)
$$

and

that is, the oscillation of current is of the amplitude of full-load current, and the oscillation of condenser potential difference is of the amplitude $i_0\sqrt{x\, x_c}$.

$x\, x_c$ is the ratio of inductance voltage to condenser current, in fractions of full-load voltage and current. We have, therefore,

$$
i_0\sqrt{x\, x_c} = i_0\sqrt{\frac{L}{C}}.
$$

Thus in circuits of very high inductance L and relatively low capacity C, $i_0\sqrt{x\, x_c}$ may be much higher than the impressed e.m.f., and a serious rise of potential occur when opening the circuit under load, while in low inductance cables of high capacity $i_0\sqrt{x\, x_c}$ is moderate; that is, the inductance, by tending to maintain the current, generates an e.m.f., producing a rise in potential, while capacity exerts a cushioning effect. Low inductance and high capacity thus are of advantage when breaking full-load current in a circuit.

68. (*b*) If a transmission line containing resistance, inductance, and capacity is short-circuited, and the short-circuit suddenly opened at time $t = 0$, we have, for $t < 0$,

$$
\left.\begin{aligned}
e_0 &= 0 \\[2mm]
i &= \frac{E}{z} \cos (\theta - \theta_0 - \gamma), \\[2mm]
z &= \sqrt{r^2 + x^2} \\[2mm]
\tan \gamma &= \frac{x}{r} ;
\end{aligned}\right\} \quad (19)
$$

and

where

and

thus, at time $t = 0$,

$$i_0 = \frac{E}{z} \cos (\theta_0 + \gamma). \tag{20}$$

Substituting these values of e_0 and i_0 in equations (9) gives

$$i''' = \frac{E}{z} \cos (\theta_0 + \gamma) \, \varepsilon^{-\frac{r}{2x}\theta} \left\{ \cos \sqrt{\frac{x_c}{x}} \theta - \frac{r}{2\sqrt{x x_c}} \sin \sqrt{\frac{x_c}{x}} \theta \right\}$$

and

$$e_1''' = \frac{E}{z} \cos (\theta_0 + \gamma) \varepsilon^{-\frac{r}{2x}\theta} \sqrt{x x_c} \sin \sqrt{\frac{x_c}{x_c}} \theta,$$

or, neglecting terms of secondary magnitude,

$$i''' = \frac{E}{z} \varepsilon^{-\frac{r}{2x}\theta} \cos (\theta_0 + \gamma) \cos \sqrt{\frac{x_c}{x}} \theta$$

and

$$e_1''' = \frac{E\sqrt{x\, x_c}}{z} \varepsilon^{-\frac{r}{2x}\theta} \cos (\theta_0 + \gamma) \sin \sqrt{\frac{x_c}{x}} \theta; \tag{21}$$

that is, i''' is of the magnitude of short-circuit current, and e_1''' of higher magnitude than the impressed e.m.f., since z is small compared with $\sqrt{x x_c}$.

The total values of current and condenser potential difference, from equation (11), are

$$i = -\frac{E}{x_c} \sin (\theta - \theta_0) + E\varepsilon^{-\frac{r}{2x}\theta} \left\{ \frac{\cos (\theta_0 + \gamma)}{z} \right.$$

$$\cos \sqrt{\frac{x_c}{x}} \theta + \frac{\cos \theta_0}{\sqrt{x x_c}} \sin \sqrt{\frac{x_c}{x}} \theta \right\}$$

and

$$e_1 = E \cos (\theta - \theta_0) - E\varepsilon^{-\frac{r}{2x}\theta} \left\{ \cos \theta_0 \cos \sqrt{\frac{x_c}{x}} \theta \right.$$

$$- \frac{\sqrt{x x_c} \cos (\theta_0 + \gamma)}{z} \sin \sqrt{\frac{x_c}{x}} \theta \right\}, \tag{22}$$

or approximately, since all terms are negligible compared with i''' and e_1''',

$$i = \frac{E}{z} \varepsilon^{-\frac{r}{2x}\theta} \cos(\theta_0 + \gamma) \cos\sqrt{\frac{x_c}{x}}\,\theta$$

and

$$e_1 = \frac{E\sqrt{xx_c}}{z} \varepsilon^{-\frac{r}{2x}\theta} \cos(\theta_0 + \gamma) \sin\sqrt{\frac{x_c}{x}}\,\theta.$$

$$(23)$$

These values are a maximum, if the circuit is opened at the moment $\theta_0 = -\gamma$, that is, at the maximum value of the short-circuit current, and are then

$$i = \frac{E}{z} \varepsilon^{-\frac{r}{2x}\theta} \cos\sqrt{\frac{x_c}{x}}\,\theta$$

and

$$e_1 = \frac{\sqrt{xx_c}}{z} E\varepsilon^{-\frac{r}{2x}\theta} \sin\sqrt{\frac{x_c}{x}}\,\theta.$$

$$(24)$$

The amplitude of oscillation of the condenser potential difference is

$$\frac{\sqrt{xx_c}}{z} E,$$

or, neglecting the line resistance, as rough approximation, $x = z$,

$$\sqrt{\frac{x_c}{x}}\,E;$$

that is, the potential difference at the condenser is increased above the impressed e.m.f. in the proportion of the square root of the ratio of condensive reactance to inductive reactance, or inversely proportional to the square root of inductance voltage times capacity current, as fraction of the impressed voltage and the full-load current. Thus, in this case, the rise of voltage is excessive.

The minimum intensity of the oscillation due to rupturing short-circuit occurs if the circuit is broken at the moment

$\theta_0 = 90° - r$, that is, at the zero value of the short-circuit current. Then we have

$$i = -\frac{E}{x_c}\cos(\theta + \gamma) + \frac{E}{\sqrt{x\,x_c}}\,\varepsilon^{-\frac{r}{2x}\theta}\sin\gamma\sin\sqrt{\frac{x_c}{x}}\theta$$

and

$$e_1 = E\sin(\theta + \gamma) - E\varepsilon^{-\frac{r}{2x}\theta}\sin\gamma\cos\sqrt{\frac{x_c}{x}}\theta;$$

$$(25)$$

that is, the potential difference at the condenser is less than twice the impressed e.m.f.; therefore is moderate. Hence, a short-circuit can be opened safely only at or near the zero value of the short-circuit current.

The phenomenon ceases to be oscillating, and becomes an ordinary logarithmic discharge, if $\sqrt{r^2 - 4\,x\,x_c}$ is real, or

$$r > 2\sqrt{x\,x_c}.$$

Some examples may illustrate the phenomena discussed in the preceding paragraphs.

69. Let, in a transmission line carrying 100 amperes at full load, under an impressed e.m.f. of 20,000 volts, the resistance drop = 8 per cent, the inductance voltage = 15 per cent of the impressed voltage, and the charging current = 8 per cent of full-load current. Assuming 1 per cent resistance drop in the step-up transformers, and a reactance voltage of $2\frac{1}{2}$ per cent, the resistance drop between the constant potential generator terminals and the middle of the transmission line is then 5 per cent, or $r = 10$ ohms, and the inductance voltage is 10 per cent, or $x = 20$ ohms. The charging current of the line is 8 amperes, thus the condensive reactance $x_c = 2500$ ohms.

Then, assuming a sine wave of impressed e.m.f., we have

$$E = 20{,}000\sqrt{2} = 28{,}280 \text{ volts};$$
$$i' = -11.3\sin(\theta - \theta_0);$$
$$e_1' = 28{,}280\cos(\theta - \theta_0);$$
$$i'' = -11.3\,\varepsilon^{-0.25\,\theta}[\sin\theta_0\cos 11.2\,\theta - 11.2\cos\theta_0\sin 11.2\,\theta],$$

and $\quad e_1'' = -28{,}280\,\varepsilon^{-0.25\,\theta}[\cos\theta_0\cos 11.2\,\theta + (0.0222\cos\theta_0$
$$+\,0.0283\sin\theta_0)\sin 11.2\,\theta]$$
$$\cong -28{,}280\,\varepsilon^{-0.25\,\theta}\cos\theta_0\cos 11.2\,\theta.$$

Therefore the oscillations produced in starting the transmission line are

$$i = -11.3 \left[\sin (\theta - \theta_0) + \varepsilon^{-0.25\,\theta} (\sin \theta_0 \cos 11.2\,\theta \right.$$
$$\left. - 11.2 \cos \theta_0 \sin 11.2\,\theta) \right]$$

and
$$e_1 = 28{,}280 \left\{ \cos (\theta - \theta_0) - \varepsilon^{-0.25\,\theta} [\cos \theta_0 \cos 11.2\,\theta \right.$$
$$\left. + (0.0222 \cos \theta_0 + 0.0283 \sin \theta_0) \sin 11.2\,\theta] \right\}$$
$$\cong 28{,}280 \left[\cos (\theta - \theta_0) - \varepsilon^{-0.25\,\theta} \cos \theta_0 \cos 11.2\,\theta \right].$$

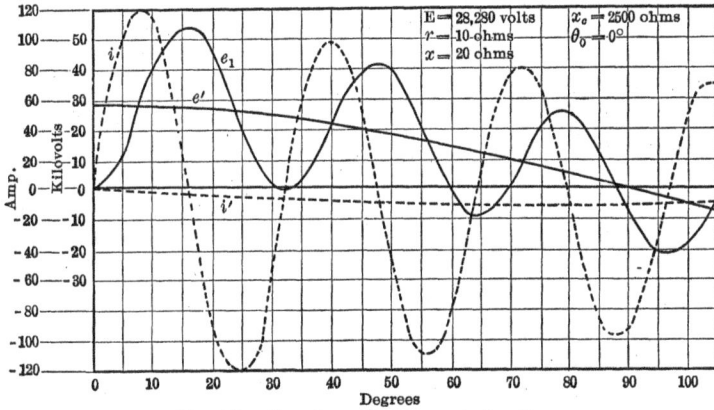

Fig. 27. Starting of a transmission line.

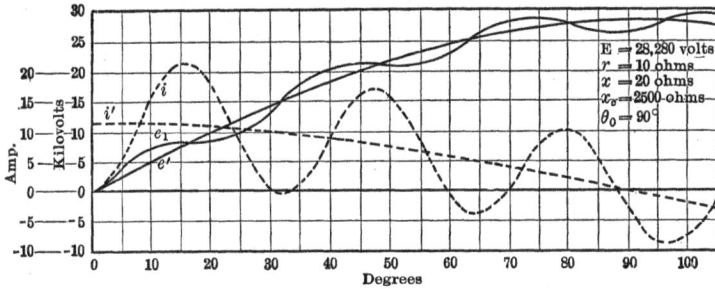

Fig. 28. Starting of a transmission line.

Hence the maximum values for $\theta_0 = 0$, are

$$i = -11.3 \left(\sin \theta - 11.2\, \varepsilon^{-0.25\,\theta} \sin 11.2\,\theta \right)$$

and
$$e_1 = 28{,}280 \left[\cos \theta - \varepsilon^{-0.25\,\theta} (\cos 11.2\,\theta + 0.0222 \sin 11.2\,\theta) \right]$$
$$\cong 28{,}280 \left(\cos \theta - \varepsilon^{-0.25\,\theta} \cos 11.2\,\theta \right),$$

and the minimum values, for $\theta_0 = 90°$, are

$$i = 11.3 \left(\cos \theta - \varepsilon^{-0.25\,\theta} \cos 11.2\,\theta \right)$$

and
$$e_1 = 28{,}280 \left(\sin \theta - 0.0283\, \varepsilon^{-0.25\,\theta} \sin 11.2\,\theta \right)$$
$$\cong 28{,}280 \sin \theta$$

These values are plotted in Figs. 27 and 28, with the current, i, in dotted and the potential difference, e_1, in drawn line. The stationary values are plotted also, in thin lines, i and e', respectively.

(a) Opening the circuit under full load, we have

$$i = -11.3 \sin (\theta - \theta_0) + i_0 \varepsilon^{-0.25\theta} \cos 11.2\,\theta$$

and $e_1 = 28{,}280 \cos (\theta - \theta_0) + 224\, i_0 \varepsilon^{-0.25\theta} \sin 11.2\,\theta.$

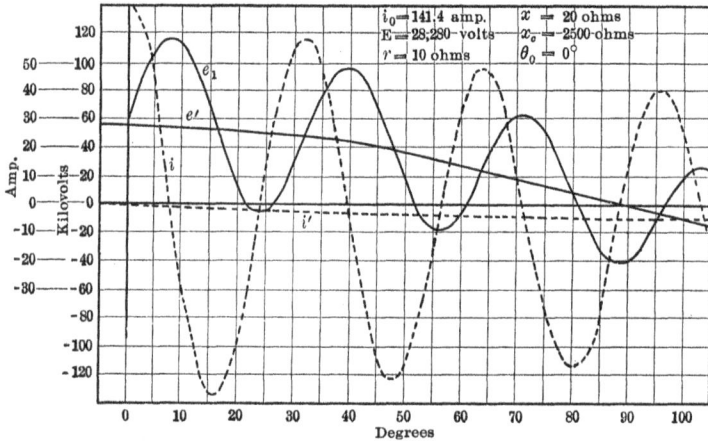

Fig. 29. Opening a loaded transmission line.

These values are maximum for $\theta_0 = 0$ and non-inductive circuit, or $i_0 = 141.4$, and are

$$i = -11.3 \sin \theta + 141.4\, \varepsilon^{-0.25\theta} \cos 11.2\,\theta$$

and $\qquad e_1 = 28{,}280 \cos \theta + 31{,}600\, \varepsilon^{-0.25\theta} \sin 11.2\,\theta.$

These values are plotted, in Fig. 29, in the same manner as Figs. 27 and 28.

(b) Rupturing the line under short-circuit, we have

$$z = 22.4$$

and $\qquad\qquad i_0 = 1265 \cos (\theta_0 + \gamma);$

and therefore

$$i = -11.3 \sin (\theta - \theta_0) + 1265\, \varepsilon^{-0.25\theta} [\cos (\theta_0 + \gamma)$$
$$\cos 11.2\,\theta + 0.1 \cos \theta_0 \sin 11.2\,\theta]$$

and $\quad e_1 = 28{,}280 \{\cos(\theta - \theta_0) - \varepsilon^{-0.25\,\theta}[\cos\theta_0 \cos 11.2\,\theta$
$\qquad\qquad - 10 \cos(\theta_0 + \gamma)\sin 11.2\,\theta]\}.$

These values are a maximum for $\theta_0 = -\gamma = -63°$, thus

$\qquad i = -11.3 \sin(\theta + 63°) + 1265\,\varepsilon^{-0.25\,\theta}(\cos 11.2\,\theta$
$\qquad\qquad + 0.044 \sin 11.2\,\theta)$

and $\quad e_1 = 28{,}280 \cos(\theta + 63°) - 282{,}800\,\varepsilon^{-0.25\,\theta}(0.044 \cos 11.2\,\theta$
$\qquad\qquad - \sin 11.2\,\theta);$

that is, the potential difference rises about tenfold, to 282,800 volts. These values are plotted in Fig. 30.

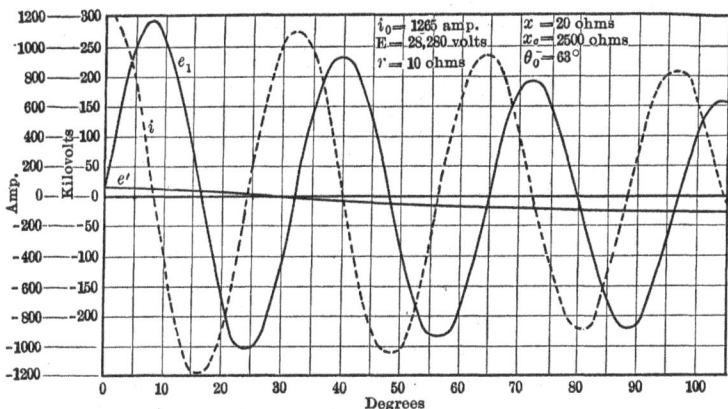

Fig. 30. Opening a short-circuited transmission line.

70. On an experimental 10,000-volt, 40-cycle line, when a destructive e.m.f. was produced by a short-circuiting arc, the author observed a drop in generator e.m.f. to about 5000 volts, due to the limited machine capacity. The resistance of the system was very low, about $r = 1$ ohm, while the inductive reactance may be estimated as $x = 10$ ohms, and the condensive reactance as $x_c = 20{,}000$ ohms. Therefore $\tan\gamma = 10$, or approximately, $\gamma = 90°$.

Herefrom it follows that

$$i = 707\,\varepsilon^{-0.05\,\theta}\cos 44.7\,\theta$$

and

$$e_1 = 316{,}000\,\varepsilon^{-0.05\,\theta}\sin 44.7\,\theta;$$

that is, the oscillation has a frequency of about 1800 cycles per second and a maximum e.m.f. of nearly one-third million volts, which fully accounts for its disruptive effects.

71. As conclusion, it follows herefrom:

1. A most important source of destructive high voltage phenomena in high potential circuits containing inductance and capacity are the electric oscillations produced by a change of circuit conditions, as starting, opening circuit, etc.

2. These phenomena are essentially independent of the frequency and the wave shape of the impressed e.m.f., but depend upon the conditions under which the circuit is changed, as the manner of change and the point of the impressed e.m.f. and current wave at which the change occurs.

3. The electric oscillations occurring in connecting a transmission line to the generator are not of dangerous potential, but the oscillations produced by opening the transmission circuit under load may reach destructive voltages, and the oscillations caused by interrupting a short-circuit are liable to reach voltages far beyond the strength of any insulation. Thus special precautions should be taken in opening a high potential circuit under load. But the most dangerous phenomenon is a low resistance short-circuit in open space.

4. The voltages produced by the oscillations in open-circuiting a transmission line under load or under short-circuit are moderate if the opening of the circuit occurs at a certain point of the e.m.f. wave. This point approximately coincides with the moment of zero current.

CHAPTER IX.

DIVIDED CIRCUIT.

72. A circuit consisting of two branches or multiple circuits 1 and 2 may be supplied, over a line or circuit 3, with an impressed e.m.f., e_0.

Let, in such a circuit, shown diagrammatically in Fig 31, r_1, L_1, C_1 and r_2, L_2, C_2 = resistance, inductance, and capacity, respectively, of the two branch circuits 1 and 2; r_0, L_0, C_0 =

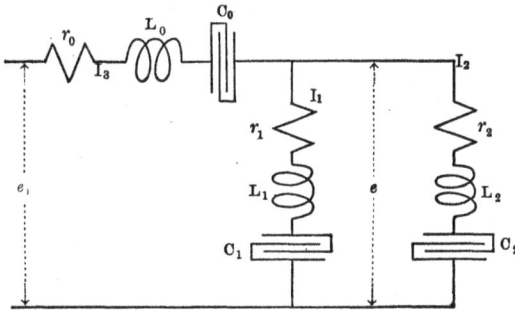

Fig. 31.   Divided circuit.

resistance, inductance, and capacity of the undivided part of the circuit, 3.   Furthermore let e = potential difference at terminals of branch circuits 1 and 2, i_1 and i_2 respectively = currents in branch circuits 1 and 2, and i_3 = current in undivided part of circuit, 3.

Then
$$i_3 = i_1 + i_2 \tag{1}$$

and e.m.f. at the terminals of circuit 1 is

$$e = r_1 i_1 + L_1 \frac{di_1}{dt} + \frac{1}{C_1} \int i_1 \, dt, \tag{2}$$

of circuit 2 is

$$e = r_2 i_2 + L_2 \frac{di_2}{dt} + \frac{1}{C_2} \int i_2 \, dt, \tag{3}$$

and of circuit 3 is

$$e_0 = e + r_0 i_3 + L_0 \frac{di_3}{dt} + \frac{1}{C_0} \int i_3 \, dt. \tag{4}$$

Instead of the inductances, L, and capacities, C, it is usually preferable, even in direct-current circuits, to introduce the reactances, $x = 2\pi fL$ = inductive reactance, $x_c = \dfrac{1}{2\pi fC}$ = condensive reactance, referred to a standard frequency, such as $f = 60$ cycles per second. Instead of the time t, then, an angle

$$\theta = 2\pi ft \tag{5}$$

is introduced, and then we have

and

$$\left. \begin{aligned} L\frac{di}{dt} &= \frac{x}{2\pi f} \frac{di}{d\theta} \frac{d\theta}{dt} = x \frac{di}{d\theta} \\ \frac{1}{C} \int i \, dt &= 2\pi f x_c \int \frac{i}{\dfrac{d\theta}{dt}} d\theta = x_c \int i \, d\theta, \end{aligned} \right\} \tag{6}$$

since

$$\frac{d\theta}{dt} = 2\pi f.$$

Hereby resistance, inductance, and capacity are expressed in the same units, ohms.

Time is expressed by an angle θ so that 360 degrees correspond to $\frac{1}{60}$ of a second, and the time effects thus are directly comparable with the phenomena on a 60-cycle circuit.

A better conception of the size or magnitude of inductance and capacity is secured. Since inductance and capacity are mostly observed and of importance in alternating-current circuits, a reactor having an inductive reactance of x ohms and i amperes conveys to the engineer a more definite meaning as regards size: it has a volt-ampere capacity of $i^2 x$, that is, the approximate size of a transformer of half this capacity, or of a $\dfrac{i^2 x}{2}$-watt transformer. A reactor having an inductance of L henrys and i amperes, however, conveys very little meaning to the engineer who is mainly familiar with the effect of inductance in alternating-current circuits.

Substituting therefore (5) and (6) in equations (2), (3), (4), gives the e.m.f. in circuit 1 as

$$e = r_1 i_1 + x_1 \frac{d i_1}{d\theta} + x_{c_1} \int i_1 \, d\theta; \qquad (7)$$

in circuit 2 as

$$e = r_2 i_2 + x_2 \frac{d i_2}{d\theta} + x_{c_2} \int i_2 \, d\theta; \qquad (8)$$

in circuit 3 as

$$e_0 = e + r_0 i_3 + x_0 \frac{d i_3}{d\theta} + x_{c_0} \int i_3 \, d\theta; \qquad (9)$$

hence, the potential differences at the condenser terminals are

$$e_1 = x_{c_1} \int i_1 \, d\theta = e - r_1 i_1 - x_1 \frac{d i_1}{d\theta}, \qquad (10)$$

$$e_2 = x_{c_2} \int i_2 \, d\theta = e - r_2 i_2 - x_2 \frac{d i_2}{d\theta}, \qquad (11)$$

and $\qquad e_3 = x_{c_0} \int i_3 \, d\theta = e_0 - e - r_0 i_3 - x_0 \frac{d i_3}{d\theta}. \qquad (12)$

Differentiating equations (7), (8), and (9), to eliminate the integral, gives as differential equations of the divided circuit:

$$x_1 \frac{d^2 i_1}{d\theta^2} + r_1 \frac{d i_1}{d\theta} + x_{c_1} i_1 = \frac{de}{d\theta}, \qquad (13)$$

$$x_2 \frac{d^2 i_2}{d\theta^2} + r_2 \frac{d i_2}{d\theta} + x_{c_2} i_2 = \frac{de}{d\theta}, \qquad (14)$$

and $\qquad x_0 \frac{d^2 i_3}{d\theta^2} + r_0 \frac{d i_3}{d\theta} + x_{c_0} i_3 = \frac{de_0}{d\theta} - \frac{de}{d\theta}. \qquad (15)$

Subtracting (14) from (13) gives

$$\left(x_1 \frac{d^2 i_1}{d\theta^2} + r_1 \frac{d i_1}{d\theta} + x_{c_1} i_1 \right) - \left(x_2 \frac{d^2 i_2}{d\theta^2} + r_2 \frac{d i_2}{d\theta} + x_{c_2} i_2 \right) = 0. \quad (16)$$

Multiplying (15) by 2, and adding thereto (13) and (14), gives, by substituting (1), $i_3 = i_1 + i_2$,

$$(2 x_0 + x_1) \frac{d^2 i_1}{d\theta^2} + (2 r_0 + r_1) \frac{d i_1}{d\theta} + (2 x_{c_0} + x_{c_1}) i_1 +$$

$$(2 x_0 + x_2) \frac{d^2 i_2}{d\theta^2} + (2 r_0 + r_2) \frac{d i_2}{d\theta} + (2 x_{c_0} + x_{c_2}) i_2 = 2 \frac{d e_0}{d\theta}. \quad (17)$$

These two differential equations (16) and (17) are integrated by the functions

$$\left.\begin{array}{c} i_1 = i_1' + A_1 \varepsilon^{-a\theta} \\[2mm] \text{and} \qquad\qquad\qquad\qquad\qquad \\[2mm] i_2 = i_2' + A_2 \varepsilon^{-a\theta}, \end{array}\right\} \quad (18)$$

where i_1' and i_2' are the permanent values of current, and $i_1'' = A_1 \varepsilon^{-a\theta}$ and $i_2'' = A_2 \varepsilon^{-a\theta}$ are the transient current terms.

Substituting (18) in (16) and (17) gives

$$\left(x_1 \frac{d^2 i_1'}{d\theta^2} + r_1 \frac{d i_1'}{d\theta} + x_{c_1} i_1' \right) - \left(x_2 \frac{d^2 i_2'}{d\theta^2} + r_2 \frac{d i_2'}{d\theta} + x_{c_2} i_2' \right)$$

$$+ A_1 \varepsilon^{-a\theta} (a^2 x_1 - a r_1 + x_{c_1}) - A_2 \varepsilon^{-a\theta} (a^2 x_2 - a r_2 + x_{c_2}) = 0 \quad (19)$$

and

$$(2 x_0 + x_1) \frac{d^2 i_1'}{d\theta^2} + (2 r_0 + r_1) \frac{d i_1'}{d\theta} + (2 x_{c_0} + x_{c_1}) i_1' + (2 x_0 + x_2)$$

$$\frac{d^2 i_2'}{d\theta^2} + (2 r_0 + r_2) \frac{d i_2'}{d\theta} + (2 x_{c_0} + x_{c_2}) i_2 + A_1 \varepsilon^{-a\theta} \{ a^2 (2 x_0 + x_1)$$

$$- a (2 r_0 + r_1) + (2 x_{c_0} + x_{c_1}) \} + A_2 \varepsilon^{-a\theta} \{ a^2 (2 x_0 + x_2)$$

$$- a (2 r_0 + r_2) + (2 x_{c_0} + x_{c_2}) \} = 2 \frac{d e_0}{d\theta}. \quad (20)$$

73. For $\theta = \infty$, the exponential terms eliminate, and there remain the differential equations of the permanent terms i_1' and i_2', thus

$$\left(x_1 \frac{d^2 i_1'}{d\theta^2} + r_1 \frac{d i_1'}{d\theta} + x_{c_1} i_1' \right) - \left(x_2 \frac{d^2 i_2'}{d\theta^2} + r_2 \frac{d i_2'}{d\theta} + x_{c_2} i_2' \right) = 0 \quad (21)$$

and

$$(2 x_0 + x_1) \frac{d^2 i_1'}{d\theta^2} + (2 r_0 + r_1) \frac{d i_1'}{d\theta} + (2 x_{c_0} + x_{c_1}) i_1' + (2 x_0 + x_2)$$

$$\frac{d^2 i_2'}{d\theta^2} + (2 r_0 + r_2) \frac{d i_2'}{d\theta} + (2 x_{c_0} + x_{c_2}) i_2' = 2 \frac{d e_0}{d\theta}. \quad (22)$$

The solution of these equations (21) and (22) is the usual equation of electrical engineering, giving i_1' and i_2' as sine waves if the e.m.f., e_0, is a sine wave; giving i_1' and i_2' as constant quantities if e_0 is constant and x_{c_0} and either x_{c_1} or x_{c_2} or both vanish, and giving i_1' and $i_2' = 0$ if either x_{c_0} or both x_{c_1} and x_{c_2} differ from zero.

Subtracting (21) and (22) from (19) and (20) leaves as differential equations of the transient terms i_1'' and i_2'',

$$\varepsilon^{-a\theta} \{A_1 (a^2 x_1 - a r_1 + x_{c_1}) - A_2 (a^2 x_2 - a r_2 + x_{c_2})\} = 0 \quad (23)$$

and

$$\varepsilon^{-a\theta} \{A_1 [a^2 (2 x_0 + x_1) - a (2 r_0 + r_1) + (2 x_{c_0} + x_{c_1})] + A_2$$
$$[a^2 (2 x_0 + x_2) - a (2 r_0 + r_2) + (2 x_{c_0} + x_{c_2})]\} = 0. \quad (24)$$

Introducing a new constant B, these equations give, from (23),

$$A_1 = B (a^2 x_2 - a r_2 + x_{c_2})$$

and

$$A_2 = B (a^2 x_1 - a r_1 + x_{c_1}); \quad (25)$$

then substituting (25) in (24) gives

$$(a^2 x_2 - a r_2 + x_{c_2}) [a^2 (2 x_0 + x_1) - a (2 r_0 + r_1) + (2 x_{c_0} + x_{c_1})]$$
$$+ (a^2 x_1 - a r_1 + x_{c_1})[a^2 (2 x_0 + x_2) - a (2 r_0 + r_2) + (2 x_{c_0}$$
$$+ x_{c_2})] = 0, \quad (26)$$

while B remains indeterminate as integration constant.

Quartic equation (26) gives four values of a, which may be all real, or two real and two conjugate imaginary, or two pairs of conjugate imaginary roots.

Rearranged, equation (26) gives

$$a^4 (x_0 x_1 + x_0 x_2 + x_1 x_2) - a^3 \{r_0 (x_1 + x_2) + r_1 (x_0 + x_2)$$
$$+ r_2 (x_0 + x_1)\} + a^2 \{(r_0 r_1 + r_0 r_2 + r_1 r_2) + x_{c_0} (x_1 + x_2)$$
$$+ x_{c_1} (x_0 + x_2) + x_{c_2} (x_0 + x_1)\} - a \{x_{c_0}(r_1 + r_2) + x_{c_1} (r_0 + r_2)$$
$$+ x_{c_2} (r_0 + r_1)\} + (x_{c_0} x_{c_1} + x_{c_0} x_{c_2} + x_{c_1} x_{c_2}) = 0. \quad (27)$$

Let a_1, a_2, a_3, a_4 be the four roots of this quartic equation (27);

then

$$i_1 = i_1' + B_1 \left(a_1^2 x_2 - a_1 r_2 + x_{c_2}\right) \varepsilon^{-a_1\theta} + B_2 \left(a_2^2 x_2 - a_2 r_2 + x_{c_2}\right) \varepsilon^{-a_2\theta}$$
$$+ B_3 \left(a_3^2 x_2 - a_3 r_2 + x_{c_2}\right) \varepsilon^{-a_3\theta} + B_4 \left(a_4^2 x_2 - a_4 r_2 + x_{c_2}\right) \varepsilon^{-a_4\theta} \quad (28)$$

and

$$i_2 = i_2' + B_1 \left(a_1^2 x_1 - a_1 r_1 + x_{c_1}\right) \varepsilon^{-a_1\theta} + B_2 \left(a_2^2 x_1 - a_2 r_1 + x_{c_1}\right) \varepsilon^{-a_2\theta}$$
$$+ B_3 \left(a_3^2 x_1 - a_3 r_1 + x_{c_1}\right) \varepsilon^{-a_3\theta} + B_4 \left(a_4^2 x_1 - a_4 r_1 + x_{c_1}\right) \varepsilon^{-a_4\theta} \quad (29)$$

where the integration constants B_1, B_2, B_3 and B_4 are determined by the terminal conditions: the currents and condenser potentials at zero time, $\theta = 0$.

The quartic equation (27) usually has to be solved by approximation.

74. *Special Cases:* Continuous-current divided circuit, with resistance and inductance but no capacity, $e_0 =$ constant.

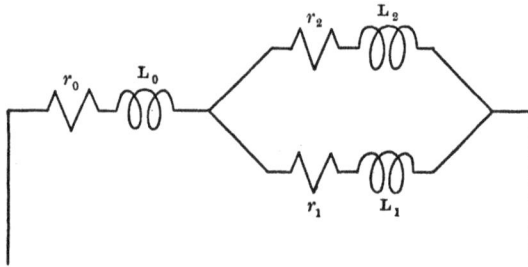

Fig. 32.   Divided continuous-current circuit without capacity.

In such a circuit, shown diagrammatically in Fig. 32, equations (7), (8), and (9) are greatly simplified by the absence of the integral, and we have

$$e = r_1 i_1 + x_1 \frac{di_1}{d\theta}, \quad (30)$$

$$e = r_2 i_2 + x_2 \frac{di_2}{d\theta}, \quad (31)$$

and
$$e_0 = e + r_0 i_3 + x_0 \frac{di_3}{d\theta}. \quad (32)$$

(30) and (31) combined give

$$r_1 i_1 - r_2 i_2 + x_1 \frac{di_1}{d\theta} - x_2 \frac{di_2}{d\theta} = 0. \quad (33)$$

Substituting (1), $i_3 = i_1 + i_2$, in (32), multiplying it by 2 and adding thereto (30) and (31), gives

$$2\,e_0 = (2\,r_0 + r_1)\,i_1 + (2\,r_0 + r_2)\,i_2 + (2\,x_0 + x_1)\,\frac{di_1}{d\theta} \\ + (2\,x_0 + x_2)\,\frac{di_2}{d\theta}. \tag{34}$$

Equations (33) and (34) are integrated by

$$i_1 = i_1' + A_1 \varepsilon^{-a\theta}$$

and

$$i_2 = i_2' + A_2 \varepsilon^{-a\theta}. \tag{35}$$

Substituting (35) in (33) and (34) gives

$$(r_1 i_1' - r_2 i_2') + \varepsilon^{-a\theta}\{A_1(r_1 - ax_1) - A_2(r_2 - ax_2)\} = 0$$

and

$$2\,e_0 = (2\,r_0 + r_1)\,i_1' + (2\,r_0 + r_2)\,i_2' + \varepsilon^{-a\theta}\{A_1[(2\,r_0 + r_1) \\ - a\,(2\,x_0 + x_1)] + A_2[(2\,r_0 + r_2) - a\,(2\,x_0 + x_2)]\}.$$

These equations resolve into the equations of permanent state, thus

$$r_1 i_1' - r_2 i_2' = 0$$

and

$$(2\,r_0 + r_1)\,i_1' + (2\,r_0 + r_2)\,i_2' = 2\,e_0.$$

Hence,

$$i_1' = e_0 \frac{r_2}{r^2}$$

and

$$i_2' = e_0 \frac{r_1}{r^2}, \tag{36}$$

where

$$r^2 = r_0 r_1 + r_0 r_2 + r_1 r_2, \tag{37}$$

and the transient equations having the coefficients

$$A_1(r_1 - ax_1) - A_2(r_2 - ax_2) = 0$$

and

$$A_1[(2\,r_0 + r_1) - a\,(2\,x_0 + x_1)] + A_2[(2\,r_0 + r_2) \\ - a\,(2\,x_0 + x_2)] = 0.$$

Herefrom it follows that

$$A_1 = B\,(r_2 - ax_2)$$

and

$$A_2 = B\,(r_1 - ax_1),$$

$$(38)$$

and

$$a^2\,(x_0x_1 + x_0x_2 + x_1x_2) - a\,[r_0\,(x_1 + x_2) + r_1\,(x_0 + x_2) + r_2\,(x_0 + x_1)] + (r_0r_1 + r_0r_2 + r_1r_2) = 0, \qquad (39)$$

$$B = \text{indefinite.} \qquad (40)$$

Substituting the abbreviations,

$$\begin{aligned}
x_0x_1 + x_0x_2 + x_1x_2 &= x^2, \\
r_0r_1 + r_0r_2 + r_1r_2 &= r^2,
\end{aligned}$$

and

$$r_0\,(x_1 + x_2) + r_1\,(x_0 + x_2) + r_2\,(x_0 + x_1) = x_0\,(r_1 + r_2) \\ + x_1\,(r_0 + r_2) + x_2\,(r_0 + r_1) = s^2,$$

$$(41)$$

gives (39)

$$a^2x^2 - as^2 + r^2 = 0, \qquad (42)$$

hence two roots,

$$a_1 = \frac{s^2 - q^2}{2\,x^2}$$

and

$$a_2 = \frac{s^2 + q^2}{2\,x^2},$$

$$(43)$$

where

$$q^2 = \sqrt{s^4 - 4\,r^2x^2}. \qquad (44)$$

The two roots of equation (42), a_1 and a_2, are always real, since in q^2

$$s^4 > 4\,r^2x^2,$$

as seen by substituting (41) therein.

The final integral equations thus are

$$i_1 = e_0\frac{r_2}{r^2} + (r_2 - a_1x_2)\,B_1\varepsilon^{-\frac{s^2-q^2}{2\,x^2}\theta} + (r_2 - a_2x_2)\,B_2\varepsilon^{-\frac{s^2+q^2}{2\,x^2}\theta}$$

and

$$i_2 = e_0\frac{r_1}{r^2} + (r_1 - a_1x_1)\,B_1\varepsilon^{-\frac{s^2-q^2}{2\,x^2}\theta} + (r_1 - a_2x_1)\,B_2\varepsilon^{-\frac{s^2+q^2}{2\,x^2}\theta}.$$

$$(45)$$

B_1 and B_2 are determined by the terminal conditions, as the currents i_1 and i_2 at the start, $\theta = 0$.

Let, at zero time, or $\theta = 0$,

$$\left. \begin{array}{c} i_1 = i_1{}^0 \\ \\ i_2 = i_2{}^0; \end{array} \right\} \tag{46}$$

and

then, substituting in (45), we have

$$\left. \begin{array}{c} i_1{}^0 = e_0 \dfrac{r_2}{r^2} + (r_2 - a_1 x_2) B_1 + (r_2 - a_2 x_2) B_2 \\ \\ i_2{}^0 = e_0 \dfrac{r_1}{r^2} + (r_1 - a_1 x_1) B_1 + (r_1 - a_2 x_1) B_2; \end{array} \right\} \tag{47}$$

and

and herefrom calculate B_1 and B_2.

75. For instance, in a continuous-current circuit, let the impressed e.m.f., $e_0 = 120$ volts; the resistance of the undivided part of the circuit, $r_0 = 20$ ohms; the reactance, $x_0 = 20$ ohms; the resistance of one of the branches, $r_1 = 20$ ohms; the reactance, $x_1 = 40$ ohms, and the resistance of the other branch, $r_2 = 5$ ohms, the reactance, $x_2 = 200$ ohms.

Thus one of the branches is of low resistance and high reactance, the other of high resistance and moderate reactance.

The permanent values of the currents, ($r^2 = 600$), are

$$\left. \begin{array}{c} i_1{}' = 1 \text{ amp.} \\ \\ i_2{}' = 4 \text{ amp.} \end{array} \right\}$$

and

(*a*) Assuming now the resistance r_0 suddenly decreased from $r_0 = 20$ ohms to $r_0 = 15$ ohms, we have the permanent values of current as

$$\left. \begin{array}{c} i_1{}' = 1.265 \text{ amp.} \\ \\ i_2{}' = 5.06 \text{ amp.} \end{array} \right\}$$

and

The previous values of currents, and thus the values of currents at the moment of start, $\theta = 0$, are

$$\left. \begin{array}{c} i_1{}^0 = 1 \text{ amp.} \\ \\ i_2{}^0 = 4 \text{ amp,} \end{array} \right\}$$

and

therefrom follow the equations of currents, by substitution in the preceding,

$$i_1 = 1.265 + 0.455 \, \varepsilon^{-0.0633 \, \theta} - 0.720 \, \varepsilon^{-0.586 \, \theta}$$

and

$$i_2 = 5.06 - 1.038 \, \varepsilon^{-0.0633 \, \theta} - 0.022 \, \varepsilon^{-0.586 \, \theta}.$$

(b) Assuming now the resistance r_0 suddenly raised again from $r_0 = 15$ ohms to $r_0 = 20$ ohms, leaving everything else the same, we have

$$i_1{}^0 = 1.265 \text{ amp.}$$

and

$$i_2{}^0 = 5.06 \text{ amp.};$$

and then

$$i_1 = 1 - 0.528 \, \varepsilon^{-0.0697 \, \theta} + 0.793 \, \varepsilon^{-0.674 \, \theta}$$

and

$$i_2 = 4 + 1.018 \, \varepsilon^{-0.0697 \, \theta} + 0.042 \, \varepsilon^{-0.674 \, \theta}.$$

(c) Assuming now the resistance r_0 suddenly raised from $r_0 = 20$ ohms to $r_0 = 25$ ohms, gives

$$i_1 = 0.828 - 0.374 \, \varepsilon^{-0.0743 \, \theta} + 0.546 \, \varepsilon^{-0.764 \, \theta}$$

and

$$i_2 = 3.312 + 0.649 \, \varepsilon^{-0.0743 \, \theta} + 0.039 \, \varepsilon^{-0.764 \, \theta}.$$

(d) Assuming now the resistance r_0 lowered again from $r_0 = 25$ ohms to $r_0 = 20$ ohms, gives

$$i_1 = 1 + 0.342 \, \varepsilon^{-0.0697 \, \theta} - 0.514 \, \varepsilon^{-0.674 \, \theta}$$

and

$$i_2 = 4 - 0.660 \, \varepsilon^{-0.0697 \, \theta} - 0.028 \, \varepsilon^{-0.674 \, \theta}.$$

76. In Fig. 33 are shown the variations of currents i_1 and i_2, resultant from a sudden variation of the resistance r_0 from 20 to 15, back to 20, to 25, and back again to 20 ohms. As seen, the readjustment of current i_2, that is, the current in the inductive branch of the circuit, to its permanent condition, is very slow and gradual. Current i_1, however, not only changes very rapidly with a change of r_0, but overreaches greatly; that is, a decrease of r_0 causes i_1 to increase rapidly to a temporary value far in excess of the permanent increase, and then gradually i_1

falls back to its normal, and inversely with an increase of r_0. Hence, any change of the main current is greatly exaggerated in the temporary component of current i_1; a permanent change of about 20 per cent in the total current results in a practically instantaneous change of the branch current i_1, by about 50 per cent in the present instance.

Thus, where any effect should be produced by a change of current, or of voltage, as a control of the circuit effected thereby, the action is made far more sensitive and quicker by shunting the operating circuit i_1, of as low inductance as possible, across

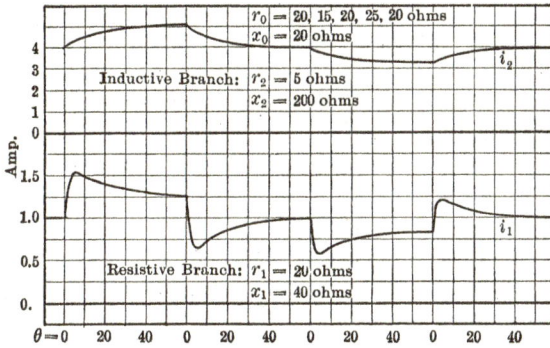

Fig. 33. Current in divided continuous-current circuit resulting from sudden variations in resistance.

a high inductance of as low resistance as possible. The sudden and temporary excess of the change of current i_1 takes care of the increased friction of rest in setting the operating mechanism in motion, and gives a quicker reaction than a mechanism operated directly by the main current.

This arrangement has been proposed for the operation of arc lamps of high arc voltage from constant potential circuits. The operating magnet, being in the circuit i_1, more or less anticipates the change of arc resistance by temporarily over-reaching.

77. The temporary increase of the voltage, e, across the branch circuit, i_1, corresponding to the temporary excess current of this circuit, may, however, result in harmful effects, as destruction of measuring instruments by the temporary excess voltage.

Let, for instance, in a circuit of impressed continuous e.m.f., $e_0 = 600$ volts, as an electric railway circuit, the resistance of the circuit equal 25 ohms, the inductive reactance 44 ohms. This gives a permanent current of $i' = 24$ amperes.

Let now a small part of the circuit, of resistance $r_2 = 1$ ohm, but including most of the reactance $x_2 = 40$ ohms — as a motor series field winding — be shunted by a voltmeter, and $r_1 = 1000$ ohms = resistance, $x_1 = 40$ ohms = reactance of the volt-meter circuit.

In permanent condition the voltmeter reads $\frac{1}{25} \times 600 = 24$ volts, but any change of circuit condition, as a sudden decrease or increase of supply voltage e_0, results in the appearance of a temporary term which may greatly increase the voltage impressed upon the voltmeter.

In this divided circuit, the constants are: undivided part of the circuit, $r_0 = 24$ ohms; $x_0 = 4$ ohms; first branch, voltmeter (practically non-inductive), $r_1 = 1000$ ohms, $x_1 = 40$ ohms; second branch, motor field, highly inductive, $r_2 = 1$ ohm, $x_2 = 40$ ohms.

(a) Assuming now the impressed e.m.f., e_0, suddenly dropped from $e_0 = 600$ volts to $e_0 = 540$ volts, that is, by 10 per cent, gives the equations

and

$$\left. \begin{aligned} i_1 &= 0.0216 - 0.0806 \, \varepsilon^{-0.832 \, \theta} + 0.0830 \, \varepsilon^{-23.1 \, \theta} \\ i_2 &= 21.6 + 2.407 \, \varepsilon^{-0.832 \, \theta} - 0.007 \, \varepsilon^{-23.1 \, \theta}. \end{aligned} \right\}$$

(b) Assuming now the voltage, e_0, suddenly raised again from $e_0 = 540$ volts to $e_0 = 600$ volts, gives the equations

and

$$\left. \begin{aligned} i_1 &= 0.024 + 0.0806 \, \varepsilon^{-0.832 \, \theta} - 0.0830 \, \varepsilon^{-23.1 \, \theta} \\ i_2 &= 24 - 2.407 \, \varepsilon^{-0.832 \, \theta} + 0.007 \, \varepsilon^{-23.1 \, \theta}. \end{aligned} \right\}$$

The voltage, e, across the voltmeter, or on circuit 1, is

$$e = r_1 i_1 + x_1 \frac{di_1}{d\theta} = 1000 \, i_1' \mp 77.9 \, \varepsilon^{-0.832 \, \theta} \pm 6.2 \, \varepsilon^{-23.1 \, \theta},$$

where

$$i_1' = e \frac{r_1}{r^2}.$$

Hence, in case (*a*), drop of impressed voltage, e_0, by 10 per cent,

$$e = 21.6 - 77.9 \, \varepsilon^{-0.832\,\theta} + 6.2 \, \varepsilon^{-23.1\,\theta},$$

and in (*b*), rise of impressed voltage,

$$e = 24.0 + 77.9 \, \varepsilon^{-0.832\,\theta} - 6.2 \, \varepsilon^{-23.1\,\theta}.$$

This voltage, *e*, in the two cases, is plotted in Fig. 34. As seen, during the transition of the voltmeter reading from 21.6 to 24.0 volts, the voltage momentarily rises to 95.7 volts, or

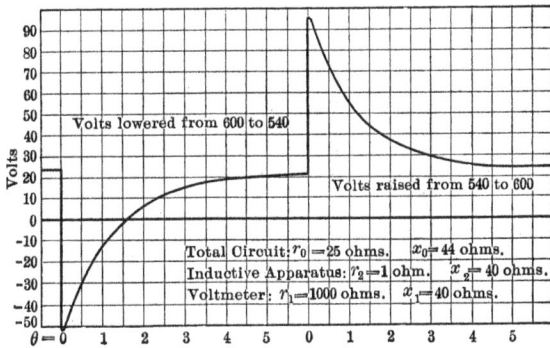

Fig. 34. Voltage across inductive apparatus in series with circuit of high resistance.

four times its permanent value, and during the decrease of permanent voltage from 24.0 to 21.6 volts the voltmeter momentarily reverses, going to 50.1 volts in reverse direction.

In a high voltage direct-current circuit, a voltmeter shunted across a low resistance, if this resistance is highly inductive, is in danger of destruction by any sudden change of voltage or current in the circuit, even if the permanent value of the voltage is well within the safe range of the voltmeter.

CAPACITY SHUNTING A PART OF A CONTINUOUS-CURRENT CIRCUIT.

78. A circuit of resistance r_1 and inductive reactance x_1 is shunted by the condensive reactance x_c, and supplied over the resistance r_0 and the inductive reactance x_0 by a continuous impressed e.m.f., e_0, as shown diagrammatically in Fig. 35.

In the undivided circuit,

$$e_0 = e + r_0 \left(i_1 + i_2\right) + x_0 \left(\frac{di_1}{d\theta} + \frac{di_2}{d\theta}\right). \qquad (48)$$

In the inductive branch,

$$e = r_1 i_1 + x_1 \frac{di_1}{d\theta}. \qquad (49)$$

In the condenser branch,

$$e = x_c \int i_2 \, d\theta. \qquad (50)$$

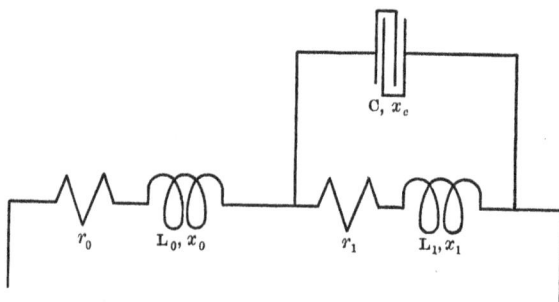

Fig. 35. Suppression of pulsations in direct-current circuits by series inductance and shunted capacity.

Eliminating *e* gives, from (48) and (49),

$$e_0 = \left(r_0 + r_1\right) i_1 + \left(x_0 + x_1\right)\frac{di_1}{d\theta} + r_0 i_2 + x_0 \frac{di_2}{d\theta}, \qquad (51)$$

and from (49) and (50),

$$x_c \int i_2 \, d\theta = r_1 i_1 + x_1 \frac{di_1}{d\theta}. \qquad (52)$$

Differentiating (52), to eliminate the integral,

$$x_c i_2 = r_1 \frac{di_1}{d\theta} + x_1 \frac{d^2 i_1}{d\theta^2}. \qquad (53)$$

Substituting (53) in (51), and rearranging,

$$e_0 = (r_0 + r_1)\, i_1 + \frac{1}{x_c} \left\{ (r_0 r_1 + x_c x_0 + x_c x_1) \frac{di_1}{d\theta} \right.$$

$$\left. + (r_0 x_1 + r_1 x_0) \frac{d^2 i_1}{d\theta^2} + x_0 x_1 \frac{d^3 i_1}{d\theta^3} \right\} ; \qquad (54)$$

a differential equation of third order.

This resolves into the permanent term

$$e_0 = (r_0 + r_1)\, i_1',$$

hence, $\qquad\qquad i_1' = \dfrac{e_0}{r_0 + r_1}, \qquad\qquad (55)$

and a transient term

$$i_1'' = A\,\varepsilon^{-a\theta}; \qquad (56)$$

that is,

$$i_1 = i_1' + A\varepsilon^{-a\theta} = \frac{e_0}{r_0 + r_1} + A\,\varepsilon^{-a\theta}. \qquad (57)$$

Equation (57) substituted in (54) gives as equation of a,

$$x_c\,(r_0 + r) - a\,(r_0 r_1 + x_c x_0 + x_c x_1) + a^2\,(r_0 x_1 + r_1 x_0) - a^3 x_0 x_1 = 0,$$

or

$$a^3 - a^2 \left(\frac{r_0}{x_0} + \frac{r_1}{x_1} \right) + a \left(\frac{r_0 r_1}{x_0 x_1} + \frac{x_c}{x_0} + \frac{x_c}{x_1} \right) - \frac{x_c\,(r_0 + r_1)}{x_0 x_1} = 0, \quad (58)$$

while A remains indefinite as integration constant.

Equation (58) has three roots, a_1, a_2, and a_3, which either are all three real, when the phenomenon is logarithmic, or, one real and two imaginary, when the phenomenon is oscillating.

The integral equation for the current in branch 1 is

$$i_1 = \frac{e_0}{r_0 + r_1} + A_1 \varepsilon^{-a_1\theta} + A_2 \varepsilon^{-a_2\theta} + A_3 \varepsilon^{-a_3\theta} ; \qquad (59)$$

the current in branch 2 is by (53)

$$i_2 = \frac{1}{x_c} \left(r_1 \frac{di_1}{d\theta} + x_1 \frac{d^2 i_1}{d\theta} \right)$$

$$= \frac{1}{x_c} \left\{ - a_1\,(r_1 - a_1 x_1)\, A_1 \varepsilon^{-a_1\theta} - a_2\,(r_1 - a_2 x_1)\, A_2 \varepsilon^{-a_2\theta} \right.$$

$$\left. - a_3\,(r_1 - a_3 x_1)\, A_3 \varepsilon^{-a_3\theta} \right\}, \qquad (60)$$

and the potential difference at the condenser is

$$e = x_c \int i_2 \, d\theta = r_1 i_1 + x_1 \frac{di_1}{d\theta}$$

$$= \frac{r_1 e_0}{r_0 + r_1} + (r_1 - a_1 x_1) A_1 \varepsilon^{-a_1 \theta} + (r_1 - a_2 x_1) A_2 \varepsilon^{-a_2 \theta}$$

$$+ (r_1 - a_3 x_1) A_3 \varepsilon^{-a_3 \theta}. \tag{61}$$

In the case of an oscillatory change, equations (59), (60), and (61) appear in complex imaginary form, and therefore have to be reduced to trigonometric functions.

The three integration constants, A_1, A_2, and A_3, are determined by the three terminal conditions, at $\theta = 0$, $i_1 = i_1^0$, $i_2 = i_2^0$, $e = e^0$.

79. As numerical example may be considered a circuit having the constants, $e_0 = 110$ volts; $r_0 = 1$ ohm; $x_0 = 10$ ohms; $r_1 = 10$ ohms; $x_1 = 100$ ohms, and $x_c = 10$ ohms.

In other words, a continuous e.m.f. of 110 volts supplies, over a line of $r_0 = 1$ ohm resistance, a circuit of $r_1 = 10$ ohms resistance. An inductive reactance $x_0 = 10$ ohms is inserted into the line, and an inductive reactance $x_1 = 100$ ohms in the load circuit, and the latter shunted by a condensive reactance of $x_c = 10$ ohms.

Then, substituting in equation (58),

$$a^3 - 0.2 \, a^2 + 1.11 \, a - 0.11 = 0.$$

This cubic equation gives by approximation one root, $a_1 = 0.1$, and, divided by $(a - 0.1)$, leaves the quadratic equation

$$a^2 - 0.1 \, a + 1.1 = 0,$$

which gives the complex imaginary roots $a_2 = 0.05 - 1.047 \, j$ and $a_3 = 0.05 + 1.047 \, j$; then from the equation of current, by substituting trigonometric functions for the exponential functions with imaginary exponent, we get the equation for the *load current* as

$$i_1 = i_1' + A_1 \varepsilon^{-0.1 \theta} + \varepsilon^{-0.05 \theta} \, (B_1 \cos 1.047 \, \theta + B_2 \sin 1.047 \, \theta),$$

the *condenser potential* as

$$e = 10 \, i_1' + \varepsilon^{-0.05 \theta} \{ (5 \, B_1 + 104.7 \, B_2) \cos 1.047 \, \theta - (104.7 \, B_1 - 5 \, B_2) \sin 1.047 \, \theta \},$$

and the *condenser current* as

$$i_2 = 10.9 \, \varepsilon^{-0.05\,\theta} \{B_1 \cos 1.047 \, \theta + B_2 \sin 1.047 \, \theta\}.$$

At $e_0 = 110$ volts impressed, the permanent current is $i_1' = 10$ amp., the permanent condenser potential is $e' = 100$ volts, and the permanent condenser current is $i_2' = 0$.

Assuming now the voltage, e_0, suddenly dropped by 10 per cent, from $e_0 = 110$ volts to $e_0 = 99$ volts, gives the permanent current as $i_1' = 9$ amp. At the moment of drop of voltage, $\theta = 0$, we have, however, $i_1 = i_1{}^0 = 10$ amp.; $e = e' = 100$ volts, and $i_2 = 0$; hence, substituting these numerical values into the above equations of i_1, e, i_2, gives the three integration constants:

$$A_1 = 1; \ B_1 = 0, \text{ and } B_2 = 0.0955;$$

therefore the load current is

$$i_1 = 9 + \varepsilon^{-0.1\,\theta} + 0.0955 \, \varepsilon^{-0.05\,\theta} \sin 1.047 \, \theta,$$

the condenser current is

$$i_2 = 1.05 \, \varepsilon^{-0.05\,\theta} \sin 1.047 \, \theta,$$

and the condenser, or load, voltage is

$$e = 90 + \varepsilon^{-0.05\,\theta} (10 \cos 1.047 \, \theta + 0.48 \sin 1.047 \, \theta).$$

Without the condenser, the equation of current would be

$$i = 9 + \varepsilon^{-0.1\,\theta}.$$

In this combination of circuits with shunted condensive reactance x_c, at the moment of the voltage drop, or $\theta = 0$, the rate of change of the load current is, approximately,

$$\frac{di_1}{d\theta} = [-0.1\varepsilon^{-0.1\,\theta} + 0.0955 \times 1.047\,\varepsilon^{-0.05\,\theta} \cos 1.047 \, \theta]_0 = 0,$$

while without the condenser it would be

$$\frac{di}{d\theta} = [-0.1 \, \varepsilon^{-0.1\,\theta}]_0 = -0.1.$$

80. By shunting the circuit with capacity, the current in the circuit does not instantly begin to change with a change or fluctuation of impressed e.m.f.

In Fig. 36 is plotted, with θ as abscissas, the change of the current, i_1, in per cent, resulting from an instantaneous change of impressed e.m.f., e_0, of 10 per cent, with condenser in shunt to the load circuit, and without condenser.

As seen, at $\theta = 172° = 3.0$ radians, both currents, i_1 with the condenser and i without condenser, have dropped by the same

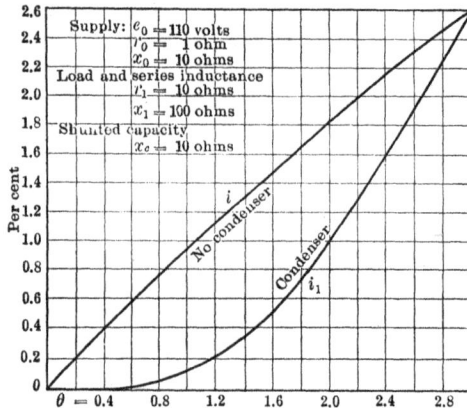

Fig. 36. Suppression of pulsations in direct-current circuits by series inductance and shunted capacity. Effect of 10 per cent drop of voltage.

amount, 2.6 per cent. But at $\theta = 57.3° = 1.0$ radian, i_1 has dropped only $\frac{1}{8}$ per cent., and i nearly 1 per cent, and at $\theta = 24°$, i_1 has not yet dropped at all, while i has dropped by 0.38 per cent.

That is, without condenser, all pulsations of the impressed e.m.f., e_0, appear in the load circuit as pulsations of the current, i, of a magnitude reduced the more the shorter the duration of the pulsation. After $\theta = 60°$, or $t = 0.00275$ seconds, the pulsation of the current has reached 10 per cent of the pulsation of impressed e.m.f.

With a condenser in shunt to the load circuit, the pulsation of current in the load circuit is still zero after $\theta = 24°$, or after 0.001 seconds, and reaches 1.25 per cent of the pulsation of impressed e.m.f., e_0, after $\theta = 60°$, or $t = 0.00275$ seconds.

A pulsation of the impressed e.m.f., e_0, of a frequency higher than 250 cycles, practically cannot penetrate to the load circuit, that is, does not appear at all in the load current i, regardless of how much a pulsation of the impressed e.m.f., e_0, it is, and a pulsation of impressed e.m.f., e_0, of a frequency of 120 cycles reappears in the load current i_1, reduced to 1 per cent of its value.

In cases where from a source of e.m.f., e_0, which contains a slight high frequency pulsation — as the pulsation corresponding to the commutator segments of a commutating machine — a current is desired showing no pulsation whatever, as for instance for the operation of a telephone exchange, a very high inductive reactance in series with the circuit, and a condensive reactance in shunt therewith, entirely eliminates all high frequency pulsations from the current, passing only harmless low frequency pulsations at a greatly reduced amplitude.

81. As a further example is shown in Fig. 37 the pulsation of a non-inductive circuit, $x_1 = 0$, of the resistance $r_1 = 4$ ohms, shunted by a condensive reactance $x_c = 10$ ohms, and supplied over a line of resistance $r_0 = 1$ ohm and inductive reactance $x_0 = 10$ ohms, by an impressed e.m.f., $e_0 = 110$ volts.

Due to $x_1 = 0$ equation (58) reduces to

$$a^2 - a\left(\frac{x_c}{r_1} + \frac{r_0}{x_0}\right) + \frac{x_c}{x_0}\left(1 + \frac{r_0}{r_1}\right) = 0;$$

or, substituting numerical values,

$$a^2 - 2.6\,a + 1.25 = 0$$

and $\qquad a_1 = 0.637, \qquad a_2 = 1.963;$

that is, both roots are real, or the phenomenon is logarithmic.

We now have

$$i_1 = i_1' + A_1\varepsilon^{-0.637\,\theta} + A_2\varepsilon^{-1.963\,\theta},$$
$$i_2 = -0.255\,A_1\varepsilon^{-0.637\,\theta} - 0.785\,A_2\varepsilon^{-1.963\,\theta},$$

and $\qquad e = r_1 i_1 = 4\,(i_1' + A_1\varepsilon^{-0.637\,\theta} + A_2\varepsilon^{-1.963\,\theta}).$

The load current is

$$i_1' = 22 \text{ amp.}$$

A reduction of the impressed e.m.f., e_0, by 10 per cent, or from 110 to 99 volts, gives the integration constants $A_1 = 3.26$ and $A_2 = -1.06$; hence,

$$i_1 = 19.8 + 3.26\,\varepsilon^{-0.637\,\theta} - 1.06\,\varepsilon^{-1.963\,\theta},$$
$$i_2 = -0.83\,(\varepsilon^{-0.637\,\theta} + \varepsilon^{-1.963\,\theta}),$$

and $\qquad e = 4\,i_1.$

Without a condenser, the equation of current would be

$$i = 19.8 + 2.2\, \varepsilon^{-0.5\,\theta}.$$

In Fig. 37 is shown, with θ as abscissas, the drop of current i_1 and i, in per cent.

Although here the change is logarithmic, while in the former paragraph it was trigonometric, the result is the same — a very great reduction, by the condenser, of the drop of current immediately after the change of e.m.f.   However, in the present case

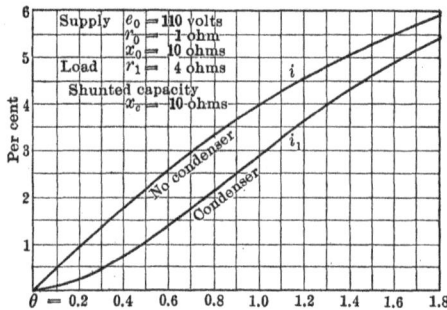

Fig. 37.   Suppression of pulsations in non-inductive direct-current circuits by series inductance and shunted capacity.   Effect of 10 per cent drop of voltage.

the change of the circuit is far more rapid than in the preceding case, due to the far lower inductive reactance of the present case. For instance, after $\theta = 0.1$, the drop of current, with condenser, is 0.045 per cent, without condenser, 0.5 per cent.   At $\theta = 0.2$, the drop of current is 0.23 and 0.95 per cent respectively.   For longer times or larger values of θ, the difference produced by the condenser becomes less and less.

This effect of a condenser across a direct-current circuit, of suppressing high frequency pulsations from reaching the circuit, requires a very large capacity.

CHAPTER X.

MUTUAL INDUCTANCE.

82. In the preceding chapters, circuits have been considered containing resistance, self-inductance, and capacity, but no mutual inductance; that is, the phenomena which take place in the circuit have been assumed as depending upon the impressed e.m.f. and the constants of the circuit, but not upon the phenomena taking place in any other circuit.

Of the magnetic flux produced by the current in a circuit and interlinked with this circuit, a part may be interlinked with a second circuit also, and so by its change generate an e.m.f. in the second circuit, and part of the magnetic flux produced by

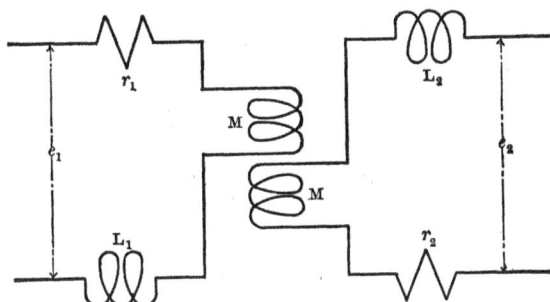

Fig. 38. Mutual inductance between circuits.

the current in a second circuit and interlinked with the second circuit may be interlinked also with the first circuit, and a change of current in the second circuit, that is, a change of magnetic flux produced by the current in the second circuit, then generates an e.m.f. in the first circuit.

Diagrammatically the mutual inductance between two circuits can be sketched as shown by M in Fig. 38, by two *coaxial* coils, while the self-inductance is shown by a single coil L, and the resistance by a *zigzag* line.

The presence of mutual inductance, with a second circuit, introduces into the equation of the circuit a term depending upon the current in the second circuit.

If i_1 = the current in the circuit and r_1 = the resistance of the circuit, then $r_1 i_1$ = the e.m.f. consumed by the resistance of the circuit. If L_1 = the inductance of the circuit, that is, total number of interlinkages between the circuit and the number of lines of magnetic force produced by unit current in the circuit, we have

$$L_1 \frac{di_1}{dt} = \text{e.m.f. consumed by the inductance,}$$

where, t = time.

If instead of time t an angle $\theta = 2\pi f t$ is introduced, where f is some standard frequency, as 60 cycles,

$$x_1 \frac{di_1}{d\theta} = \text{e.m.f. consumed by the inductance,}$$

where $x_1 = 2\pi f L_1$ = inductive reactance.

If now M = mutual inductance between the circuit and another circuit, that is, number of interlinkages of the circuit with the magnetic flux produced by unit current in the second circuit, and i_2 = the current in the second circuit, then

$$M \frac{di_2}{dt} = \text{e.m.f. consumed by mutual inductance in the first}$$
$$\text{circuit,}$$

$$M \frac{di_1}{dt} = \text{e.m.f. consumed by mutual inductance in the second}$$
$$\text{circuit.}$$

Introducing $x_m = 2\pi f M$ = mutual reactance between the two circuits, we have

$$x_m \frac{di_2}{d\theta} = \text{e.m.f. consumed by mutual inductance in the first}$$
$$\text{circuit,}$$

$$x_m \frac{di_1}{d\theta} = \text{e.m.f. consumed by mutual inductance in the second}$$
$$\text{circuit.}$$

If now e_1 = the e.m.f. impressed upon the first circuit and e_2 = the e.m.f. impressed upon the second circuit, the equations of the circuits are

$$e_1 = r_1 i_1 + x_1 \frac{di_1}{d\theta} + x_m \frac{di_2}{d\theta} + x_{c_1} \int i_1 \, d\theta \qquad (1)$$

and

$$e_2 = r_2 i_2 + x_2 \frac{di_2}{d\theta} + x_m \frac{di_1}{d\theta} + x_{c_2} \int i_2 \, d\theta, \qquad (2)$$

where r_1 = the resistance, $x_1 = 2\pi f L_1$ = the inductive reactance, and $x_{c_1} = \dfrac{1}{2\pi f C_1}$ = the condensive reactance of the first circuit; r_2 = the resistance, $x_2 = 2\pi f L_2$ = the inductive reactance, $x_{c_2} = \dfrac{1}{2\pi f C_2}$ = the condensive reactance of the second circuit, and $x_m = 2\pi f M$ = mutual inductive reactance between the two circuits.

83. In these equations, x_1 and x_2 are the total inductive reactance, L_1 and L_2 the total inductance of the circuit, that is, the number of magnetic interlinkages of the circuit with the total flux produced by unit current in the circuit, the self-inductive flux as well as the mutual inductive flux, and not merely the self-inductive reactance and inductance respectively.

In induction apparatus, such as transformers and induction machines, it is usually preferable to separate the total reactance x, into the self-inductive reactance x_s, referring to the magnetic flux interlinked with the inducing circuit only, but with no other circuit, and the mutual inductive reactance, x_m, usually represented as a susceptance, which refers to the mutual inductive component of the total inductance; in which case $x = x_s + x_m$. This is not done in the present case.

Furthermore it is assumed that the circuits are inductively related to each other symmetrically, or reduced thereto; that is, where the mutual inductance is due to coils enclosed in the first circuit, interlinked magnetically with coils enclosed in the second circuit, as the primary and the secondary coils of a transformer, or a shunt and a series field winding of a generator, the two coils are assumed as of the same number of turns, or reduced thereto.

If $a = \dfrac{n_2}{n_1} = \dfrac{\text{No. turns second circuit}}{\text{No. turns first circuit}}$, the currents in the second circuit are multiplied, the e.m.fs. divided by a, the resistances and reactances divided by a^2, to reduce the second circuit to the first circuit, in the manner customary in dealing with transformers and especially induction machines.*

If the ratio of the number of turns is introduced in the equations, that is, in the first equation $\dfrac{n_2}{n_1} x_m$ substituted for x_m, in the second equation $\dfrac{n_1}{n_2} x_m$ for x_m, and the equations then are

$$e_1 = r_1 i_1 + x_1 \frac{di_1}{d\theta} + \frac{n_2}{n_1} x_m \frac{di_2}{d\theta} + x_{c_1} \int i_1 \, d\theta \qquad (3)$$

and

$$e_2 = r_2 i_2 + x_2 \frac{di_2}{d\theta} + \frac{n_1}{n_2} x_m \frac{di_1}{d\theta} + x_{c_2} \int i_2 \, d\theta. \qquad (4)$$

Since the solution and further investigation of these equations (3), (4) are the same as in the case of equations (1) and (2), except that n_1 and n_2 appear as factors, it is preferable to eliminate n_1 and n_2 by reducing one circuit to the other by the ratio of turns $a = \dfrac{n_2}{n_1}$, and then use the simpler equations (1), (2).

(A) CIRCUITS CONTAINING RESISTANCE, INDUCTANCE, AND MUTUAL INDUCTANCE BUT NO CAPACITY.

84. In such a circuit, shown diagrammatically in Fig. 38, we have

$$e_1 = r_1 i_1 + x_1 \frac{di_1}{d\theta} + x_m \frac{di_2}{d\theta} \qquad (5)$$

and

$$e_2 = r_2 i_2 + x_2 \frac{di_2}{d\theta} + x_m \frac{di_1}{d\theta}. \qquad (6)$$

Differentiating (6) gives

$$\frac{de_2}{d\theta} = r_2 \frac{di_2}{d\theta} + x_2 \frac{d^2 i_2}{d\theta^2} + x_m \frac{d^2 i_1}{d\theta^2}; \qquad (7)$$

* See the chapters on induction machines, etc., in " Theory and Calculation of Alternating Current Phenomena."

from (5) follows

$$\frac{di_2}{d\theta} = \frac{e_1 - r_1 i_1 - x_1 \dfrac{di_1}{d\theta}}{x_m},$$ (8)

and, differentiated,

$$\frac{d^2 i_2}{d\theta^2} = \frac{1}{x_m} \left\{ \frac{de_1}{d\theta} - r_1 \frac{di_1}{d\theta} - x_1 \frac{d^2 i_1}{d\theta^2} \right\}.$$ (9)

Substituting (8) and (9) in (7) gives

$$r_2 e_1 + x_2 \frac{de_1}{d\theta} - x_m \frac{de_2}{d\theta} = r_1 r_2 i_1 + (r_1 x_2 + r_2 x_1) \frac{di_1}{d\theta}$$

$$+ (x_1 x_2 - x_m^2) \frac{d^2 i_1}{d\theta^2},$$ (10)

and analogously,

$$r_1 e_2 + x_1 \frac{de_2}{d\theta} - x_m \frac{de_1}{d\theta} = r_1 r_2 i_2 + (r_1 x_2 + r_2 x_1) \frac{di_2}{d\theta}$$

$$+ (x_1 x_2 - x_m^2) \frac{d^2 i_2}{d\theta^2}.$$ (11)

Equations (10) and (11) are the two differential equations of second order, of currents i_1 and i_2.

If e_1', i_1' and e_2', i_2' are the permanent values of impressed e.m.fs. and of currents in the two circuits, and e_1'', i_1'' and e_2'', i_2'' are their transient terms, we have,

$$e_1 = e_1' + e_1'', \qquad\qquad i_1 = i_1' + i_1'',$$
$$e_2 = e_2' + e_2'', \qquad\qquad i_2 = i_2' + i_2''.$$

Since the permanent terms must fulfill the differential equations (10) and (11),

$$r_2 e_1' + x_2 \frac{de_1'}{d\theta} - x_m \frac{de_2'}{d\theta} = r_1 r_2 i_1' + (r_1 x_2 + r_2 x_1) \frac{di_1'}{d\theta}$$

$$+ (x_1 x_2 - x_m^2) \frac{d^2 i_1'}{d\theta^2}$$ (12)

and

$$r_1 e_2' + x_1 \frac{de_2'}{d\theta} - x_m \frac{de_1'}{d\theta} = r_1 r_2 i_2' + (r_1 x_2 + r_2 x_1) \frac{di_2'}{d\theta}$$

$$+ (x_1 x_2 - x_m^2) \frac{d^2 i_2'}{d\theta^2},$$ (13)

subtracting equations (12) and (13) from (10) and (11) gives the differential equations of the transient terms,

$$r_2 e_1'' + x_2 \frac{d e_1''}{d\theta} - x_m \frac{d e_2''}{d\theta} = r_1 r_2 i_1'' + (r_1 x_2 + r_2 x_1) \frac{d i_1''}{d\theta}$$
$$+ (x_1 x_2 - x_m^2) \frac{d^2 i_1''}{d\theta^2} \tag{14}$$

and

$$r_1 e_2'' + x_1 \frac{d e_2''}{d\theta} - x_m \frac{d e_1''}{d\theta} = r_1 r_2 i_2'' + (r_1 x_2 + r_2 x_1) \frac{d i_2''}{d\theta}$$
$$+ (x_1 x_2 - x_m^2) \frac{d^2 i_2''}{d\theta^2}. \tag{15}$$

These differential equations of the transient terms are the same as the general differential equations (10) and (11) and the differential equations of the permanent terms (12) and (13).

85. If, as is usually the case, the impressed e.m.fs. contain no transient term, that is, the transient terms of current do not react upon the sources of supply of the impressed e.m.fs. and affect them, we have

$$e_1'' = 0 \quad \text{and} \quad e_2'' = 0;$$

hence, the differential equations of the transient terms are

$$0 = r_1 r_2 i + (r_1 x_2 + r_2 x_1) \frac{di}{d\theta} + (x_1 x_2 - x_m^2) \frac{d^2 i}{d\theta^2} \tag{16}$$

and are the same for both currents i_1'' and i_2'', that is, the transient terms of currents differ only by their integration constants, or the terminal conditions.

Equation (16) is integrated by the function

$$i = A \varepsilon^{-a\theta}. \tag{17}$$

Substituting (17) in (16) gives

$$A \varepsilon^{-a\theta} \{ r_1 r_2 - a (r_1 x_2 + r_2 x_1) + a^2 (x_1 x_2 - x_m^2) \} = 0;$$

hence,

A = indefinite, as integration constant, and

$$a^2 - \frac{r_1 x_2 + r_2 x_1}{x_1 x_2 - x_m^2} a + \frac{r_1 r_2}{x_1 x_2 - x_m^2} = 0. \tag{18}$$

The exponent a is given by a quadratic equation (18). This quadratic equation (18) always has two real roots, and in this respect differs from the quadratic equation appearing in a circuit containing capacity, which latter may have two imaginary roots and so give rise to an oscillation.

Mutual induction in the absence of capacity thus always gives a logarithmic transient term; thus,

$$a = \frac{(r_1 x_2 + r_2 x_1) \pm \sqrt{(r_1 x_2 - r_2 x_1)^2 + 4\,r_1 r_2 x_m^2}}{2\,(x_1 x_2 - x_m^2)}. \quad (19)$$

As seen, the term under the radical in (19) is always positive, that is, the two roots a_1 and a_2 always real and always positive, since the square root is smaller than the term outside of it.

Herefrom then follows the integral equation of one of the currents, for instance i_1, as

$$i_1 = i_1' + A_1 \varepsilon^{-a_1 \theta} + A_2 \varepsilon^{-a_2 \theta}, \quad (20)$$

and eliminating from the two equations (5) and (6) the term $\dfrac{di_2}{d\theta}$ gives

$$i_2 = \frac{1}{r_2 x_m} \left\{ r_1 x_2 i_1 + (x_1 x_2 - x_m^2) \frac{di_1}{d\theta} + x_m e_2 - x_2 e_1 \right\}, \quad (21)$$

leaving the two integration constants A_1 and A_2 to be determined by the terminal conditions, as $\theta = 0$,

$$i_1 = i_1^0 \quad \text{and} \quad i_2 = i_2^0.$$

86. If the impressed e.m.fs. e_1 and e_2 are constant, we have

$$\frac{de_1}{d\theta} = 0 \quad \text{and} \quad \frac{de_2}{d\theta} = 0;$$

hence, the equations of the permanent terms (12) and (13) give

$$i_1' = \frac{e_1}{r_1} \quad \text{and} \quad i_2' = \frac{e_2}{r_2}; \quad (22)$$

thus:

$$i_1 = \frac{e_1}{r_1} + A_1 \varepsilon^{-a_1 \theta} + A_2 \varepsilon^{-a_2 \theta}$$

and

$$i_2 = \frac{e_2}{r_2} + A_1' \varepsilon^{-a_1 \theta} + A_2' \varepsilon^{-a_2 \theta},$$

$$(23)$$

where, A_1' and A_2' follow from A_1 and A_2 by equation (21).

If the mutual inductance between the two circuits is perfect, that is,

$$x_m{}^2 = x_1 x_2, \tag{24}$$

equation (18) becomes, by multiplication with $\dfrac{x_1 x_2 - x_m{}^2}{r_1 x_2 + r_2 x_1}$,

$$a = \frac{r_1 r_2}{r_1 x_2 + r_2 x_1}; \tag{25}$$

that is, only one transient term exists.

As example may be considered a circuit having the following constants: $e_1 = 100$ volts; $e_2 = 0$; $r_1 = 5$ ohms; $r_2 = 5$ ohms; $x_1 = 100$ ohms; $x_2 = 100$ ohms, and $x_m = 80$ ohms. This gives

$$i_1' = 20 \text{ amp. and } i_2' = 0,$$

and

$$a^2 - 0.278\, a + 0.00695 = 0;$$

the roots are $a_1 = 0.0278$ and $a_2 = 0.251$

and

$$i_1 = 20 + A_1 \varepsilon^{-0.0278\,\theta} + A_2 \varepsilon^{-0.251\,\theta}.$$

By equation (21),

$$i_2 = -25 + 1.25\, i_1 + 9\frac{di_1}{d\theta};$$

hence,

$$i_2 = A_1 \varepsilon^{-0.0278\,\theta} - A_2 \varepsilon^{-0.251\,\theta}.$$

For $\theta = 0$ let $i_1{}^0 = 18$ amp., or the current 10 per cent below the normal, and $i_2{}^0 = 0$; then substituted, gives:

$$18 = 20 + A_1 + A_2 \quad \text{and} \quad 0 = A_1 - A_2,$$

hence, $A_1 = A_2 = -1;$

and we have

$$i_1 = 20 - (\varepsilon^{-0.0278\,\theta} + \varepsilon^{-0.251\,\theta})$$

and $i_2 = -(\varepsilon^{-0.0278\,\theta} - \varepsilon^{-0.251\,\theta}).$

87. An interesting application of the preceding is the investigation of the building up of an overcompounded direct-current generator, with sudden changes of load, or the building up, or down, of a compound wound direct-current booster.

While it would be desirable that a generator or booster, under sudden changes of load, should instantly adjust its voltage to the change so as to avoid a temporary fluctuation of voltage, actually an appreciable time must elapse.

A 600-kw. 8-pole direct-current generator overcompounds from 500 volts at no load to 600 volts at terminals at full load of 1000 amperes. The circuit constants are: resistance of armature winding, $r_0 = 0.01$ ohm; resistance of series field winding, $r_2' = 0.003$ ohm; number of turns per pole in shunt field winding, $n_1 = 1000$, and magnetic flux per pole at 500 volts, $\Phi = 10$ megalines. At 600 volts full load *terminal voltage* (or voltage from brush to brush) the generated e.m.f. is $e + ir_0 = 610$ volts.

From the saturation curve or magnetic characteristics of the machine, we have:

At no load and 500 volts:

> 5000 ampere-turns, 10 megalines and 5 amp. in shunt field circuit.

At no load and 600 volts:

> 7000 ampere-turns and 12 megalines.

At no load and 610 volts:

> 7200 ampere-turns and 12.2 megalines.

At full load and 600 volts:

> 8500 ampere-turns, 12.2 megalines and 6 amp. in shunt field.

Hence the demagnetizing force of the armature, due to the shift of brushes, is 1300 ampere-turns per pole.

At 600 volts and full load the shunt field winding takes 6 amperes, and gives 6000 ampere-turns, so that the series field winding has to supply 2500 ampere-turns per pole, of which 1300 are consumed by the armature reaction and 1200 magnetize.

At 1000 amp. full load the series field winding thus has 2.5 turns per pole, of which 1.3 neutralize the armature reaction and $n_2 = 1.2$ turns are effective magnetizing turns.

The ratio of effective turns in series field winding and in shunt field winding is $a = \dfrac{n_2}{n_1} = 1.2 \times 10^{-3}$. This then is the reduction factor of the shunt circuit to the series circuit.

It is convenient to reduce the phenomena taking place in the shunt field winding to the same number of turns as the series field winding, by the factors a and a^2 respectively.

If then e = terminal voltage of the armature, or voltage impressed upon the main circuit consisting of series field winding and external circuit, the same voltage is impressed upon the shunt field winding and reduced to the main circuit by factor a, gives
$$e_1 = ae = 1.2 \times 10^{-3} e.$$

Since at 500 volts impressed the shunt field current is 5 amperes, the field rheostat must be set so as to give to the shunt field circuit the total resistance of $r_1' = \dfrac{500}{5} = 100$ ohms.

Reduced to the main circuit by the square of the ratio of turns, this gives the resistance,
$$r_1 = a^2 r_1' = 144 \times 10^{-6} \text{ ohms.}$$

An increase of ampere-turns from 5000 to 7000, corresponding to an increase of current in the shunt field winding by 2 amperes, increases the generated e.m.f. from 500 to 600 volts, and the magnetic flux from 10 to 12, or by 2 megalines per pole. In the induction range covered by the overcompounding from 500 to 600 volts, 1 ampere increase in the shunt field increases the flux by 1 megaline per pole, and so, with $n_1 = 1000$ turns, gives 10^9 magnetic interlinkages per pole, or 8×10^9 interlinkages with 8 poles, per ampere, hence 80×10^9 interlinkages per unit current or 10 amperes, that is, an inductance of 80 henrys. Reduced to the main circuit this gives an inductance of $1.2^2 \times 10^{-6} \times 80 = 115.2 \times 10^{-6}$ henrys. This is the inductance due to the magnetic flux in the field poles, which interlinks with shunt and series coil, or the mutual inductance, $M = 115.2 \times 10^{-6}$ henrys.

Assuming the total inductance L_1 of the shunt field winding as 10 per cent higher than the mutual inductance M, that is, assuming 10 per cent stray flux, we have
$$L_1 = 1.1 M = 126.7 \times 10^{-6} \text{ henrys.}$$

In the main circuit, full load is 1000 amp. at 600 volts. This gives the effective resistance of the main circuit as $r = 0.6$ ohm.

The quantities referring to the main circuit may be denoted without index.

The total inductance of the main circuit depends upon the character of the load. Assuming an average railway motor load, the inductance may be estimated as about $L = 2000 \times 10^{-6}$ henrys.

In the present problem the impressed e.m.fs. are not constant but depend upon the currents, that is, the sum $i + i_1$, where $i_1 =$ shunt field current reduced to the main circuit by the ratio of turns.

The impressed e.m.f., e, is approximately proportional to the magnetic flux Φ, hence less than proportional to the current, in consequence of magnetic saturation. Thus we have

$$e = 500 \text{ volts for } 5000 \text{ ampere-turns,}$$

$$\text{or } i + i_1 = \frac{5000}{1.2} = 4170 \text{ amp. and}$$

$$e = 600 \text{ volts for } 7200 \text{ ampere-turns,}$$

$$\text{or } i + i_1 = \frac{7200}{1.2} = 6000 \text{ amp.;}$$

hence, 1830 amp. produce a rise of voltage of 100, or 1 amp. raises the voltage by $\dfrac{100}{1830} = \dfrac{1}{18.3}$.

At 6000 amp. the voltage is $\dfrac{6000}{18.3} = 328$ volts higher than at 0 amp., that is, the voltage in the range of saturation between 500 and 600 volts, when assuming the saturation curve in this range as straight line, is given by the equation

$$e = 272 + \frac{i + i_1}{18.3}.$$

The impressed e.m.f. of the shunt field is the same, hence, reduced to the main circuit by the ratio of turns, $a = 1.2 \times 10^{-3}$, is

$$e_1 = \left(272 + \frac{i + i_1}{18.3}\right) 1.2 \times 10^{-3}.$$

Assuming now as standard frequency, $f = 60$ cycles per sec., the constants of the two mutually inductive circuits shown diagrammatically in Fig. 38 are:

	Main Circuit.	Shunt Field Circuit.
Current..........	i amp.	i_1 amp.
Impressed e.m.f...	$e = 272 + \dfrac{i + i_1}{18.3}$ volts	$e_1 = \left(272 + \dfrac{i+i_1}{18.3}\right) 1.2 \times 10^{-3}$ volts
Resistance........	$r = .6$ ohms	$r_1 = 0.144 \times 10^{-3}$ ohms
Inductance.......	$L = 2000 \times 10^{-6}$ henrys	$L_1 = 126.7 \times 10^{-6}$ henrys
Reactance, $2\pi f L$..	$x = 755 \times 10^{-3}$ ohms	$x_1 = 47.8 \times 10^{-3}$ ohms
Mutual inductance	$M = 115.2 \times 10^{-6}$ henrys	
Mutual reactance.	$x_m = 43.5 \times 10^{-3}$ ohms	

This gives the differential equations of the problem as

$$272 + \frac{i + i_1}{18.3} = 0.6\,i + 0.755\frac{di}{d\theta} + 0.0435\frac{di_1}{d\theta} \qquad (26)$$

and

$$1.2\left(272 + \frac{i + i_1}{18.3}\right) = 0.144\,i_1 + 47.8\frac{di_1}{d\theta} + 43.5\frac{di}{d\theta}. \qquad (27)$$

88. Eliminating $\dfrac{di}{d\theta}$ from equations (26) and (27) gives

$$\frac{di_1}{d\theta} = 0.695\,i - 0.0712\,i_1 - 338. \qquad (28)$$

Equation (28) substituted in (26) gives

$$i_1 = 13.07\frac{di}{d\theta} + 9.95\,i - 4950. \qquad (29)$$

Equation (29) substituted in (28) gives

$$\frac{di_1}{d\theta} = -0.93\frac{di}{d\theta} - 0.015\,i + 15. \qquad (30)$$

Equation (29) differentiated, and equated with (30), gives

$$\frac{d^2i}{d\theta^2} + 0.828\frac{di}{d\theta} + 0.00115\,i - 1.15 = 0. \qquad (31)$$

Equation (31) is integrated by

$$i = i_0 + A\varepsilon^{-a\theta}.$$

Substituting this in (31) gives

$$A\varepsilon^{-a\theta}\{a^2 - 0.828\,a + 0.00115\} + \{0.00115\,i_0 - 1.15\} = 0,$$

hence, $i_0 = 1000$, A is indefinite, as integration constant, and

$$a^2 - 0.828\,a + 0.00115 = 0;$$

thus $$a = 0.414 \pm 0.4126,$$

and the roots are

$$a_1 = 0.0014 \quad \text{and} \quad a_2 = 0.827.$$

Therefore

$$i = 1000 + A_1\varepsilon^{-0.0014\,\theta} + A_2\varepsilon^{-0.827\,\theta}. \tag{32}$$

Substituting (32) in (29) gives

$$i_1 = 5000 + 9.932\,A_1\varepsilon^{-0.0014\,\theta} - 0.85\,A_2\varepsilon^{-0.827\,\theta}. \tag{33}$$

Substituting in (32) and (33) the terminal conditions $\theta = 0$, $i = 0$, and $i_1 = 4170$, gives

$$A_1 + A_2 = -1000 \quad \text{and} \quad 9.932\,A_1 - 0.85\,A_2 = -830,$$

that is,

$$A_1 = -156 \quad \text{and} \quad A_2 = -844.$$

Therefore

$$i = 1000 - 156\,\varepsilon^{-0.0014\,\theta} - 844\,\varepsilon^{-0.827\,\theta} \tag{34}$$

and

$$i_1 = 5000 - 1550\,\varepsilon^{-0.0014\,\theta} + 720\,\varepsilon^{-0.827\,\theta}; \tag{35}$$

or the shunt field current i_1 reduced back to the number of turns of the shunt field by the factor $a = 1.2 \times 10^{-3}$ is

$$i_1' = 6 - 1.86\,\varepsilon^{-0.0014\,\theta} + 0.86\,\varepsilon^{-0.827\,\theta}, \tag{36}$$

and the terminal voltage of the machine is

$$e = 272 + \frac{i + i_1}{18.3},$$

or, $e = 600 - 93.2\,\varepsilon^{-0.0014\,\theta} - 6.8\,\varepsilon^{-0.827\,\theta}.$ (37)

As seen, of the two exponential terms one disappears very quickly, the other very slowly.

Introducing now instead of the angle $\theta = 2\pi f t$ the time, t, gives the *main current* as

$$i = 1000 - 156\,\varepsilon^{-0.53\,t} - 844\,\varepsilon^{-311\,t},$$

the *shunt field current* as

$$i_1' = 6 - 1.86\,\varepsilon^{-0.53\,t} + 0.86\,\varepsilon^{-311\,t},$$ (38)

and the *terminal voltage* as

$$e = 600 - 93.2\,\varepsilon^{-0.53\,t} - 6.8\,\varepsilon^{-311\,t}.$$

89. Fig. 39 shows these three quantities, with the time, t, as abscissas.

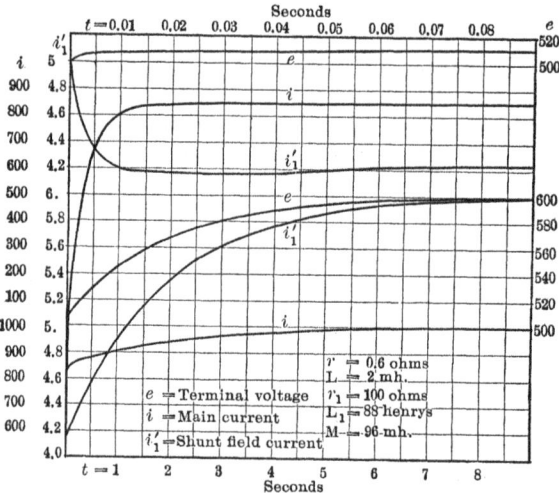

Fig. 39.   Building-up of over-compounded direct-current generator from 500 volts no load to 600 volts load.

The upper part of Fig. 39 shows the first part of the curve with 100 times the scale of abscissas as the lower part. As seen, the transient phenomenon consists of two distinctly different periods: first a very rapid change covering a part of the range of current or e.m.f., and then a very gradual adjustment to the final condition.

So the main current rises from zero to 800 amp. in 0.01 sec., but requires for the next 100 amp., or to rise to a total of 900 amp., about a second, reaching 95 per cent of full value in 2.25 sec. During this time the shunt field current first falls very rapidly, from 5 amp. at start to 4.2 amp. in 0.01 sec., and then, after a minimum of 4.16 amp., at $t = 0.015$, gradually and very slowly rises, reaching 5 amp., or its starting point, again after somewhat more than a second. After 2.5 sec. the shunt field current has completed half of its change, and after 5.5 sec. 90 per cent of its change.

The terminal voltage first rises quickly by a few volts, and then rises slowly, completing 50 per cent of its change in 1.2 sec., 90 per cent in 4.5 sec., and 95 per cent in 5.5 sec.

Physically, this means that the terminal voltage of the machine rises very slowly, requiring several seconds to approach stationary conditions. First, the main current rises very rapidly, at a rate depending upon the inductance of the external circuit, to the value corresponding to the resistance of the external circuit and the initial or no load terminal voltage, and during this period of about 0.01 sec. the magnetizing action of the main current is neutralized by a rapid drop of the shunt field current. Then gradually the terminal voltage of the machine builds up, and the shunt field current recovers to its initial value in 1.15 sec., and then rises, together with the main current, in correspondence with the rising terminal voltage of the machine.

It is interesting to note, however, that a very appreciable time elapses before approximately constant conditions are reached.

90. In the preceding example, as well as in the discussion of the building up of shunt or series generators in Chapter II, the e.m.fs. and thus currents produced in the iron of the magnetic field by the change of the field magnetization have not been considered. The results therefore directly apply to a machine with laminated field, but only approximately to one with solid iron poles.

In machines with solid iron in the magnetic circuit, currents produced in the iron act as a second electric circuit in inductive relation to the field exciting circuit, and the transition period thus is slower.

As example may be considered the excitation of a series booster with solid and with laminated poles; that is, a machine with series field winding, inserted in the main circuit of a feeder, for the purpose of introducing into the circuit a voltage proportional to the load, and thus to compensate for the increasing drop of voltage with increase of load.

Due to the production of eddy currents in the solid iron of the field magnetic circuit, the magnetic flux density is not uniform throughout the whole field section during a change of the magnetic field, since the outer shell of the field iron is magnetized by the field coil only, while the central part of the iron is acted upon by the impressed m.m.f. of the field coil and the m.m.f. of the eddy currents in the outer part of the iron, and the change of magnetic flux density in the interior thus lags behind that of the outside of the iron. As result hereof the eddy currents in the different layers of the structure differ in intensity and in phase.

A complete investigation of the distribution of magnetism in this case leads to a transient phenomenon in space, and is discussed in Section III. For the present purpose, where the total m.m.f. of the eddy currents is small compared with that of the main field, we can approximate the effect of eddy currents in the iron by a closed circuit secondary conductor, that is, can assume uniform intensity and phase of secondary currents in an outer layer of the iron, that is, consider the outer

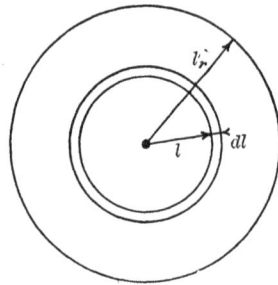

Fig. 40. Section of a magnetic circuit.

layer of the iron, up to a certain depth, as a closed circuit secondary.

Let Fig. 40 represent a section of the magnetic circuit of the machine, and assume uniform flux density. If Φ = the total magnetic flux, l_r = the radius of the field section, then at a distance l from the center, the magnetic flux enclosed by a circle with radius l is $\left(\dfrac{l}{l_r}\right)^2 \Phi$, and the e.m.f. generated in the zone at distance l from the center is proportional to $\left(\dfrac{l}{l_r}\right)^2 \Phi$, that

is, $e = a\left(\dfrac{l}{l_r}\right)^2 \Phi$. The current density of the eddy currents in

this zone, which has the length $2\,\pi l$, is therefore proportional to

$\dfrac{e}{2\,\pi l}$, or is $i = \dfrac{bl}{l_r^2}\Phi$. This current density acts as a m.m.f. upon

the space enclosed by it, that is, upon $\left(\dfrac{l}{l_r}\right)^2$ of the total field

section, and the magnetic reaction of the secondary current at

distance l from the center therefore is proportional to $i\left(\dfrac{l}{l_r}\right)^2$, or

is $\mathcal{F} = \dfrac{cl^3}{l_r^4}\Phi$, and therefore the total magnetic reaction of the

eddy currents is

$$i_0 = \int_0^{l_r} \mathcal{F}\, dl = \frac{c}{4}\Phi.$$

At the outer periphery of the field iron, the generated e.m.f.

is $e_1 = a\Phi$, the current density therefore $i_1 = \dfrac{b}{l_r}\Phi$, and the

magnetic reaction $\mathcal{F}_1 = \dfrac{c}{l_r}\Phi$, and therefore

$$i_0 = \frac{l_r}{4}\mathcal{F}_1;$$

that is, the magnetic reaction of the eddy currents, assuming uniform flux density in the field poles, is the same as that of the currents produced in a closed circuit of a thickness $\dfrac{l_r}{4}$, or one-fourth the depth of the pole iron, of the material of the field pole and surrounding the field pole, that is, fully induced and fully magnetizing.

The eddy currents in the solid material of the field poles thus can be represented by a closed secondary circuit of depth $\dfrac{l_r}{4}$ surrounding the field poles.

The magnitude of the depth of the field copper on the spools is probably about one-fourth the depth of the field poles. Assuming then the width of the band of iron which represents the eddy current circuit as about twice the width of the field coils — since eddy currents are produced also in the yoke of the machine, etc. — and the conductivity of the iron as about 0.1 that of the field copper, the effective resistance of the eddy current circuit, reduced to the field circuit, approximates five times that of the field circuit.

Hence, if r_2 = resistance of main field winding, $r_1 = 5 r_2$ = resistance of the secondary short circuit which represents the eddy currents.

Since the eddy currents extend beyond the space covered by the field coils, and considerably down into the iron, the self-inductance of the eddy current circuit is considerably greater than its mutual inductance with the main field circuit, and thus may be assumed as twice the latter.

91. As example, consider a 200-kw. series booster covering the range of voltage from 0 to 200, that is, giving a full load value of 1000 amperes at 200 volts. Making the assumptions set forth in the preceding paragraph, the following constants are taken: the armature resistance = 0.008 ohms and the series field winding resistance = 0.004 ohm; hence, the short circuit — or eddy current resistance — r_1 = 0.02 ohm. Furthermore let $M = 900 \times 10^{-6}$ henry = mutual inductance between main field and short-circuited secondary; hence, $x_m = 0.34$ ohm = mutual reactance, and therefore, assuming a leakage flux of the secondary equal to the main flux, $L_1 = 1800 \times 10^{-6}$ henry and $x_1 = 0.68$ ohm.

The booster is inserted into a constant potential circuit of 550 volts, so as to raise the voltage from 550 volts no load to 750 volts at 1000 amperes.

The total resistance of the circuit at full load, including main circuit and booster, therefore is $r = 0.75$ ohm.

The inductance of the external circuit may be assumed as $L = 4500 \times 10^{-6}$ henrys; hence, the reactance at $f = 60$ cycles per sec. is $x = 1.7$ ohms. The impressed e.m.f. of the circuit is $e = 550 + e'$, e' being the e.m.f. generated in the booster. Since at no load, for $i = 0$, $e' = 0$, and at full load, for $i = 1000$, $e' = 200$, assuming a straight line magnetic characteristic or saturation curve, that is, assuming the effect of magnetic saturation as negligible within the working range of the booster, we have

$$e = 550 + 0.2 \, (i + i_1).$$

This gives the following constants:

	Main Circuit.	Eddy Current Circuit.
Current................	i amp.	i_1 amp.
Impressed e.m.f......	$e = 550 + 0.2\,(i + i_1)$ volts.	0 volts.
Resistance..........	$r = 0.75$ ohm.	$r_1 = 0.02$ ohm.
Inductance.........	$L = 4500 \times 10^{-6}$ henrys.	$L_1 = 1800 \times 10^{-6}$ henrys.
Reactance..........	$x = 1.7$ ohms.	$x_1 = 0.68$ ohm.
Mutual inductance...	$M = 900 \times 10^{-6}$ henrys.	
Mutual reactance....	$x_m = 0.34$ ohm.	

This gives the differential equations of the problem as

$$550 - 0.55 \, i + 0.2 \, i_1 - 1.7 \, \frac{di}{d\theta} - 0.34 \, \frac{di_1}{d\theta} = 0 \qquad (39)$$

and

$$0.02 \, i_1 + 0.34 \, \frac{di}{d\theta} + 0.68 \, \frac{di_1}{d\theta} = 0. \qquad (40)$$

Adding 2 times (39) to (40) gives

$$1100 - 1.1 \, i + 0.42 \, i_1 - 3.06 \, \frac{di}{d\theta} = 0, \qquad (41)$$

or

$$i_1 = 7.28 \, \frac{di}{d\theta} + 2.62 \, i - 2620, \qquad (42)$$

herefrom:

$$0.02 \, i_1 = 0.1456 \, \frac{di}{d\theta} + 0.0524 \, i - 52.4, \qquad (43)$$

and

$$0.68 \, \frac{di_1}{d\theta} = 4.95 \, \frac{d^2i}{d\theta^2} + 1.78 \, \frac{di}{d\theta}, \qquad (44)$$

substituting the last two equations into (40),

$$\frac{d^2i}{d\theta^2} + 0.458 \, \frac{di}{d\theta} + 0.0106 \, i - 10.6 = 0. \qquad (45)$$

If

$$i = i_0 + A\varepsilon^{-a\theta}, \qquad (46)$$

then

$$A\varepsilon^{-a\theta}(a^2 - 0.458 \, a + 0.0106) + 0.0106 \, i_9 - 10.6 = 0.$$

As transient and permanent terms must each equal zero,

$$i_0 = 1000 \quad \text{and} \quad a^2 - 0.458\,a + 0.0106 = 0,$$

wherefrom $\qquad a = 0.229 \pm 0.205;$

the roots are $\qquad a_1 = 0.024 \quad \text{and} \quad a_2 = 0.434;$

then we have

$$i = 1000 + A_1\varepsilon^{-0.024\,\theta} + A_2\varepsilon^{-0.434\,\theta} \tag{47}$$

and

$$i_1 = 2.45\,A_1\varepsilon^{-0.024\,\theta} - 0.55\,A_2\varepsilon^{-0.434\,\theta}. \tag{48}$$

With terminal conditions $\theta = 0$, $i = 0$, and $i_1 = 0$,

$$A_1 = -183 \quad \text{and} \quad A_2 = -817.$$

If $\theta = 2\,\pi f t = 377.5$, we have

$$\left.\begin{aligned}
i &= 1000 - 183\,\varepsilon^{-9.07\,t} - 817\,\varepsilon^{-164\,t}, \\
i_1 &= -450\left\{\varepsilon^{-9.07\,t} - \varepsilon^{-164\,t}\right\}, \\
e &= 750 - 127\,\varepsilon^{-9.07\,t} - 73^{-164\,t}.
\end{aligned}\right\} \tag{49}$$

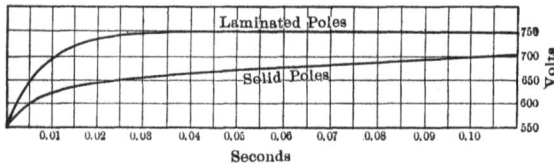

Fig. 41. Building up of feeder voltage by series booster.

In the absence of a secondary circuit, or with laminated field poles, equation (39) would assume the form $i_1 = 0$, or

$$550 + 0.2\,i = 0.75\,i + 1.7\,\frac{di}{d\theta}; \tag{50}$$

hence, $\qquad \dfrac{di}{d\theta} = 0.323\,(1000 - i)$

and $\qquad i = 1000\,(1 - \varepsilon^{-0.323\,\theta});$

or $\qquad i = 1000\,(1 - \varepsilon^{-122\,t})$

and $\qquad \left.\begin{aligned}e &= 750 - 200\,\varepsilon^{-122\,t};\end{aligned}\right\} \tag{51}$

that is, the e.m.f., e, approaches final conditions at a more rapid rate.

Fig. 41 shows the curves of the e.m.f., e, for the two conditions, namely, solid field poles, (49), and laminated field poles, (51).

(B) MUTUAL INDUCTANCE IN CIRCUITS CONTAINING SELF-INDUCTANCE AND CAPACITY.

92. The general eqations of such a pair of circuits, (1) and (2), differentiated to eliminate the integral give

$$\frac{de_1}{d\theta} = x_{c_1}i_1 + r_1\frac{di_1}{d\theta} + x_1\frac{d^2i_1}{d\theta^2} + x_m\frac{d^2i_2}{d\theta^2} \qquad (52)$$

and

$$\frac{de_2}{d\theta} = x_{c_2}i_2 + r_2\frac{di_2}{d\theta} + x_2\frac{d^2i_2}{d\theta^2} + x_m\frac{d^2i_1}{d\theta^2}, \qquad (53)$$

and the potential differences at the condensers, from (1) and (2), are

$$e_1' = x_{c_1}\int i_1 \, d\theta = e_1 - r_1i_1 - x_1\frac{di_1}{d\theta} - x_m\frac{di_2}{d\theta} \qquad (54)$$

and

$$e_2' = x_{c_2}\int i_2 \, d\theta = e_2 - r_2i_2 - x_2\frac{di_2}{d\theta} - x_m\frac{di_1}{d\theta}. \qquad (55)$$

If now the impressed e.m.fs., e_1 and e_2, contain no transient term, that is, if the transient values of currents i_1 and i_2 exert no appreciable reaction on the source of e.m.f., and if i_1' and i_2' are the permanent terms of current, then, substituting i_1' and i_2' in equations (52) and (53), and subtracting the result of this substitution from (52) and (53), gives the equations of the transient terms of the currents i_1 and i_2, thus:

$$0 = x_{c_1}i_1 + r_1\frac{di_1}{d\theta} + x_1\frac{d^2i_1}{d\theta^2} + x_m\frac{d^2i_2}{d\theta^2} \qquad (56)$$

and

$$0 = x_{c_2}i_2 + r_2\frac{di_2}{d\theta} + x_2\frac{d^2i_2}{d\theta^2} + x_m\frac{d^2i_1}{d\theta^2}. \qquad (57)$$

If the impressed e.m.fs., e_1 and e_2, are constant, $\dfrac{de_1}{d\theta}$ and $\dfrac{de_2}{d\theta}$

equal zero, and equations (52) and (53) assume the form (56) and (57); that is, equations (56) and (57) are the differential equations of the transient terms, for the general case of any e.m.fs., e_1 and e_2, which have no transient terms, and are the general differential equations of the case of constant impressed e.m.fs., e_1 and e_2.

From (56) it follows that

$$x_m \frac{d^2 i_2}{d\theta^2} = - x_{c_1} i_1 - r_1 \frac{d i_1}{d\theta} - x_1 \frac{d^2 i_1}{d\theta^2} . \qquad (58)$$

Differentiating equation (57) twice, and substituting therein (58), gives

$$(x_1 x_2 - x_m{}^2) \frac{d^4 i}{d\theta^4} + (r_1 x_2 + r_2 x_1) \frac{d^3 i}{d\theta^3} + (x_{c_1} x_2 + x_{c_2} x_1 + r_1 r_2) \frac{d^2 i}{d\theta^2}$$

$$+ (x_{c_1} r_2 + x_{c_2} r_1) \frac{d i}{d\theta} + x_{c_1} x_{c_2} i = 0. \qquad (59)$$

This is a differential equation of fourth order, symmetrical in $r_1 x_1 x_{c_1}$ and $r_2 x_2 x_{c_2}$, which therefore applies to both currents, i_1 and i_2.

The expressions of the two currents i_1 and i_2 therefore differ only by their integration constants, as determined by the terminal conditions.

Equation (59) is integrated by

$$i = A \varepsilon^{-a\theta} \qquad (60)$$

and substituting (60) in (59) gives for the determination of the exponent a the quartic equation

$$(x_1 x_2 - x_m{}^2) a^4 - (r_1 x_2 + r_2 x_1) a^3 + (x_{c_1} x_2 + x_{c_2} x_1 + r_1 r_2) a^2$$
$$- (x_{c_1} r_2 + x_{c_2} r_1) a + x_{c_1} x_{c_2} = 0,$$

or

$$a^4 - \frac{r_1 x_2 + r_2 x_1}{x_1 x_2 - x_m{}^2} a^3 + \frac{x_{c_1} x_2 + x_{c_2} x_1 + r_1 r_2}{x_1 x_2 - x_m{}^2} a^2$$

$$- \frac{x_{c_1} r_2 + x_{c_2} r_1}{x_1 x_2 - x_m{}^2} a + \frac{x_{c_1} x_{c_2}}{x_1 x_2 - x_m{}^2} = 0. \qquad (61)$$

The solution of this quartic equation gives four values of a, and thus gives

$$i = A_1 \varepsilon^{-a_1\theta} + A_2 \varepsilon^{-a_2\theta} + A_3 \varepsilon^{-a_3\theta} + A_4 \varepsilon^{-a_4\theta}. \qquad (62)$$

The roots, a, may be real, or two real and two imaginary, or all imaginary, and the solution of the equation by approximation therefore is difficult.

In the most important case, where the resistance, r, is small compared with the reactances x and x_c — and which is the only case where the transient terms are prominent in intensity and duration, and therefore of interest — as in the transformer and the induction coil or Ruhmkorff coil, the equation (61) can be solved by a simple approximation.

In this case, the roots, a, are two pairs of conjugate imaginary numbers, and the phenomenon oscillatory.

The real components of the roots, a, must be positive, since the exponential $\varepsilon^{-a\theta}$ must decrease with increasing θ.

The four roots thus can be written:

$$\left. \begin{aligned} a_1 &= \alpha_1 - j\beta_1, \\ a_2 &= \alpha_1 + j\beta_1, \\ a_3 &= \alpha_2 - j\beta_2, \\ a_4 &= \alpha_2 + j\beta_2, \end{aligned} \right\} \tag{63}$$

where α and β are positive numbers.

In the equation (61), the coefficients of a^3 and a are small, since they contain the resistances as factor, and this equation thus can be approximated by

$$a^4 + \frac{x_{c_1} x_2 + x_{c_2} x_1}{x_1 x_2 - x_m^2} a^2 + \frac{x_{c_1} x_{c_2}}{x_1 x_2 - x_m^2} = 0; \tag{64}$$

hence,

$$a^2 = -\frac{1}{2} \left\{ \frac{x_{c_1} x_2 + x_{c_2} x_1}{x_1 x_2 - x_m^2} \pm \sqrt{\left(\frac{x_{c_1} x_2 + x_{c_2} x_1}{x_1 x_2 - x_m^2}\right)^2 - \frac{4\, x_{c_1} x_{c_2}}{x_1 x_2 - x_m^2}} \right\}; \tag{65}$$

that is, a^2 is negative, having two roots,

$$b_1 = -\beta_1^2 \quad \text{and} \quad b_2 = -\beta_2^2.$$

This gives the four imaginary roots of a as first approximation:

$$\left. \begin{aligned} a &= \pm j\beta_1 \\ &\quad \pm j\beta_2. \end{aligned} \right\} \tag{66}$$

If a_1, a_2, a_3, a_4 are the four roots of equation (61), this equation can be written

$$f(a) = (a - a_1)(a - a_2)(a - a_3)(a - a_4) = 0;$$

or, substituting (63),

$$f(a) = \{(a - \alpha_1)^2 + \beta_1^2\}\{(a - \alpha_2)^2 + \beta_2^2\} = 0, \quad (67)$$

and comparing (67) with (61) gives as coefficients of a^3 and of a,

and

$$\left.\begin{array}{c} 2(\alpha_1 + \alpha_2) = \dfrac{r_1 x_2 + r_2 x_1}{x_1 x_2 - x_m^2} \\[4mm] 2(\alpha_1\beta_2^2 + \alpha_2\beta_1^2) = \dfrac{x_{c_1} r_2 + x_{c_2} r_1}{x_1 x_2 - x_m^2}, \end{array}\right\} \quad (68)$$

and since β_1^2 and β_2^2 are given by (65) and (66) as roots of equation (64), α_1, α_2, β_1, β_2, and hereby the four roots a_1, a_2, a_3, a_4 of equation (61) are approximated by (64), (65), (66), (68).

The integration constants A_1, A_2, A_3, A_4 now follow from the terminal conditions.

93. As an example may be considered the operation of an inductorium, or Ruhmkorff coil, by make and break of a direct-current battery circuit, with a condenser shunting the break, in the usual manner.

Let $e_1 = 10$ volts = impressed e.m.f.; $r_1 = 0.4$ ohm = resistance of primary circuit, giving a current, at closed circuit and in stationary condition, of $i_0 = 25$ amp.; $r_2 = 0.2$ ohm = resistance of secondary circuit, reduced to the primary by the square of the ratio of primary $\div$ secondary turns; $x_1 = 10$ ohms = primary inductive reactance; $x_2 = 10$ ohms = secondary inductive reactance, reduced to primary; $x_m = 8$ ohms = mutual inductive reactance; $x_{c_1} = 4000$ ohms = primary condensive reactance of the condenser shunting the break of the interrupter in the battery circuit, and $x_{c_2} = 6000$ ohms = secondary condensive reactance, due to the capacity of the terminals and the high tension winding.

Substituting these values, we have

$$\left.\begin{array}{lll} e_1 = 10 \text{ volts} & i_0 = 25 \text{ amp.} & \\[2mm] r_1 = 0.4 \text{ ohm} & x_1 = 10 \text{ ohms} & x_{c_1} = 4000 \text{ ohms} \\[2mm] r_2 = 0.2 \text{ ohm} & x_2 = 10 \text{ ohms} & x_{c_2} = 6000 \text{ ohms} \\[2mm] & x_m = 8 \text{ ohms.} & \end{array}\right\} \quad (69)$$

These values in equation (61) give

$$f(a) = a^4 - 0.167\, a^3 + 2780\, a^2 - 89\, a + 667{,}000 = 0, \quad (70)$$

and in equation (64) they give

$$f_1(a) = a^4 + 2780\, a^2 + 667{,}000 = 0$$

and
$$a^2 = -(1390 \pm 1125)$$
$$= -2515,$$

or
$$= -265;$$

hence,
$$\beta_1 = 50.15$$

and
$$\beta_2 = 16.28.$$

From (68) it follows that

$$\alpha_1 + \alpha_2 = 0.0833$$

and
$$265\, \alpha_1 + 2515\, \alpha_2 = 44.5;$$

hence,
$$\alpha_1 = 0.073,$$
$$\alpha_2 = 0.010.$$

Introducing for the exponentials with imaginary exponents the trigonometric functions give

$$
\left.
\begin{aligned}
i_1 &= \varepsilon^{-0.073\,\theta}\left\{ A_1 \cos 50.15\,\theta + A_2 \sin 50.15\,\theta \right\} \\
&\quad + \varepsilon^{-0.010\,\theta}\left\{ B_1 \cos 16.28\,\theta + B_2 \sin 16.28\,\theta \right\} \\
i_2 &= \varepsilon^{-0.073\,\theta}\left\{ C_1 \cos 50.15\,\theta + C_2 \sin 50.15\,\theta \right\} \\
&\quad + \varepsilon^{-0.010\,\theta}\left\{ D_1 \cos 16.28\,\theta + D_2 \sin 16.28\,\theta \right\},
\end{aligned}
\right\}
\quad (71)
$$

where the constants C and D depend upon A and B by equations (56), (57), or (58), thus:

Substituting (71) into (58),

$$8\frac{d^2 i_2}{d\theta^2} + 4000\, i_1 + 0.4\frac{d i_1}{d\theta} + 10\frac{d^2 i_1}{d\theta^2} = 0 \quad (58)$$

gives an identity, from which, by equating the coefficients of $\varepsilon^{-a\theta}\cos b\theta$ and $\varepsilon^{-a\theta}\sin b\theta$ to zero, result four equations, in the coefficients

$$A_1\ B_1\ C_1\ D_1$$
$$A_2\ B_2\ C_2\ D_2,$$

from which follows, with sufficient approximation,

$$\left.\begin{array}{l} A_1 = -0.95\,C_1 \\ A_2 = -0.98\,C_2 \\ B_1 = +1.57\,D_1 \\ B_2 = +1.57\,D_2; \end{array}\right\} \quad (72)$$

hence,

$$\begin{aligned} i_1 = {} & -0.96\,\varepsilon^{-0.073\,\theta}\{C_1\,\cos 50.15\,\theta + C_2\,\sin 50.15\,\theta\} \\ & + 1.57\,\varepsilon^{-0.010\,\theta}\{D_1\,\cos 16.28\,\theta + D_2\,\sin 16.28\,\theta\} \end{aligned} \quad (73)$$

and substituting (71) and (73) in the equations of the condenser potential, (54) and (55), gives

$$\left.\begin{aligned} e_1' = {} & 10 + 79\,\varepsilon^{-0.073\,\theta}\{C_2\,\cos 50.15\,\theta - C_1\,\sin 50.15\,\theta\} \\ & - 385\,\varepsilon^{-0.010\,\theta}\{D_2\,\cos 16.28\,\theta - D_1\,\sin 16.28\,\theta\} \\[2mm] e_2' = {} & 118\,\varepsilon^{-0.073\,\theta}\{C_2\,\cos 50.15\,\theta - C_1\,\sin 50.15\,\theta\} \\ & + 367\,\varepsilon^{-0.010\,\theta}\{D_2\,\cos 16.28\,\theta - D_1\,\sin 16.28\,\theta\} \end{aligned}\right\} \quad (74)$$

94. Substituting now the terminal conditions of the circuit: At the moment where the interrupter opens the primary circuit the current in this circuit is $i_1 = \dfrac{e_0}{r_1} = 25$ amp. The condenser in the primary circuit, which is shunted across the break, was short-circuited before the break, hence of zero potential difference. The secondary circuit was dead. This then gives the conditions $\theta = 0$; $i_1 = 25$, $i_2 = 0$, $e_1' = 0$, and $e_2' = 0$. Substituting these values in equations (71), (73), (74) gives

$$\left.\begin{array}{l} 25 = -0.95\ C_1 + 1.58\,D_1 \\ 0 = \qquad\ C_1 + \quad\ D_1 \\ 0 = 10 + 79\,C_2 - 385\,D_2 \\ 0 = 118\,C_2 \qquad + 367\,D_2 \end{array}\right\}$$

hence

$$\begin{array}{l} C_1 = -10 \\ C_2 = -0.05 \cong 0 \\ D_1 = +10 \\ D_2 = +0.016 \cong 0, \end{array}$$

and

$$i_1 = 9.6 \, \varepsilon^{-0.073\,\theta} \cos 50.15 \, \theta + 15.7 \, \varepsilon^{-0.010\,\theta} \cos 16.28 \, \theta$$

$$i_2 = -10 \, \varepsilon^{-0.073\,\theta} \cos 50.15 \, \theta + 10 \, \varepsilon^{-0.010\,\theta} \cos 16.28 \, \theta$$

$$e_1' = 10 + 790 \, \varepsilon^{-0.073\,\theta} \sin 50.15 \, \theta + 3850 \, \varepsilon^{-0.010\,\theta} \sin 16.28 \, \theta$$

$$e_2' = 1180 \, \varepsilon^{-0.073\,\theta} \sin 50.15 \, \theta - 3670 \, \varepsilon^{-0.010\,\theta} \sin 16.28 \, \theta$$

$$(75)$$

Approximately therefore we have

$$i_1 = 9.6 \, \varepsilon^{-0.073\,\theta} \cos 50.15 \, \theta + 15.7 \, \varepsilon^{-0.010\,\theta} \cos 16.28 \, \theta$$

$$i_2 = -10 \left\{ \varepsilon^{-0.073\,\theta} \cos 50.15 \, \theta - \varepsilon^{-0.010\,\theta} \cos 16.28 \, \theta \right\}$$

$$e_1' = 3850 \, \varepsilon^{-0.010\,\theta} \sin 16.28 \, \theta$$

$$e_2' = -3670 \, \varepsilon^{-0.010\,\theta} \sin 16.28 \, \theta.$$

The two frequencies of oscillation are 3009 and 977 cycles per sec., hence rather low.

The secondary terminal voltage has a maximum of nearly 4000, reduced to the primary, or 400 times as large as corresponds to the ratio of turns.

In this particular instance, the frequency 3009 is nearly suppressed, and the main oscillation is of the frequency 977.

CHAPTER XI.

GENERAL SYSTEM OF CIRCUITS.

(A) CIRCUITS CONTAINING RESISTANCE AND INDUCTANCE ONLY.

95. Let, upon a general system or network of circuits connected with each other directly or inductively, and containing resistance and inductance, but no capacity, a system of e.m.fs., e_γ be impressed. These e.m.fs. may be of any frequency or wave shape, or may be continuous or anything else, but are supposed to be given by their equations. They may be free of transient terms, or may contain transient terms depending upon the currents in the system. In the latter case, the dependency of the e.m.f. upon the currents must obviously be given.

Then, in each branch circuit,

$$e - ri - L\frac{di}{dt} - \sum_\delta M_\delta \frac{di_\delta}{dt} = 0, \qquad (1)$$

where e = total impressed e.m.f.; r = resistance; L = inductance, of the circuit or branch of circuit traversed by current i, and M_δ = mutual inductance of this circuit with any circuit in inductive relation thereto and traversed by current i_δ.

The currents in the different branch circuits of the system depend upon each other by Kirchhoff's law,

$$\sum i = 0 \qquad (2)$$

at every branching point of the system.

By equation (2) many of the currents can be eliminated by expressing them in terms of the other currents, but a certain number of independent currents are left.

Let n = the number of independent currents, denoting these currents by i_κ, where $\kappa = 1, 2, \ldots n$. $\qquad$ (3)

Usually, from physical considerations, the number of independent currents of the system, n, can immediately be given.

For these n currents i_κ, n independent differential equations of form (1) can be written down, between the impressed e.m.fs. e_γ or their combinations, and currents which are expressed by the n independent currents i_κ. They are given by applying equation (1) to a closed circuit or ring in the system.

These equations are of the form

$$e_q - \sum_1^n {}_\kappa\, b_\kappa{}^q\, i_\kappa - \sum_1^n {}_\kappa\, c_\kappa{}^q \frac{di_\kappa}{dt} = 0,$$

where $\qquad\qquad q = 1, 2, \ldots n,$ $\qquad\qquad$ (4)

where the n^2 coefficients $b_\kappa{}^q$ are of the dimension of *resistance* and the n^2 coefficients $c_\kappa{}^q$ of the dimension of *inductance*. $\qquad$ (5)

These n simultaneous differential equations of n variables i_κ are integrated by the equations

$$i_\kappa = i_\kappa{}' + \sum_1^m {}_i\, A_i{}^\kappa \varepsilon^{-a_i t}, \qquad\qquad (6)$$

where $i_\kappa{}'$ is the stationary value of current i_κ, reached for $t = \infty$.

Substituting (6) in (4) gives

$$e_q - \sum_1^n {}_\kappa\, b_\kappa{}^q\, i_\kappa{}' - \sum_1^n {}_\kappa\, c_\kappa{}^q \frac{di_\kappa{}'}{dt} - \sum_1^n {}_\kappa\, b_\kappa{}^q \sum_1^m {}_i\, A_i{}^\kappa \varepsilon^{-a_i t} + \sum_1^n {}_\kappa\, c_\kappa{}^q$$

$$\sum_1^m {}_i\, a_i A_i{}^\kappa \varepsilon^{-a_i t} = 0. \qquad\qquad (7)$$

For $t = \infty$, this equation becomes

$$e_q - \sum_1^n {}_\kappa\, b_\kappa{}^q\, i_\kappa{}' - \sum_1^n {}_\kappa\, c_\kappa{}^q \frac{di_\kappa{}'}{dt} = 0. \qquad\qquad (8)$$

These n equations (8) determine the stationary components of the n currents, $i_\kappa{}'$.

Subtracting (8) from (7) gives, for the transient components of currents i_κ,

$$i_\kappa{}'' = \sum_1^m {}_i\, A_i{}^\kappa \varepsilon^{-a_i t}, \qquad\qquad (9)$$

the n equations

$$\sum_1^n {}_\kappa\, b_\kappa{}^q \sum_1^m {}_i\, A_i{}^\kappa \varepsilon^{-a_i t} - \sum_1^n {}_\kappa\, c_\kappa{}^q \sum_1^m {}_i\, a_i A_i{}^\kappa \varepsilon^{-a_i t} = 0. \qquad (10)$$

Reversing the order of summation in (10) gives

$$\sum_{1}^{m} i \; \varepsilon^{-a_i t} \sum_{1}^{n} \kappa \; A_i^{\kappa} \, (b_{\kappa}{}^{q} - a_i c_{\kappa}{}^{q}) = 0. \tag{11}$$

The n equations (11) must be identities, that is, the coefficients of $\varepsilon^{-a_i t}$ must individually disappear. Each equation (11) thus gives m equations between the constants a, A, b, c, for $i = 1$, $2, \ldots m$, and since n equations (11) exist, we get altogether mn equations of the form

$$\left. \begin{aligned} \sum_{1}^{n} \kappa \; A_i^{\kappa} \, (b_{\kappa}{}^{q} - a_i c_{\kappa}{}^{q}) = 0, \\[6pt] \text{where} \qquad\qquad\qquad\qquad\qquad\qquad \\[-18pt] q = 1, 2, 3, \ldots n \quad \text{and} \quad i = 1, 2, 3, \ldots m. \end{aligned} \right\} \tag{12}$$

In addition hereto, the n terminal conditions, or values of current i_{κ}'' for $t = 0$: $i_{\kappa}{}^{0}$, give by substitution in (9) n further equations,

$$\dot{i}_{\kappa}{}^{0} = \sum_{1}^{m} i \; A_i^{\kappa} . \tag{13}$$

There thus exist $(mn + n)$ equations for the determination of the mn constants A_i^{κ} and the m constants a_i, or altogether $(mn + m)$ constants. That is,

$$m = n \tag{14}$$

and

$$\dot{i}_{\kappa} = \dot{i}_{\kappa}' + \sum_{1}^{n} i \; A_i^{\kappa} \, \varepsilon^{-a_i t}, \tag{15}$$

where

$$\left. \sum_{1}^{n} \kappa \; A_i^{\kappa} \, (b_{\kappa}{}^{q} - a_i c_{\kappa}{}^{q}) = 0; \right. \tag{16}$$

$$\left. \sum_{1}^{n} i \; A_i^{\kappa} = \dot{i}_{\kappa}{}^{0}; \right. \tag{17}$$

$$\left. \begin{aligned} q = 1, 2, \ldots n, \\ \kappa = 1, 2, \ldots n, \\ i = 1, 2, \ldots n. \end{aligned} \right\} \tag{18}$$

and

Each of the n sets of n linear homogeneous equations in A_i^κ (16) which contains the same index i gives by elimination of A_i^κ the same determinant:

$$\|b_\kappa^q - a_i c_\kappa^q\| = \begin{vmatrix} b_1^1 - a_i c_1^1, & b_1^2 - a_i c_1^2, & b_1^3 - a_i c_1^3 & \ldots & b_1^n - a_i c_1^n \\ b_2^1 - a_i c_2^1, & b_2^2 - a_i c_2^2, & b_2^3 - a_i c_2^3 & \ldots & b_2^n - a_i c_2^n \\ b_3^1 - a_i c_3^1, & b_3^2 - a_i c_3^2, & b_3^3 - a_i c_3^3 & \ldots & b_3^n - a_i c_3^n \\ \cdot & \cdot & \cdot & \cdots & \cdot \\ b_n^1 - a_i c_n^1, & b_n^2 - a_i c_n^2, & b_n^3 - a_i c_n^3 & \ldots & b_n^n - a_i c_n^n \end{vmatrix} = 0. \quad (19)$$

Thus the n values of a_i are the n roots of the equation of nth degree (19), and determined by solving this equation.

Substituting these n values of a_i in the equations (16) gives n^2 linear homogeneous equations in A_i^κ, of which $n\,(n-1)$ are independent equations, and these $n\,(n-1)$ independent equations together with the n equations (17) give the n^2 linear equations required for the determination of the n^2 constants A_i^κ.

The problem of determining the equations of the phenomena in starting, or in any other way changing the circuit conditions, in a general system containing only resistance and inductance, with n independent currents and such impressed e.m.fs., e_γ, that the equations of stationary condition,

$$i_\kappa' = f_\kappa(t),$$

can be solved, still depends upon the solution of an equation of nth degree, in the exponents a_i of the exponential functions which represent the transient term.

96. As an example of the application of this method may be considered the following case, sketched diagrammatically in Fig. 42:

An alternator of e.m.f. $E \cos(\theta - \theta_0)$ feeds over resistance r_1 the primary of a transformer of mutual reactance x_m. The secondary of this transformer feeds over resistances r_2 and r_3 the primary of a second transformer of mutual reactance x_{m_0}, and the secondary of this second transformer is closed by resistance r_4. Across the circuit between the two transformers and the two resistances r_2 and r_3, is connected a continuous-current e.m.f., e_0, as a battery, in series with an inductive reactance x. The transformers obviously must be such as not to be saturated magnetically by the component of continuous current which traverses them, must for instance be open core transformers.

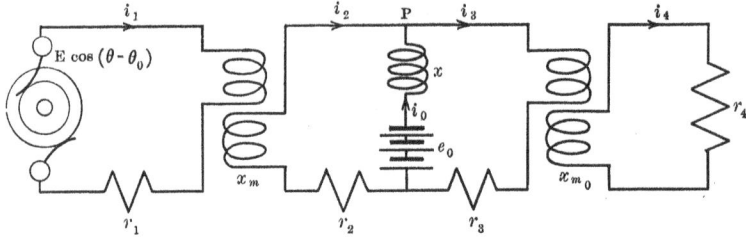

Fig. 42.   Alternating-current circuit containing mutual and self-inductive reactance, resistance and continuous e.m.f.

Let i_1, i_2, i_0, i_3, i_4 = currents in the different circuits; then, at the dividing point P, by equation (2) we have

$$i_0 + i_2 - i_3 = 0;$$

hence, $$i_0 = i_3 - i_2,$$ (20)

leaving four independent currents i_1, i_2, i_3, i_4.

This gives four equations (4):

$$E \cos (\theta - \theta_0) - r_1 i_1 - x_m \frac{di_2}{d\theta} = 0,$$

$$- e_0 - r_2 i_2 - x_m \frac{di_1}{d\theta} + x \left(\frac{di_3}{d\theta} - \frac{di_2}{d\theta}\right) = 0,$$

$$e_0 - r_3 i_3 - x_{m_0} \frac{di_4}{d\theta} - x \left(\frac{di_3}{d\theta} - \frac{di_2}{d\theta}\right) = 0,$$ (21)

and

$$- r_4 i_4 - x_{m_0} \frac{di_3}{d\theta} = 0.$$

If now i_1', i_2', i'_3, i_4' are the permanent terms of current, by substituting these into (21) and subtraction, the equations of the transient terms rearranged are:

$q:$ $\kappa = 1$ 2   3   4

$$
\left.
\begin{array}{l}
1 \quad r_1 i_1 \quad + x_m \dfrac{di_2}{d\theta} \qquad\qquad\qquad = 0, \\[2mm]
2 \quad x_m \dfrac{di_1}{d\theta} + r_2 i_2 + x \dfrac{di_2}{d\theta} \quad - x \dfrac{di_3}{d\theta} \qquad = 0, \\[2mm]
3 \qquad\qquad - x \dfrac{di_2}{d\theta} + r_3 i_3 + x \dfrac{di_3}{d\theta} \quad + x_m \dfrac{di_4}{d\theta} = 0, \\[2mm]
4 \qquad\qquad\qquad\qquad x_m \dfrac{di_3}{d\theta} + r_4 i_4 \qquad = 0.
\end{array}
\right\} \qquad (22)
$$

These equations integrated by

$$
i_\kappa{}'' = \sum_1^4 i\, A_i^\kappa \, \varepsilon^{-\alpha_i \theta} \tag{23}
$$

give for the determination of the exponents a_i the determinant (19):

$$
\begin{vmatrix}
r_1 & -ax_m & 0 & 0 \\
-ax_m & r_2 - ax & ax & 0 \\
0 & ax & r_3 - ax & -ax_{m_0} \\
0 & 0 & -ax_{m_0} & r_4
\end{vmatrix} = 0; \tag{24}
$$

or, resolved,

$$
f = a^4 x_m{}^2 x_{m_0}{}^2 + a^3 x \, (x_m{}^2 r_4 + x_{m_0}{}^2 r_1) - a^2 \, (x_m{}^2 r_3 r_4 + x_{m_0}{}^2 r_1 r_2)
$$
$$
- a x r_1 r_4 \, (r_2 + r_3) + r_1 r_2 r_3 r_4 = 0. \tag{25}
$$

Assuming now the numerical values,

$$
\left.
\begin{array}{ll}
r_1 = 1 & x_m = 10 \\
r_2 = 1 & x_{m_0} = 10 \\
r_3 = 1 & x = 100 \\
r_4 = 10 &
\end{array}
\right\} \tag{26}
$$

equation (25) gives

$$
f = a^4 + 11\, a^3 - 0.11\, a^2 - 0.2\, a + 0.001 = 0. \tag{27}
$$

The sixteen coefficients,

$$
A_i^\kappa, \quad i = 1, 2, 3, 4, \quad k = 1, 2, 3, 4,
$$

are now determined by the 16 independent linear equations (12) and (13).

(B) CIRCUITS CONTAINING RESISTANCE, SELF-INDUCTANCE, MUTUAL INDUCTANCE AND CAPACITY.

97. The general method of dealing with such a system is the same as in (A).

Kirchhoff's equation (1) is of the form

$$e - ri - L\frac{di}{dt} - \sum_\delta M_\delta \frac{di_\delta}{dt} - \frac{1}{C}\int i\, dt = 0. \qquad (28)$$

Eliminating now all the currents which can be expressed in terms of other currents, by means of equation (2), leaves n independent currents:

$$i_\kappa, \quad \kappa = 1, 2, \ldots n.$$

Substituting these currents i_κ in equations (28) gives n independent equations of the form

$$e_q - \sum_1^n {}_\kappa\, b_\kappa{}^q i_\kappa - \sum_1^n {}_\kappa\, c_\kappa{}^q \frac{di_\kappa}{dt} - \sum_1^n {}_\kappa\, g_\kappa{}^q \int i_\kappa\, dt = 0. \qquad (29)$$

Resolving these equations for $\int i_\kappa\, dt$ gives

$$e_\kappa' = \frac{1}{C_\kappa}\int i_\kappa\, dt = \sum e + \sum bi + \sum c\frac{di}{dt} \qquad (30)$$

as the equations of the potential differences at the condensers.

Differentiating (29) gives

$$\frac{de_q}{dt} - \sum_1^n {}_\kappa\, g_\kappa{}^q\, i_\kappa - \sum_1^n {}_\kappa\, b_\kappa{}^q \frac{di_\kappa}{dt} - \sum_1^n {}_\kappa\, c_\kappa{}^q \frac{d^2 i_\kappa}{dt^2} = 0, \qquad (31)$$

where

$$q = 1, 2, \ldots n.$$

By the same reasoning as before, the solution of these equations (31) can be split into two components, a permanent term,

$$i_\kappa' = f(t), \qquad (32)$$

and a transient term, which disappears for $t = \infty$, and is given by the n simultaneous differential equations of second order, thus:

$$\sum_1^n {}_\kappa \left\{ g_\kappa{}^q i_\kappa + b_\kappa{}^q \frac{di_\kappa}{dt} + c_\kappa{}^q \frac{d^2 i_\kappa}{dt^2} \right\} = 0. \qquad (33)$$

These equations are integrated by

$$i_\kappa = \sum_1^m{}^i A_i{}^\kappa \varepsilon^{-a_i t}. \qquad (34)$$

Substituting (34) in (33) gives

$$\sum_1^n{}^\kappa \sum_1^m{}^i A_i{}^\kappa \varepsilon^{-a_i t} \left\{ g_\kappa{}^q - a_i b_\kappa{}^q + a_i{}^2 c_\kappa{}^q \right\}, \qquad (35)$$

where
$$\left. \begin{array}{l} q = 1, 2, \ldots n, \\ \kappa = 1, 2, \ldots n, \\ i = 1, 2, \ldots m. \end{array} \right\} \qquad (36)$$
and

Reversing in these n equations the order of summation,

$$\sum_1^m{}^i \varepsilon^{-a_i t} \sum_1^n{}^\kappa A_i{}^\kappa \left\{ g_\kappa{}^q - a_i b_\kappa{}^q + a_i{}^2 c_\kappa{}^q \right\} = 0, \qquad (37)$$

and this gives, as identity, the mn equations for the determination of the constants:

$$\left. \begin{array}{c} \sum_1^n{}^\kappa A_i{}^\kappa \left\{ g_\kappa{}^q - a_i b_\kappa{}^q + a_i{}^2 c_\kappa{}^q \right\} = 0, \\ q = 1, 2, \ldots n \quad \text{and} \quad i = 1, 2, \ldots m. \end{array} \right\} \qquad (38)$$
where

In addition to these mn equations (38), two sets of terminal conditions exist, depending respectively on the instantaneous current and the instantaneous condenser potential at the moment of start.

The current is

$$i_\kappa = i_\kappa{}' + \sum_1^m{}^i A_i{}^\kappa \varepsilon^{-a_i t} \qquad (39)$$

and the condenser potential of the circuit q is

$$e_q{}' = \sum_1^n{}^\kappa g_\kappa{}^q \int i_\kappa \, dt = e_q - \sum_1^n{}^\kappa b_\kappa{}^q i_\kappa - \sum_1^n{}^\kappa c_\kappa{}^q \frac{d i_\kappa}{dt} ; \qquad (40)$$

hence, for $t = 0$,

$$i_\kappa{}^0 = i_\kappa{}' + \sum_1^m{}^i A_i{}^\kappa, \qquad (41)$$

where $\qquad\kappa = 1, 2, \ldots n,$

and
$$e_q^0 = e_q - \sum_1^n {}_\kappa \, b_\kappa^{\,q} i_\kappa^0 - \sum_1^n {}_\kappa \, c_\kappa^{\,q} \frac{di_\kappa^0}{dt}, \qquad (42)$$

where, $\qquad q = 1, 2 \ldots n;$

or, substituting (39) in (40), and then putting $t = 0$,

$$e_q^0 = \left[e_q - \sum_1^n {}_\kappa \left\{ b_\kappa^{\,q} i_\kappa' - c_\kappa^{\,q} \frac{di_\kappa'}{dt} \right\} \right]_{t=0}$$
$$- \sum_1^n {}_\kappa \sum_1^m {}_i \, A_i^\kappa (b_\kappa^{\,q} - a_i c_\kappa^{\,q}). \qquad (43)$$

As seen, in (41) and (43), the first term is the instantaneous value of the permanent current i'_κ and condenser potential e_q'.

These two sets of n equations each, given by the terminal conditions of the current, $i'_\kappa = i_\kappa^0$ (42), and condenser potential, $e_q' = e_q^0$ (43), together with the mn equations (38), give a total of $(mn + 2n)$ equations for the determination of the mn constants A_i^κ and the m constants a_i, that is, a total of $(mn + m)$ constants.

From
$$mn + 2n = mn + m$$

it follows that
$$m = 2n. \qquad (44)$$

We have, then, $2n$ constants, a_i, giving the coefficients in the exponents of the $2n$ exponential transient terms, and $2n^2$ coefficients, A_i^κ, and for their determination $2n^2$ equations,

$$\sum_1^n {}_\kappa \, A_i^\kappa \, (g_\kappa^{\,q} - a_i b_\kappa^{\,q} + a_i^2 c_\kappa^{\,q}) = 0, \qquad (45)$$

n equations,

$$\sum_1^{2n} {}_i \, A_i^\kappa = i_\kappa^0, \qquad (46)$$

and n equations,

$$\sum_1^n {}_\kappa \sum_1^{2n} {}_i \, A_i^\kappa \, (b_\kappa^{\,q} - a_i c_\kappa^{\,q}) = k_q^0, \qquad (47)$$

where
$$k_q^{\,0} = \left[e_q - \sum_1^n {}_\kappa \, g_\kappa^{\,q} \int i_\kappa \, dt \right]_{t=o} ; \qquad (48)$$

or the difference between the condenser potential required by the permanent term and the actual condenser potential at time $t = 0$, where

$$\left. \begin{aligned} q &= 1, 2, 3, \ldots n, \\ \kappa &= 1, 2, 3, \ldots n, \\ i &= 1, 2, 3, \ldots 2\,n. \end{aligned} \right\} \qquad (49)$$

and

Eliminating $A_i^{\,\kappa}$ from the equations (45) gives for each of the $2\,n$ sets of n equations which have the same a_i the determinant:

$$\left\| g_\kappa^{\,q} - a_i b_\kappa^{\,q} + a_i^2 c_\kappa^{\,q} \right\| =$$

$$\begin{vmatrix} g_1^{\,1} - a_i b_1^{\,1} + a_i^2 c_1^{\,1}, & g_1^{\,2} - a_i b_1^{\,2} + a_i^2 c_1^{\,2}, & \ldots & g_1^{\,n} - a_i b_1^{\,n} + a_i^2 c_1^{\,n} \\ g_2^{\,1} - a_i b_2^{\,1} + a_i^2 c_2^{\,1}, & g_2^{\,2} - a_i b_2^{\,2} + a_i^2 c_2^{\,2}, & \ldots & g_2^{\,n} - a_i b_2^{\,n} + a_i^2 c_2^{\,n} \\ g_3^{\,1} - a_i b_3^{\,1} + a_i^2 c_3^{\,1}, & g_3^{\,2} - a_i b_3^{\,2} + a_i^2 c_3^{\,2}, & \ldots & g_3^{\,n} - a_i b_3^{\,n} + a_i^2 c_3^{\,n} \\ \cdot & \cdot & \cdot & \cdot \\ \cdot & \cdot & \cdot & \cdot \\ g_n^{\,1} - a_i b_n^{\,1} + a_i^2 c_n^{\,1}, & g_n^{\,2} - a_i b_n^{\,2} + a_i^2 c_n^{\,2}, & \ldots & g_n^{\,n} - a_i b_n^{\,n} + a_i^2 c_n^{\,n} \end{vmatrix} = 0. (50)$$

The $2\,n$ values of a_i thus are the roots of an equation of $2\,n$th order.

Substituting these values of a_i in equations (45), (46), (47), leaves $2\,n\,(n-1)$ independent equations (45) and $2\,n$ independent equations (46) and (47), or a total of $2\,n^2$ linear equations, for the determination of the $2\,n^2$ constants $A_i^{\,\kappa}$, which now can easily be solved.

The roots of equation (50) may either be real or may be complex imaginary, and in the latter case each pair of conjugate roots gives by elimination of the imaginary form an electric oscillation.

That is, the solution of the problem of n independent circuits leads to n transient terms, each of which may be either an oscillation or a pair of exponential functions.

98. The preceding discussion gives the general method of the determination of the transient phenomena occurring in any system or net work of circuits containing resistances, self-inductances and mutual inductances and capacities, and impressed and counter e.m.fs. of any frequency or wave shape, alternating or continuous.

It presupposes, however,

(1) That the solution of the system for the permanent terms of currents and e.m.fs. is given.

(2) That, if the impressed e.m.fs. contain transient terms depending upon the currents in the system, these transient terms of impressed or counter e.m.fs. are given as linear functions of the currents or of their differential coefficients, that is, the rate of change of the currents.

(3) That resistance, inductance, and capacity are constant quantities, and for instance magnetic saturation does not appear.

The determination of the transient terms requires the solution of an equation of $2\,n$th degree, which is lowered by one degree for every independent circuit which contains no capacity.

Thus, for instance, a divided circuit having capacity in either branch leads to a quartic equation. A transmission line loaded with inductive or non-inductive load, when representing the capacity of the line by a condenser shunted across its middle, leads to a cubic equation.

CHAPTER XII.

MAGNETIC SATURATION AND HYSTERESIS IN ALTERNATING-CURRENT CIRCUITS.

99. If an alternating e.m.f. is impressed upon a circuit containing resistance and inductance, the current and thereby the magnetic flux produced by the current assume their final or permanent values immediately only in case the circuit is closed at that point of the e.m.f. wave at which the permanent current is zero. Closing the circuit at any other point of the e.m.f. wave produces a transient term of current and of magnetic flux. So for instance, if the circuit is closed when the current i should have its negative maximum value $-I_0$, and therefore the magnetic flux and the magnetic flux density also be at their negative maximum value $-\Phi_0$ and $-\mathfrak{B}_0$ — that is, in an inductive circuit, near the zero value of the decreasing e.m.f. wave — during the first half wave of e.m.f. the magnetic flux, which generates the counter e.m.f., should vary from $-\Phi_0$ to $+\Phi_0$, or by $2\Phi_0$; hence, starting with 0, to generate the same counter e.m.f., it must rise to $+2\Phi_0$, that is, twice its permanent value, and so the current i also rises, at constant inductance L, from zero to twice its maximum permanent value, $2I_0$. Since the e.m.f. consumed by the resistance during the variation from 0 to $2I_0$ is greater than during the normal variation from $-I_0$ to $+I_0$, less e.m.f. is to be generated by the change of magnetic flux, that is, the magnetic flux does not quite rise to $2\Phi_0$, but remains below this value the more, the higher the resistance of the circuit. During the next half wave the e.m.f. has reversed, but the current is still mostly in the previous direction, and the generated e.m.f. thus must give the resistance drop, that is, the total variation of magnetic flux must be greater than $2\Phi_0$, the more, the higher the resistance. That is, starting at a value somewhat below $2\Phi_0$, it decreases below zero, and reaches a negative value. During the third half wave the magnetic flux, starting not at zero as in the first half wave, but at a negative

179

value, thus reaches a lower positive maximum, and thus gradually, at a rate depending upon the resistance of the circuit, the waves of magnetic flux Φ, and thereby current i, approach their final permanent or symmetrical cycles.

100. In the preceding, the assumption has been made that the magnetic flux, Φ, or the flux density, $\mathfrak{B}$, is proportional to the current, or in other words, that the inductance, L, is constant. If the magnetic circuit interlinked with the electric circuit contains iron, and especially if it is an iron-clad or closed magnetic circuit, as that of a transformer, the current is not proportional to the magnetic flux or magnetic flux density, but increases for high values of flux density more than proportional, that is, the flux density in the iron reaches a finite limiting value. In the case illustrated above, the current corresponding to double the normal maximum magnetic flux, Φ_0, or flux density, $\mathfrak{B}_0$, may be many times greater than twice the normal maximum current, I_0. For instance, if the maximum permanent current is $I_0 = 4.5$ amperes, the maximum permanent flux density, $\mathfrak{B}_0 = 10,000$, and the circuit closed, as above, at that point of the e.m.f. wave where the flux density should have its negative maximum, $-\mathfrak{B}_0 = -10,000$, but the actual flux density is 0, during the first half wave of e.m.f., the flux density, when neglecting the resistance of the electric circuit, should rise from 0 to $2\mathfrak{B}_0 = 20,000$, and at this high value of saturation the corresponding current maximum would be, by the magnetic cycle, Fig. 43, 200 amperes, that is, not twice but 44.5 times the normal value. With such excessive values of current, the e.m.f. consumed by resistance would be in general considerable, and the e.m.f. consumed by inductance, and therefore the variation of magnetic flux density, considerably decreased, that is, the maximum magnetic flux density would not rise to 20,000, but remain considerably below this value. The maximum current, however, would be still very much greater than twice the normal maximum. That is, in an iron-clad circuit, in starting, the transient term of current may rise to values relatively very much higher than in air magnetic circuits. While in the latter it is limited to twice the normal value, in the iron-clad circuit, if the magnetic flux density reaches into the range of magnetic saturation, very much higher values of transient current are found. Due to the far greater effect of the resistance with such excessive values of current, the transient term of current during the first half waves decreases at a more rapid rate; due to the lack of proportionality between current and magnetic flux density, the transient term does not follow the exponential law any more.

101. In an iron-clad magnetic circuit, the current is not only not proportional to the magnetic flux density, but the same magnetic flux density can be produced by different currents, or with the same current the flux density can have very different values, depending on the point of the hysteresis cycle. Therefore the magnetic flux density for zero current may equal zero, or, on the decreasing branch of the hysteresis cycle, Fig. 43, may be $+ 7600$, or, on the increasing branch, $- 7600$. Thus, when closing the electric circuit energizing an iron-clad magnetic circuit, as a transformer, at the moment of zero current, the magnetic flux density may not be zero, but may still have a high value, as remanent magnetism. For instance, closing the circuit at the point of the e.m.f. wave where the permanent wave of magnetic flux density would have its negative maximum value, $- \mathcal{B}_0 = - 10,000$, the actual density at this moment may be $\mathcal{B}_r = + 7600$, the remanent magnetism of the cycle. During the first half wave of impressed e.m.f. the variation of flux density by $2\,\mathcal{B}_0$, as required to generate the counter e.m.f., when neglecting the resistance, would bring the positive maximum of flux density up to $\mathcal{B}_r + 2\,\mathcal{B}_0 = 27,600$, requiring 1880 amperes maximum current, or 420 times the normal current. Obviously, no such rise could occur, since the resistance of the circuit would consume a considerable part of the e.m.f., and so lower the flux density by reducing the e.m.f. consumed by inductance.

It is obvious, however, that excessive values of transient current may occur in transformers and other iron-clad magnetic circuits.

102. When disconnecting a transformer, its current becomes zero, that is, the magnetic flux density is left at the value of the remanent magnetism $\pm \mathcal{B}_r$, and during the period of rest more or less decreases spontaneously towards zero. Hence, in connecting a transformer into circuit its flux density may be anywhere between $+ \mathcal{B}_r$ and $- \mathcal{B}_r$. The maximum magnetic flux density during the first half cycle of impressed e.m.f. therefore is produced if the circuit is closed at the moment where the permanent value of the flux density should be a maximum, $\pm\ \mathfrak{B}_0$, and the actual density in this moment is the remanent magnetism in opposite direction, $\mp\ \mathfrak{B}_r$, and the maximum value of density which could occur then is $\pm\ (\mathfrak{B}_r + 2\ \mathfrak{B}_0)$. If therefore the maximum magnetic flux density $\mathfrak{B}_0$ in the transformer is such that $\mathfrak{B}_r + 2\ \mathfrak{B}_0$ is still below saturation, the transient term of current cannot reach abnormal values. At $\mathfrak{B} = 16,000$, the flux density is about at the bend of the saturation curve, and the current still moderate. Estimating $\mathfrak{B}_r = 0.75\ \mathfrak{B}_0$ as approximate value, $\mathfrak{B}_r + 2\ \mathfrak{B}_0 = 16,000$ thus gives $\mathfrak{B} = 5800$, or 37,500 lines of magnetic flux per square inch.

Such low maximum density is uneconomical. However, for $B_r = 0$, which probably more nearly represents the starting conditions of a transformer, which has been disconnected for some time, the limit is $B_0 = 8,000$ or 51,600 lines per square inch, and at least, at 60 cycles, in well designed transformers, the maximum densities do not very much exceed this value. With a large starting current, not only the resistance of the current consumes voltage, but the self-inductive or leakage flux of the transformer, which is essentially an air flux, and as such not limited by saturation, also consumes voltage. Furthermore, the terminal voltage usually, more or less, drops by the impedance between transformer and generating system, and as a result, at least in 60 cycle circuits, this phenomenon is not serious.

103. Since the relation between the current, i, and the magnetic flux density, $\mathfrak{B}$, is empirically given by the magnetic cycle of the material, and cannot be expressed with sufficient accuracy by a mathematical equation, the problem of determining the transient starting current of a transformer is investigated by constructing the curves of current and magnetic flux density.

Let the normal magnetic cycle of a transformer be represented by the dotted curve in Figs. 43 and 44; the characteristic points are: the maximum values, $\pm\ \mathfrak{B}_0 = \pm\ 10,000$; the remanent values, $\pm\ \mathfrak{B}_r = \pm\ 7600$, and the maximum exciting current, $i_m = \pm\ 4.5$ amp.

At very high values of flux density an appreciable part of the total magnetic flux Φ may be carried through space, outside of the iron, depending on the construction of the transformer. The most convenient way of dealing with such a case is to resolve the magnetic flux density, $\mathfrak{B}$, in the iron into the "metallic

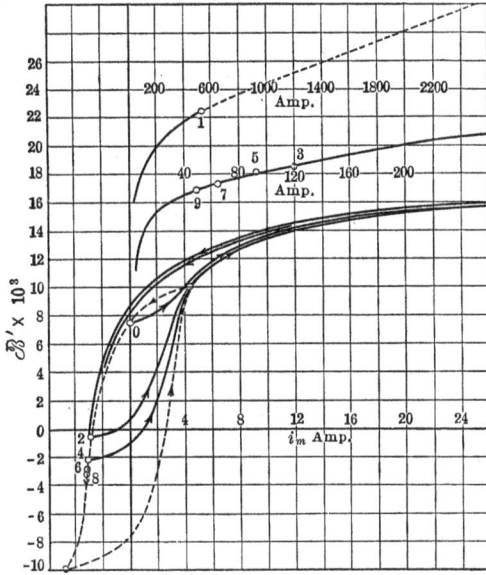

Fig. 43. Magnetic cycle of a transformer starting with low stray field.

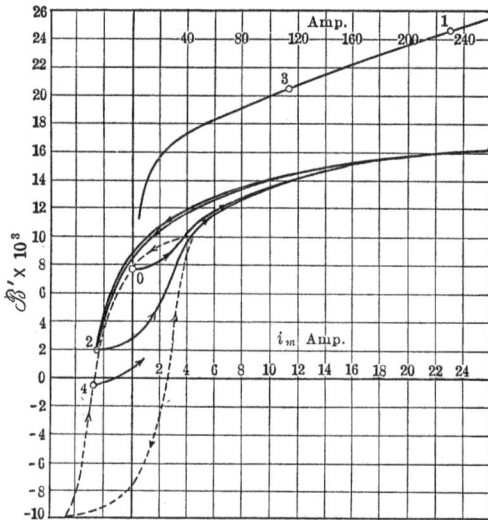

Fig. 44. Magnetic cycle of a transformer starting with high stray field.

flux density," $\mathcal{B}' = \mathcal{B} - \mathcal{H}$, which reaches a finite limiting value, and the density in space, $\mathcal{H}$. The total magnetic flux then consists of the flux carried by the molecules of the iron, $\Phi' = A'\mathcal{B}'$, where A' is the section of the iron circuit, and the space flux, $\Phi'' = A''\mathcal{H}$, where A'' is the total section interlinked with the electric circuit, including iron as well as other space.

$$\Phi = \Phi' + \Phi'' = A'\mathcal{B}' + A''\mathcal{H}.$$

If then $A'' = kA'$, that is, the total space inside of the coil is k times the space filled by the iron, we have

$$\Phi = A' (\mathcal{B}' + k\mathcal{H}),$$

or the total magnetic flux even in a case where considerable stray field exists, that is, magnetic flux can pass also outside of

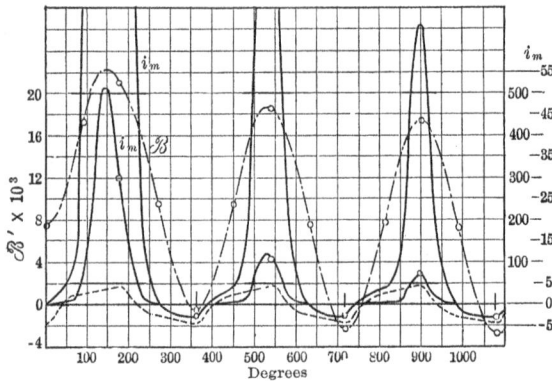

Fig. 45. Starting current of a transformer. Low stray field.

the iron, can be calculated by considering only the iron section as carrying magnetic flux, but using as curve of magnetic flux density not the usual curve,

$$\mathcal{B} = \mathcal{B}' + \mathcal{H},$$

but a curve derived therefrom,

$$\mathcal{B} = \mathcal{B}' + k\mathcal{H},$$

where k = ratio of total section to iron section.

This, for instance, is the usual method of calculating the m.m.f. consumed in the armature teeth of commutating machines at very high saturations.

In investigating the transient transformer starting current, the magnetic density curve thus is corrected for the stray field.

Figs. 43 and 45 correspond to $k = 3$, or a total effective air section equal to three times the iron section, that is, $\mathscr{B} = \mathscr{B}' + 3\mathscr{K}$.

Figs. 44 and 46 correspond to $k = 25$, or a section of stray field equal to 25 times the iron section, that is, $\mathscr{B} = \mathscr{B}' + 25\mathscr{K}$.

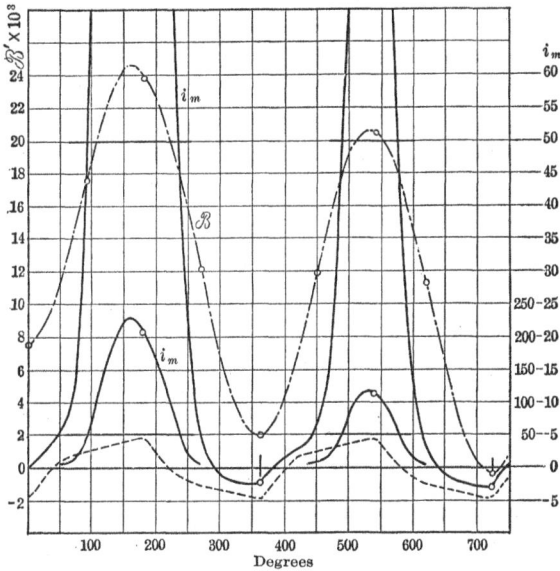

Fig. 46.   Starting current of a transformer.   High stray field.

104. At very high values of current the resistance consumes a considerable voltage, and thus reduces the e.m.f. generated by the magnetic flux, and thereby the maximum magnetic flux and transient current.   The resistance, which comes into consideration here, is the total resistance of the transformer primary circuit plus leads and supply lines, back to the point where the voltage is kept constant, as generator, busbars, or supply main.

Assuming then at full load of $i_m = 50$ amperes effective in the transformer, a resistance drop of 8 per cent, or the voltage consumed by the resistance, as $e_r = 0.08$ of the impressed e.m.f.

Let now the remanent magnetic flux density be $\mathscr{B}_r = + 7600$, and the circuit be closed at the moment $\theta = 0$, where the flux density should be $\mathcal{B} = -\mathcal{B}_0 = -10,000$; then the impressed e.m.f. is given by

$$e = -E \sin \theta = E \frac{d}{d\theta} (\cos \theta). \qquad (1)$$

It is, however,

$$e = A \frac{d\mathcal{B}}{d\theta} + Ci, \qquad (2)$$

where A and C are constants; that is, the impressed e.m.f., e, is consumed by the self-inductance, or the e.m.f. generated by the changing magnetic density, which is proportional to $\dfrac{d\mathcal{B}}{d\theta}$, and by the voltage consumed by the resistance, which is proportional to the current i.

Combining (1) and (2) gives

$$A \frac{d\mathcal{B}}{d\theta} + Ci = E \frac{d \cos \theta}{d\theta}. \qquad (3)$$

However, at full load, we have $\dfrac{E}{\sqrt{2}} = $ effective impressed e.m.f. and $i_m = 50$ amperes $= $ effective current; hence

$$Ci_m = 50\,C = \text{e.m.f. consumed by resistance,}$$

and since this equals $e_r = 0.08$ of impressed e.m.f.,

$$Ci_m = \frac{e_r}{\sqrt{2}},$$

or

$$50\,C = \frac{0.08\,E}{\sqrt{2}},$$

or

$$\frac{C}{E} = \frac{e_r}{Ei_m\sqrt{2}} = \frac{0.08}{50\sqrt{2}} = 0.00113. \qquad (4)$$

From (3) follows

$$d\mathcal{B} = \frac{E}{A} d \cos \theta - \frac{Ci}{A} d\theta \qquad (5)$$

and

$$\int_0^{\frac{\pi}{2}} d\mathcal{B} = \frac{E}{A} \int_0^{\frac{\pi}{2}} d \cos \theta - \frac{C}{A} \int_0^{\frac{\pi}{2}} i\, d\theta;$$

hence, for $i = 0$, or negligible resistance drop, that is, permanent condition,

$$\mathcal{B}_0 = \frac{E}{A} = 10{,}000. \tag{6}$$

Multiplying (4) and (6) gives

$$\frac{C}{A} = \frac{e_r \mathcal{B}_0}{E i_m \sqrt{2}} = 11.3, \tag{7}$$

and substituting (6) and (7) in (5) gives

$$d\mathcal{B} = \mathcal{B}_0 d \cos \theta - i \frac{e_r \mathcal{B}_0}{E i_m \sqrt{2}} d\theta$$

$$= 10{,}000 \, d \cos \theta - 11.3 \, i \, d\theta. \tag{8}$$

Changing now from differential to difference, that is, replacing, as approximation, d by Δ, gives

$$\Delta\mathcal{B} = \mathcal{B}_0 \Delta \cos \theta - i \frac{e_r \mathcal{B}_0}{E i_m \sqrt{2}} \Delta\theta$$

$$= 10{,}000 \, \Delta \cos \theta - 11.3 \, i\Delta\theta. \tag{9}$$

Assuming now

$$\Delta\theta = 10° = 0.175 \tag{10}$$

gives for the increment of magnetic flux density during 10° change of angle the value

$$\Delta\mathcal{B} = 10{,}000 \, \Delta \cos \theta - 2 \, i \tag{11}$$

and

$$\mathcal{B} = \mathcal{B}' + \Delta\mathcal{B}$$

$$= \mathcal{B}' + 10{,}000 \, \Delta \cos \theta - 2 \, i. \tag{12}$$

From equation (12) the instantaneous values of magnetic flux density $\mathcal{B}$, and therefrom, by the magnetic cycles, Figs. 43 and 44, respectively, the values of current i are calculated, by starting, for $\theta = 0$, with the remanent density $\mathcal{B}' = \mathcal{B}_r = 7600$, adding thereto the change of cosine, $10{,}000 \, \Delta \cos \theta$, which gives a value $\mathcal{B}_1 = \mathcal{B}' + 10{,}000 \, \Delta \cos \theta$, taking the corresponding value of i from the hysteresis cycle, Figs. 43 and 44, subtracting $2 \, i$ from $\mathcal{B}_1$, and then correcting i for the value corresponding to $\mathcal{B} = \mathcal{B}_1 - 2 \, i$.

The quantity $2 \, i$ is appreciable only during the range of the curve where i is very large.

105. The following table is given to illustrate the beginning of the calculation of the curve for low stray field.

STARTING CURRENT OF A TRANSFORMER.

$\theta°$	$\cos\theta$	$\Delta\mathcal{B}_1 = 10\,\Delta\cos\theta$	$\mathcal{B}_1 = \mathcal{B} + \Delta\mathcal{B}_1$	i	$D_1 = 2i \times 10^{-3}$	$\mathcal{B}$
0	+1.00	....	7.6	0	....	....
10	0.98	+0.2	7.8	1.0	....	....
20	0.94	0.4	8.2	1.9	....	....
30	0.87	0.7	8.9	2.9	....	....
40	0.77	1.0	9.9	3.8	....	....
50	0.64	1.3	11.2	5.2	....	....
60	0.50	1.4	12.6	7.3	....	....
70	0.34	1.6	14.2	12.0	....	....
80	+0.17	1.7	15.9	27	....	....
90	0	1.7	17.6	70	0.1	17.5
100	−0.17	1.7	19.2	138	0.3	18.9
110	0.34	1.7	20.6	220	0.45	20.15
120	0.50	1.6	21.75	270	0.55	21.2
130	0.64	1.4	22.6	450	0.9	21.7
140	0.77	1.3	23.0	510	1.0	22.0
150	0.87	1.0	23.0	510	1.0	22.0
160	0.94	0.7	22.7	440	0.9	21.8
170	0.98	0.4	22.2	350	0.7	21.5
180	1.00	+0.2	21.7	250	0.5	21.2
190	0.98	−0.2	21.0	200	0.4	20.6
200	0.94	0.4	20.2	180	0.35	19.85
210	0.87	0.7	19.15	130	0.25	18.9
220	0.77	1.0	17.9	76	0.15	17.75
230	0.64	1.3	16.45	40	0.2	16.25
240	0.50	1.4	14.85	14	0.05	14.8
250	0.34	1.6	13.2	7	....	....
260	−0.17	1.7	11.5	3.2	....	....
270	0	1.7	9.8	−1.0	....	....
280	+0.17	1.7	8.1	− .4	....	....
290	0.34	1.7	6.4	−1.3	....	....
300	0.50	1.6	4.8	−1.9	....	....
310	0.64	1.4	3.4	−2.2	....	....
320	0.77	1.3	2.1	−2.5	....	....
330	0.87	1.0	1.1	−2.6	....	....
340	0.94	0.7	.4	−2.75	....	....
350	0.98	0.4	0	−2.8	....	....
360	+1.00	−0.2	− .2	−2.8	....	....
370	0.98	+0.2	0	−1.3	....	....
380	0.94	0.4	+0.4	−0.5	....	....
390	0.87	0.7	1.1	+0.3	....	....
400	0.77	1.0	2.1	1.0	....	....
410	0.64	1.3	3.5	1.7	....	....
420	0.50	1.4	4.9	2.2	....	....
430	0.34	1.6	6.5	2.7	....	....
440	+0.17	1.7	8.2	3.3	....	....
450	0	1.7	9.9	4.3	....	....
460	−0.17	1.7	11.6	6.2	....	....
470	−0.34	1.7	13.3	9.5	....	....
480	−0.50	1.6	14.9	16.5	....	....

The first column gives angle θ,

The second column gives cos θ,

The third column gives $\Delta\mathfrak{B}_1 = 10 \,\Delta \cos \theta$, in kilolines per sq. cm.,

The fourth column gives $\mathfrak{B}_1 = \mathfrak{B} + \Delta\mathfrak{B}_1$,

The fifth column gives i,

The sixth column gives $D_1 = 2\,i \times 10^{-3}$, and

The seventh column gives $\mathfrak{B} = \mathfrak{B}_1 - D_1$,

i in the fifth column being chosen, by trial, so as to correspond, on the hysteresis cycles, not to $\mathfrak{B}_1$, but to $\mathfrak{B} = \mathfrak{B}_1 - D_1$.

These values are recorded as magnetic cycles on Figs. 43 and 44, and as waves of flux density, current, etc., in Figs. 45 and 46.

The maximum values of successive half waves are:

A. Low Stray Field. $k = 3$			B. High Stray Field. $k = 25$		
θ	$\mathfrak{B}$	i_m	θ	$\mathfrak{B}$	i_m
0	7.6	0	0	7.6	0
145°	22.0	510	160°	24.6	230
360°	− .2	−2.8	360°	+2.0	−2.4
530°	18.6	120	530°	20.7	117
720°	−2.2	−2.9	720°	− .4	−2.7
900°	18.0	92	....	....	....
1080°	−2.8	−3.0	....	....	....
1260°	17.4	66	....	....	....
1440°	−3.1	−3.0	....	....	....
1620°	16.9	50	....	....	....
Permanent	±10.0	±4.5	....	±10.0	±4.5

As seen, the maximum value of current during the first cycle, 510, is more than one hundred times the final value 4.5, and more than 7 times the maximum value of the full-load current, $50\sqrt{2} = 70.7$ amperes, and the transient current falls below full-load current only in the fourth cycle. That is, the excessive value of transient current in an ironclad circuit lasts for a considerable number of cycles.

In the presence of iron in the magnetic field of electric circuits, transient terms of current may thus occur which are very large compared with the transient terms in ironless reactors, which do not follow the exponential curve, can usually not be calculated by general equations, but require numerical investigation by the use of the magnetic cycles of the iron.

These transient terms lead to excessive current values only if the normal magnetic flux density exceeds half the saturation value of the iron, and so are most noticeable in 25-cycle circuits.

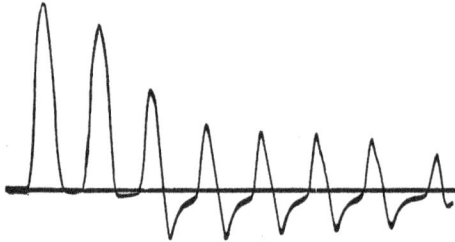

Fig. 47.   Starting current of a 25-cycle transformer.

As illustration is shown, in Fig. 47, an oscillogram of the starting current of a 25-cycle transformer having a resistance in the supply circuit somewhat smaller than in the above instance, thus causing a still longer duration of the transient term of excessive current.

These starting transients of the ironclad inductance at high density are unsymmetrical waves, that is, successive half waves have different shapes, and when resolved into a trigonometric series, would give even harmonics as well as the odd harmonics.

Thus the first wave of Fig. 45 can, when neglecting the transient factor, be represented by the series:

$$i = + 108.3 - 183.8 \cos (\theta + 28.0°)$$
$$+ 112.4 \cos 2 (\theta + 29.8°) - 53.1 \cos 3 (\theta + 33.3°)$$
$$+ 27.2 \cos 4 (\theta + 39.1°) - 18.4 \cos 5 (\theta + 38.1°)$$
$$+ 13.6 \cos 6 (\theta + 33.4°) - 8.1 \cos 7 (\theta + 32.7°)$$

or, substituting: $\theta = \beta + 150°$, gives:

$$e = E \sin (\beta + 150°)$$
$$i = 108.3 + 183.8 \cos (\beta - 2.0°) + 112.4 \cos 2 (\beta - 0.2°)$$
$$+ 53.1 \cos 3 (\beta + 3.3°) + 27.2 \cos 4 (\beta + 9.1°)$$
$$+ 18.4 \cos 5 (\beta + 8.1°) + 13.6 \cos 6 (\beta + 3.4°)$$
$$+ 8.1 \cos 7 (\beta + 2.7°).$$

106. An approximate estimate of the initial value of the starting current of the transformer, at least of its magnitude under

conditions where it is very large, can be made by separately considering the iron flux, that is the flux B_0 carried by the iron proper,
which reaches a finite saturation value of about $S = 21,000$ lines
per cm.[2], and the air flux or space flux, which is proportional to
the current. Also neglecting the remanent magnetism—which
usually is small—that is, assuming the initial magnetic cycle
performed between zero and a maximum of flux density.

Let

$\qquad i =$ maximum value of current

$\qquad \Phi =$ total maximum magnetic flux

$\qquad n =$ number of turns of transformer circuit

$\qquad Z_1 = r_1 + jx_1 =$ impedance of transformer circuit, where

$\qquad x_1 =$ reactance of leakage flux

$\qquad Z_2 = r_2 + jx_2 =$ impedance of circuit between transformer and source of constant voltage e_0.

$\qquad e_0 =$ effective value of this constant voltage.

The voltage consumed by a variation of current between 0
and i, corresponding to an effective value of

$$i' = \frac{i}{2\sqrt{2}}$$

then is:

in the supply lines:

$$E_2 = i'\,(r_2 + jx_2)$$
$$= \frac{i}{2\sqrt{2}}\,(r_2 + jx_2)$$

in the transformer impedance:

$$E_1 = i'\,(r_1 + jx_1)$$
$$= \frac{i}{2\sqrt{2}}\,(r_1 + jx_1)$$

by the magnetic flux Φ, at frequency f:

$$E = j\frac{\pi f n \Phi}{\sqrt{2}}10^{-8}$$

thus the total supply voltage:

$$E_0 = E + E_1 + E_2$$
$$= (r_1 + r_2)\frac{i}{2\sqrt{2}} + j\left[(x_1 + x_2)\frac{i}{2\sqrt{2}} + \frac{\pi f n \Phi}{\sqrt{2}}10^{-8}\right]$$

and, absolute:

$$e_0{}^2 = \frac{1}{8}\{(r_1 + r_2)^2\, i^2 + [(x_1 + x_2)\, i + 2\, \pi f n \Phi 10^{-8}]^2\} \qquad (1)$$

If now:

$$s_1 = \text{magnetic iron section, in cm.}^2,$$

the magnetic flux in the iron proper is:

$$\Phi_1 = s_1 S = 21{,}000 s$$

if

$$s_2 = \text{section, in cm.}^2,$$

$$l_2 = \text{length, in cm.,}$$

of the total magnetic circuit, including iron and air space, the magnetic space flux is

$$\Phi_2 = 0.4\ \pi n i \frac{s_2}{l_2}$$

hence, the total magnetic flux:

$$\Phi = \Phi_1 + \Phi_2$$

$$= s_1 S + 0.4\ \pi n i \frac{s_2}{l_2} \qquad (2)$$

Substituting (2) into (1) gives:

$$e_0{}^2 = \frac{1}{8}\left\{ (r_1 + r_2)^2 i^2 + \left[\left(x_1 + x_2 + 0.8\ \pi^2 f n^2 \frac{s_2}{l_2}\, 10^{-8} \right) i + \right.\right.$$
$$\left.\left. 2\ \pi f n s_1 S \right]^2 \right\} \qquad (3)$$

From equation (3) follows the value of i, the maximum initial or starting current of the transformer or reactor.

107. An approximate calculation, giving an idea of the shape of the transient of the ironclad magnetic circuit, can be made by neglecting the difference between the rising and decreasing magnetic characteristic, and using the approximation of the magnetic characteristic given by Fröhlich's formula:

$$\mathcal{B} = \frac{\mathcal{3C}}{\alpha + \sigma \mathcal{3C}}, \qquad (1)$$

which is usually represented in the form given by Kennelly:

$$\rho = \frac{\mathcal{3C}}{\mathcal{B}} = \alpha + \sigma \mathcal{3C}; \qquad (2)$$

that is, the reluctivity is a linear function of the field intensity. It gives a fair approximation for higher magnetic densities.

This formula is based on the fairly rational assumption that the permeability of the iron is proportional to its remaining magnetizability. That is, the magnetic-flux density $\mathcal{B}$ consists of a component $\mathcal{3C}$, the field intensity, which is the flux density in space, and a component $\mathcal{B}' = \mathcal{B} - \mathcal{3C}$, which is the additional flux density carried by the iron. $\mathcal{B}'$ is frequently called the "metallic-flux density." With increasing $\mathcal{3C}$, $\mathcal{B}'$ reaches a finite, limiting value, which in iron is about

$$\mathcal{B}_\infty' = 20{,}000 \text{ lines per cm}^2.$$

At any density $\mathcal{B}'$, the remaining magnetizability then is $\mathcal{B}_\infty' - \mathcal{B}'$, and, assuming the (metallic) permeability as proportional hereto, gives

$$\mu = c(\mathcal{B}_\infty' - \mathcal{B}'),$$

and, substituting

$$\mu = \frac{\mathcal{B}'}{\mathcal{3C}},$$

gives

$$\mathcal{B}' = \frac{c\mathcal{B}_\infty'\mathcal{3C}'}{1 + c\mathcal{3C}'},$$

or, substituting

$$\frac{1}{c\mathcal{B}_\infty'} = \alpha, \frac{1}{\mathcal{B}_\infty'} = \sigma,$$

gives equation (1).

For $\mathcal{3C} = 0$ in equation (1), $\dfrac{\mathcal{B}}{\mathcal{3C}} = \dfrac{1}{\alpha}$; for $\mathcal{3C} = \infty$, $\mathcal{B} = \dfrac{1}{\sigma}$; that is, in equation (1), $\dfrac{1}{\alpha} = $ initial permeability, $\dfrac{1}{\sigma} = $ saturation value of magnetic density.

If the magnetic circuit contains an air gap, the reluctance of the iron part is given by equation (2), that of the air part is constant, and the total reluctance thus is

$$\rho = \beta + \sigma\mathcal{3C},$$

where $\beta = \alpha$ plus the reluctance of the air gap. Equation (1), therefore, remains applicable, except that the value of α is increased.

In addition to the metallic flux given by equation (1), a greater or smaller part of the flux always passes through the air or through space in general, and then has constant permeance, that is, is given by

$$\mathcal{B} = c\mathcal{3C}.$$

In general, the flux in an ironclad magnetic circuit can, there-fore, be represented as function of the current by an expression of the form

$$\Phi = \frac{ai}{1 + bi} + ci, \tag{3}$$

where $\frac{ai}{1 + bi} = \Phi'$ is that part of the flux which passes through the iron and whatever air space may be in series with the iron, and ci is the part of the flux passing through nonmagnetic material.

Denoting now

$$\left. \begin{array}{l} L_1 = na \ 10^{-8}, \\ L_2 = nc \ 10^{-8}, \end{array} \right\} \tag{4}$$

where n = number of turns of the electric circuit, which is inter-linked with the magnetic circuit, L_2 is the inductance of the air part of the magnetic circuit, L_1 the (virtual) initial inductance, that is, inductance at very small currents, of the iron part of the magnetic circuit, and $\frac{a}{b}$ the saturation value of the flux in the iron. That is, for $i = 0, \frac{n\Phi'}{i} = L_1$; and for $i = \infty$, $\Phi' = \frac{a}{b}$.

If r = resistance, the duration of the component of the tran-sient resulting from the air flux would be

$$T_2 = \frac{L_2}{r} = \frac{nc \ 10^{-8}}{r}, \tag{5}$$

and the duration of the transient which would result from the initial inductance of the iron flux would be

$$T_1 = \frac{L_1}{r} = \frac{na \ 10^{-8}}{r}. \tag{6}$$

108. The differential equation of the transient is: induced voltage plus resistance drop equal zero; that is,

$$n \frac{d\Phi}{dt} 10^{-8} + ri = 0.$$

Substituting (3) and differentiating gives

$$\frac{na \ 10^{-8}}{(1 + bi)^2} \frac{di}{dt} + nc \ 10^{-8} \frac{di}{dt} + ri = 0,$$

and, substituting (5) and (6),

$$\left\{ \frac{T_1}{(1 + bi)^2} + T_2 \right\} \frac{di}{dt} + i = 0;$$

hence, separating the variables,

$$\frac{T_1 di}{i(1+bi)^2} + \frac{T_2 di}{i} + dt = 0. \tag{7}$$

The first term is integrated by resolving into partial fractions:

$$\frac{1}{i(1+bi)^2} = \frac{1}{i} - \frac{b}{1+bi} - \frac{b}{(1+bi)^2},$$

and the integration of differential equation (7) then gives

$$T_1 \log \frac{i}{1+bi} + T_2 \log i + \frac{T_1}{1+bi} + t + C = 0. \tag{8}$$

If, then, for the time $t = t_0$, the current is $i = i_0$, these values substituted in (8) give the integration constant C:

$$T_1 \log \frac{i_0}{1+bi_0} + T_2 \log i_0 + \frac{T_1}{1+bi_0} + t_0 + C = 0, \tag{9}$$

and, subtracting (8) from (9), gives

$$t - t_0 = T_1 \log \frac{i_0(1+bi)}{i(1+bi_0)} + T_2 \log \frac{i_0}{i} + T_1 \left\{ \frac{1}{1+bi_0} - \frac{1}{1+bi} \right\} \tag{10}$$

This equation is so complex in i that it is not possible to calculate from the different values of t the corresponding values of i; but inversely, for different values of i the corresponding values of t can be calculated, and the corresponding values of i and t, derived in this manner, can be plotted as a curve, which gives the single-energy transient of the ironclad magnetic circuit.

Such is done in Fig. 48, for the values of the constants:

$$r = .3,$$
$$a = 4 \times 10^5,$$
$$c = 4 \times 10^4,$$
$$b = .6,$$
$$n = 300.$$

This gives

$$T_1 = 4,$$
$$T_2 = .4.$$

Assuming $i_0 = 10$ amperes for $t_0 = 0$, gives from (10) the equation:

$$T = 2.92 - \left\{ 9.21 \log^{10} \frac{i}{1+0.6\,i} + 0.921 \log^{10} i + \frac{4}{1+0.6\,i} \right\}.$$

Herein, the logarithms have been reduced to the base 10 by division with $log^{10}\epsilon = 0.4343$.

For comparison is shown, in dotted line, in Fig. 48, the transient of a circuit containing no iron, and of such constants as to give about the same duration:

$$t = 1.085 \log^{10} i - 0.507.$$

As seen, in the ironclad transient the current curve is very much steeper in the range of high currents, where magnetic sat-

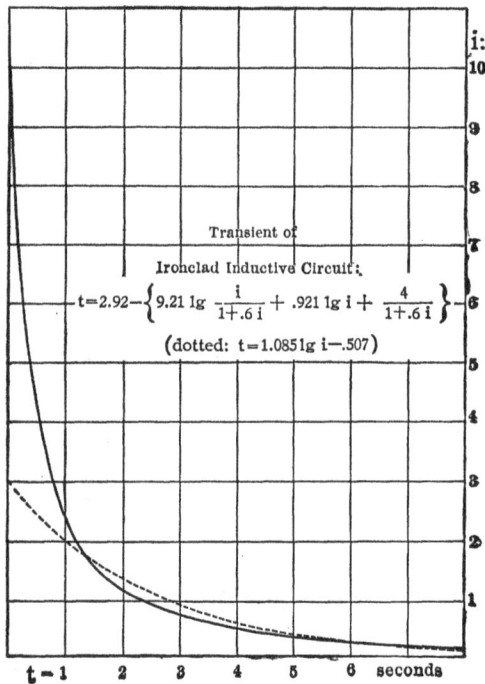

Fig. 48.

uration is reached, but the current is lower in the range of medium magnetic densities.

Thus, in ironclad transients very high-current values of short duration may occur, and such transients, as those of the starting current of alternating-current transformers, may therefore be of serious importance by their excessive current values.

CHAPTER XIII.

TRANSIENT TERM OF THE ROTATING FIELD.

109. The resultant of n_p equal m.m.fs. equally displaced from each other in space angle and in time-phase is constant in intensity, and revolves at constant synchronous velocity. When acting upon a magnetic circuit of constant reluctance in all directions, such a polyphase system of m.m.fs. produces a revolving magnetic flux, or a rotating field. ("Theory and Calculation of Alternating Current Phenomena.") That is, if n_p equal magnetizing coils are arranged under equal space angles of $\dfrac{360}{n_p}$ electrical degrees, and connected to a symmetrical n_p phase electrical degrees, and connected to a symmetrical n_p phase system, that is, to n_p equal e.m.fs. displaced in time-phase by $\dfrac{360}{n_p}$ degrees, the resultant m.m.f. of these n_p coils is a constant and uniformly revolving m.m.f., of intensity $\mathfrak{F}_0 = \dfrac{n_p}{2}\mathfrak{F}$, where $\mathfrak{F}$ is the maximum value $\left(\text{hence } \dfrac{\mathfrak{F}}{\sqrt{2}} \text{ the effective value}\right)$ of the m.m.f. of each coil.

In starting, that is, when connecting such a system of magnetizing coils to a polyphase system of e.m.fs., a transient term appears, as the resultant magnetic flux first has to rise to its constant value. This transient term of the rotating field is the resultant of the transient terms of the currents and therefore the m.m.fs. of the individual coils.

If, then, $\mathfrak{F} = nI$ = maximum value of m.m.f. of each coil, where n = number of turns, and I = maximum value of current, and τ = space-phase angle of the coil, the instantaneous value of the m.m.f. of the coil, under permanent conditions, is

$$f' = \mathfrak{F} \cos (\theta - \tau), \tag{1}$$

197

and if the time θ is counted from the moment of closing the circuit, the transient term is, by Chapter IV,

$$f'' = - \mathfrak{F}\varepsilon^{-\frac{r}{x}\theta} \cos \tau, \tag{2}$$

where $$Z = r - jx.$$

The complete value of m.m.f. of one coil is

$$f_1 = f' + f'' = \mathfrak{F} \left\{ \cos (\theta - \tau) - \varepsilon^{-\frac{r}{x}\theta} \cos \tau \right\}. \tag{3}$$

In an n_p-phase system, successive e.m.fs. and therefore currents are displaced from each other by $\dfrac{1}{n_p}$ of a period, or an angle $\dfrac{2\pi}{n_p}$, and the m.m.f. of coil, i, thus is

$$f_i = \mathfrak{F} \left\{ \cos \left(\theta - \tau - \frac{2\pi}{n_p} i \right) - \varepsilon^{-\frac{r}{x}\theta} \cos \left(\tau + \frac{2\pi}{n_p} i \right) \right\}. \tag{4}$$

The resultant of n_p such m.m.fs. acting together in the same direction would be

$$\sum_1^{np} i\, f_i = \mathfrak{F} \sum_1^{n_p} i\, \cos \left(\theta - \tau - \frac{2\pi}{n_p} i \right)$$

$$- \mathfrak{F}\varepsilon^{-\frac{r}{x}\theta} \sum_1^{n_p} i\, \cos \left(\tau + \frac{2\pi}{n_p} i \right) = 0; \tag{5}$$

that is, the sum of the instantaneous values of the permanent terms as well as the transient terms of all the phases of a symmetrical polyphase system equals zero.

In the polyphase field, however, these m.m.fs. (4) do not act in the same direction, but in directions displaced from each other by a space angle $\dfrac{2\pi}{n_p}$ equal to the time angle of their phase displacement.

The component of the m.m.f., f_1, acting in the direction $(\theta_0 - \tau)$, thus is

$$f_1' = f_1 \cos \left(\theta_0 - \tau - \frac{2\pi}{n_p} i \right), \tag{6}$$

and the sum of the components of all the n_p m.m.fs., in the direction $(\theta_0 - \tau)$, that is, the component of the resultant m.m.f. of the polyphase field, in the direction $(\theta_0 - \tau)$, is

$$f = \sum_1^{n_p} i \ f_i'$$

$$= \mathfrak{F} \sum_1^{n_p} i \left\{ \cos \left(\theta - \tau - \frac{2\,\pi}{n_p} i \right) - \varepsilon^{-\frac{r}{x}\theta} \cos \left(\tau + \frac{2\,\pi}{n_p} i \right) \right\},$$

$$\cos \left(\theta_0 - \tau - \frac{2\,\pi}{n_p} i \right). \tag{7}$$

Transformed, this gives

$$f = \frac{\mathfrak{F}}{2} \left\{ \sum_1^{n_p} i \ \cos \left(\theta + \theta_0 - 2\,\tau - \frac{4\,\pi}{n_p} i \right) + \sum_1^{n_p} i \ \cos \left(\theta - \theta_0 \right) \right.$$

$$\left. - \varepsilon^{-\frac{r}{x}\theta} \sum_1^{n_p} i \cos \theta_0 - \varepsilon^{-\frac{r}{x}\theta} \sum_1^{n_p} i \cos \left(\theta_0 - 2\,\tau - \frac{4\,\pi}{n_p} i \right) \right\},$$

and as the sums containing $\dfrac{4\,\pi}{n_p} i$ equal zero, we have

$$f = \frac{n_p}{2} \mathfrak{F} \left\{ \cos \left(\theta - \theta_0 \right) - \varepsilon^{-\frac{r}{x}\theta} \cos \theta_0 \right\}, \tag{8}$$

and for $\theta = \infty$, that is as permanent term, this gives

$$f_0 = \frac{n_p}{2} \mathfrak{F} \cos \left(\theta - \theta_0 \right); \tag{9}$$

hence, a maximum, and equal to $\dfrac{n_p}{2} \mathfrak{F}$, that is, constant, for $\theta_0 = \theta$, that is, uniform synchronous rotation. That is, the resultant of a polyphase system of m.m.fs., in permanent condition, rotates at constant intensity and constant synchronous velocity.

Before permanent condition is reached, however, the resultant m.m.f. in the direction $\theta_0 = \theta$, that is, in the direction of the synchronously rotating vector, in which in permanent condition the m.m.f. is maximum and constant, is given during the transient period, from equation (8), by

$$f_0 = \frac{n_p}{2} \mathfrak{F} \left\{ 1 - \varepsilon^{-\frac{r}{x}\theta} \cos \theta \right\}, \tag{10}$$

that is, it is not constant but periodically varying. ·

As example is shown, in Fig. 49, the resultant m.m.f. f_0 in the direction of the synchronously revolving vector, $\theta_0 = \theta$, for the

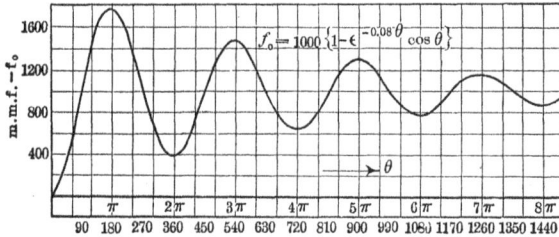

Fig. 49. Transient term of polyphase magnetomotive force.

constants $n_p = 3$, or a three-phase system; $\mathfrak{F} = 667$, and $Z = r - jx = 0.32 - 4\,j$; hence,

$$f_0 = 1000 \left(1 - \varepsilon^{-1.08\,\theta} \cos \theta \right),$$

with θ as abscissas, showing the gradual oscillatory approach to constancy.

110. The direction, $\theta_0 = \theta$, is, however, not the direction in which the resultant m.m.f. in equation (8) is a maximum, but the maximum is given by

$$\frac{df}{d\theta_0} = 0, \tag{11}$$

this gives

$$\sin \left(\theta - \theta_0 \right) + \varepsilon^{-\frac{r}{x}\theta} \sin \theta_0 = 0, \tag{12}$$

hence,

$$\cot \theta_0 = \frac{\cos \theta - \varepsilon^{-\frac{r}{x}\theta}}{\sin \theta}; \tag{13}$$

that is, the resultant maximum m.m.f. of the polyphase system does not revolve synchronously, in the starting condition, but revolves with a varying velocity, alternately running ahead and dropping behind the position of uniform synchronous rotation, by equation (13), and only for $\theta = \infty$, equation (12) becomes $\cot \theta_0 = \cot \theta$, or $\theta_0 = \theta$, that is, uniform synchronous rotation.

The speed of rotation of the maximum m.m.f. is given from equation (12) by differentiation as

$$S = \frac{d\theta_0}{d\theta} = -\frac{\dfrac{dQ}{d\theta}}{\dfrac{dQ}{d\theta_0}},$$

where

$$Q = \sin (\theta - \theta_0) + \varepsilon^{-\frac{r}{x}\theta} \sin \theta_0;$$

hence,

$$S = \frac{\cos (\theta - \theta_0) - \dfrac{r}{x}\varepsilon^{-\frac{r}{x}\theta} \sin \theta_0}{\cos (\theta - \theta_0) - \varepsilon^{-\frac{r}{x}\theta} \cos \theta_0}, \tag{14}$$

or approximately,

$$S = \frac{1 - \dfrac{r}{x}\varepsilon^{-\frac{r}{x}\theta} \sin \theta_0}{1 - \varepsilon^{-\frac{r}{x}\theta} \cos \theta_0}. \tag{15}$$

For $\theta = \infty$, equation (14) becomes $S = 1$, or uniform synchronous rotation, but during the starting period the speed alternates between below and above synchronism.

From (13) follows

and

$$\left.\begin{array}{l} \cos \theta_0 = \dfrac{\cos \theta - \varepsilon^{-\frac{r}{x}\theta}}{\mathfrak{R}} \\[3mm] \sin \theta_0 = \dfrac{\sin \theta}{\mathfrak{R}}, \end{array}\right\} \tag{16}$$

where

$$\mathfrak{R} = \sqrt{\left(\cos \theta - \varepsilon^{-\frac{r}{x}\theta}\right)^2 + \sin^2 \theta} = \sqrt{1 - 2\varepsilon^{-\frac{r}{x}\theta} \cos \theta + \varepsilon^{-2\frac{r}{x}\theta}}. \tag{17}$$

The maximum value of the resultant m.m.f., at time-phase θ, and thus of direction θ_0 as given by equation (13) or (16), (17), is derived by substituting (16), (17) into (8), as:

$$f_m = \frac{n_p}{2} \mathfrak{F} \, \mathfrak{R}$$

$$= \frac{n_p}{2} \mathfrak{F} \sqrt{1 - 2 \, \varepsilon^{-\frac{r}{x}\theta} \cos \theta + \varepsilon^{-2\frac{r}{x}\theta}}, \tag{18}$$

hence is not constant, but pulsates periodically, with gradually decreasing amplitude of pulsation, around the mean value $\frac{n_p}{2} \mathfrak{F}$.

For $\theta = 0$, or at the moment of start, it is, by (13),

$$\cot \theta_0' = \frac{\cos \theta - \varepsilon^{-\frac{r}{x}\theta}}{\sin \theta} = \frac{0}{0},$$

hence, differentiating numerator and denominator,

$$\cot \theta_0' = \frac{-\sin \theta + \frac{r}{x} \varepsilon^{-\frac{r}{x}\theta}}{\cos \theta} = \frac{r}{x}$$

and

$$\tan \theta_0' = \frac{x}{r} \, ;$$

that is, the position of maximum resultant m.m.f. starts from angle θ_0' ahead of the permanent position, where θ_0' is the time-phase angle of the electric magnetizing circuit. The initial value of the resultant m.m.f., for $\theta = 0$, is $f_m = 0$, that is, the revolving m.m.f. starts from zero.

Substituting (16) in (15) gives the speed as function of time

$$S = \frac{1 - \varepsilon^{-\frac{r}{x}\theta} \left(\cos \theta - \frac{r}{x} \sin \theta \right)}{1 + \varepsilon^{-2\frac{r}{x}\theta} - 2 \varepsilon^{-\frac{r}{x}\theta} \cos \theta} \tag{19}$$

for $\theta = 0$ this gives the starting speed of the rotating field

$$S_0 = \frac{0}{0}, \text{ or, indefinite;}$$

hence, after differentiating numerator and denominator twice, this value becomes definite.

$$S_0 = \frac{1}{2};$$ (20)

that is, the rotating field starts at half speed.

As illustration are shown, in Fig. 50, the maximum value of the resultant polyphase m.m.f., f_m, and its displacement in

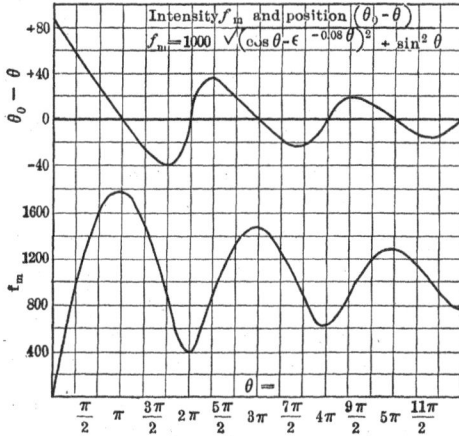

Fig. 50.   Start of rotating field.

position from that of uniform synchronous rotation, $\theta_0 - \theta$, for the same constants as before, namely: $n_p = 3$; $\mathfrak{F} = 667$, and $Z = r - jx = 0.32 - 4j$; hence,

$$f_m = 1000 \sqrt{1 - 2\,\varepsilon^{-0.08\theta}\,\cos\theta + \varepsilon^{-0.16\theta}},$$

with the time-phase angle θ as abscissas, for the first three cycles.

111. As seen, the resultant maximum m.m.f. of the polyphase system, under the assumed condition, starting at zero in the moment of closing the three-phase circuit, rises rapidly — within 60 time-degrees — to its normal value, overreaches and exceeds it by 78 per cent, then drops down again below normal, by 60 per cent, rises 47 per cent above normal, drops 37 per cent below normal, rises 28 per cent above normal, and thus by a series of oscillations approaches the normal value. The maximum value of the resultant m.m.f. starts in position

85 time-degrees ahead, in the direction of rotation, but has in half a period dropped back to the normal position, that is, the position of uniform synchronous rotation, then drops still further back to the maximum of 40 deg., runs ahead to 34 deg., drops 23 deg. behind, etc.

It is interesting to note that the transient term of the rotating field, as given by equations (10), (13), (18), does not contain the phase angle, that is, does not depend upon the point of the wave, $\theta = \tau$, at which the circuit is closed, while in all preceding investigations the transient term depended upon the point of the wave at which the circuit was closed, and that this transient term is oscillatory. In the preceding chapter, in circuits containing only resistance and inductance, the transient term has always been gradual or logarithmic, and oscillatory phenomena occurred only in the presence of capacity in addition to inductance. In the rotating field, or the polyphase m.m.f., we thus have a case where an oscillatory transient term occurs in a circuit containing only resistance and inductance but not capacity, and where this transient term is independent of the point of the wave at which the circuits were closed, that is, is always the same, regardless of the moment of start of the phenomenon.

The transient term of the polyphase m.m.f. thus is independent of the moment of start, and oscillatory in character, with an amplitude of oscillation depending only on the reactance factor, $\dfrac{x}{r}$, of the circuit.

CHAPTER XIV.

SHORT-CIRCUIT CURRENTS OF ALTERNATORS.

112. The short-circuit current of an alternator is limited by armature reaction and armature self-inductance; that is, the current in the armature represents a m.m.f. which with lagging current, as at short circuit, is demagnetizing or opposing the impressed m.m.f. of field excitation, and by combining therewith to a resultant m.m.f. reduces the magnetic flux from that corresponding to the field excitation to that corresponding to the resultant of field excitation and armature reaction, and thus reduces the generated e.m.f. from the nominal generated e.m.f., e_0, to the virtual generated e.m.f., e_1. The armature current also produces a local magnetic flux in the armature iron and pole-faces which does not interlink with the field coils, but is a true self-inductive flux, and therefore is represented by a reactance x_1. Combined with the effective resistance, r_1, of the armature winding, this gives the self-inductive impedance $Z_1 = r_1 - jx_1$, or $z_1 = \sqrt{r_1^2 + x_1^2}$. Vectorially subtracted from the virtual generated e.m.f., e_1, the voltage consumed by the armature current in the self-inductive impedance Z_1 then gives the terminal voltage, e.

At short circuit, the virtual generated e.m.f., e_1, is consumed by the armature self-inductive impedance, z_1. As the effective armature resistance, r_1, is very small compared with its self-inductive reactance, x_1, it can be neglected compared thereto, and the short-circuit current of the alternator, in permanent condition, thus is

$$i = \frac{e_1}{x_1}.$$

As shown in Chapter XXII, "Theory and Calculation of Alternating Current Phenomena," the armature reaction can be represented by an equivalent, or effective reactance, x_2, and the self-inductive reactance, x_1, and the effective reactance of

armature reaction, x_2, combine to form the synchronous reactance, $x_0 = x_1 + x_2$, and the short-circuit current of the alternator, in permanent condition, therefore can be expressed by

$$i = \frac{e_0}{x_0},$$

·where e_0 = nominal generated e.m.f.

113. The effective reactance of armature reaction, x_2, differs,
however, essentially from the true self-inductive reactance, x_1,
in that x_1 is instantaneous in its action, while the effective
reactance of armature reaction, x_2, requires an appreciable time
to develop: x_2 represents the change of the magnetic field flux
produced by the armature m.m.f. The field flux, however, cannot change instantaneously, as it interlinks with the field exciting
coil, and any change of the field flux generates an e.m.f. in the
field coils, changing the field current so as to retard the change
of the field flux. Hence, at the first moment after a change of
armature current, the current change meets only the reactance,
x_1, but not the reactance x_2. Thus, when suddenly short-circuiting an alternator from open circuit, in the moment before
the short circuit, the field flux is that corresponding to the
impressed m.m.f. of field excitation and the voltage in the armature, i.e., the nominal generated e.m.f., e_0 (corrected for magnetic saturation). At the moment of short circuit, a counter
m.m.f., that of the armature reaction of the short-circuit
current, is opposed to the impressed m.m.f. of the field excitation,
and the magnetic flux, therefore, begins to decrease at such a
rate that the e.m.f. generated in the field coils by the decrease
of field flux increases the field current and therewith the m.m.f.
so that when combined with the armature reaction it gives a
resultant m.m.f. producing the instantaneous value of field flux.
Immediately after short circuit, while the field flux still has full
value, that is, before it has appreciably decreased, the field m.m.f.
thus must have increased by a value equal to the counter m.m.f.
of armature reaction. As the field is still practically unchanged,
the generated e.m.f. is the nominal generated voltage, e_0, and
the short-circuit current is

$$i'^0 = \frac{e_0}{x_1},$$

and from this value gradually dies down, with a decrease of the field flux and of the generated e.m.f., to

$$i = \frac{e_1}{x_1} = \frac{e_0}{x_0}.$$

Hence, approximately, when short-circuiting an alternator, in the first moment the short-circuit current is

$$i'^0 = \frac{e_0}{x_1},$$

while the field current has increased from its normal value i_0 to the value

$$i_0 \times \frac{\text{Field excitation} + \text{Armature reaction}}{\text{Field excitation}};$$

gradually the armature current decreases to

$$i = \frac{e_1}{x_1} = \frac{e_0}{x_0},$$

and the field current again to the normal value i_0.

Therefore, the momentary short-circuit current of an alternator bears to the permanent short-circuit current the ratio

$$\frac{i'^0}{i} = \frac{x_0}{x_1} = \frac{x_1 + x_2}{x_1},$$

that is,

$$\frac{\text{Armature self-inductance} + \text{Armature reaction}}{\text{Armature self-inductance}}.$$

In machines of high self-inductance and low armature reaction, as uni-tooth high frequency alternators, this increase of the momentary short-circuit current over the permanent short-circuit current is moderate, but may reach enormous values in machines of low self-inductance and high armature reaction, as large low frequency turbo alternators.

114. Superimposed upon this transient term, resulting from the gradual adjustment of the field flux to a change of m.m.f., is the transient term of armature reaction. In a polyphase alternator, the resultant m.m.f. of the armature in permanent conditions is constant in intensity and revolves with regard to the armature at uniform synchronous speed, hence is stationary

with regard to the field.   In the first moment, however, the
resultant armature m.m.f. is changing in intensity and in velocity,
approaching its constant value by a series of oscillations, as
discussed in Chapter XIII.   Hence, with regard to the field, the
transient term of armature reaction is pulsating in intensity and
oscillating in position, and therefore generates in the field coils

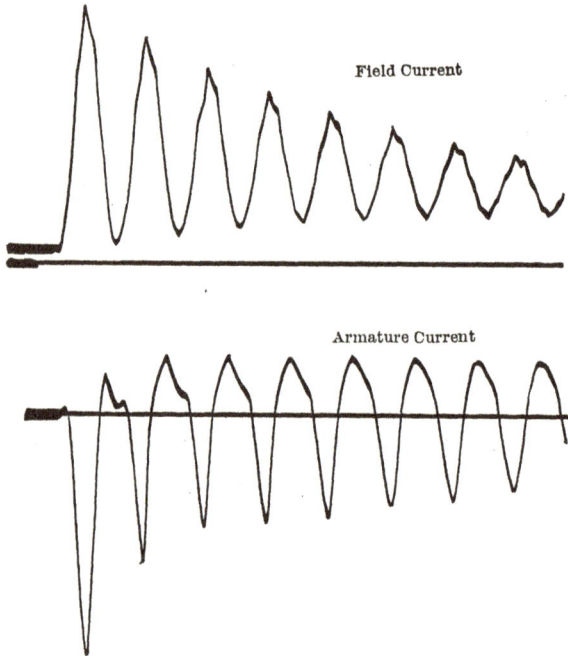

Field Current

Armature Current

Fig. 51.   Three-phase short-circuit current of a turbo-alternator.

an e.m.f. and causes a corresponding pulsation in the field
current and field terminal voltage, of the same frequency as
the armature current, as shown by the oscillogram of such a
three-phase short-circuit, in Fig. 51.   This pulsation of field
current is independent of the point in the wave, at which the
short-circuit occurs, and dies out gradually, with the dying out
of the transient term of the rotating m.m.f.

In a single-phase alternator, the armature reaction is alternating with regard to the armature, hence pulsating, with double
frequency, with regard to the field, varying between zero and its

maximum value, and therefore generates in the field coils a double frequency e.m.f., producing a pulsation of field current of double frequency. This double-frequency pulsation of the field current and voltage at single-phase short-circuit is proportional to the armature current, and does not disappear with the disappearance of the transient term, but persists also after the permanent condition of short-circuit has been reached,

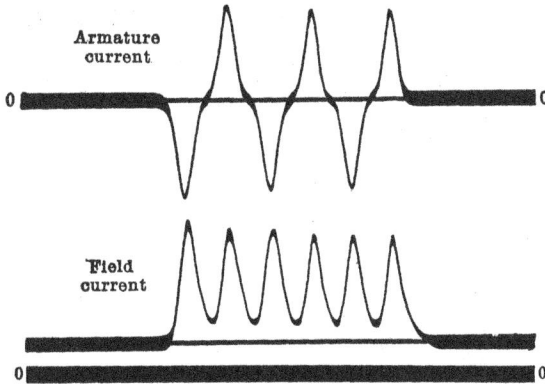

Fig. 52.   Single-phase short-circuit current of a three-phase turbo-alternator.

merely decreasing with the decrease of the armature current. It is shown in the oscillogram of a single-phase short-circuit on a three-phase alternator, Fig. 52.

Superimposed on this double frequency pulsation is a single-frequency pulsation due to the transient term of the armature current, that is, the same as on polyphase short-circuit. With single-phase short-circuit, however, this normal frequency pulsation of the field depends on the point of the wave at which the short-circuit occurs, and is zero, if the circuit is closed at the moment when the short-circuit current is zero, as in Fig. 51, and a maximum when the short-circuit starts at the maximum point of the current wave. As this normal frequency pulsation gradually disappears, it causes the successive waves of the double frequency pulsation to be unequal in size at the beginning of the transient term, and gradually become equal, as shown in the oscillogram, Fig. 53.

The calculation of the transient term of the short-circuit current of alternators thus involves the transient term of the armature and the field current, as determined by the self-inductance of armature and of field circuit, and the mutual inductance between the armature circuits and the field circuit, and the impressed or generated voltage; therefore is rather complicated; but a simpler approximate calculation can be

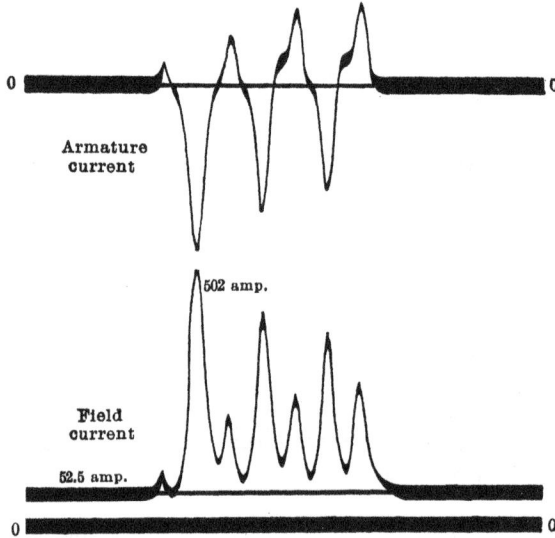

Fig. 53. Single-phase short-circuit current of a three-phase turbo-alternator.

given by considering that the duration of the transient term is short compared with that of the armature reaction on the field.

(A) *Polyphase alternator.*

115. Let n_p = number of phases; $\theta = 2 \pi f t$ = time-phase angle; n_0 = number of field turns in series per pole; n_1 = number of armature turns in series per pole; $Z_0 = r_0 - j x_0$ = self-inductive impedance of field circuit; $Z_1 = r_1 - j x_1$ = self-inductive impedance of armature circuit; p = permeance of field magnetic circuit; $a = 2 \pi f n_1 \, 10^{-8}$ = induction coefficient of armature; E_0 = exciter voltage; $I_0 = \dfrac{E_0}{r_0}$ = field exciting current, in permanent condition; i_0 = field exciting current at time θ; i_0^0 = field exciting current immediately after short-circuit; i = armature current at time θ, and $k_t = \dfrac{2}{n_p} \dfrac{n_0}{n_1}$ = transformation ratio of field to resultant armature. Counting the time angle θ from the moment of short circuit, $\theta = 0$, and letting $\theta' =$ time-phase angle of one of the generator circuits at the moment of short circuit, we have,

$\mathfrak{F}_0 = n_0 I_0 =$ field excitation, in permanent or stationary condition, (1)

$\Phi_0 = p\mathfrak{F}_0 = pn_0 I_0 =$ magnetic flux corresponding thereto,

and

$e_0{}^0 = ap\mathfrak{F}_0 = apn_0 I_0.$ (2)
 = nominal generated voltage, maximum value, at $\theta = 0$.

Hence, $$I^0 = \frac{e_0{}^0}{x_1} = \frac{apn_0}{x_1} I_0$$ (3)

 = momentary short-circuit current at time $\theta = 0$, and

$$\mathfrak{F}_1{}^0 = \frac{n_p}{2} n_1 I^0 = \frac{n_p apn_1 n_0 I_0}{2\,x_1}$$ (4)

 = resultant armature reaction thereof.

Assume this armature reaction as opposite to the field excitation,

$$\mathfrak{F}_0{}^0 = n_0 \dot{i}_0{}^0,$$ (5)

as is the case at short circuit.

The resultant m.m.f. of the magnetic circuit at the moment of short-circuit is

$$\mathfrak{F}^0 = \mathfrak{F}_0{}^0 - \mathfrak{F}_1{}^0.$$ (6)

At this moment, however, the field flux is still Φ_0, and the resultant m.m.f. is given by (1) as

$$\mathfrak{F}^0 = \mathfrak{F}_0 = n_0 I_0.$$ (7)

Substituting (4), (5), (7) in (6) gives

$$n_0 I_0 = n_0 \dot{i}_0{}^0 - \frac{n_p apn_1 n_0 I_0}{2\,x_1},$$

hence, $$\dot{i}_0{}^0 = \frac{x_1 + \dfrac{n_p apn_1}{2}}{x_1} I_0.$$ (8)

Writing
$$x_2 = \frac{n_p a p n_1}{2},\qquad(9)$$

we have
$$\dot{i_0}^0 = \frac{x_1 + x_2}{x_1} I_0;\qquad(10)$$

that is, at the moment of short circuit the field exciting current rises from I_0 to $\dot{i_0}^0$, and then gradually dies down again to I_0 at a rate depending on the field impedance Z_0, that is, by $\varepsilon^{-\frac{r_0}{x_0}\theta}$, as discussed in preceding chapters. Hence, it can be represented by

$$\dot{i_0} = \frac{x_1 + x_2\, \varepsilon^{-\frac{r_0}{x_0}\theta}}{x_1} I_0.\qquad(11)$$

The resultant armature m.m.f., or armature reaction, is

$$\frac{n_p n_1 I^0}{2};$$

thus the magnetic flux which would be produced by it is

$$\frac{p n_p n_1 I^0}{2},$$

and therefore the voltage generated by this flux is

$$\frac{a p n_p n_1 I^0}{2};$$

hence,

$$x_2 = \frac{a p n_p n_1}{2}$$

$$= \frac{\text{Voltage corresponding to the m.m.f. of armature current}}{\text{Armature current}};$$

that is, x_2 is the equivalent or effective reactance of armature reaction.

In equations (10) and (11) the external self-inductance of the field circuit, that is, the reactance of the field circuit outside of the machine field winding, has been neglected. This would introduce a negative transient term in (11), thus giving equation (11) the approximate form

$$i_0 = \frac{x_1 + x_2\left(\varepsilon^{-\frac{r_0}{x_0}\theta} - \varepsilon^{-\frac{r_0}{x_3}\theta}\right)}{x_1} I_0, \tag{12}$$

where x_3 = self-inductive reactance of the field circuit outside of alternator field coils.

The more complete expression requires consideration when x_3 is very large, as when an external reactive coil is inserted in the field circuit.

In reality, x_2 is a mutual inductive reactance, and x_3 can be represented approximately by a corresponding increase of x_1.

116. If I = maximum value of armature current, we have

$$\mathfrak{F}_1 = \frac{n_p n_1 I}{2} = \text{armature m.m.f.,}$$

hence,

$$\mathfrak{F} = n_0 i_0 - \frac{n_p n_1 I}{2} \tag{13}$$

$$= \text{resultant m.m.f.,}$$

and

$$E = ap\mathfrak{F}$$

$$= \text{e.m.f. maximum generated thereby,}$$

and

$$I = \frac{E}{x_1} = \frac{ap}{x_1}\mathfrak{F} \tag{14}$$

$$= \text{armature current, maximum.}$$

Substituting (13) in (14) gives

$$x_1 I = apn_0 i_0 - \frac{n_p apn_1}{2} I$$

and

$$I = \frac{apn_0 i_0}{x_1 + \dfrac{n_p apn_1}{2}} = \frac{apn_0 i_0}{x_1 + x_2}; \tag{15}$$

or, by (9),

$$I = \frac{2}{n_p}\frac{n_0}{n_1} i_0 \frac{x_2}{x_1 + x_2}, \tag{16}$$

where $\dfrac{2}{n_p} \dfrac{n_0}{n_1} = k_t = $ transformation ratio of field turns to resultant armature turns; hence,

$$I = k_t i_0 \frac{x_2}{x_1 + x_2} . \tag{17}$$

Substituting (11) in (17) thus gives the *maximum value of the armature current* as

$$I = k_t I_0 \frac{x_2}{x_1 + x_2} \frac{x_1 + x_2 \varepsilon^{-\frac{r_0}{x_0}\theta}}{x_1}, \tag{18}$$

the *instantaneous value of the armature current* as

$$i = k_t I_0 \frac{x_2 \left(x_1 + x_2 \varepsilon^{-\frac{r_0}{x_0}\theta}\right)}{x_1 (x_1 + x_2)} \left\{ \cos (\theta - \theta') - \varepsilon^{-\frac{r_1}{x_1}\theta} \cos \theta' \right\}, \tag{19}$$

and by equation (10) of Chapter XIII, the *armature reaction* as

$$f = \frac{n_p n_1}{2} I_0 \frac{x_2 \left(x_1 + x_2 \varepsilon^{-\frac{r_0}{x_0}\theta}\right)}{(x_1 + x_2)} \left\{ 1 - \varepsilon^{-\frac{r_1}{x_1}\theta} \cos \theta \right\}, \tag{20}$$

where $x_1 + x_2 = x_0$ is the synchronous reactance of the alternator.

For $\theta = \infty$, or in permanent condition, equations (18), (19), (20) assume the usual form:

$$\left. \begin{aligned} I &= k_t I_0 \frac{x_2}{x_0}, \\[2mm] i &= k_t I_0 \frac{x_2}{x_0} \cos (\theta - \theta') \\[2mm] f &= \frac{n_p n_1}{2} k_t I_0 \frac{x_2}{x_0} . \end{aligned} \right\} \tag{21}$$

and

117. As an example is calculated, the instantaneous value of the transient short-circuit current of a three-phase alternator, with the time angle θ as abscissas, and for the constants: the field turns, $n_0 = 100$; the normal field current, $I_0 = 200$ amp.; the field impedance, $Z_0 = r_0 - jx_0 = 1.28 - 160\,j$ ohms; the armature turns, $n_1 = 25$, and the armature impedance, $Z_1 = $

$r_1 - jx_1 = 0.4 - 5j$ ohms. For the phase angle, $\theta' = 0$, the transformation ratio then is

$$k_t = \frac{2}{n_p} \frac{n_0}{n_1} = \frac{8}{3} = 2.67,$$

and the equivalent impedance of armature reaction is

$$x_2 = \frac{n_p}{2} \left(\frac{n_1}{n_0}\right)^2 x_0$$
$$= 15,$$

and we have

$$I = \quad 400 \; (1 + 3 \, \varepsilon^{-0.008\,\theta}), \tag{18}$$
$$i = \quad 400 \; (1 + 3 \, \varepsilon^{-0.008\,\theta}) \; (\cos\theta - \varepsilon^{-0.08\,\theta}), \tag{19}$$

and $\quad f = 15,000 \; (1 + 3 \, \varepsilon^{-0.008\,\theta}) \; (1 - \varepsilon^{-0.08\,\theta} \cos\theta). \tag{20}$

(B) Single-phase alternator.

118. In a single-phase alternator, or in a polyphase alternator with one phase only short-circuited, the armature reaction is pulsating.

The m.m.f. of the armature current,

$$i = I \cos (\theta - \theta'), \tag{22}$$

of a single-phase alternator, is, with regard to the field,

$$f_1 = n_1 I \cos (\theta - \theta') \cos (\theta_0 - \theta');$$

hence, for position angle $\theta_0 =$ time angle θ, or synchronous rotation,

$$f_1 = \frac{n_1}{2} I \left\{1 + \cos 2 \, (\theta - \theta')\right\}; \tag{23}$$

that is, of double frequency, with the average value,

$$\mathcal{F}_1 = \frac{n_1}{2} \, I, \tag{24}$$

pulsating between 0 and twice the average value.

The average value (24) is the same as the value of the polyphase machine, for $n_p = 1$.

Using the same denotations as in (A), we have:

(1) $\quad \mathcal{F}_0 = n_0 I_0, \tag{25}$

(4) $\quad \mathcal{F}_1{}^0 = \dfrac{apn_1 n_0}{2 \, x_1} I_0 \left\{1 + \cos 2 \, (\theta - \theta')\right\} = \dfrac{apn_1 n_0}{2 \, x_1} I_0 \left\{1 + \cos 2 \, \theta'\right\}.$

$$\tag{26}$$

Denoting the effective reactance of armature reaction thus:

$$x_2 = \frac{apn_1}{2},\qquad(27)$$

and substituting (27) in (26) we obtain

$$\mathcal{F}_1{}^0 = \frac{x_2}{x_1} n_0 I_0 \{1 + \cos 2\,(\theta - \theta')\} = \frac{x_2}{x_1} n_0 I_0 \{1 + \cos 2\,\theta'\};\quad(28)$$

hence, by (6),

$$n_0 I_0 = n_0 i_0{}^0 - \frac{x_2}{x_1} n_0 I_0 \{1 + \cos 2\,\theta'\}$$

and

$$i_0{}^0 = \frac{x_1 + x_2}{x_1} I_0 \left\{1 + \frac{x_2}{x_1 + x_2} \cos 2\,\theta'\right\},\qquad(29)$$

and the *field current,*

$$i_0 = \frac{x_1 + x_2 \epsilon^{-\frac{r_0}{x_0}\theta}}{x_1} I_0 \left\{1 + \frac{x_2}{x_1 + x_2} \cos 2\,(\theta - \theta_0)\right\}\qquad(30)$$

119. If I = maximum value of armature current,

$$\mathcal{F}_1 = \frac{n_1}{2} I \{1 + \cos 2\,(\theta - \theta')\}\qquad(31)$$

$$= \text{armature m.m.f.;}$$

hence,

$$\mathcal{F} = n_0 i_0 - \mathcal{F}_1\qquad(32)$$

$$= \text{resultant m.m.f.}$$

Since, however,

$$\Phi = p\mathcal{F},$$

$$e = a\Phi = ap\mathcal{F},$$

$$I = \frac{e}{x_1} = \frac{ap\mathcal{F}}{x_1},\qquad(33)$$

and, by (27),

$$ap = \frac{2}{n_1} x_2,$$

we have, by (33)

$$\mathfrak{F} = \frac{x_1}{ap} I = \frac{n_1}{2} \frac{x_1}{x_2} I. \tag{34}$$

Substituting (30), (31), and (34) into (32) gives

$$\frac{n_1}{2} \frac{x_1}{x_2} I = n_0 I_0 \frac{x_1 + x_2 \varepsilon^{-\frac{r_0}{x_0} \theta}}{x_1} \left\{ 1 + \frac{x_2}{x_1 + x_2} \cos 2 (\theta - \theta') \right\}$$

$$- \frac{n_1}{2} I \left\{ 1 + \cos 2 (\theta - \theta') \right\};$$

or, substituting,

$$k_t = 2 \frac{n_0}{n_1} = \text{transformation ratio}, \tag{35}$$

and rearranging, gives

$$I = k_t I_0 \frac{x_2}{x_1 + x_2} \frac{x_1 + x_2 \varepsilon^{-\frac{r_0}{x_0} \theta}}{x_1} \tag{36}$$

as the *maximum value of the armature current.*

This is the same expression as found in (18) for the poly-phase machine, except that now the reactances have different values.

Herefrom it follows that the *instantaneous value of the armature current* is

$$i = k_t I_0 \frac{x_2 \left(x_1 + x_2 \varepsilon^{-\frac{r_0}{x_0} \theta} \right)}{x_1 (x_1 + x_2)} \left\{ \cos (\theta - \theta') - \varepsilon^{-\frac{r_1}{x_1} \theta} \cos \theta' \right\}, \tag{37}$$

and, by (31), the *armature reaction* is

$$\mathfrak{F}_1 = \frac{n_1}{2} k_t I_0 \frac{x_2 \left(x_1 + x_2 \varepsilon^{-\frac{r_0}{x_0} \theta} \right)}{x_1 (x_1 + x_2)} \left\{ 1 + \cos 2 (\theta - \theta') \right\}. \tag{38}$$

For $\theta = \infty$, or permanent condition, equations (30), (36), (37), and (38) give

$$\left.\begin{array}{l} i_0 = I_0 \left\{ 1 + \dfrac{x_2}{x_1 + x_2} \cos 2 \, (\theta - \theta') \right\}, \\[2ex] I = k_t I_0 \dfrac{x_2}{x_1 + x_2}, \\[2ex] i = k_t I_0 \dfrac{x_2}{x_1 + x_2} \cos (\theta - \theta'), \\[2ex] \mathfrak{F}_1 = \dfrac{n_1 k_t}{2} I_0 \dfrac{x_2}{x_1 + x_2} \left\{ 1 + \cos 2 \, (\theta - \theta') \right\}. \end{array}\right\} \qquad (39)$$

As seen, the field current i_0 is pulsating even in permanent condition, the more so the higher the armature reaction x_2 compared with the armature self-inductive reactance x_1.

120. Choosing the same example as in Fig. 53, paragraph 117, but assuming only one phase short-circuited, that is, a single-phase short circuit between two terminals, we have the effective armature series turns, $n_1 = 25 \sqrt{3} = 43.3$; the armature impedance, $Z_1 = r_1 - jx_1 = 0.8 - 10 \, j$; $\theta' = 0$; the transformation ratio, $k_t = 4.62$, and the effective reactance of armature reaction

$$x_2 = \frac{3}{32} x_0 = 15; \text{ herefrom,}$$

$$I = 555 \, (1 + 1.5 \, \varepsilon^{-0.008 \, \theta}), \qquad (36)$$

$$i = 555 \, (1 + 1.5 \, \varepsilon^{-0.008 \, \theta}) \, (\cos \theta - \varepsilon^{-0.08 \, \theta}), \qquad (37)$$

and $\qquad f = 12,000 \, (1 + 1.5 \, \varepsilon^{-0.008 \, \theta}) \, (1 + \cos 2 \, \theta); \qquad (38)$

and the field current is

$$i_0 = 200 \, (1 + 1.5 \, \varepsilon^{-0.008 \, \theta}) \, (1 + 0.6 \cos 2 \, \theta). \qquad (30)$$

In this case, in the open-circuited phase of the machine, a high third harmonic voltage is generated by the double frequency pulsation of the field, and to some extent also appears in the short-circuit current.

121. It must be considered, however, that the effective self-inductive reactance of the armature under momentary short-circuit conditions is not the same as the self-inductive reactance under permanent short circuit conditions, and is not constant, but varying, is a *transient reactance*.

The armature magnetic circuit is in inductive relation with the field magnetic circuit. Under permanent conditions, the resultant m.m.f. of the armature currents with regard to the field is constant, and the mutual inductance between field and armature circuits thus exerts no inductive effect. In the moment of short circuit, however—and to a lesser extent in the moment of any change of armature condition—the resultant armature m.m.f. is pulsating in intensity and direction with regard to the field, as seen in the preceding chapter, and appears as a transient term of the armature self-inductance.

The relations between armature magnetic circuit, field magnetic circuit and mutual magnetic circuit in alternators are similar as the relations in the alternating current transformer, between primary leakage flux, secondary leakage flux and mutual magnetic flux, except that in the transformer the inductive action of the mutual magnetic flux is permanent, while in the alternator it exists only in the moment of change of armature condition, and gradually disappears, thus must be represented by a transient effective reactance. See the chapters on "Reactance of Apparatus" in "Theory and Calculation of Electric Circuits."

For a more complete study, such as required for the predetermination of short circuit currents by the numerical calculation of the constants from the design of the machine, the reader must be referred to the literature.*

SECTION II

PERIODIC TRANSIENTS

PERIODIC TRANSIENTS

CHAPTER I.

INTRODUCTION.

1. Whenever in an electric circuit a sudden change of the circuit conditions is produced, a transient term appears in the circuit, that is, at the moment when the change begins, the circuit quantities, as current, voltage, magnetic flux, etc., correspond to the circuit conditions existing before the change, but do not, in general, correspond to the circuit conditions brought about by the change, and therefore must pass from the values corresponding to the previous condition to the values corresponding to the changed condition. This transient term may be a gradual approach to the final condition, or an approach by a series of oscillations of gradual decreasing intensities.

Gradually — after indefinite time theoretically, after relatively short time practically — the transient term disappears, and permanent conditions of current, of voltage, of magnetism, etc., are established. The numerical values of current, of voltage, etc., in the permanent state reached after the change of circuit conditions, in general, are different from the values of current, voltage, etc., existing in the permanent state before the change, since they correspond to a changed condition of the circuit. They may, however, be the same, or such as can be considered the same, if the change which gives rise to the transient term can be considered as not changing the permanent circuit conditions. For instance, if the connection of one part of a circuit, with regard to the other part of the circuit, is reversed, a transient term is produced by this reversal, but the final or permanent condition after the reversal is the same as before, except that the current, voltage, etc., in the part of the circuit which has been reversed, are now in opposite direction. In this latter case, the same change can be produced again and again after equal intervals of time t_0, and thus the transient term made to recur periodically. The electric quantities i, e, etc., of the circuit, from time $t = 0$ to $t = t_0$, have the same values as from time $t = t_0$ to $t = 2 t_0$, from $t = 2 t_0$ to $t = 3 t_0$, etc., and it is sufficient to investigate one cycle, from $t = 0$ to $t = t_0$.

In this case, the starting values of the electrical quantities during each period are the end values of the preceding period, or, in other words, the terminal values at the moment of start of the transient term, $t = 0$, $i = i_0$ and $e = e_0$, are the same as the values at the end of the period $t = t_0$, $i = i'$ and $e = e'$; that is, $i_0 = \pm i'$, $e_0 = \pm e'$, etc.; where, the plus sign applies for the unchanged, and the minus sign for the reversed part of the circuit.

2. With such periodically recurrent changes of circuit conditions, the period of recurrence t_0 may be so long, that the transient term produced by a change has died out, the permanent conditions reached, before the next change takes place. Or, at the moment where a change of circuit conditions starts a transient term, the transient term due to the preceding change has not yet disappeared, that is, the time, t_0, of a period is shorter than the duration of the transient term.

In the first case, the terminal or starting values, that is, the values at the moment when the change begins, are the same as the permanent values, and periodic recurrence has no effect on the character of the transient term, but the phenomenon is calculated as discussed in Section I, as single transient term, which gradually dies out.

If, however, at the moment of change, the transient term of the preceding change has not yet vanished, then the starting or terminal values of the electric quantities, as i_0 and e_0, also contain a transient term, namely, that existing at the end of the preceding period. The same term then exists also at the end of the period, or at $t = t_0$. Hence in this case, the terminal conditions are given, not as fixed numerical values, but as an equation between the electric quantities at time $t = 0$ and at time $t = t_0$; or, at the beginning and at the end of the period, and the integration constants, thus, are calculated from this equation.

3. In general, the permanent values of electric quantities after a change are not the same as before, and therefore at least two changes are required before the initial condition of the circuit is restored, and the cycle can be repeated. Periodically recurring transient phenomena, thus usually consist of two or more successive changes, at the end of which the original condition of the circuit is reproduced, and therefore the series of changes can be repeated. For instance, increasing the resistance of a circuit brings about a change. Decreasing this resistance again to its original value brings about a second change, which restores the condition existing before the first change, and thus completes the cycle. In this case, then, the starting values of the electric quantities during the first part of the period equal the end values during the second part of the period, and the starting values of the second part of the period equal the end values of the first part of the period. That is, if a resistor is inserted at time $t = 0$, short circuited at time $t = t_1$, and inserted again at time $t = t_0$, and e and i are voltage and current respectively during the first, e_1 and i_1 during the second part of the period, we have

$$/e/_{t=0} = /e_1/_{t=t_0}; \quad /e_1/_{t=t_1} = /e/_{t=t_1},$$

and

$$/i/_{t=0} = /i_1/_{t=t_0}; \quad /i_1/_{t=t_1} = /i/_{t=t_1}.$$

If during the times t_1 and $t_0 - t_1$ the transient terms have already vanished, and permanent conditions established, so that the transient terms of each part of the period depend only upon the permanent values during the other part of the period, the length of time t_1 and t_0 has no effect on the transient term, that is, each change of circuit conditions takes place and is calculated independently of the other change, or the periodic recurrence. A number of such cases have been discussed in Section I, as for instance, the effect of cutting a resistor in and out of a divided inductive circuit, paragraph 75, Fig. 33. In this case, four successive changes are made before the cycle recurs: a resistor is cut in, in two steps, and cut out again in two steps, but at each change, sufficient time elapses to reach practically permanent condition.

In general, and especially in those cases of periodic transient phenomena, which are of engineering importance, successive changes occur before the permanent condition is reached, or even approximated after the preceding change, so that frequently the values of the electric quantities are very different throughout the whole cycle from the permanent values which they would gradually assume; that is, the transient term preponderates in the values of current, voltage, etc., and the permanent term occasionally is very small compared with the transient term.

4. Periodic transient phenomena are of engineering importance mainly in three cases: (1) in the control of electric circuits; (2) in the production of high frequency currents, and (3) in the rectification of alternating currents.

1. In controlling electric circuits, etc., by some operating mechanism, as a potential magnet increasing and decreasing the resistance of the circuit, or a clutch shifting brushes, etc., the main objections are due to the excess of the friction of rest over the friction while moving. This results in a lack of sensitiveness, and an overreaching of the controlling device. To overcome the friction of rest, the deviation of the circuit from normal must become greater than necessary to maintain the motion of the operating mechanism, and when once started, the mechanism overreaches. This objection is eliminated by never allowing the operating mechanism to come to rest, but arranging it in unstable equilibrium, as a "floating system," so that the condition of the circuit is never normal, but continuously and periodically varies between the two extremes, and the resultant effect is the average of the transient terms, which rapidly and periodically succeed each other. By changing the relative duration of the successive transient terms, any resultant intermediary between the two extremes can thus be produced. On this principle, for instance, operated the controlling solenoid of the Thomson-Houston arc machine, and also numerous automatic potential regulators.

2. Production of high frequency oscillating currents by periodically recurring condenser discharges has been discussed under "oscillating current generator," in Section I, paragraph 44.

High frequency alternating currents are produced by an arc, when made unstable by shunting it with a condenser, as discussed before.

The Ruhmkorff coil or inductorium also represents an application of periodically recurring transient phenomena, as also does Prof. E. Thomson's dynamostatic machine.

3. By reversing the connections between a source of alter-

nating voltage and the receiver circuit, synchronously with the alternations of the voltage, the current in the receiver circuit is made unidirectional (though more or less pulsating) and therefore rectified.

In rectifying alternating voltages, either both half waves of voltage can be taken from the same source, as the same transformer coil, and by synchronous reversal of connections sent in the same direction into the receiver circuit, or two sources of voltage, as the two secondary coils of a transformer, may be used, and the one half wave taken from the one source, and sent into the receiver circuit, the other half wave taken from the other source, and sent into the receiver circuit in the same direction as the first half wave. The latter arrangement has the disadvantage of using the alternating current supply source less economically, but has the advantage that no reversal, but only an opening and closing of connections, is required, and is therefore the method commonly applied in stationary rectifying apparatus.

5. In rectifying alternating voltages, the change of connections between the alternating supply and the unidirectional receiving circuit can be carried out as outlined below:

(a) By a synchronously moving commutator or contact maker, in mechanical rectification. Such mechanical rectifiers may again be divided, by the character of the alternating supply voltage, into single phase and polyphase, and by the character of the electric circuit, into constant potential and constant current rectifiers. Mechanical rectification by a commutator driven by a separate synchronous motor has not yet found any extensive industrial application. Rectification by a commutator driven by the generator of the alternating voltage has found very extended and important industrial use in the excitation of the field, or a part of the field (the series field) of alternators and synchronous motors, and especially in the constant-current arc machine. The Brush arc machine is a quarter-phase alternator connected to a rectifying commutator on the armature shaft, and the Thomson-Houston arc machine is a star-connected three-phase alternator connected to a rectifying commutator on the armature shaft. The reason for using rectification in these machines, which are intended to produce constant direct current at very high voltage, is that the ordinary commutator of the continuous-current machine cannot safely commutate, even at limited current, more than 30 to 50 volts per commutator segment, while the rectifying commutator of the constant-current arc machine can control from 2000 to 3000 volts per segment, and therefore rectification is superior to commutation for very high voltages at limited current, as explained by the character of this phenomenon, discussed in Chapter III.

(b) The synchronous change of circuit connection required by the rectification of alternating e.m.fs. can be brought about without any mechanical motion in so-called "arc rectifiers," by the characteristic properties of the electric arc, to be a good conductor in one, an insulator in the opposite direction. By thus inserting an arc in the path of the alternating circuit, current can exist and thus a circuit be established for that half wave of alternating voltage, which sends the current in the same direction as the current in the arc, while for the reversed half wave of voltage the arc acts as open circuit. As seen, the arc cannot reverse, but only open and close the circuit, and so can rectify only one half wave, that is, two separate sources of alternating voltage, or two rectifiers with the same source of voltage, are required to rectify both half waves of alternating voltage.

(c) Some electrolytic cells, as those containing aluminum as one terminal, offer a low resistance to the passage of current in one direction, but a very high resistance, or practically interrupt the current, in opposite direction, due to the formation of a non-conducting film on the aluminum, when it is the positive terminal. Such electrolytic cells can therefore be used for rectification in a similar manner as arcs.

The three main classes of rectifiers thus are: (a) mechanical rectifiers; (b) arc rectifiers; (c) electrolytic rectifiers.

Still other methods of rectification, as by the unidirectional character of vacuum discharges, of the conduction in some crystals, etc., are not yet of industrial importance.

CHAPTER II.

CIRCUIT CONTROL BY PERIODIC TRANSIENT PHENOMENA.

6. As an example of a system of periodic transient phenomena, used for the control of electric circuits, may be considered an automatic potential regulator operating in the field circuit of the exciter of an alternating current system.

Let, $r_0 = 40$ ohms = resistance and $L = 400$ henrys = inductance of the exciter field circuit.

A resistor, having a resistance, $r_1 = 24$ ohms, is inserted in series to r_0, L in the exciter field, and a potential magnet, controlled by the alternating current system, is arranged so as to short circuit resistance, r_1, if the alternating potential is below, to throw resistance r_1 into circuit again, if the potential is above normal.

With a single resistance step, r_1, in the one position of the regulator, with r_1 short circuited, and only r_0 as exciter field winding resistance, the alternating potential would be above normal, that is, the regulator cannot remain in this position, but as soon after short circuiting resistance r_1 as the potential has risen sufficiently, the regulator must change its position and cut resistance r_1 into the circuit, increasing the exciter field circuit resistance to $r_0 + r_1$. This resistance now is too high, would lower the alternating potential too much, and the regulator thus cuts resistance r_1 out again. That is, the regulator continuously oscillates between the two positions, corresponding to the exciter field circuit resistances r_0 and $(r_0 + r_1)$ respectively, at a period depending on the momentum of the moving mass, the force of the magnets, etc., that is, approximately constant. The time of contact in each of the two positions, however, varies: when requiring a high field excitation, the regulator remains a longer time in position r_0, hence a shorter time in position $(r_0 + r_1)$, before the rising potential throws it over into the next position; while at light load, requiring low field excitation, the duration of the period of high resistance,

$(r_0 + r_1)$, is greater, and that of the period of low resistance, r_0, less.

7. Let, t_1 = the duration of the short circuit of resistance r_1; t_2 = the time during which resistance r_1 is in circuit, and $t_0 = t_1 + t_2$.

During each period t_0, the resistance of the exciter field, therefore, is r_0 for the time t_1, and $(r_0 + r_1)$ for the time t_2.

Furthermore, let, i_1 = the current during time t_1, and i_2 = the current during time t_2.

During each of the two periods, let the time be counted anew from zero, that is, the transient current i_1 exists during the time $0 < t < t_1$, through the resistance r_0, the transient current, i_2, during the time $0 < t < t_2$, through the resistance $(r_0 + r_1)$.

This gives the terminal conditions:

and
$$\left. \begin{aligned} /i_1/_{t=0} &= /i_2/_{t=t_2} \\[2mm] /i_2/_{t=0} &= /i_1/_{t=t_1}; \end{aligned} \right\} \tag{1}$$

that is, the starting point of the current, i_1, is the end value of the current, i_2, and inversely.

If now, e = voltage impressed upon the exciter field circuit, the differential equations are:

and
$$\left. \begin{aligned} e &= r_0 i_1 + L \frac{di_1}{dt} \\[4mm] e &= (r_0 + r_1)\, i_2 + L \frac{di_2}{dt}; \end{aligned} \right\} \tag{2}$$

or,

$$\left. \begin{aligned} \frac{di_1}{i_1 - \dfrac{e}{r_0}} &= -\frac{r_0}{L}\, dt, \\[6mm] \frac{di_2}{i_2 - \dfrac{e}{r_0 + r_1}} &= -\frac{r_0 + r_1}{L}\, dt. \end{aligned} \right\} \tag{3}$$

Integrated,

$$i_1 = \frac{e}{r_0} + c_1 \varepsilon^{-\frac{r_0}{L}t}$$

and

$$i_2 = \frac{e}{r_0 + r_1} + c_2 \varepsilon^{-\frac{r_0 + r_1}{L}t}. \qquad (4)$$

Substituting the terminal conditions (1) in equations (4), gives for the integration constants c_1 and c_2 the equations,

$$\frac{e}{r_0} + c_1 = \frac{e}{r_0 + r_1} + c_2 \varepsilon^{-\frac{r_0 + r_1}{L}t_2}$$

and

$$\frac{e}{r_0 + r_1} + c_2 = \frac{e}{r_0} + c_1 \varepsilon^{-\frac{r_0}{L}t_2};$$

herefrom,

$$c_1 = - \frac{er_1\left\{1 - \varepsilon^{-\frac{r_0 + r_1}{L}t_2}\right\}}{r_0(r_0 + r_1)\left\{1 - \varepsilon^{-\frac{r_0}{L}t_1 - \frac{r_0 + r_1}{L}t_2}\right\}}$$

and

$$c_2 = + \frac{er_1\left\{1 - \varepsilon^{-\frac{r_0}{L}t_1}\right\}}{r_0(r_0 + r_1)\left\{1 - \varepsilon^{-\frac{r_0}{L}t_1 - \frac{r_0 + r_1}{L}t_2}\right\}}. \qquad (5)$$

Substituting (5) in (4),

$$i_1 = \frac{e}{r_0}\left\{1 - \frac{r_1\left\{1 - \varepsilon^{-\frac{r_0 + r_1}{L}t_2}\right\}}{(r_0 + r_1)\left\{1 - \varepsilon^{-\frac{r_0}{L}t_1 - \frac{r_0 + r_1}{L}t_2}\right\}} \varepsilon^{-\frac{r_0}{L}t}\right\}$$

and

$$i_2 = \frac{e}{r_0 + r_1}\left\{1 + \frac{r_1\left\{1 - \varepsilon^{-\frac{r_0}{L}t_1}\right\}}{r_0\left\{1 - \varepsilon^{-\frac{r_0}{L}t_1 - \frac{r_0 + r_1}{L}t_2}\right\}} \varepsilon^{-\frac{r_0 + r_1}{L}t}\right\}. \qquad (6)$$

If, $e = 250$ volts; $t_0 = 0.2$ sec., or 5 complete cycles per sec.; $t_1 = 0.15$, and $t_2 = 0.05$ sec.; then

$$i_1 = 6.25\left\{1 - 0.128\,\varepsilon^{-0.1\,t}\right\}$$

and

$$i_2 = 3.91\left\{1 + 0.391\,\varepsilon^{-0.16\,t}\right\}. \qquad (7)$$

8. The mean value of current in the circuit is

$$i = \frac{1}{t_1 + t_2} \left\{ \int_0^{t_1} i_1 dt + \int_0^{t_2} i_2 dt \right\}. \tag{8}$$

This integrated gives,

$$i = \frac{t_1 \dfrac{e}{r_0} + t_2 \dfrac{e}{r_0 + r_1}}{t_1 + t_2}; \tag{9}$$

and, if

$$i_1' = \frac{e}{r_0}$$

and

$$i_2' = \frac{e}{r_0 + r_1} \tag{10}$$

are the two extreme values of permanent current, corresponding respectively to the resistances r_0 and $(r_0 + r_1)$, we have

$$i = \frac{t_1 i_1' + t_2 i_2'}{t_1 + t_2} = \frac{t_1}{t_0} i_1' + \frac{t_2}{t_0} i_2'; \tag{11}$$

that is, the current, i, varies between i_1' and i_2' as linear function of the durations of contact, t_1 and t_2.

The maximum variation of current during the periodic change is given by the ratio of maximum current and minimum current; or,

$$\left|\frac{i_2}{i_1}\right|_{t=0} = q, \tag{12}$$

and is

$$q = \frac{r_0 \left(1 - \varepsilon^{-s_1 - s_2}\right) + r_1 \left(1 - \varepsilon^{-s_1}\right)}{r_0 \left(1 - \varepsilon^{-s_1 - s_2}\right) + r_1 \varepsilon^{-s_2} \left(1 - \varepsilon^{-s_1}\right)}; \tag{13}$$

where,

$$s_1 = \frac{r_0}{L} t_1$$

and

$$s_2 = \frac{r_0 + r_1}{L} t_2. \tag{14}$$

Substituting

$$1 - \varepsilon^{-x} = x - \frac{x^2}{2} + \frac{x^3}{6} - + \dots , \qquad (15)$$

by using only term of first order;

gives
$$\left. \begin{array}{c} 1 - \varepsilon^{-x} = x, \\[2mm] q = 1; \end{array} \right\} \qquad (16)$$

that is, the primary terms eliminate, and the difference between i_1 and i_2 is due to terms of secondary order only, hence very small.

Substituting

$$1 - \varepsilon^{-x} = x - \frac{x^2}{2}; \qquad (17)$$

that is, using also terms of second order, gives

$$q = \frac{\{r_0 (s_1 + s_2) + r_1 s_1\} - \frac{1}{2}\{r_0 (s_1 + s_2)^2 + r_1 s_1^2\}}{\{r_0 (s_1 + s_2) + r_1 s_1\} - \frac{1}{2}\{r_0 (s_1 + s_2)^2 + r_1 s_1^2 + 2 r_1 s_1 s_2\}}; \qquad (18)$$

or, approximately,

$$q = 1 + \frac{r_1 s_1 s_2}{r_0 (s_1 + s_2) + r_1 s_1}, \qquad (19)$$

and, substituting (14),

$$q = 1 + \frac{r_1 t_1 t_2}{L (t_1 + t_2)}; \qquad (20)$$

that is, the percentage variation of current is

$$q - 1 = \frac{r_1 t_1 t_2}{L (t_1 + t_2)}. \qquad (21)$$

Equation (21) is a maximum for

$$t_1 = t_2 = \frac{t_0}{2}, \qquad (22)$$

and, then, is

$$q - 1 = \frac{r_1 t_0}{4L}; \qquad (23)$$

or, in the above example, $(r_1 = 24; \ L = 400; \ t_0 = 0.2)$;

$$q - 1 = 0.003;$$

that is, 0.3 per cent.

The time t_0 of a cycle, which gives 1 per cent variation of current, $q - 1 = 0.01$, is

$$t_0 = \frac{4L}{r_1} (q - 1), \tag{24}$$

$$= \tfrac{2}{3} \text{ sec.}$$

The pulsation of current, 0.3 per cent respectively 1 per cent, thus is very small compared with the pulsation of the resistance, $r_1 = 24$ ohms, which is 46 per cent of the average resistance $r_0 + \dfrac{r_1}{2} = 52$ ohms.

CHAPTER III.

MECHANICAL RECTIFICATION.

9. If an alternating-current circuit is connected, by means of a synchronously operated circuit breaker or rectifier, with a second circuit in such a manner, that the connection between the two circuits is reversed at or near the moment when the alternating voltage passes zero, then in the second circuit current and voltage are more or less unidirectional, although they may not be constant, but pulsating.

If i = instantaneous value of alternating current, and i_0 = instantaneous value of rectified current, then we have, before reversal, $i_0 = i$, and after reversal, $i_0 = -i$; that is, during the reversal of the circuit one of the currents must reverse. Since, however, due to the self-inductance of the circuits, neither current can reverse instantly, the reversal occurs gradually, so that for a while during rectification the instantaneous value of the alternating and of the rectified current differ from each other. Thus means have to be provided either to shunt the difference between the two currents through a non-inductive bypath, or, the difference of the two currents exists as arc over the surface of the rectifying commutator.*

The general phenomenon of single-phase rectification thus is: The alternating and the rectified circuit are in series. Both circuits are closed upon themselves at the rectifier, by the resistances, r and r_0, respectively. The terminals are reversed. The shunt-resistance circuits are opened, leaving the circuits in series in opposite direction.

Special cases hereof are:

1. If $r = r_0 = 0$, that is, during rectification both circuits are short circuited. Such short-circuit rectification is feasible only in limited-current circuits, as on arc lighting machines, or circuits of high self-inductance, or in cases where the voltage of the recti-

---

* If the circuit is reversed at the moment when the alternating current passes zero, due to self-inductance of the rectified circuit its current differs from zero, and an arc still appears at the rectifier

fied circuit is only a small part of the total voltage, and thus the current not controlled thereby, as when rectifying for the supply of series fields of alternators.

2. $r = r_0 = \infty$, or open circuit rectification. This is feasible only if the rectified circuit contains practically no self-inductance, but a constant counter e.m.f., e, (charging storage batteries), so that in the moment when the alternating impressed e.m.f. falls to e, and the current disappears, the circuit is opened, and closed again in opposite direction when after reversal the alternating impressed e.m.f. has reached the value, e.

In polyphase rectification, the rectified circuit may be fed successively by the successive phases of the system, that is shifted over from a phase of falling e.m.f. to a phase of rising e.m.f., by shunting the two phases with each other during the time the current changes from the one to the next phase. Thus the Thomson-Houston arc machine is a star-connected three-phase constant-current alternator with rectifying commutator. The Brush arc machine is a quarter-phase machine with rectifying commutator.

In rectification frequently the sine wave term of the current is entirely overshadowed by the transient exponential term, and thus the current in the rectified circuit is essentially of an exponential nature.

As examples, three cases will be discussed:

1. Single-phase constant-current rectification; that is, a rectifier is inserted in an alternating-current circuit, and the voltage consumed by the rectified circuit is small compared with the total circuit voltage; the current thus is not noticeably affected by the rectifier. In other words, a sine wave of current is sent over a rectifying commutator.

2. Single-phase constant-potential rectification; that is, a constant-potential alternating e.m.f. is rectified, and the impedance between the alternating voltage and the rectifying commutator is small, so that the rectified circuit determines the current wave shape.

3. Quarter-phase constant-current rectification as occurring in the Brush arc machine.

1. Single-phase constant-current rectification.

10. A sine wave of current, $i_0 \sin \theta$, derived from an e.m.f. very large compared with the voltage consumed in the rectified circuit, feeds, after rectification, a circuit of impedance $Z = r - jx$. This circuit is permanently shunted by a circuit of resistance r_1.

Rectification takes place over short-circuit from the moment $\pi - \theta_2$ to $\pi + \theta_1$; that is, at $\pi - \theta_2$ the rectified and the alternating circuit are closed upon themselves at the rectifier, and this short-circuit opened, after reversal, at $\pi + \theta_1$, as shown by the diagrammatic representation of a two-pole model of such a rectifier in Fig. 54. In this case the space angles $\pi + \tau_1$ and $\pi - \tau_2$ and the time angles $\pi + \theta_1$ and $\pi - \theta_2$ are identical.

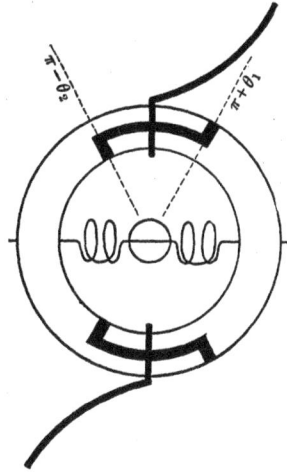

This represents the conditions existing in compound-wound alternators, that is, alternators feeding a series field winding through a rectifier.

Fig. 54.   Single-phase current rectifier commutator.

Let, during the period from θ_1 to $\pi - \theta_2$, i = current in impedance Z, and i_1 = current in resistance r_1, then:

$$i + i_1 = i_0 \sin \theta. \tag{1}$$

However,

$$i_1 r_1 = ir + x \frac{di}{d\theta} \tag{2}$$

and substituting (1) in (2) gives the differential equation:

$$i (r + r_1) + x \frac{di}{d\theta} - i_0 r_1 \sin \theta = 0, \tag{3}$$

which is integrated by the function:

$$i = A\varepsilon^{-a\theta} + B \sin (\theta - \delta). \tag{4}$$

Substituting (4) in (3) and arranging, gives:

$$A (r + r_1 - ax) \varepsilon^{-a\theta} + [B ([r + r_1] \cos \delta + x \sin \delta) - i_0 r_1] \sin \theta$$
$$- [(r + r_1) \sin \delta - x \cos \delta] B \cos \theta = 0, \tag{5}$$

which equation must be an identity, thus:

$$r + r_1 - ax = 0,$$
$$B\left([r + r_1]\cos\delta + x\sin\delta\right) - i_0 r_1 = 0$$

and
$$(r + r_1)\sin\delta - x\cos\delta = 0,$$

and herefrom:

$$a = \frac{r + r_1}{x},$$

$$\tan\delta = \frac{x}{r + r_1} \tag{6}$$

and
$$B = i_0 \frac{r_1}{\sqrt{(r + r_1)^2 + x^2}} = \frac{i_0 r_1}{z},$$

where

$$z = \sqrt{(r + r_1)^2 + x^2}; \tag{7}$$

hence:

$$i = A\,\varepsilon^{-\frac{r + r_1}{x}\theta} + i_0 \frac{r_1}{z}\sin(\theta - \delta). \tag{8}$$

During the time of short-circuit, from $\pi - \theta_2$ to $\pi + \theta_1$, if $i' = $ current in impedance Z, we have

$$i'r + x\frac{di'}{d\theta} = 0, \tag{9}$$

hence:

$$i' = A'\varepsilon^{-\frac{r}{x}\theta}. \tag{10}$$

The condition of sparkless rectification is, that no sudden change of current occur anywhere in the system. In consequence hereof we must have:

$$i = i' = i_0\sin\theta \text{ at the moment } \theta = \pi - \theta_2,$$

and, at the moment $\theta = \pi + \theta_1$, i' must have reached the same value as i and $i_0\sin\theta$ at the moment $\theta = \theta_1$.

This gives the two double equations:

$$\left.\begin{array}{c} i_{\pi-\theta_2} = i'_{\pi-\theta_2} = i_0 \sin(\pi - \theta_2) \\ i_{\theta_1} = i'_{\pi+\theta_1} = i_0 \sin \theta_1; \end{array}\right\} \quad (11)$$

or, substituting (8) and (9),

$$A\,\varepsilon^{-\frac{r+r_1}{x}(\pi-\theta_2)} + i_0\frac{r_1}{z}\sin(\delta+\theta_2) = A'\varepsilon^{-\frac{r}{x}(\pi-\theta_2)} = i_0 \sin \theta_2 \quad (12)$$

and

$$A\,\varepsilon^{-\frac{r+r_1}{x}\theta_1} - i_0\frac{r_1}{z}\sin(\delta-\theta_1) = A'\varepsilon^{-\frac{r}{x}(\pi+\theta_1)} = i_0 \sin \theta_1. \quad (13)$$

These four equations (12) (13) determine four of the five quantities, A, A', θ_1, θ_2, r_1, leaving one indeterminate.

Thus, one of these five quantities can be chosen. The determination of the four remaining quantities, however, is rather difficult, due to the complex character of equations (12) (13), and is feasible only by approximation, in a numerical example.

11. EXAMPLE: Let an alternating current of effective value of 100 amp., that is, of maximum value $i_0 = 141.4$, be rectified for the supply of a circuit of impedance $Z = 0.2 - 2\,j$, shunted by a non-inductive circuit of resistance r_1.

Let the series connection of the rectified and alternating circuits be established 30 time-degrees after the zero value of alternating current, that is, $\theta_1 = 30$ deg. $= \dfrac{\pi}{6}$ chosen.

Then, from equation (13), we have

$$A'\varepsilon^{-\frac{r}{x}(\pi+\theta_1)} = i_0 \sin \theta_1,$$

hence, substituting r, x, θ_1, i_0, gives

$$A' = 102.$$

From equation (12),

$$A'\varepsilon^{-\frac{r}{x}(\pi-\theta_2)} = i_0 \sin \theta_2,$$

and, substituting,

$$\sin \theta_2 = 0.527\ \varepsilon^{0.1\theta_2};$$

approximately

$$\sin \theta_2 = 0.527 \text{ and } \theta_2 = 32°;$$

thus, more closely

$$\sin \theta_2 = 0.527 \; \epsilon^{3.20°} = 0.558, \text{ and } \theta_2 = 34°;$$

thus, more closely

$$\sin \theta_2 = 0.527 \; \epsilon^{3.4°} = 0.559, \text{ and } \theta_2 = 34°.$$

From equations (12) and (13) it follows:

$$A\epsilon^{-\frac{r+r_1}{x}(\pi-\theta_2)} + i_0 \frac{r_1}{z} \sin(\delta + \theta_2) = i_0 \sin \theta_2,$$

$$A\epsilon^{-\frac{r+r_1}{x}\theta_1} - i_0 \frac{r_1}{z} \sin(\delta - \theta_1) = i_0 \sin \theta_1;$$

eliminating A gives

$$\epsilon^{-\frac{r}{x}(\pi-\theta_1-\theta_2)} \epsilon^{-\frac{r_1}{x}(\pi-\theta_1-\theta_2)} = \frac{z \sin \theta_2 - r_1 \sin(\delta + \theta_2)}{z \sin \theta_1 + r_1 \sin(\delta - \theta_1)};$$

substituting $\sin \delta = \dfrac{x}{z}$, $\cos \delta = \dfrac{r+r_1}{z}$, $z^2 = (r+r_1)^2 + x^2$, and

substituting for r, x, θ_1, θ_2, gives after some changes:

$$\epsilon^{-58 r_1°} = \frac{1.5 - 1.04\, r_1}{1.1 - r_1};$$

calculating by approximation,

assuming $r_1 = 0.5$,
$$0.603 = 0.612;$$

assuming $r_1 = 0.51$,
$$0.597 = 0.602;$$

assuming $r_1 = 0.52$,
$$0.591 = 0.592;$$

hence, $r_1 = 0.52$,

and $z = 2.124$,

$$\delta = 70°.$$

Substituting these values in (12) or (13) gives

$$A = 114;$$

hence, as final equations, we have

$$i = 112 \, \varepsilon^{-0.36\,\theta} + 34.6 \sin(\theta - 70°),$$
$$i' = 102 \, \varepsilon^{-0.1\,\theta},$$
$$i_0 = 141.4 \sin \theta,$$

and
$$i_1 = i_0 - i;$$

which gives the following results:

Quantity.	Instantaneous Values.										Effective Value.	Arithmetic Mean Value.
$\theta° =$	30	50	70	90	110	130	146	170	190	210	. . . . . .	. . . .
$i =$	70.8	70.5	72.6	74.9	79.0	80.4	79.0	. . . . .	. . . . .	. . . . .	75.2	75.2
$i' =$							79.0	75.8	73.2	70.8		
$i_0 \sin\theta =$	70.8	108	133	141.4	133	108	79.0	24.7	−24.7	−70.8	100.0	. . . . .
$i_1 =$	0	37.5	60.4	66.5	54.0	27.6	0	(−51.1	−48.5)	0	38.2	27.3
	. . . .	. . . .	. . . .	. . . .	. . . .	. . . .	. . . .	. . . .			(44.9)	

Curves of these quantities are plotted in Fig. 55, for $i_0 = 100 \sin \theta$.

The effective value of the rectified current is 75.2 amp., and this current is fairly constant, pulsating only between 70.5 and 80.4 amp., or by 6.6 per cent from the mean; that is, due to the self-inductance, the fluctuations of current are practically suppressed, and taken up by the non-inductive shunt, and the arithmetic mean value of this current is therefore equal to its effective value. The effective value of the shunt current is 38.2 amp., and this current is unidirectional also, but very fluctuating. Its arithmetic mean value is only 27.3 amp.; that is, in this circuit a continuous-current ammeter would record 27.3, an alternating ammeter 38.2 amperes. The effective value of the total difference between alternating and rectified current (shunt plus short-circuit current) is 44.9 amp.

The current divides between the inductive rectified circuit and its non-inductive shunt, not in proportion to their respective impedances, but more nearly, though not quite, in proportion to the resistances; that is, in a rectified circuit, self-inductance does not greatly affect the intensity of the current, but only its character as regards fluctuations.

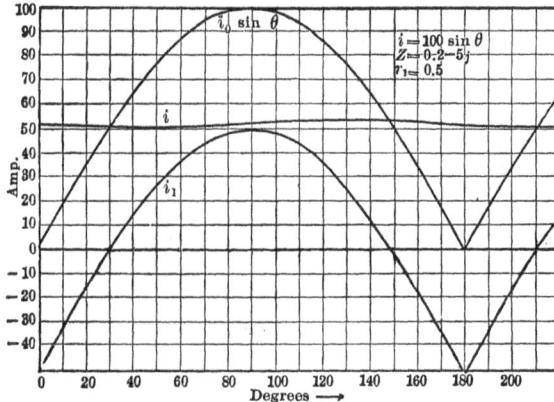

Fig. 55.   Single-phase current rectification.

2. Single-phase constant-potential rectification.

12. Let the alternating e.m.f. $e_0 \sin \theta$ of the alternating circuit of impedance $Z_0 = r_0 - jx_0$ be rectified by connecting it at the moment θ_1 with the direct-current receiver circuit of impedance $Z = r - jx$ and continuous counter e.m.f. e, disconnecting it therefrom at the moment $\pi - \theta_2$, and closing during the time from $\pi - \theta_2$ to $\pi + \theta_1$ the alternating circuit by the resistance r_1, the direct-current circuit by the resistance r_2, then connecting the circuits again in series in opposite direction, at $\pi + \theta_1$, etc., as shown diagrammatically by Fig. 56, where

$$ r_1 = \cfrac{1}{\cfrac{1}{r' + r''} + \cfrac{1}{r''' + r''''}} \, ; $$

$$ r_2 = \cfrac{1}{\cfrac{1}{r' + r''''} + \cfrac{1}{r'' + r'''}} \, . $$

1. Then, during the time from θ_1 to $\pi - \theta_2$, if $i_1 =$ current, the differential equation is

$$ e_0 \sin \theta - e - i_1 (r + r_0) - (x + x_0) \frac{di_1}{d\theta} = 0, \tag{1} $$

which is integrated by

$$i_1 = A_1 + B_1 \varepsilon^{-a_1 \theta} + C_1 \sin (\theta - \delta_1) \qquad (2)$$

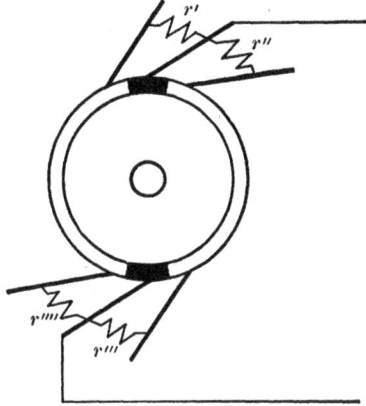

Fig. 56. Single-phase constant-potential rectifying commutator.

Equation (2) substituted in (1) gives

$$e_0 \sin \theta - e - (r + r_0) [A_1 + B_1 \varepsilon^{-a_1 \theta} + C_1 \sin (\theta - \delta_1)]$$
$$- (x + x_0) [- a_1 B_1 \varepsilon^{-a_1 \theta} + C_1 \cos (\theta - \delta_1)] = 0;$$

or, transposing,

$$- [+ e + (r + r_0) A_1] + B_1 \varepsilon^{-a\theta} [a_1 (x + x_0) - (r + r_0)]$$
$$+ \sin \theta [e_0 - (r + r_0) C_1 \cos \delta_1 - (x + x_0) C_1 \sin \delta_1]$$
$$+ C_1 \cos \theta [(r + r_0) \sin \delta_1 - (x + x_0) \cos \delta_1] = 0;$$

herefrom it follows that

$$e + (r + r_0) A_1 = 0,$$
$$a_1 (x + x_0) - (r + r_0) = 0,$$
$$e_0 - (r + r_0) C_1 \cos \delta_1 - (x + x_0) C_1 \sin \delta_1 = 0,$$

and

$$(r + r_0) \sin \delta_1 - (x + x_0) \cos \delta_1 = 0;$$

hence

$$A_1 = -\frac{e}{r+r_0},$$

$$a_1 = \frac{r+r_0}{x+x_0},$$

$$\tan \delta_1 = \frac{x+x_0}{r+r_0},$$

and

$$C_1 = \frac{e_0}{\sqrt{(r+r_0)^2 + (x+x_0)^2}},$$

$$\tag{3}$$

and, substituting in (2),

$$i_1 = -\frac{e}{r+r_0} + B_1 \varepsilon^{-\frac{r+r_0}{x+x_0}\theta} + \frac{e_0 \sin(\theta - \delta_1)}{\sqrt{(r+r_0)^2 + (x+x_0)^2}},$$

$$= -\frac{e}{r+r_0} + B_1 \varepsilon^{-\frac{r+r_0}{x+x_0}\theta} + \frac{e_0[(r+r_0)\sin\theta - (x+x_0)\cos\theta]}{(r+r_0)^2 + (x+x_0)^2},$$

$$\tan \delta_1 = \frac{x+x_0}{r+r_0}.$$

$$\tag{4}$$

2. During the time from $\pi - \theta_2$ to $\pi + \theta_1$, if i_2 = current in the direct circuit, i_3 = current in alternating circuit, we have

Alternating-current circuit:

$$e_0 \sin \theta - i_3 (r_0 + r_1) - x_0 \frac{di_3}{d\theta} = 0, \tag{5}$$

which is integrated the same as in (1), by

$$i_3 = B_3 \varepsilon^{-\frac{r_0+r_1}{x_0}\theta} + \frac{e_0 \sin(\theta - \delta_3)}{\sqrt{(r_0+r_1)^2 + x_0^2}}$$

$$= B_3 \varepsilon^{-\frac{r_0+r_1}{x_0}\theta} + \frac{e_0[(r_0+r_1)\sin\theta - x_0\cos\theta]}{(r_0+r_1)^2 + x_0^2},$$

$$\tan \delta_3 = \frac{x_0}{r_0+r_1}.$$

$$\tag{6}$$

Direct-current circuit:

$$- e - i_2 (r + r_2) - x \frac{d i_2}{d \theta} = 0, \tag{7}$$

integrated by

$$i_2 = - \frac{e}{r + r_2} + B_2 \varepsilon^{- \frac{r + r_2}{x} \theta} \tag{8}$$

At $\theta = \pi - \theta_2$, however, we must have

$$i_1 = i_2 = i_3$$

and $\quad i_2$ at $\theta = \pi + \theta_1$ must be equal to

$\quad i_1$ at $\theta = \theta_1$, and opposite to i_3 at $\theta = \pi + \theta_1$;

$\quad i_1 [\theta = \theta_1] = i_2 [\theta = \pi + \theta_1] = - i_3 [\theta = \pi + \theta_1].$ $\qquad$ (9)

These terminal conditions represent four equations, which suffice for the determination of the three remaining integration constants, B_1, B_2, B_3, and one further constant, as θ_1 or θ_2, or r_1 or r_2, or e; that is, with the circuit conditions Z_0, Z, r_1, r_2, e_0, e chosen, the moment θ_1 depends on θ_2 and inversely.

13. *Special case:*

$$Z_0 = 0, \quad r_2 = 0, \quad e = 0; \tag{10}$$

that is, the alternating e.m.f. $e_0 \sin \theta$ is connected to the circuit of impedance $Z = r - jx$ during time θ_1 to $\pi - \theta_2$, and closed by resistance r_1, while the rectified circuit is short-circuited, during time $\pi - \theta_2$ to $\pi + \theta_1$.

The equations are:

1. Time θ_1 to $\pi - \theta_2$:

$$i_1 = B_1 \varepsilon^{- \frac{r}{x} \theta} + \frac{e_0}{r^2 + x^2} [r \sin \theta - x \cos \theta].$$

2. Time $\pi - \theta_2$ to $\pi + \theta_1$:

$$i_2 = B_2 \varepsilon^{- \frac{r}{x} \theta}$$

$$i_3 = \frac{e_0 \sin \theta}{r_1}. \qquad (11)$$

The terminal conditions now assume the following forms:

At
$$\theta = \pi - \theta_2,$$

$$\dot{i}_1 = \dot{i}_2 = \dot{i}_3$$

$$B_1 \varepsilon^{-\frac{r}{x}(\pi - \theta_2)} + \frac{e_0}{r^2 + x^2}(r \sin \theta_2 + x \cos \theta_2) = B_2 \varepsilon^{-\frac{r}{x}(\pi - \theta_2)}$$

$$= \frac{e_0}{r_1} \sin \theta_2;$$

at $\theta = \pi + \theta_1$ and θ_1 respectively

$$B_1 \varepsilon^{-\frac{r}{x}\theta_1} + \frac{e_0}{r^2 + x^2}(r \sin \theta_1 - x \cos \theta_1) = B_2 \varepsilon^{-\frac{r}{x}(\pi + \theta_1)}$$

$$= \frac{e_0}{r_1} \sin \theta_1.$$

$$(12)$$

These four equations suffice for the determination of the two integration constants B_1 and B_2, and two of the three rectification constants, θ_1, θ_2, r_1, so that one of the latter may be chosen.

Choosing θ_2, the moment of beginning reversal, the equations (12) transposed and expanded give

$$\varepsilon^{-\frac{r}{x}(\theta_1 + \theta_2)} = \frac{\sin \theta_1}{\sin \theta_2},$$

$$\cot \theta_1 + \varepsilon^{\frac{r\pi}{x}} \cot \theta_2 = \left(\frac{r^2 + x^2}{r_1 x} - \frac{r}{x}\right)\left(\varepsilon^{\frac{r\pi}{x}} - 1\right),$$

$$B_2 = \frac{e_0 \sin \theta_2}{r_1} \varepsilon^{+\frac{r}{x}(\pi - \theta_2)},$$

and

$$B_1 = B_2 - \frac{e_0}{r^2 + x^2}(r \sin \theta_2 + x \cos \theta_2)\varepsilon^{\frac{r}{x}(\pi - \theta_2)},$$

$$(13)$$

which give θ_1, r_1, B_2, B_1: θ_1 is calculated by approximation.

Assuming, as an example,

$$e_0 = 156 \sin \theta \text{ (corresponding to 110 volts effective),}$$

$$Z = 10 - 30\,j,$$

and

$$\theta_2 = \frac{\pi}{6} = 30°,$$

$$(14)$$

by equations (13) we have:

$$\log \sin \theta_1 = -0.3765 - 0.1448\,\theta_1,$$

$$\text{and} \quad \theta_1 = 21.7°,$$

$$r_1 = 7.63,$$

$$B_2 = 24.4,$$

and $\quad\quad\quad B_1 = 12.8;$ \hfill (15)

thus

$$i_1 = 12.8\,\varepsilon^{-\frac{\theta}{3}} + 1.56\,(\sin\theta - 3\cos\theta),$$

$$i_2 = 24.4\,\varepsilon^{-\frac{\theta}{3}},$$

and $\quad\quad\quad i_3 = 20.5\sin\theta,$ \hfill (16)

which gives:

$\theta°$.	i_1.	i_2.	i_3.	$\theta°$.	i_1.	i_2.	i_3.
21.7	7.55			135	10.27		
30	7.47			150	10.20	10.2	10.2
45	7.7			165		9.4	5.3
60	8.02			180		8.6	0
75	8.56			195		7.9	−5.3
90	9.18			201.7		7.55	−7.55
105	9.67						
120	10.09						

The mean value of the rectified current is derived herefrom as 8.92 amp., while without rectification the effective value of alternating current would be $\dfrac{110}{\sqrt{r^2 + x^2}} = 3.48$. 110 volts effective corresponds to $\dfrac{2\sqrt{2}}{\pi}\,110 = 99$ volts mean, which in $r = 10$ would give the current as 9.9 amp.

Thus, in a rectified circuit, self-inductance has little effect besides smoothing out the fluctuations of current, which in this case varies between 7.47 and 10.27, with 8.92 as mean, while without self-inductance it would vary between 0 and 15.6, with 9.9 as mean, and without rectification the current would be $4.95\sin(\theta - 71.6°)$.

As seen, in this case the exponential or transient term of current largely preponderates over the permanent or sinusoidal term.

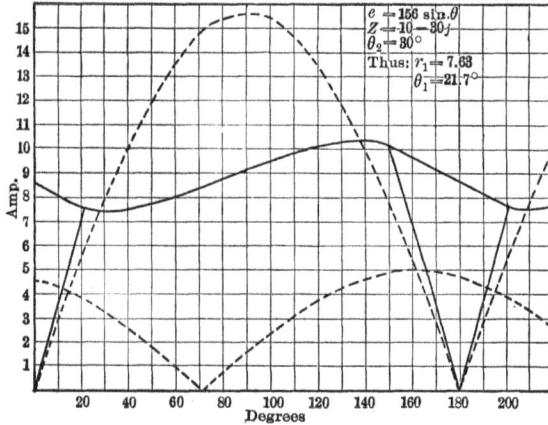

Fig. 57.   Single-phase e.m.f. rectification.

In Fig. 57 is shown the rectified current in drawn line, the value it would have without self-inductance, and the value the alternating current would have, in dotted lines.

3. Quarter-phase constant-current rectification.

14. In the quarter-phase constant-current arc machine, as the Brush machine, two e.m.fs., $E_1 = e \cos \theta$ and $E_2 = e \sin \theta$, are connected to a rectifying commutator, so that while the first E_1 is in circuit E_2 is open-circuited.   At the moment θ_1, E_2 is connected in parallel, as shown diagrammatically in Fig. 58, with E_1, and the rising e.m.f. in E_2 gradually shifts the current i_0 away from E_1 into E_2, until at the moment θ_2, E_1 is disconnected and E_2 left in circuit.

Assume that, due to the superposition of a number of such quarter-phase e.m.fs., displaced in time-phase from each other, and rectified by a corresponding number of commutators offset against each other, and due to self-inductance in the external circuit, the rectified current is practically steady and has the value i_0.   Thus up to the moment θ_1 the current in E_1 is i_0, in

E_2 is 0. From θ_1 to θ_2 the current in E_2 may be i; thus in E_1 it is $i_2 = i_0 - i$. After θ_2, the current in E_1 is 0, in E_2 it is i_0.

A change of current occurs only during the time from θ_1 to θ_2, and it is only this time that needs to be considered.

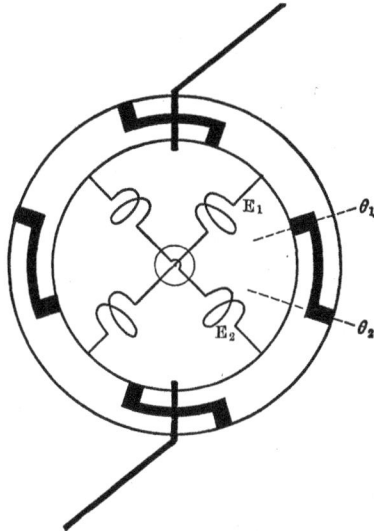

Fig. 58. Quarter-phase constant-current rectifying commutator.

Let $Z = r - jx =$ impedance per phase, where $x = 2\pi f L$; then at the time t and the corresponding angle $\theta = 2\pi f t$ the difference of potential in E_1 is

$$\left.\begin{array}{c} e \cos \theta - (i_0 - i)\, r - L\dfrac{d\,(i_0 - i)}{dt} \\[2mm] = e \cos \theta - (i_0 - i)\, r + x\dfrac{di}{d\theta}; \\[4mm] \text{the difference of potential in } E_2 \text{ is} \\[2mm] e \sin \theta - ir - x\dfrac{di}{d\theta}; \end{array}\right\} \qquad (1)$$

the difference of potential in E_2 is

$$e \sin \theta - ir - x\frac{di}{d\theta};$$

and, since these two potential differences are connected in parallel, they are equal

$$e\,(\sin \theta - \cos \theta) + i_0 r - 2\,ir - 2\,x\frac{di}{d\theta} = 0. \qquad (2)$$

The differential equation (2) is integrated by

$$i = A + B\varepsilon^{-a\theta} + C \cos(\theta - \delta); \qquad (3)$$

thus
$$\frac{di}{d\theta} = -aB\varepsilon^{-a\theta} - C\sin(\theta - \delta),$$

and substituting in (2),

$$e(\sin\theta - \cos\theta) + i_0 r - 2Ar - 2Br\varepsilon^{-a\theta} - 2Cr\cos(\theta - \delta)$$
$$+ 2aBx\varepsilon^{-a\theta} + 2Cx\sin(\theta - \delta) = 0 \cdot$$

or, transposed,

$$(i_0 - 2A)r + 2B\varepsilon^{-a\theta}(ax - r) + \sin\theta\,[e - 2Cr\sin\delta$$
$$+ 2Cx\cos\delta] - \cos\theta\,[e + 2Cr\cos\delta + 2Cx\sin\delta] = 0;$$

thus
$$i_0 - 2A = 0,$$
$$ax - r = 0,$$
$$e - 2Cr\sin\delta + 2Cx\cos\delta = 0,$$

and
$$e + 2Cx\sin\delta + 2Cr\cos\delta = 0,$$

and herefrom, letting $\dfrac{x}{r} = \tan\sigma$, we have

$$\left.\begin{aligned}
e &= -2Cz\sin(\sigma - \delta), \\
e &= -2Cz\cos(\sigma - \delta), \\
A &= \frac{i_0}{2}, \\
a &= \frac{r}{x}, \\
\tan\tau &= \frac{x - r}{x + r}, \\
C &= -\frac{e}{\sqrt{2(x^2 + r^2)}}, \\
\tan(\sigma - \delta) &= 1,
\end{aligned}\right\} \qquad (4)$$

and
$$C = \frac{e}{\sqrt{2z}} \cdot$$

These values substituted in (3) give

$$i = \frac{i_0}{2} + B\varepsilon^{-\frac{r}{x}\theta} - \frac{e}{\sqrt{2\,(x^2 + r^2)}}\cos(\theta - \delta),$$

$$\tan \delta = \frac{x - r}{x + r}. \qquad (5)$$

At $\theta = \theta_1$, $i = 0$, and we have

$$0 = \frac{i_0}{2} + B\varepsilon^{-\frac{r}{x}\theta_1} - \frac{e}{\sqrt{2\,(x^2 + r^2)}}\cos(\theta_1 - \delta);$$

hence,

$$B\varepsilon^{-\frac{r}{x}\theta_1} = \frac{e}{\sqrt{2\,(x^2 + r^2)}}\cos(\theta_1 - \delta) - \frac{i_0}{2}; \qquad (6)$$

substituting in (5), we have the equations of current in the two coils as follows:

$$
\left.
\begin{aligned}
i &= \frac{i_0}{2} + \left(\frac{e}{\sqrt{2\,(x^2 + r^2)}}\cos(\theta_1 - \delta) - \frac{i_0}{2}\right)\varepsilon^{-\frac{r}{x}(\theta - \theta_1)} \\
&\quad - \frac{e}{\sqrt{2\,(x^2 + r^2)}}\cos(\theta - \delta) \\
&= \frac{i_0}{2}\left(1 - \varepsilon^{-\frac{r}{x}(\theta - \theta_1)}\right) \\
&\quad - \frac{e}{\sqrt{2\,(x^2 + r^2)}}\left(\cos(\theta - \delta) - \cos(\theta_1 - \delta)\,\varepsilon^{-\frac{r}{x}(\theta - \theta_1)}\right)\theta_1; \\
\text{and} \\
i_0 - i &= \frac{i_0}{2} - \left(\frac{e}{\sqrt{2\,(x^2 + r^2)}}\cos(\theta_1 - \delta) - \frac{i_0}{2}\right)\varepsilon^{-\frac{r}{x}(\theta - \theta_1)} \\
&\quad + \frac{e}{\sqrt{2\,(x^2 + r^2)}}\cos(\theta - \delta).
\end{aligned}
\right\} (7)
$$

At $\theta = \theta_2$, $i = i_0$; thus

$$\frac{i_0}{2} - \left(\frac{e}{\sqrt{2\,(x^2 + r^2)}} \cos (\theta_1 - \delta) - \frac{i_0}{2}\right) \varepsilon^{-\frac{r}{x}(\theta_2 - \theta_1)}$$

$$+ \frac{e}{\sqrt{2\,(x^2 + r^2)}} \cos (\theta_2 - \delta) = 0;$$

or, multiplied by $\varepsilon^{-\frac{r}{x}\theta_1}$ and rearranged, we have the condition connecting moments θ_1 and θ_2, as follows:

$$\frac{i_0}{2}\left(\varepsilon^{-\frac{r}{x}\theta_1} + \varepsilon^{-\frac{r}{x}\theta_2}\right) + \frac{e}{\sqrt{2\,(x^2 + r^2)}}\left\{\varepsilon^{-\frac{r}{x}\theta_1} \cos (\theta_2 - \delta)\right.$$

$$\left. - \varepsilon^{-\frac{r}{x}\theta_2} \cos (\theta_1 - \delta)\right\} = 0$$

and

$$\frac{i_0}{2}\left(\varepsilon^{\frac{r}{x}\theta_1} + \varepsilon^{\frac{r}{x}\theta_2}\right) = \frac{e}{\sqrt{2\,(x^2 + r^2)}}\left\{\varepsilon^{\frac{r}{x}\theta_1} \cos (\theta_1 - \delta)\right.$$

$$\left. - \varepsilon^{\frac{r}{x}\theta_2} \cos (\theta_2 - \delta)\right\}. \tag{8}$$

Rearranged equation (8) gives

$$f(\theta_2) = \varepsilon^{\frac{r}{x}\theta_2}\left[1 + \frac{2\,e}{i_0 \sqrt{2\,(x^2 + r^2)}} \cos (\theta_2 - \delta)\right]$$

$$= \varepsilon^{\frac{r}{x}\theta_1}\left[\frac{2\,e}{i_0 \sqrt{2\,(x^2 + r^2)}} \cos (\theta_1 - \delta) - 1\right], \tag{9}$$

where

$$\tan \delta = \frac{x - r}{x + r}.$$

By approximation, from this equation the value of θ_2, corresponding to a given θ_1, is derived.

15. Example:

$$e = 2000, \quad i_0 = 10, \quad \text{and} \quad Z = 10 - 40\,j.$$

Thus $\qquad \delta = 31° = 0.54$ radians

and
$$i = 5 + [34.3 \cos (\theta_1 - 31°) - 5]\, \varepsilon^{-0.25\,(\theta - \theta_1)}$$
$$- 34.3 \cos (\theta - 31°),$$
$$f(\theta_2) = \varepsilon^{0.25\,\theta_2}\,[1 + 6.86 \cos (\theta_2 - 31°)]$$
$$= \varepsilon^{0.25\,\theta_1}\,[6.86 \cos (\theta_1 - 31°) - 1].$$

Substituting for θ_1, $30° = \dfrac{\pi}{6}$, $45° = \dfrac{\pi}{4}$, and $60° = \dfrac{\pi}{3}$, respectively, gives:

$$\theta_1 = \frac{\pi}{6},\ i = 5 + 29.3\,\varepsilon^{-0.25\left(\theta - \frac{\pi}{6}\right)} - 34.3 \cos (\theta - 31°)$$

$$\frac{\pi}{4},\ i = 5 + 28.3\,\varepsilon^{-0.25\left(\theta - \frac{\pi}{4}\right)} - 34.3 \cos (\theta - 31°)$$

$$\frac{\pi}{3},\ i = 5 + 25.1\,\varepsilon^{-0.25\left(\theta - \frac{\pi}{3}\right)} - 34.3 \cos (\theta - 31°)$$

and

	$\theta_1 = \dfrac{\pi}{6}$		$\theta_1 = \dfrac{\pi}{4}$		$\theta_1 = \dfrac{\pi}{3}$			
$\theta°$	i	i_2	i	i_2	i	i_2	e_2	e_1
25	.....	.....	.....	.....	.....	.....	1810	850
30	0	10	.....	.....	.....	.....		
35	.....	.....	.....	.....	.....	.....	1640	1150
40	−0.9	10.9	.....	.....	.....	.....		
45	.....	.....	0	10	.....	.....	1410	1410
50	−0.6	10.6	.....					
55	.....	.....	.8	9.2	.....	.....	1150	1640
60	+0.6	9.4	.....	.....	0	10	.....	.....
65	.....	.....	2.5	7.5	.....	.....	850	1810
70	3.0	7.0	.....	.....	2.2	7.8	.....	.....
75	.....	.....	5.1	4.9	.....	.....	520	1930
80	6.0	4.0	.....	.....	5.4	4.6	.....	.....
85	.....	.....	8.6	1.4	.....	.....	170	1990
90	9.9	0.1	.....	.....	9.3	0.7	.....	.....
95	.....	.....	12.8	−2.8	.....	.....	−170	1990
100	14.3	−4.3	.....	.....	13.8	−3.8	.....	.....
105	.....	.....	17.3	−7.3	.....	.....	−520	1930
110	19.1	−9.1	.....	.....	18.6	−8.6	.....	.....
115	.....	.....	22.2	−12.2	.....	.....	−850	1810

These values are plotted in Fig. 59, together with e_1 and e_2. It follows then,

$$\theta_2 = \qquad 90.2° \qquad\quad 88.6° \qquad\quad 91.7°$$

The actual curves of an arc machine differ, however, **very** greatly from those of Fig. 59. In the arc machine, inherent regulation for constant current is produced by opposing a very high armature reaction to the field excitation, so that the resultant m.m.f., or m.m.f. which produces the effective magnetic flux, is

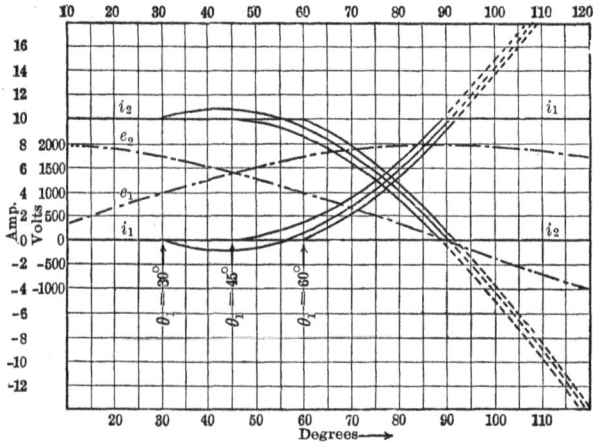

Fig. 59. Quarter-phase rectification.

small compared with the total field m.m.f. and the armature reaction, and so greatly varies with a small variation of armature current. As result, a very great distortion of the field occurs, and the magnetic flux is concentrated at the pole corner. This gives an e.m.f. wave which has a very sharp and high peak, with very long flat zero, and so cannot be approximated by an equivalent sine wave, but the actual e.m.f. curves have to be used in a more exact investigation.

CHAPTER IV.

ARC RECTIFICATION.

I. The Arc.

16. The operation of the arc rectifier is based on the characteristic of the electric arc to be a good conductor in one direction but a non-conductor in the opposite direction, and so to permit only unidirectional currents.

In an electric arc the current is carried across the gap between the terminals by a bridge of conducting vapor consisting of the material of the negative or the cathode, which is produced and constantly replenished by the cathode blast, a high velocity blast issuing from the cathode or negative terminal towards the anode or positive terminal.

An electric arc, therefore, cannot spontaneously establish itself. Before current can exist as an arc across the gap between two terminals, the arc flame or vapor bridge must exist, i.e., energy must have been expended in establishing this vapor bridge. This can be done by bringing the terminals into contact and so starting the current, and then by gradually withdrawing the terminals derive the energy of the arc flame by means of the current, from the electric circuit, as is done in practically all arc lamps. Or by increasing the voltage across the gap between the terminals so high that the electrostatic stress in the gap represents sufficient energy to establish a path for the current, i.e., by jumping an electrostatic spark across the gap, this spark is followed by the arc flame. An arc can also be established between two terminals by supplying the arc flame from another arc, etc.

The arc therefore must be continuous at the cathode, but may be shifted from anode to anode. Any interruption of the cathode blast puts out the arc by interrupting the supply of conducting vapor, and a reversal of the arc stream means stopping the cathode blast and producing a reverse cathode blast, which, in general, requires a voltage higher than the electrostatic striking voltage (at arc temperature) between the electrodes. With an alternating impressed e.m.f. the arc if established goes out at the end of the half wave, or if a cathode blast is maintained continuously by a second arc (excited by direct current or overlapping sufficiently with the first arc), only alternate half waves can pass, those for which that terminal is negative from which the continuous blast issues. The arc, with an alternating impressed voltage, therefore rectifies, and the voltage range of rectification is the range between the arc voltage and the electro-static spark voltage through the arc vapor, or the air or residual gas which may be mixed with it. Hence it is highest with the mercury arc, due to its low temperature.

The mercury arc is therefore almost exclusively used for arc rectification. It is enclosed in an evacuated glass vessel, so as to avoid escape of mercury vapor and entrance of air into the arc stream. Due to the low temperature of the boiling point of mercury, enclosure in glass is feasible with the mercury arc.

II. Mercury Arc Rectifier.

17. Depending upon the character of the alternating supply, whether a source of constant alternating potential or constant alternating current, the direct-current circuit receives from the rectifier either constant potential or constant current. Depending on the character of the system, thus constant-potential rectifiers and constant-current rectifiers can be distinguished. They differ somewhat from each other in their construction and that of the auxiliary apparatus, since the constant-potential rectifier operates at constant voltage but varying current, while the constant-current rectifier operates at varying voltage. The general character of the phenomenon of arc rectification is, how-ever, the same in either case, so that only the constant-current rectifier will be considered more explicitly in the following paragraphs.

The constant-current mercury arc rectifier system, as used for the operation of constant direct-current arc circuits from an alternating constant potential supply of any frequency, is sketched diagrammatically in Fig. 60. It consists of a constant-current transformer with a tap C brought out from the middle of the secondary coil AB. The rectifier tube has two graphite anodes *a*, *b*, and a mercury cathode *c*, and usually two auxiliary mercury anodes near the cathode *c* (not shown in diagram, Fig. 60), which are used for excitation, mainly in starting, by establishing between the cathode *c* and the two auxiliary mercury anodes, from a small low voltage constant-potential transformer, a pair of low current rectifying arcs. In the constant-potential rectifier, generally one auxiliary anode only is used, connected through a resistor *r* with one of the main anodes, and the constant-

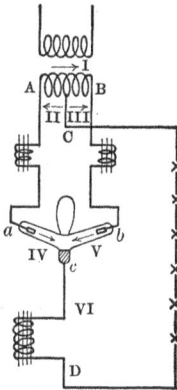

Fig. 60.   Constant-current
mercury arc rectifier.

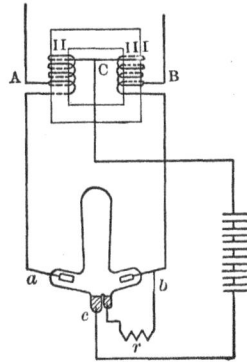

Fig. 61.   Constant-potential
mercury arc rectifier.

current transformer is replaced by a constant-potential transformer or compensator (auto-transformer) having considerable inductance between the two half coils II and III, as shown in Fig. 61. Two reactive coils are inserted between the outside terminals of the transformer and rectifier tube respectively, for the purpose of producing an overlap between the two rectifying arcs, *ca* and *cb*, and thereby the required continuity of the arc stream at *c*. Or instead of separate reactances, the two half coils II and III may be given sufficient reactance, as in Fig. 61. A reactive coil is inserted into the rectified or arc circuit, which connects between transformer neutral *C* and rectifier neutral *c*, for the purpose of reducing the fluctuation of the rectified current to the desired amount.

In the constant-potential rectifier, instead of the transformer *ACB* and the reactive coils *Aa* and *Ba*, generally a compensator or auto-transformer is used, as shown in Fig. 61, in which the two halves of the coil, AC and BC, are made of considerable self-inductance against each other, as by their location on different magnet cores, and the reactive coil at c frequently omitted. The modification of the equations resulting herefrom is obvious. Such auto-transformer also may raise or lower the impressed voltage, as shown in Fig. 61.

The rectified or direct voltage of the constant-current rectifier is somewhat less than one-half of the alternating voltage supplied by the transformer secondary AB, the rectified or direct current somewhat more than double the effective alternating current supplied by the transformer.

In the constant-potential rectifier, in which the currents are larger, and so a far smaller angle of overlap θ is permissible, the direct-current voltage therefore is very nearly the mean value of half the alternating voltage, minus the arc voltage, which is about 13 volts. That is, if $e =$ effective value of alternating voltage between rectifier terminals ab of compensator (Fig. 61),

hence $\dfrac{2\sqrt{2}}{\pi}\, e =$ mean value, the direct current voltage is

$$e_0 = \frac{\sqrt{2}}{\pi} e - 13.$$

III. MODE OF OPERATION.

18. Let, in Figs. 62 and 63, the impressed voltage between the secondary terminals AB of an alternating-current transformer be shown by curve I. Let C be the middle or center of the transformer secondary AB. The voltages from C to A and from C to B then are given by curves II and III.

If now A, B, C are connected with the corresponding rectifier terminals a, b, c and at c a cathode blast maintained, those currents will exist for which c is negative or cathode, i.e., the current through the rectifier from a to c and from b to c, under the impressed e.m.fs. II and III, are given by curves IV and V, and the current derived from c is the sum of IV and V, as shown in curve VI.

Such a rectifier as shown diagrammatically in Fig. 62 requires some outside means for maintaining the cathode blast at c, since the current in the half wave 1 in curve VI goes down to zero at the zero value of e.m.f. III before the current of the next half wave 2 starts by the e.m.f. II.

It is therefore necessary to maintain the current of the half wave 1 beyond the zero value of its propel- ling impressed e.m.f. III until the current of the next half wave 2 has started, i.e., to overlap the currents of the successive half waves. This is done by inserting reactances into the leads from the transformer to the rectifier, i.e., between A and a, B and b respec- tively, as shown in Fig. 60. The effect of this reactance is that the current of half wave 1, V, continues beyond the zero of its im- pressed e.m.f. III i.e., until the e.m.f. III has died out and reversed, and the current of the half wave 2, IV, started by e.m.f. II; that is, the two half waves of the current overlap, and each half wave lasts for more than half a period or 180 degrees.

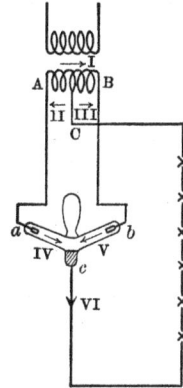

Fig. 62. Constant- current mercury arc rectifier.

The current waves then are shown in curve VII. The current half wave 1 starts at the zero value of its e.m.f. III, but rises more slowly than it would without react-

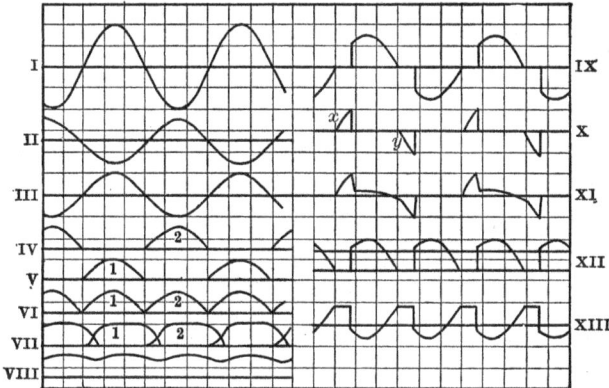

Fig. 63. E.m.f. and current waves of constant-current mercury arc rectifier.

ance, following essentially the exponential curve of a starting current wave, and the energy which is thus consumed by the reactance as counter e.m.f. is returned by maintaining the current half wave 1 beyond the e.m.f. wave, i.e., beyond 180 degrees, by θ_0 time-degrees, so that it overlaps the next half wave 2 by θ_0 time-degrees.

Hereby the rectifier becomes self-exciting, i.e., each half wave of current, by overlapping with the next, maintains the cathode blast until the next half wave is started.

The successive current half waves added give the rectified or unidirectional current curve VIII.

During a certain period of time in each half wave from the zero value of e.m.f. both arcs ca and cb exist. During the existence of both arcs there can be no potential difference between the rectifier terminals a and b, and the impressed e.m.f. between the rectifier terminals a and b therefore has the form shown in curve IX, Fig. 63, i.e., remains zero for θ_0 time-degrees, and then with the breaking of the arc of the preceding half wave jumps up to its normal value.

The generated e.m.f. of the transformer secondary, however, must more or less completely follow the primary impressed e.m.f. wave, that is, has a shape as shown in curve I, and the difference between IX and I must be taken up by the reactance. That is, during the time when both arcs exist in the rectifier, the a. c. reactive coils consume the generated e.m.f. of the transformer secondary, and the voltage across these reactive coils, therefore, is as shown in curve X. That is, the reactive coil consumes voltage at the start of the current of each half wave, at x in curve X, and produces voltage near the end of the current, at y. Between these times, the reactive coil has practically no effect and its voltage is low, corresponding to the variation of the rectified alternating current, as shown in curve XI. That is, during this intermediary time the alternating reactive coils merely assist the direct-current reactive coil.

Since the voltage at the alternating terminals of the rectifier, a, b, has two periods of zero value during each cycle, the rectified voltage between c and C must also have the same zero periods, and is indeed the same curve as IX, but reversed, as shown in curve XII.

Such an e.m.f. wave cannot satisfactorily operate arcs, since during the zero period of voltage XII the arcs go out. The voltage on the direct-current line must never fall below the "counter e.m.f." of the arcs, and since the resistance of this circuit is low, frequently less than 10 per cent, it follows that the total variation of direct-current line voltage must be below 10 per cent, i.e., the voltage practically constant, as shown by the straight line in curve XII. Hence a high reactance is inserted into the direct-current circuit, which consumes the excess voltage during that part of curve XII where the rectified voltage is above line voltage, and supplies the line voltage during the period of zero rectified voltage. The voltage across this reactive coil, therefore, is as shown by curve XIII.

IV. Constant-Current Rectifier.

19. The angle of overlap θ_0 of the two arcs is determined by the desired stability of the system. By the angle θ_0 and the impressed e.m.f. is determined the sum total of e.m.fs. which has to be consumed and returned by the a. c. reactive coil, and herefrom the size of the a. c. reactive coil.

From the angle θ_0 also follows the wave shape of the rectified voltage, and therefrom the sum total of e.m.f. which has to be given by the d. c. reactive coil, and hereby the size of the d. c. reactive coil required to maintain the d. c. current fluctuation within certain given limits.

The efficiency, power factor, regulation, etc., of such a mercury arc rectifier system are essentially those of the constant-current transformer feeding the rectifier tube.

Let f = frequency of the alternating-current supply system, i_0 = mean value of the rectified direct current, and a = the pulsation of the rectified current from the mean value, i.e., $i_0 (1 + a)$ the maximum and $i_0 (1 - a)$ the minimum value of direct current. A pulsation from a mean of 20 to 25 per cent is permissible in an arc circuit. The total variation of the rectified current then is $2 ai_0$, i.e., the alternating component of the direct current has the maximum value ai_0, hence the effective value $\dfrac{a}{\sqrt{2}} i_0$ (or for $a = 0.2$, $0.141 i_0$) and the frequency $2 f$. Hysteresis and eddy losses in the direct-current reactive coil, therefore, correspond to an alternating current of frequency $2 f$ and effective value $\dfrac{a}{\sqrt{2}} i_0$, or about $0.141 i_0$, i.e., are small even at relatively high densities.

In the alternating-current reactive coils the current varies, unidirectionally, between 0 and $i_0 (1 + a)$, i. e., its alternating component has the maximum value $\dfrac{1 + a}{2} i_0$ and the effective value $\dfrac{1 + a}{2\sqrt{2}} i_0$ (or, for $a = + 0.2$, $0.425\, i_0$) and the frequency f. The hysteresis loss, therefore, corresponds to an alternating current of frequency f and effective value $\dfrac{1 + a}{2 \sqrt{2}} i_0$, or about $0.425\, i_0$.

With decreasing load, at constant alternating-current supply, the rectified direct current slightly increases, due to the increasing overlap of the rectifying arcs, and to give constant direct current the transformer must therefore be adjusted so as to regulate for a slight decrease of alternating-current output with decrease of load.

V. Theory and Calculation.

20. In the constant-current mercury-arc rectifier shown diagrammatically in Fig. 64, let $e \sin \theta =$ sine wave of e.m.f. impressed between neutral and outside of alternating-current supply to the rectifier; that is, $2 e \sin \theta =$ total secondary generated e.m.f. of the constant-current transformer; $Z_1 = r_1 - jx_1 =$ impedance of the reactive coil in each anode circuit of the rectifier ("alternating-current reactive coil"), inclusive of the internal self-inductive impedance between the two halves of the transformer secondary coil; i_1 and $i_2 =$ anode currents, counted in the direction from anode to cathode; $e_a =$ counter e.m.f. of rectifying arc, which is constant; $Z_0 = r_0 - jx_0 =$ impedance of reactive coil in rectified circuit ("direct-current reactive coil"); $Z_2 = r_2 - jx_2 =$ impedance of load or arc-lamp circuit; $e_0' =$ counter e.m.f. in rectified circuit, which is con-

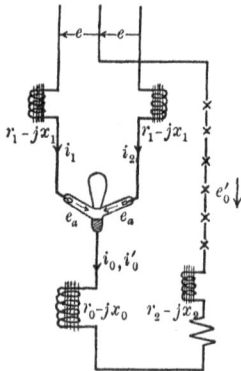

Fig. 64. Constant-current mercury arc rectifier.

stant (equal to the sum of the counter e.m.fs. of the arcs in the lamp circuit); θ_0 = angle of overlap of the two rectifying arcs, or overlap of the currents i_1 and i_2; i_0 = rectified current during the period, $0 < \theta < \theta_0$, where both rectifying arcs exist, and i_0' = rectified current during the period, $\theta_0 < \theta < \pi$, where only one arc or one anode current i_1 exists.

Let $e_0 = e_0' + e_a$ = total counter e.m.f. in the rectified circuit and $Z = r - jx = (r_1 + r_0 + r_2) - j(x_1 + x_0 + x_2)$ = total impedance per circuit; then we have

(a) During the period when both rectifying arcs exist,

$$0 < \theta < \theta_0,$$

$$i_0 = i_1 + i_2. \tag{1}$$

In the circuit between the e.m.f. $2 e \sin \theta$, the rectifier tube, and the currents i_1 and i_2, according to Kirchhoff's law, it is, Fig. **64,**

$$2 e \sin \theta - r_1 i_1 - x_1 \frac{di_1}{d\theta} + r_1 i_2 + x_1 \frac{di_2}{d\theta} = 0. \tag{2}$$

In the circuit from the transformer neutral over e.m.f. $e \sin \theta$, current i_1, rectifier arc e_a and rectified circuit i_0, back to the transformer neutral, we have

$$e \sin \theta - r_1 i_1 - x_1 \frac{di_1}{d\theta} - e_a - r_0 i_0 - x_0 \frac{di_0}{d\theta} - r_2 i_0 - x_2 \frac{di_0}{d\theta} - e_0' = 0;$$

or,

$$e \sin \theta - r_1 i_1 - x_1 \frac{di_1}{d\theta} - (r_0 + r_2) i_0 - (x_0 + x_2) \frac{di_0}{d\theta} - e_0 = 0. \tag{3}$$

(b) During the period when only one rectifying arc exists,

$$\theta_0 < \theta < \pi,$$

$$i_1 = i_0';$$

hence, in this circuit,

$$e \sin \theta - r_1 i_0' - x_1 \frac{di_0'}{d\theta} - (r_0 + r_2) i_0' - (x_0 + x_2) \frac{di_0'}{d\theta} - e_0 = 0. \tag{4}$$

Substituting (1) in (2) and combining the result (5) of this substitution with (3) gives the *differential equations of the rectifier:*

$$2\,e \sin\theta + r_1\,(i_0 - 2\,i_1) + x_1\frac{d}{d\theta}\,(i_0 - 2\,i_1) = 0, \qquad (5)$$

$$2\,e_0 + (2\,r - r_1)\,i_0 + (2\,x - x_1)\frac{di_0}{d\theta} = 0, \qquad (6)$$

and $\qquad e \sin\theta - e_0 - ri_0' - x\dfrac{di_0'}{d\theta} = 0. \qquad (7)$

In these equations, i_0 and i_1 apply for the time, $0 < \theta < \theta_0$, i_0' for the time, $\theta_0 < \theta < \pi$.

21. These differential equations are integrated by the functions

$$i_0 - 2\,i_1 = A\varepsilon^{-a\theta} + A' \sin(\theta - \beta), \qquad (8)$$

$$i_0 = B\varepsilon^{-b\theta} + B', \qquad (9)$$

and $\qquad i_0' = C\varepsilon^{-c\theta} + C' + C'' \sin(\theta - \gamma). \qquad (10)$

Substituting (8), (9), and (10) into (5), (6), and (7) gives three identities:

$$2\,e \sin\theta + A'\,[r_1 \sin(\theta - \beta) + x_1 \cos(\theta - \beta)] + A\varepsilon^{-a\theta}\,(r_1 - ax_1) = 0,$$

$$2\,e_0 + B'(2\,r - r_1) + B\varepsilon^{-b\theta}\,[(2\,r - r_1) - b\,(2\,x - x_1)] = 0,$$

and

$$e \sin\theta - e_0 - C''[r\sin(\theta - \gamma) + x \cos(\theta - \gamma)] - C'r - C\varepsilon^{-c\theta}(r - cx) = 0;$$

hence,

$$\left.\begin{array}{c}
r_1 - ax_1 = 0, \\[4pt]
(2\,r - r_1) - b\,(2\,x - x_1) = 0, \\[4pt]
r - cx = 0. \\[4pt]
2\,e_0 + B'\,(2\,r - r_1) = 0, \\[4pt]
e_0 + C'r = 0, \\[4pt]
2\,e + A'\,(r_1 \cos\beta + x_1 \sin\beta) = 0, \\[4pt]
A'\,(r_1 \sin\beta - x_1 \cos\beta) = 0, \\[4pt]
e - C''\,(r \cos\gamma + x \sin\gamma) = 0, \\[4pt]
C''\,(r \sin\gamma - x \cos\gamma) = 0.
\end{array}\right\} \qquad (11)$$

and

Writing

$$z_1 = \sqrt{r_1^2 + x_1^2},$$

$$\tan \alpha_1 = \frac{x_1}{r_1}, \tag{12}$$

and

$$z = \sqrt{r^2 + x^2},$$

$$\tan \alpha = \frac{x}{r}. \tag{13}$$

Substituting (12) and (13) gives by solving the 9 equations (11) the values of the coefficients a, b, c, A', B', C', C'', β, γ:

$$a = \frac{r_1}{x_1},$$

$$b = \frac{2r - r_1}{2x - x_1}, \tag{14}$$

$$c = \frac{r}{x},$$

$$\beta = \alpha_1,$$

$$\gamma = \alpha, \tag{15}$$

$$A' = -\frac{2e}{z_1},$$

$$B' = -\frac{2e_0}{2r - r_1}, \tag{16}$$

$$C' = -\frac{e_0}{r},$$

$$C'' = +\frac{e}{z}, \tag{17}$$

and thus the *integral equations of the rectifier* are

$$i_0 - 2i_1 = A\varepsilon^{-a\theta} - \frac{2e}{z_1}\sin(\theta - \alpha_1), \tag{18}$$

$$i_0 = B\varepsilon^{-b\theta} - \frac{2e_0}{2r - r_1}, \tag{19}$$

and

$$i_0' = C\varepsilon^{-c\theta} - \frac{e_0}{r} + \frac{e}{z}\sin(\theta - \alpha), \tag{20}$$

where a, b, c are given by equations (14), α and α_1 by equations (12) and (13), and A, B, C are integration constants given by the terminal conditions of the problem.

22. These terminal conditions are:

$$\left.|i_1|\right._{\theta=0} = 0,$$
$$\left.|i_0|\right._{\theta=0} = \left.|i_0'|\right._{\theta=\pi},$$

and

$$\left.|i_1|\right._{\theta=\theta_0} = \left.|i_0|\right._{\theta=\theta_0} = \left.|i_0'|\right._{\theta=\theta_0}. \tag{21}$$

That is, at $\theta = 0$ the anode current $i_1 = 0$. After half a period, or $\pi = 180°$, the rectified current repeats the same value. At $\theta = \theta_0$, all three currents i_1, i_0, i_0' are identical.

The four equations (21) determine four constants, A, B, C, θ_0.

Substituting these constants in equations (18), (19), (20) gives the equations of the rectified current i_0, i_0', and of the anode currents i_1 and $i_2 = i_0 - i_1$, determined by the constants of the system, Z, Z_1, e_0, and by the impressed e.m.f., e.

In the constant-current mercury-arc rectifier system of arc lighting, e, the secondary generated voltage of the constant-current transformer, varies with the load, by the regulation of the transformer, and the rectified current, i_0, i_0', is required to remain constant, or rather its average value.

Let then be given as condition of the problem the average value i of the rectified current, 4 amperes in a magnetite or mercury arc lamp circuit, 5 or 6.6 or 9.6 amperes in a carbon arc lamp circuit.

Assume as fair approximation that the pulsating rectified current i_0, i_0' has its mean value i at the moment, $\theta = 0$. This then gives the additional equation

$$\left.|i_0|\right._{\theta=0} = i, \tag{22}$$

and from the five equations (21) and (22) the five constants A, B, C, θ_0, e are determined.

Substituting (22), (18), (19), (20) in equations (21) gives

$$A = i - \frac{2\,e}{z_1}\sin\alpha_1,$$
$$B = i + \frac{2\,e_0}{2\,r - r_1}, \tag{23}$$
$$C = \varepsilon^{c\pi}\left\{i + \frac{e_0}{r} - \frac{e}{z}\sin\alpha\right\}$$

$$- A\varepsilon^{-a\theta_0} - \frac{2\,e}{z_1} \sin\,(\alpha_1 - \theta_0) = B\varepsilon^{-b\theta_0} - \frac{2\,e_0}{2\,r - r_1}$$

$$= C\varepsilon^{-c\theta_0} - \frac{e_0}{r} - \frac{e}{z} \sin\,(\alpha - \theta_0). \tag{24}$$

Substituting (23) in (24) gives

$$\frac{2\,e}{z_1}\left\{ \varepsilon^{-a\theta_0} \sin\,\alpha_1 - \sin\,(\alpha_1 - \theta_0)\right\} = i\left\{\varepsilon^{-a\theta_0} + \varepsilon^{-b\theta_0}\right\}$$

$$- \frac{2\,e_0}{2\,r - r_1}\left\{1 - \varepsilon^{-b\theta_0}\right\} \tag{25}$$

and

$$\frac{e}{z}\left\{\varepsilon^{c(\pi - \theta_0)} \sin\,\alpha + \sin\,(\alpha - \theta_0)\right\} = i\left\{\varepsilon^{c\,(\pi - \theta_0)} - \varepsilon^{-b\theta_0}\right\}$$

$$+ \frac{2\,e_0}{2\,r - r_1}\left\{1 - \varepsilon^{-b\theta_0}\right\} + \frac{e_0}{r}\left\{\varepsilon^{c(\pi - \theta_0)} - 1\right\}, \tag{26}$$

and eliminating e from these two equations gives

$$\frac{\varepsilon^{c(\pi - \theta_0)} \sin\,\alpha + \sin\,(\alpha - \theta_0)}{\varepsilon^{-a\theta_0} \sin\,\alpha_1 - \sin\,(\alpha_1 - \theta_0)}$$

$$= \frac{2\,z}{z_1}\,\frac{\left\{\varepsilon^{c(\pi - \theta_0)} - \varepsilon^{-b\theta_0}\right\} + \frac{2\,e_0}{i\,(2\,r - r_1)}\left\{1 - \varepsilon^{-b\theta_0}\right\} + \frac{e_0}{ir}\left\{\varepsilon^{c(\pi - \theta_0)} - 1\right\}}{\left\{\varepsilon^{-a\theta_0} + \varepsilon^{-b\theta_0}\right\} - \frac{2\,e_0}{i\,(2\,r - r_1)}\left\{1 - \varepsilon^{-b\theta_0}\right\}} \tag{27}$$

Equation (27) determines angle θ_0, and by successive substitution in (26), (23), e, A, B, C are found.

Equation (27) is transcendental, and therefore has to be solved by approximation, which however is very rapid.

As first approximation, $a\theta_0 = b\theta_0 = c\theta_0 = 0$; $\alpha = \alpha_1 = 90°$ or $\frac{\pi}{2}$ and substituting these values in (27) gives

$$\frac{\varepsilon^{c\pi} + \cos\,\theta_1}{1 - \cos\,\theta_1} = \frac{2\,z}{z_1}\,\frac{\left(\varepsilon^{c\pi} - 1\right)\left(1 + \frac{e_0}{ir}\right)}{2}$$

and

$$\cos \theta_1 = \frac{\dfrac{z}{z_1}\left(\varepsilon^{c\pi} - 1\right)\left(1 + \dfrac{e_0}{ir}\right) - \varepsilon^{c\pi}}{\dfrac{z}{z_1}\left(\varepsilon^{c\pi} - 1\right)\left(1 + \dfrac{e_0}{ir}\right) + 1}. \tag{28}$$

This value of θ_1 substituted in the exponential terms of equation (27) gives a simple trigonometric equation in θ_0, from which follows the second approximation θ_2, and, by interpolation, the final value,

$$\theta_0 = \theta_2 + \frac{(\theta_2 - \theta_1)^2}{\theta_1}. \tag{29}$$

23. For instance, let $e_0 = 950$, $i = 3.8$, the constants of the circuit being $Z_1 = 10 - 185\,j$ and $Z = 50 - 1000\,j$.

Herefrom follows

$$a = 0.054,\; b = 0.050,\; \text{and}\; c = 0.050, \tag{14}$$
$$\alpha_1 = 86.9° \text{ and } \alpha = 87.1°. \tag{15}$$

From equation (28) follows as first approximation, $\theta_1 = 47.8°$; as second approximation, $\theta_2 = 44.2°$.

Hence, by (29),

$$\theta = 44.4°.$$

Substituting a in (26) gives $e = 2100$,
hence, the effective value of transformer secondary voltage,

$$\frac{2\,e}{\sqrt{2}} = 2980 \text{ volts}$$

and, from (23),

$$A = -18.94,\; B = 24.90,\; C = 24.20.$$

Therefore, the equations of the currents are

$i_0 = 24.90\,\varepsilon^{-0.050\,\theta} - 21.10,$

$i_0' = 24.20\,\varepsilon^{-0.050\,\theta} - 19.00 + 2.11 \sin(\theta - 87.1°),$

$i_1 = 12.45\,\varepsilon^{-0.050\,\theta} + 9.47\,\varepsilon^{-0.054\,\theta} - 10.58 + 11.35 \sin(\theta - 86.9°),$

and

$$i_2 = i_0 - i_1.$$

The effective or equivalent alternating secondary current of the transformer, which corresponds to the primary load current, that is, primary current minus exciting current, is

$$i' = i_1 - i_2.$$

From these equations are calculated the numerical values of rectified current i_0, i_0', of anode current i_1, and of alternating current i', and plotted as curves in Fig. 65.

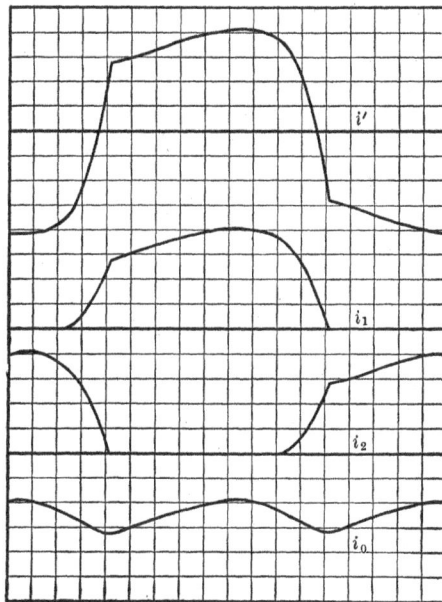

Fig. 65. Current waves of constant-current mercury arc rectifier.

24. As illustrations of the above phenomena are shown in Fig. 66 the performance curves of a small constant-current rectifier, and in Figs. 67 to 76 oscillograms of this rectifier.

Interesting to note is the high frequency oscillation at the termination of the jump of the potential difference cC (Fig. 60) which represents the transient term resulting from the electrostatic capacity of the transformer. At the end of the period of overlap of the two rectifying arcs one of the anode currents reaches

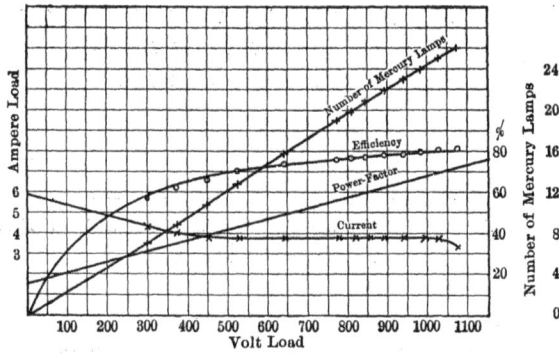

Fig. 66.　Results from tests made on a constant-current mercury arc rectifier.

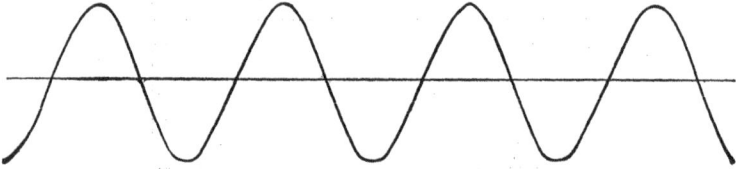

Fig. 67.　Supply e.m.f. to constant-current rectifier.

Fig. 68.　Secondary terminal e.m.f. of transformer.

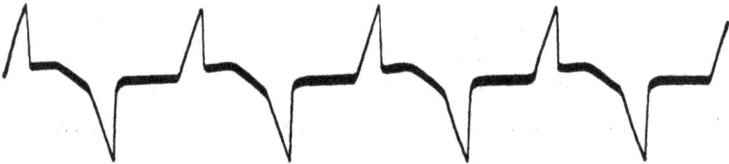

Fig. 69.　E.m.f. across a.c. reactive coils.

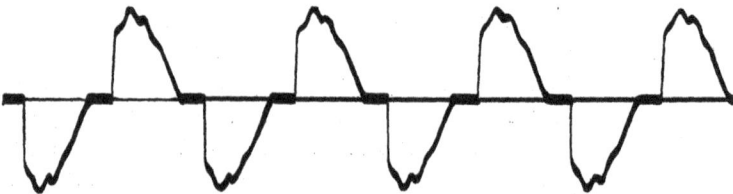

Fig. 70.　Alternating e.m.f. impressed upon rectifier tube.

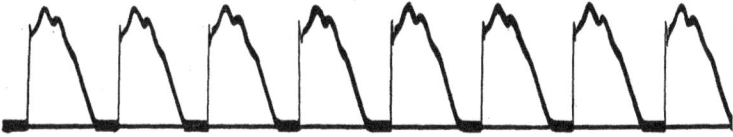

Fig. 71. Unidirectional e.m.f. produced between rectifier neutral and transformer neutral.

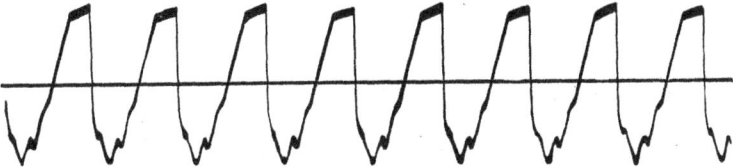

Fig. 72. E.m.f. across d.c. reactive coils.

Fig. 73. Rectified e.m.f. supplied to arc circuit.

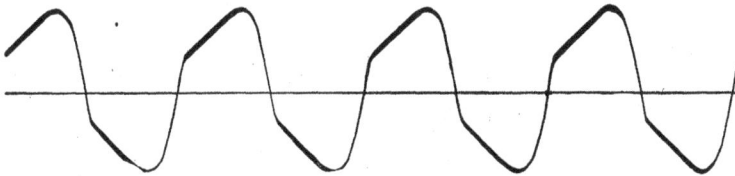

Fig. 74. Primary supply current.

Fig. 75. Current in rectifying arcs.

Fig. 76. Rectified current in arc circuit.

zero and stops, and so its $L\dfrac{di}{dt}$ abruptly changes; that is, a sudden change of voltage takes place in the circuit $aACDc$ or $bBCDc$. Since this circuit contains distributed capacity, that of the transformer coil $ACBC$ respectively, the line, etc., and inductance, an oscillation results of a frequency depending upon the capacity and inductance, usually a few thousand cycles per second, and of a voltage depending upon the impressed e.m.f.; that is, the $L\dfrac{di}{dt}$ of the circuit. An increase of inductance L increases the angle of overlap and so decreases the $\dfrac{di}{dt}$, hence does not greatly affect the amplitude, but decreases the frequency of this oscillation. An increase of $\dfrac{di}{dt}$ at constant L, as resulting from a decrease of the angle of overlap by delayed starting of the arc, caused by a defective rectifier, however increases the amplitude of this oscillation, and if the electrostatic capacity is high, and therefore the damping out of the oscillation slow, the

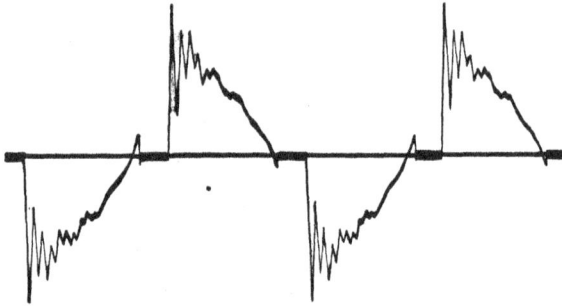

Fig. 77. E.m.f. between rectifier anodes.

oscillation may reach considerable values, as shown in oscillogram, Fig. 77, of the potential difference ab. In such cases, if the second half wave of the oscillation reaches below the zero value of the e.m.f. wave ab, the rectifying arc is blown out and a disruptive discharge may result.

VI. Equivalent Sine Waves.

25. The curves of voltage and current, in the mercury-arc rectifier system, as calculated in the preceding from the constants of the circuit, consist of successive sections of exponential or of exponential and trigonometric character.

In general, such wave structures, built up of successive sections of different character, are less suited for further calculation. For most purposes, they can be replaced by their equivalent sine waves, that is, sine waves of equal effective value and equal power.

The actual current and e.m.f. waves of the arc rectifier thus may be replaced by their equivalent sine waves, for general calculation, except when investigating the phenomena resulting from the discontinuity in the change of current, as the high frequency oscillation at the end and to a lesser extent at the beginning of the period of overlap of the rectifying arcs, and similar phenomena.

In a constant-current mercury arc rectifier system, of which the exact equations or rather groups of equations of currents and of e.m.fs. were given in the preceding, let i_0 = the mean value of direct current; e_0 = the mean value of direct or rectified voltage; i = the effective value of equivalent sine wave of secondary current of transformer feeding the rectifier; e = the effective value of equivalent sine wave of total e.m.f. generated in the transformer secondary coils, hence, $\dfrac{e}{2}$ = the effective equivalent sine wave of generated e.m.f. per secondary transformer coil, and θ_0 = the angle of overlap of rectifying arcs.

The secondary generated e.m.f., e, is then represented by a sine wave curve I, Fig. 78, with $e\sqrt{2}$ as maximum value.

Neglecting the impedance voltage of the secondary circuit during the time when only one arc exists and the current changes are very gradual, the terminal voltage between the rectifier anodes, e_1, is given by curve II, Fig. 78, with $e\sqrt{2}$ as maximum value. This curve is identical with e, except during the angle of overlap θ_0, when e_1 is zero. Due to the impedance of the reactive coils in the anode leads, curve II differs slightly from I, but the difference is so small that it can be neglected in deriving the equivalent sine wave, and this impedance considered afterwards as inserted into the equivalent sine-wave circuit.

The rectified voltage, e_2, is then given by curve III, Fig. 78, with a maximum value of $\dfrac{e}{2}\sqrt{2} = \dfrac{e}{\sqrt{2}}$ and zero value during the angle of overlap θ_0, or rather a value $= e_a$, the e.m.f. consumed by the rectifying arc (13 to 18 volts).

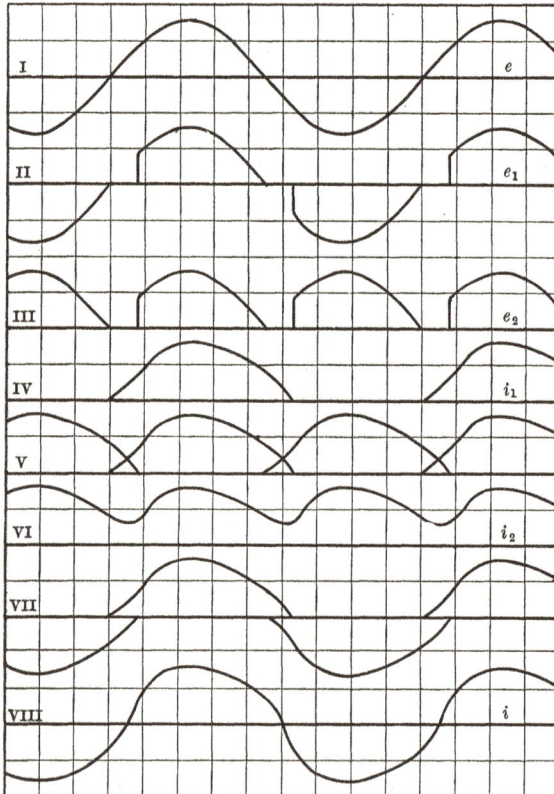

Fig. 78. E.m.f. and current curves in a mercury arc rectifier system.

The direct voltage e_0, when neglecting the effective resistance of the reactive coils, is then the mean value of the rectified voltage, e_2, of curve III, hence is

$$e_0 = \frac{e}{\sqrt{2}} \frac{1}{\pi} \int_{\theta_0}^{\pi} \sin \theta \, d\theta$$

$$= \frac{e\,(1 + \cos\theta_0)}{\sqrt{2}\,\pi};$$

or,
$$e = e_0 \frac{\pi\sqrt{2}}{1 + \cos\theta_0}.$$

If e_a = the mercury arc voltage, r_0 = the effective resistance of reactive coils and i_0 = the direct current, more correctly it is

$$e = (e_0 + e_a + r_0 i_0)\frac{\pi\sqrt{2}}{1 + \cos\theta_0}.$$

The effective alternating voltage between the rectifier anodes is the $\sqrt{\text{mean square of } e_1}$, curve II, hence is

$$e_1 = e\sqrt{2}\,\sqrt{\frac{1}{\pi}\int_{\theta_0}^{\pi}\sin^2\theta\,d\theta}$$

$$= e\sqrt{2}\,\sqrt{\frac{1}{\pi}\left[\frac{\theta}{2} - \frac{\sin 2\theta}{4}\right]_{\theta_0}^{\pi}}$$

$$= e\sqrt{2}\,\sqrt{\frac{\pi - \theta_0}{2\,\pi} + \frac{\sin 2\theta_0}{4\,\pi}}$$

$$= e\sqrt{1 - \frac{2\theta_0 - \sin 2\theta_0}{2\,\pi}},$$

and the drop of voltage in the reactive coils in the anode leads, caused by the overlap of the arcs, thus is

$$e - e_1 = e\left\{1 - \sqrt{1 - \frac{2\theta_0 - \sin 2\theta_0}{2\,\pi}}\right\}.$$

26. Let i' = the maximum variation of direct current from mean value i_0, hence, $i_2 = i_0 + i'$ = the maximum value of rectified current, and therefore also the maximum value of anode current.

The anode current thus has a maximum value i_2, and each half wave has a duration $\pi + \theta_0$, as shown by curve IV, Fig. 78.

The direct current, i_0, is then given by the superposition or addition of the two anode currents shown in curves V, and is given in curve VI.

The effective value of the equivalent alternating secondary current of the transformer is derived by the subtraction of the two anode currents, or their superposition in reverse direction, as shown by curves VII, and is given by curve VIII.

Each impulse of anode current covers an angle $\pi + \theta_0$, or somewhat more than one half wave.

Denoting, however, each anode wave by π, that is, considering each anode impulse as one half wave (which corresponds to a lower frequency $\dfrac{\pi}{\pi + \theta_0}$), then, referred to the anode impulse as half wave, the angle of overlap is

$$\theta_1 = \frac{\pi}{\pi + \theta_0} \theta_0.$$

The direct current, i_0, is the mean value of the anode current curves V, VI, and, assuming the latter as equivalent sine waves of maximum value $i_2 = i_0 + i'$, the direct current, i_0, is

$$i_0 = i_2 \frac{1}{\pi - \theta_1} \int_0^\pi \sin \theta' \, d\theta'$$

$$= \frac{2 i_2}{\pi - \theta_1}$$

$$= \frac{2 (\pi + \theta_0) i_2}{\pi^2},$$

and

$$i_2 = i_0 \frac{\pi - \theta_1}{2};$$

or,

$$i_2 = \frac{\pi^2 i_0}{2 (\pi + \theta_0)},$$

and the pulsation of the direct current, $i' = i_2 - i_0$, is

$$i' = i_0 \left\{ \frac{\pi^2}{2 (\pi + \theta_0)} - 1 \right\}.$$

The effective value of the secondary current, as equivalent sine wave in one transformer coil, is the $\sqrt{\ }$ mean square of curves VII, VIII, or, assuming this current as existing in both transformer secondary coils in series — actually it alternates, one half wave in one, the other in the other transformer coil — is half this value, or

$$i = \frac{i_2}{2} \sqrt{\frac{1}{\pi - \theta_1} \left\{ \int_{\theta_1}^{\pi - \theta_1} \sin^2 \theta' d\theta' + \int_0^{\theta_1} [\sin \theta'' + \sin (\theta' - \theta_1)]^2 d\theta' \right\}}$$

$$= \frac{i_2}{2} \sqrt{\frac{1}{\pi - \theta_1} \left\{ \int_0^{\pi} \sin^2 \theta' d\theta' + \int_0^{\theta_1} 2 \sin \theta' \sin (\theta' - \theta_1) d\theta' \right\}}$$

$$= \frac{i_2}{2} \sqrt{\frac{1}{\pi - \theta_1} \left\{ \frac{\pi}{2} + \left[\theta' \cos \theta_1 - \sin (2 \theta' - \theta_1) \right]_0^{\theta_1} \right\}}$$

$$= \frac{i_2}{2} \sqrt{\frac{1}{\pi - \theta_1} \left\{ \frac{\pi}{2} + \theta_1 \cos \theta_1 - \sin \theta_1 \right\}} ;$$

or, substituting

$$i_2 = i_0 \frac{\pi - \theta_1}{2},$$

$$i = \frac{i_0}{4} \sqrt{(\pi - \theta_1) \left\{ \frac{\pi}{2} + \theta_1 \cos \theta_1 - \sin \theta_1 \right\}}$$

$$= \frac{i_0}{2} \frac{\pi}{2\sqrt{2}} \sqrt{\left(1 - \frac{\theta_1}{\pi} \right) \left(1 + \frac{2 \theta_1}{\pi} \cos \theta_1 - \frac{2}{\pi} \sin \theta_1 \right)} ;$$

or, substituting

$$\theta_1 = \frac{\pi \theta_0}{\pi + \theta_0},$$

$$i = \frac{i_0}{2} \frac{\pi}{2\sqrt{2}} \sqrt{1 - \frac{\theta_0}{\pi + \theta_0} + \frac{2 \theta_0 \pi}{(\pi + \theta_0)^2} \cos \frac{\theta_0 \pi}{\pi + \theta_0} - \frac{2}{\pi + \theta_0} \sin \frac{\theta_0 \pi}{\pi + \theta_0}},$$

where $\dfrac{\pi}{2\sqrt{2}}$ = ratio of effective value to mean value of sine wave.

27. An approximate representation by equivalent sine waves, if e_0 = the mean value of direct terminal voltage, i_0 = the mean value of direct current, is therefore as follows:

The secondary generated e.m.f. of the transformer is

$$e = (e_0 + e_a + r_0 i_0) \frac{\pi\sqrt{2}}{1 + \cos \theta_0} ;$$

the secondary current of the transformer is

$$= \frac{i_0}{2} \frac{\pi}{2\sqrt{2}} \sqrt{1 - \frac{\theta_0}{\pi + \theta_0} + \frac{2\,\theta_0\pi}{(\pi + \theta_0)^2} \cos\frac{\theta_0\pi}{\pi + \theta_0} - \frac{2}{\pi + \theta_0} \sin\frac{\theta_0\pi}{\pi + \theta_0}} \; ;$$

the pulsation of the direct current is

$$i' = i_0 \left\{ \frac{\pi^2}{2\,(\pi + \theta_0)} - 1 \right\} ;$$

the anode voltage of the rectifier is

$$e_1 = e \sqrt{1 - \frac{2\,\theta_0 - \sin 2\,\theta_0}{2\,\pi}},$$

and herefrom follows the apparent efficiency of rectification, $\dfrac{e_0 i_0}{ei}$,

the power factor, the efficiency, etc.

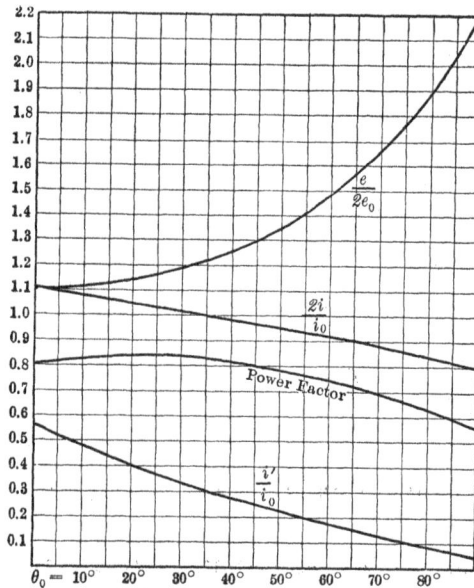

Fig. 79. E.m.f. and current ratio and secondary power factor of constant-current mercury arc rectifier.

From the equivalent sine waves, e and i, of the transformer secondary, and their phase angle, the primary impressed e.m.f. and the primary current of the transformer, and thereby the power factor, the efficiency, and the apparent efficiency of the system, are calculated in the usual manner.

In the secondary circuit, the power factor is below unity essentially due to wave shape distortion, less due to lag of current.

As example are shown, in Fig. 79, with the angle of overlap θ_0 as abscissas, the ratio of voltages, $\dfrac{e}{2\,e_0}$; the ratio of currents, $\dfrac{2\,i}{i_0}$; the current pulsation, $\dfrac{i'}{i_0}$, and the power factor of the secondary circuit.

SECTION III

TRANSIENTS IN SPACE

TRANSIENTS IN SPACE

CHAPTER I.

INTRODUCTION.

1. The preceding sections deal with transient phenomena in time, that is, phenomena occurring during the time when a change or transition takes place between one condition of a circuit and another. The time, t, then is the independent variable, electric quantities as current, e.m.f., etc., the dependent variables.

Similar transient phenomena also occur in space, that is, with space, distance, length, etc., as independent variable. Such transient phenomena then connect the conditions of the electric quantities at one point in space with the electric quantities at another point in space, as, for instance, current and potential difference at the generator end of a transmission line with those at the receiving end of the line, or current density at the surface of a solid conductor carrying alternating current, as the rail return of a single-phase railway, with the current density at the center or in general inside of the conductor, or the distribution of alternating magnetism inside of a solid iron, as a lamina of an alternating-current transformer, etc. In such transient phenomena in space, the electric quantities, which appear as functions of space or distance, are not the instantaneous values, as in the preceding chapters, but are alternating currents, e.m.fs., etc., characterized by intensity and phase, that is, they are periodic functions of time, and the analytical method of dealing with such phenomena therefore introduces two independent variables, time t and distance l, that is, the electric quantities are periodic functions of time and transient functions of space.

The introduction of the complex quantities, as representing the alternating wave by a constant algebraic number, eliminates the time t as variable, so that, in the denotation by complex quantities, the transient phenomena in space are functions of one independent variable only, distance l, and thus lead to the same equations as the previously discussed phenomena, with the difference, however, that here, in dealing with space phenomena, the dependent variables, current, e.m.f., etc., are complex quantities, while in the previous discussion they appeared as instantaneous values, that is, real quantities.

Otherwise the method of treatment and the general form of the equations are the same as with transient functions of time.

2. Some of the cases in which transient phenomena in space are of importance in electrical engineering are:

(*a*) Circuits containing distributed capacity and self-inductance, as long-distance energy transmission lines, long-distance telephone circuits, multiple spark-gaps, as used in some forms of high potential lightning arresters (multi-gap arrester), etc.

(*b*) The distribution of alternating current in solid conductors and the increase of effective resistance and decrease of effective inductance resulting therefrom.

(*c*) The distribution of alternating magnetic flux in solid iron, or the screening effect of eddy currents produced in the iron, and the apparent decrease of permeability and increase of power consumption resulting therefrom.

(*d*) The distribution of the electric field of a conductor through space, resulting from the finite velocity of propagation of the electric field, and the variation of self-inductance and mutual inductance and of capacity of a conductor without return, as function of the frequency, in its effect on wireless telegraphy.

(*e*) Conductors conveying very high frequency currents, as lightning discharges, wireless telegraph and telephone currents, etc.

Only the current and voltage distribution in the long distance transmission line can be discussed more fully in the following, and the investigation of the other phenomena only indicated in outline, or the phenomena generally discussed, as lighting conductors.

CHAPTER II.

LONG-DISTANCE TRANSMISSION LINE.

3. If an electric impulse is sent into a conductor, as a transmission line, this impulse travels along the line at the velocity of light (approximately), or 188,000 miles (3×10^{10} cm.) per second. If the line is open at the other end, the impulse there is reflected and returns at the same velocity. If now at the moment when the impulse arrives at the starting point a second impulse, of opposite direction, is sent into the line, the return of the first impulse adds itself, and so increases the second impulse; the return of this increased second impulse adds itself to the third impulse, and so on; that is, if alternating impulses succeed each other at intervals equal to the time required by an impulse to travel over the line and back, the effects of successive impulses add themselves, and large currents and high e.m.fs. may be produced by small impulses, that is, low impressed alternating e.m.fs., or inversely, when once started, even with zero impressed e.m.f., such alternating currents traverse the lines for some time, gradually decreasing in intensity by the energy consumption in the conductor, and so fading out.

The condition of this phenomenon of electrical resonance thus is that alternating impulses occur at time intervals equal to the time required for the impulse to travel the length of the line and back; that is, the time of one half wave of impressed e.m.f. is the time required by light to travel twice the length of the line, or the time of one complete period is the time light requires to travel four times the length of the line; in other words, the number of periods, or frequency of the impressed alternating e.m.fs., in resonance condition, is the velocity of light divided by four times the length of the line; or, in free oscillation or resonance condition, the length of the line is one quarter wave length.

285

If then l = length of line, S = speed of light, the frequency of oscillations or natural period of the line is

$$f_0 = \frac{S}{4\,l},\qquad(1)$$

or, with l given in miles, hence S = 188,000 miles per second, it is

$$f_0 = \frac{47,000}{l}\ \text{cycles.}\qquad(2)$$

To get a resonance frequency as low as commercial frequencies, as 25 or 60 cycles, would require l = 1880 miles for f_0 = 25 cycles, and l = 783 miles for f_0 = 60 cycles.

It follows herefrom that many existing transmission lines are such small fractions of a quarter-wave length of the impressed frequency that the change of voltage and current along the line can be assumed as linear, or at least as parabolic; that is, the line capacity can be represented by a condenser in the middle of the line, or by condensers in the middle and at the two ends of the line, the former of four times the capacity of either of the two latter (the first approximation giving linear, the second a parabolic distribution).

For further investigation of these approximations see "Theory and Calculation of Alternating-Current Phenomena."

If, however, the wave of impressed e.m.f. contains appreciable higher harmonics, some of the latter may approach resonance frequency and thus cause trouble. For instance, with a line of 150 miles length, the resonance frequency is f_0 = 313 cycles per second, or between the 5th harmonic and the 7th harmonic, 300 and 420 cycles of a 60-cycle system; fairly close to the 5th harmonic.

The study of such a circuit of distributed capacity thus becomes of importance with reference to the investigation of the effects of higher harmonics of the generator wave.

In long-distance telephony the important frequencies of speech probably range from 100 to 2000 cycles. For these frequencies the wave length varies from $\dfrac{S}{l}$ = 1880 miles down to 94 miles, and a telephone line of 1000 miles length would thus contain from about one-half to 11 complete waves of the impressed frequency. For long-distance telephony the phenomena occurring in the line thus can be investigated only by considering the complete equation of distributed capacity and inductance as so-called "wave transmission" and the phenomena thus essentially differ from those in a short energy transmission line.

4. Therefore in very long circuits, as in lines conveying alternating currents of high value at high potential over extremely long distances, by overhead conductors or underground cables, or with very feeble currents at extremely high frequency, such as telephone currents, the consideration of the *line resistance*, which consumes e.m.fs. in phase with the current, and of the *line reactance*, which consumes e.m.fs. in quadrature with the current, is not sufficient for the explanation of the phenomena taking place in the line, but several other factors have to be taken into account.

In long lines, especially at high potentials, the *electrostatic capacity* of the line is sufficient to consume noticeable currents. The charging current of the line condenser is proportional to the difference of potential and is one-fourth period ahead of the e.m.f. Hence, it either increases or decreases the main current, according to the relative phase of the main current and the e.m.f.

As a consequence the current changes in intensity, as well as in phase, in the line from point to point; and the e.m.fs. consumed by the resistance and inductance, therefore, also change in phase and intensity from point to point, being dependent upon the current.

Since no insulator has an infinite resistance, and since at high potentials not only leakage over surfaces but even direct *escape of electricity* into the air takes place by "brush discharge," or "corona," we have to recognize the existence of a current approximately proportional and in phase with the e.m.f. of the line. This current represents consumption of power, and is therefore analogous to the e.m.f. consumed by resistance, while the capacity current and the e.m.f. of inductance are wattless or reactive.

Furthermore, the alternating current passing over the line produces in all neighboring conductors secondary currents, which react upon the primary current and thereby introduce e.m.fs. of *mutual inductance* into the primary circuit. Mutual inductance is neither in phase nor in quadrature with the current, and can therefore be resolved into a *power component* of mutual inductance in phase with the current, which acts as an increase of resistance, and into a *reactive component* in quadrature with the current, which appears as a self-inductance.

This mutual inductance is not always negligible, as, for instance, its disturbing influence in telephone circuits shows.

The alternating potential of the line induces, by *electrostatic influence*, electric charges in neighboring conductors outside of the circuit, which retain corresponding opposite charges on the line wires. This electrostatic influence requires the expenditure of a current proportional to the e.m.f. and consisting of a *power component* in phase with the e.m.f. and a *reactive component* in quadrature thereto.

The alternating electromagnetic field of force set up by the line current produces in some materials a loss of power by *magnetic hysteresis*, or an expenditure of e.m.f. in phase with the current, which acts as an increase of resistance. This electromagnetic hysteresis loss may take place in the conductor proper if iron wires are used, and may then be very serious at high frequencies such as those of telephone currents.

The effect of *eddy currents* has already been referred to under "mutual inductance," of which it is a power component.

The alternating electrostatic field of force expends power in dielectrics by what is called *dielectric hysteresis*. In concentric cables, where the electrostatic gradient in the dielectric is comparatively large, the dielectric hysteresis may at high potentials consume considerable amounts of power. The dielectric hysteresis appears in the circuit as consumption of a current whose component in phase with the e.m.f. is the *dielectric power current*, which may be considered as the power component of the charging current.

Besides this there is the apparent increase of ohmic resistance due to *unequal distribution of current*, which, however, is usually not large enough to be noticeable at low frequencies.

Also, especially at very high frequency, energy is *radiated* into space, due to the finite velocity of the electric field, and can be represented by power components of current and of voltage respectively.

5. This gives, as the most general case and per unit length of line,

E.m.fs. consumed in phase with the current, I, and $= rI$, representing consumption of power, and due to *resistance,* and its apparent increase by unequal current distribution; to the power component of *mutual inductance: to secondary currents;* to the power component of *self–inductance: to electromagnetic hysteresis;* and to *electromagnetic radiation.*

E.m.fs. consumed in quadrature with the current, I, and $= xI$, reactive, and due to *self–inductance* and *mutual inductance.*

Currents consumed in phase with the e.m.f., E, and $= gE$, representing consumption of power, and due: to *leakage* through the insulating material, brush discharge or corona; to the power component of *electrostatic influence;* to the power component of *capacity,* or *dielectric hysteresis,* and to *electrostatic radiation.*

Currents consumed in quadrature with the e.m.f., E, and $= bE$, being reactive, and due to *capacity* and *electrostatic influence.*

Hence we get four constants per unit length of line, namely: Effective resistance, r; effective reactance, x; effective conductance, g, and effective susceptance, $b = -b_c$ (b_c being the absolute value of susceptance). These constants represent the coefficients per unit length of line of the following: e.m.f. consumed in phase with the current; e.m.f. consumed in quadrature with the current; current consumed in phase with the e.m.f., and current consumed in quadrature with the e.m.f.

6. This line we may assume now as supplying energy to a *receiver circuit of any description,* and determine the current and e.m.f. at any point of the circuit.

That is, an e.m.f. and current (differing in phase by any desired angle) may be given at the terminals of the receiving circuit. To be determined are the e.m.f. and current at any point of the line, for instance, at the generator terminals; or the impedance, $Z_1 = r_1 + jx_1$, or admittance, $Y_1 = g_1 - jb_1$, of the receiver circuit, and e.m.f., E_0, at generator terminals are given; the current and e.m.f. at any point of circuit to be determined, etc.

7. Counting now the distance, l, from a point 0 of the line which has the e.m.f.

$$E_1 = e_1 - je_1'$$

and the current

$$I_1 = i_1 - ji_1',$$

and counting l positive in the direction of rising power and negative in the direction of decreasing power, at any point l, in the line differential dl the leakage current is

$$Eg\ dl$$

and the capacity current is

$$jEb\ dl;$$

hence, the total current consumed by the line differential dl is

$$dI = E\ (g + jb)\ dl$$
$$= EY\ dl,$$

or
$$\frac{dI}{dl} = YE. \tag{1}$$

In the line differential dl the e.m.f. consumed by resistance is

$$I r\ dl,$$

the e.m.f. consumed by inductance is

$$jIx\ dl;$$

hence, the total e.m.f. consumed by the line differential dl is

$$dE = I\ (r + jx)\ dl$$
$$= IZ\ dl,$$

or
$$\frac{dE}{dl} = ZI. \tag{2}$$

These *fundamental differential equations* (1) and (2) are symmetrical with respect to I and E.

Differentiating these equations (1) and (2) gives

$$\left.\begin{array}{l} \dfrac{d^2I}{dl^2} = Y\dfrac{dE}{dl} \\[3mm] \dfrac{d^2E}{dl^2} = Z\dfrac{dI}{dl}, \end{array}\right\} \tag{3}$$

and

and substituting (1) and (2) in (3) gives the *differential equations of E and I*, thus:

$$\frac{d^2\dot{E}}{dl^2} = YZ\dot{E} \tag{4}$$

and

$$\frac{d^2\dot{I}}{dl^2} = YZ\dot{I}. \tag{5}$$

These differential equations are identical in form, and consequently I and E are functions differing by their integration constants or by their limiting conditions only.

These equations are of the form

$$\frac{d^2U}{dl^2} = ZY\dot{U}$$

and are integrated by

$$\dot{U} = A\varepsilon^{Vl},$$

where ε is the basis of the natural logarithms, $= 2.718283$.

Choosing equation (5), which is integrated by

$$\dot{I} = A\varepsilon^{Vl}, \tag{6}$$

and differentiating (6) twice gives

$$\frac{d^2\dot{I}}{dl^2} = V^2 A\varepsilon^{Vl},$$

and substituting (6) in (5), the factor $A\varepsilon^{Vl}$ cancels, and we have

$$V^2 = ZY,$$

or

$$V = \sqrt{ZY}, \tag{7}$$

hence, the general integral,

$$\dot{I} = A_1\varepsilon^{+Vl} - A_2\varepsilon^{-Vl}. \tag{8}$$

By equation (1),

$$\dot{E} = \frac{1}{Y}\frac{d\dot{I}}{dl},$$

and substituting herein equation (8) gives

$$\dot{E} = \frac{V}{Y}\left\{ A_1\varepsilon^{+Vl} + A_2\varepsilon^{-Vl} \right\}, \tag{9}$$

or, substituting (7),

$$E = \sqrt{\frac{Z}{Y}} \left\{ A_1 \varepsilon^{+Vl} + A_2 \varepsilon^{-Vl} \right\}. \tag{10}$$

The integration constants A_1 and A_2 in (8), (9), (10), in general, are complex quantities. The coefficient of the exponent, V, as square root of the product of two complex quantities, also is a complex quantity, therefore may be written

$$V = \alpha + j\beta, \tag{11}$$

and substituting for V, Z and Y gives

$$(\alpha + j\beta)^2 = (r + jx)(g + jb),$$

or

$$(\alpha^2 - \beta^2) + 2 j\alpha\beta = (rg - xb) + j(rb + gx),$$

and this resolves into the two separate equations

$$\left. \begin{array}{l} \alpha^2 - \beta^2 = rg - xb \\ 2\,\alpha\beta = rb + gx, \end{array} \right\} \tag{12}$$

since, when two complex quantities are equal, their real terms as well as their imaginary terms must be equal.

Equations (12) squared and added give

$$\begin{aligned} (\alpha^2 + \beta^2)^2 &= (rg - xb)^2 + (rb + xg)^2 \\ &= (r^2 + x^2)(g^2 + b^2) \\ &= z^2 y^2; \end{aligned}$$

hence,

$$\alpha^2 + \beta^2 = zy, \tag{13}$$

and from (12) and (13),

$$\begin{array}{l} \alpha = \sqrt{\tfrac{1}{2}(zy + rg - xb)} \\ \beta = \sqrt{\tfrac{1}{2}(zy - rg + xb)}. \end{array} \right\} \tag{14}$$

and

Equations (8) and (10) now assume the form

$$\left. \begin{array}{l} I = A_1 \varepsilon^{+(\alpha + j\beta)l} - A_2 \varepsilon^{-(\alpha + j\beta)l} \\ \\ E = \sqrt{\dfrac{Z}{Y}} \left\{ A_1 \varepsilon^{+(\alpha + j\beta)l} + A_2 \varepsilon^{-(\alpha + j\beta)l} \right\}. \end{array} \right\} \tag{15}$$

and

Substituting for the exponential function with an imaginary exponent the trigonometric expression

$$\varepsilon^{\pm j\beta l} = \cos \beta l \pm j \sin \beta l, \tag{16}$$

equations (15) assume the form

$$\left.\begin{aligned} I &= A_1 \varepsilon^{+al}(\cos \beta l + j \sin \beta l) - A_2 \varepsilon^{-al}(\cos \beta l - j \sin \beta l) \\ \text{and} & \\ E &= \sqrt{\frac{Z}{Y}} \Big\{ A_1 \varepsilon^{+al}(\cos \beta l + j \sin \beta l) + A_2 \varepsilon^{-al}(\cos \beta l - j \sin \beta l) \Big\}, \end{aligned}\right\} \tag{17}$$

where A_1 and A_2 are the constants of integration.

The distribution of current I and voltage E along the circuit, therefore, is represented by the sum of two products of exponential and trigonometric functions of the distance l. Of these terms, the one, with factor $A\varepsilon^{+al}$, increases with increasing distance l, that is, increases towards the generator, while the other, with factor $A_2 \varepsilon^{-al}$, decreases towards the generator and thus increases with increasing distance from the generator. The phase angle of the former decreases, that of the latter increases towards the generator, and the first term thus can be called the main wave, the second term the reflected wave.

At the point $l = 0$, by equations (17) we have

$$I_0 = A_1 - A_2,$$

$$E_0 = \sqrt{\frac{Z}{Y}} \Big\{ A_1 + A_2 \Big\},$$

and the ratio

$$\frac{A_2}{A_1} = m (\cos \tau - j \sin \tau),$$

where τ may be called the angle of reflection, and m the ratio of amplitudes of reflected and main wave at the reflection point.

8. The general integral equations of current and voltage distribution (17) can be written in numerous different forms.

Substituting $- A_2$ instead of $+ A_2$, the sign between the terms reverses, and the current appears as the sum, the voltage as difference of main and reflected wave.

Rearranging (17) gives

$$I = (A_1 \varepsilon^{+al} - A_2 \varepsilon^{-al}) \cos \beta l + j (A_1 \varepsilon^{+al} + A_2 \varepsilon^{-al}) \sin \beta l$$
and
$$E = \sqrt{\frac{Z}{Y}} \left\{ (A_1 \varepsilon^{+al} + A_2 \varepsilon^{-al}) \cos \beta l + j (A_1 \varepsilon^{+al} - A_2 \varepsilon^{-al}) \sin \beta l \right\} \cdot \tag{18}$$

Substituting (7) gives

$$\sqrt{\frac{Z}{Y}} = \frac{Z}{V} = \frac{V}{Y}, \tag{19}$$

and substituting

and
$$\frac{A_1}{V} = B_1$$
$$\frac{A_2}{V} = B_2,$$

or

and
$$\frac{A_1}{Y} = C_1$$
$$\frac{A_2}{Y} = C_2,$$

changes equations (17) to the forms,

$$I = V \left\{ B_1 \varepsilon^{+al} (\cos \beta l + j \sin \beta l) - B_2 \varepsilon^{-al} (\cos \beta l - j \sin \beta l) \right\}$$
and
$$E = Z \left\{ B_1 \varepsilon^{+al} (\cos \beta l + j \sin \beta l) + B_2 \varepsilon^{-al} (\cos \beta l - j \sin \beta l), \right\} \tag{20}$$

or

$$I = Y \left\{ C_1 \varepsilon^{+al} (\cos \beta l + j \sin \beta l) - C_2 \varepsilon^{-al} (\cos \beta l - j \sin \beta l) \right\}$$
and
$$E = V \left\{ C_1 \varepsilon^{+al} (\cos \beta l + j \sin \beta l) + C_2 \varepsilon^{-al} (\cos \beta l - j \sin \beta l) \right\} \tag{21}$$

Substituting in (17)

$$\frac{A_1}{\sqrt{Y}} = D_1 \text{ and } \frac{A_2}{\sqrt{Y}} = D_2$$

gives

$$\left. \begin{aligned} I &= \sqrt{Y} \left\{ D_1 \varepsilon^{+al} (\cos\beta l + j\sin\beta l) - D_2 \varepsilon^{-al} (\cos\beta l - j\sin\beta l) \right\} \\ \text{and} \\ E &= \sqrt{Z} \left\{ D_1 \varepsilon^{+al} (\cos\beta l + j\sin\beta l) + D_2 \varepsilon^{-al} (\cos\beta l - j\sin\beta l) \right\}. \end{aligned} \right\} \quad (22)$$

Reversing the sign of l, that is, counting the distance in the opposite direction, or positive for decreasing power, from the generator towards the receiving circuit, and not, as in equations (17) to (22), from the receiving circuit towards the generator, exchanges the position of the two terms; that is, the first term, or the main wave, decreases with increasing distance, and lags; the second term, or the reflected wave, increases with the distance, and leads.

Equations (17) thus assume the form

$$\left. \begin{aligned} I &= A_1 \varepsilon^{-al} (\cos\beta l - j\sin\beta l) - A_2 \varepsilon^{+al} (\cos\beta l + j\sin\beta l) \\ \text{and} \\ E &= \sqrt{\frac{Z}{Y}} \left\{ A_1 \varepsilon^{-al} (\cos\beta l - j\sin\beta l) + A_2 \varepsilon^{+al} (\cos\beta l + j\sin\beta l) \right\}, \end{aligned} \right\} \quad (23)$$

and correspondingly equations (18) to (22) modify.

9. The two integration constants contained in equations (17) to (23) require two conditions for their determination, such as current and voltage at one point of the circuit, as at the generator or at the receiving end; or current at one point, voltage at the other; or voltage at one point, as at the generator, and ratio of voltage and current at the other end, as the impedance of the receiving circuit.

Let the current and voltage (in intensity as well as phase, that is, as complex quantities) be given at one point of the circuit, and counting the distance l from this point, the terminal conditions are

$$\left. \begin{aligned} l &= 0, \\ I &= I_0 = i_0 - ji_0', \\ E &= E_0 = e_0 - je_0'. \end{aligned} \right\} \quad (24)$$

and

Substituting (24) in (17) gives

and

$$I_0 = A_1 - A_2$$

$$E_0 = \sqrt{\frac{Z}{Y}}\left(A_1 + A_2\right);$$

hence,

and

$$A_1 = \frac{1}{2}\left\{I_0 + E_0\sqrt{\frac{Y}{Z}}\right\}$$

$$A_2 = -\frac{1}{2}\left\{I_0 - E_0\sqrt{\frac{Y}{Z}}\right\},$$

and substituted in (17) gives

and

$$
\begin{aligned}
I &= \frac{1}{2}\left\{\left(I_0 + E_0\sqrt{\frac{Y}{Z}}\right)\varepsilon^{+al}\left(\cos \beta l + j \sin \beta l\right)\right.\\
&\quad + \left.\left(I_0 - E_0\sqrt{\frac{Y}{Z}}\right)\varepsilon^{-al}\left(\cos \beta l - j \sin \beta l\right)\right\}\\
E &= \frac{1}{2}\left\{\left(E_0 + I_0\sqrt{\frac{Z}{Y}}\right)\varepsilon^{+al}\left(\cos \beta l + j \sin \beta l\right)\right.\\
&\quad + \left.\left(E_0 - I_0\sqrt{\frac{Z}{Y}}\right)\varepsilon^{-al}\left(\cos \beta l - j \sin \beta l\right)\right\}.
\end{aligned}
\tag{25}
$$

If then I_0 and E_0 are the current and voltage respectively at the receiving end or load end of a circuit of length l_0, equations (25) represent current and voltage at any point of the circuit, from the receiving end $l = 0$ to the generator end $l = l_0$.

If I_0 and E_0 are the current and voltage at the generator terminals, since in equations (17) l is counted towards rising power, in the present case the receiving end of the line is represented by $l = -l_0$; that is, the negative values of l represent the distance from the generator end, along the line. In this case it is more convenient to reverse the sign of l, that is, use equations (22) and the distribution of current and voltage at distance l from the generator terminals. I, E are then given by

$$I = \frac{1}{2}\left\{\left(I_0 + E_0\sqrt{\frac{Y}{Z}}\right)\varepsilon^{-al}(\cos\beta l - j\sin\beta l)\right.$$

$$\left. + \left(I_0 - E_0\sqrt{\frac{Y}{Z}}\right)\varepsilon^{+al}(\cos\beta l + j\sin\beta l)\right\}$$

and
$$\hspace{6cm}(26)$$

$$E = \frac{1}{2}\left\{\left(E_0 + I_0\sqrt{\frac{Z}{Y}}\right)\varepsilon^{-al}(\cos\beta l - j\sin\beta l)\right.$$

$$\left. + \left(E_0 - I_0\sqrt{\frac{Z}{Y}}\right)\varepsilon^{+al}(\cos\beta l + j\sin\beta l)\right\}.$$

10. Assume that the character of the load, that is, the impedance, $\frac{E_1}{I_1} = Z_1 = r_1 + jx_1$, or admittance, $\frac{I_1}{E_1} = Y_1 = \frac{1}{Z_1} = g_1 - jb_1$, of the receiving circuit and the voltage E_0 at the generator end of the circuit be given.

Let l_0 = length of circuit, and counting distance l from the generator end, for $l = 0$ we have

$$E = E_0;$$

this substituted in equation (23) gives

$$E_0 = \sqrt{\frac{Z}{Y}}(A_1 + A_2). \hspace{2cm}(27)$$

However, for $l = l_0$,

$$\frac{E}{I} = Z_1;$$

substituting (23) herein gives

$$Z_1 = \sqrt{\frac{Z}{Y}}\frac{A_1\varepsilon^{-al_0}(\cos\beta l_0 - j\sin\beta l_0) + A_2\varepsilon^{+al_0}(\cos\beta l_0 + j\sin\beta l_0)}{A_1\varepsilon^{-al_0}(\cos\beta l_0 - j\sin\beta l_0) - A_2\varepsilon^{+al_0}(\cos\beta l_0 + j\sin\beta l_0)};$$

hence, substituting (19) and expanding,

$$\frac{A_1\varepsilon^{-al_0}(\cos\beta l_0 - j\sin\beta l_0)}{A_2\varepsilon^{+al_0}(\cos\beta l_0 + j\sin\beta l_0)} = \frac{VZ_1 + Z}{VZ_1 - Z},$$

or
$$A_2 = A_1\frac{VZ_1 - Z}{VZ_1 + Z}\varepsilon^{-2al_0}(\cos 2\beta l_0 - j\sin 2\beta l_0),$$

and denoting the complex factor by

$$\dot{C} = \frac{VZ_1 - Z}{VZ_1 + Z}\varepsilon^{-2\,al_0}(\cos 2\,\beta l_0 - j\sin \beta l_0), \qquad (28)$$

which may be called the reflection constant, we have

$$\dot{A}_2 = \dot{C}\dot{A}_1,$$

and by (27),

$$\left.\begin{aligned}\dot{A}_1 &= \frac{E_0}{1+\dot{C}}\sqrt{\frac{Y}{Z}}\\[2mm]\text{and}\qquad \dot{A}_2 &= \frac{E_0\,\dot{C}}{1+\dot{C}}\sqrt{\frac{Y}{Z}};\end{aligned}\right\} \qquad (29)$$

hence, substituted in (17),

$$\left.\begin{aligned}\dot{I} &= \frac{E_0}{1+\dot{C}}\sqrt{\frac{Y}{Z}}\left\{\varepsilon^{-al}(\cos \beta l - j\sin \beta l) - \dot{C}\varepsilon^{+al}(\cos \beta l + j\sin \beta l)\right\}\\[2mm]\text{and}\\[1mm]\dot{E} &= \frac{E_0}{1+\dot{C}}\left\{\varepsilon^{-al}(\cos \beta l - j\sin \beta l) + \dot{C}\varepsilon^{+al}(\cos \beta l + j\sin \beta l)\right\}.\end{aligned}\right\} \qquad (30)$$

11. As an example, consider the problem of delivering, in a three-phase system, 200 amperes per phase, at 90 per cent power factor lag at 60,000 volts per phase (or between line and neutral) and 60 cycles, at the end of a transmission line 200 miles in length, consisting of two separate circuits in multiple, each consisting of number 00 B. and S. wire with 6 feet distance between the conductors.

Number 00 B. and S. wire has a resistance of 0.42 ohms per mile, and at 6 feet distance from the return conductor an inductance of 2.4 mh. and capacity of 0.015 mf. per mile.

The two circuits in multiple give, at 60 cycles, the following line constants per mile: $r = 0.21$ ohm, $L = 1.2 \times 10^{-3}$ henry, and $C = 0.03 \times 10^{-6}$ farad; hence,

$$\begin{aligned}x &= 2\,\pi fL = 0.45,\\ Z &= 0.21 + 0.45\,j,\\ z &= 0.50,\end{aligned}$$

and, neglecting the conductance $(g = 0)$,

$$\begin{aligned}b &= 2\,\pi fC = 11\times 10^{-6},\\ Y &= 11 \times 10^{-6}\,j,\\ y &= 11 \times 10^{-6},\end{aligned}$$

and
$$\left.\begin{array}{l} \alpha = 0.524 \times 10^{-3}, \\ \beta = 2.285 \times 10^{-3}, \end{array}\right\} \quad (31)$$
$$V = (0.524 + 2.285\,j)\,10^{-3},$$

$$\sqrt{\frac{Y}{Z}} = \frac{V}{Z} = (4.53 + 0.9\,j)\,10^{-3}$$

and
$$\sqrt{\frac{Z}{Y}} = \frac{V}{Y} = (0.208 - 0.047\,j)\,10^{+3}.$$

Counting the distance l from the receiving end, and choosing the receiving voltage as zero vector, we have

$$l = 0,$$

$$E = E_0 = e_0 = 60,000 \text{ 'volts,}$$

and the current of 200 amperes at 90 per cent power factor,

$$I = I_0 = i_0 + j i_0' = 180 - 87\,j,$$

and substituting these values in equations (25) gives

$$\left.\begin{array}{l} I = (226 - 14.4\,j)\,\varepsilon^{+\alpha l}\,(\cos \beta l + j \sin \beta l) - (46 + 72.6\,j)\,\varepsilon^{-\alpha l} \\ \qquad (\cos \beta l - j \sin \beta l), \text{ in amperes,} \\ \text{and} \\ E = (46.7 - 13.3\,j)\,\varepsilon^{+\alpha l}(\cos \beta l + j \sin \beta l) + (13.3 + 13.3\,j)\,\varepsilon^{-\alpha l} \\ \qquad (\cos \beta l - j \sin \beta l), \text{ in kilovolts,} \end{array}\right\} \quad (32)$$

where α and β are given by above equations (31).

From equations (32) the following results are obtained.

Receiving end of line, $l = 0$
   $I = 180 - 87\,j$ $i = 200$ amp. $\tan \theta_1 = 0.483$ $\theta_1 = 26°$
   $E = 60 \times 10^3$ $e = 60,000$ volts $\theta_2 = 0$

                                        power factor, $\overline{0.90}$ lag

Middle of line, $l = 100$
   $I = 177 - 18\,j$ $i = 178$ amp. $\tan \theta_1 = +0.102$ $\theta_1 = 6°$
   $E = (66.2 + 6.9\,j)\,10^3$ $e = 66,400$ volts $\tan \theta_2 = -0.104$ $\theta_2 = -6°$

                                    $\theta_1 - \theta_2 = \theta = 12°$

                           power factor, $\cos \theta = 0.979$ lag.

Generator end of line, $l = 200$
   $I = 165.7 + 56\,j$ $i = 175$ amp. $\tan \theta_1 = -0.338$ $\theta_1 = -19°$
   $E = (69 - 15\,j)\,10^3$ $e = 70,700$ volts $\tan \theta_2 = -0.218$ $\theta_2 = -12°$

                                    $\theta_1 - \theta_2 = \theta = -7°$

                           power factor, $\cos \theta = 0.993$ lead.

As seen, the current decreases from the receiving end to the middle of the line, but from there to the generator remains practically constant. The voltage increases more in the receiving half of the line than in the generator half. The power-factor is practically unity from the middle of the line to the generator.

12. It is interesting to compare with above values the values derived by neglecting the distributed character of resistance, inductance, and capacity.

From above constants per mile it follows, for the total line of 200 miles length, $r_0 = 42$ ohms, $x_0 = 90$ ohms, and $b_0 = 2.2 \times 10^{-3}$ mho; hence,

$$Z_0 = 42 + 90\,j$$
$$\text{and } Y_0 = 2.2\,j\,10^{-3}.$$

(1) Neglecting the line capacity altogether, with I_0 and E_0 at the receiver terminals, at the generator terminals we have

$$I_1 = I_0$$

and

$$E_1 = E_0 + Z_0 I_0;$$

hence,

$I_1 = 180 - 87\,j$	$i_1 = 200$ amp.	$\tan \theta_1 = 0.483$	$\theta_1 = + 26°$
$E_1 = (75.4 + 12.6\,j)\,10^3$	$e_1 = 76,400$ volts	$\tan \theta_2 = -0.167$	$\theta_2 = -9°$
		$\theta_1 - \theta_2 = \theta = +34°$	
		power factor, $\cos \theta = 0.83$ lag.	

These values are extremely inaccurate, voltage and current at generator too high and power factor too low.

(2) Representing the line capacity by a condenser at the generator end, that is, adding the condenser current at the generator end,

$$I_1 = I_0 + Y_0 E_1$$

and

$$E_1 = E_0 + Z_0 I_0;$$

hence,

$I_1 = 152 + 89\,j$	$i_1 = 176$ amp.	$\tan \theta_1 = -0.585$	$\theta_1 = -30°$
$E_1 = (75.4 + 12.6\,j)\,10^3$	$e_1 = 76,400$ volts	$\tan \theta_2 = -0.167$	$\theta_2 = -9°$
		$\theta_1 - \theta_2 = \theta = -21°$	
		power factor, $\cos \theta = 0.93$ lead.	

As seen, the current is approximately correct, but the voltage is far too high and the power factor is still low, but now leading.

(3) Representing the line capacity by a condenser at the receiving end, that is, adding the condenser current at the load,

$$I_1 = I_0 + Y_0 E_0$$

and

$$E_1 = E_0 + Z_0 I_1;$$

hence,

$I_1 = 180 + 45 j$ $i_1 = 186$ amp. $\tan \theta_1 = -\,.250$ $\theta_1 = -14°$
$E_1 = (63.5 + 18.1\,j)\ 10^3$ $e_1 = 66{,}000$ volts $\tan \theta_2 = -\,.285$ $\theta_2 = -16°$
$$\theta_1 - \theta_2 = \overline{\theta} = +\ 2°$$
$$\text{power factor, } \cos \theta = 1.00$$

In this case the voltage e_1 is altogether too low, the current somewhat high, but the power factor fairly correct.

(4) Taking the average of the values of (2) and of (3) gives

$I_1 = 166 + 67 j$ $i_1 = 179$ amp. $\tan \theta_1 = -\,0.403$ $\theta_1 = -22°$
$E_1 = (69.4 + 15.3\,j)\ 10^3$ $e_1 = 71{,}100$ volts $\tan \theta_2 = -\,0.220$ $\theta_2 = -12°$
$$\theta_1 - \theta_2 = \theta = -\ 10°$$
$$\text{power-factor, } \cos \theta = 0.985 \text{ lead.}$$

As seen by comparing these average values with the exact result as derived above, these values are not very different, but constitute a fair approximation in the present case. Such a close coincidence of this approximation with the exact result can, however, not be counted upon in all instances.

13. In the equations (17) to (23) the length

$$l_w = \frac{2\,\pi}{\beta} \tag{33}$$

is a complete wave length, which means that in the distance $\dfrac{2\,\pi}{\beta}$ the phases of the components of current and of e.m.f. repeat, and that in half this distance they are just opposite.

Hence, the remarkable condition exists that in a very long line at different points the currents are simultaneously in opposite directions and the e.m.fs. are opposite.

The difference of space phase τ between current I and e.m.f. E at any point l of the line is determined by the equation

$$m (\cos \tau - j \sin \tau) = \frac{\dot{E}}{\dot{I}}, \tag{34}$$

where m is a constant.

Hence, τ varies from point to point, oscillating around a medium position, τ_∞, which it approaches at infinity.

This difference of phase, τ_∞, towards which current and e.m.f. tend at infinity, is determined by the expression

$$m (\cos \tau_\infty - j \sin \tau_\infty) = \left[\frac{\dot{E}}{\dot{I}}\right]_{l=\infty},$$

or, substituting for $\dot{E}$ and $\dot{I}$ their values from equations (23), and since $\varepsilon^{-al} = 0$, and $A_1 \varepsilon^{al} (\cos \beta l + j \sin \beta l)$ cancels,

$$m (\cos \tau_\infty + j \sin \tau_\infty) = \sqrt{\frac{Z}{Y}} = \frac{V}{Y} = \frac{\alpha + j\beta}{g + jb}$$

$$= \frac{(\alpha g + \beta b) - j (\alpha b - \beta g)}{b^2 + g^2};$$

hence,
$$\tan \tau_\infty = \frac{+ \alpha b - \beta g}{\alpha g + \beta b}. \tag{35}$$

14. This angle, $\tau_\infty = 0$; that is, current and e.m.f. come more and more in phase with each other when

$$\alpha b - \beta g = 0; \text{ that is,}$$
$$\alpha \div \beta = g \div b, \text{ or}$$
$$\frac{\alpha^2 - \beta^2}{2 \alpha\beta} = \frac{g^2 - b^2}{2 gb};$$

substituting (12) gives

$$\frac{gr - bx}{gx + br} = \frac{g^2 - b^2}{2 gb};$$

hence, expanding, $\quad r \div x = g \div b;$ \hfill (36)

that is, *the ratio of resistance to inductance equals the ratio of leakage to capacity.*

This angle, $\tau_\infty = 45°$; that is, current and e.m.f. differ by one-eighth period if $+ \alpha b - \beta g = \alpha g + \beta b$, or

$$\frac{\alpha}{\beta} = \frac{b + g}{b - g},$$

which gives $\qquad\qquad rg + xb = 0, \qquad\qquad (37)$

which means that two of the four line-constants, either g and x or g and b, must be zero.

The case where $g = 0 = x$, that is, a line having only resistance and distributed capacity but no self-inductance, is approximately realized in concentric or multiple-conductor cables, and in these the space-phase angle tends towards 45 degrees lead for infinite length.

15. As an example are shown the characteristic curves of a transmission line of the relative constants,

$$r : x : g : b = 8 : 32 : 1.25 \times 10^{-4} : 25 \times 10^{-4} \text{ and } e = 25,000,$$
$$i = 200 \text{ at the receiving circuit, for the conditions}$$

(a) Non-inductive load in the receiving circuit, Fig. 80.

(b) Wattless receiving circuit of 90 time-degrees lag, Fig. 81.

(c) Wattless receiving circuit of 90 time-degrees lead, Fig. 82.

These curves are determined graphically by constructing the topographic circuit characteristics in polar coördinates as explained in "Theory and Calculation of Alternating-Current Phenomena," and deriving corresponding values of current, potential difference, and phase angle therefrom.

As seen from these diagrams, for wattless receiving circuit, current and e.m.f. oscillate in intensity inversely to each other, with an amplitude of oscillation gradually decreasing when passing from the receiving circuit towards the generator, while the space-phase angle between current and e.m.f. oscillates between lag and lead with decreasing amplitude. Approximately maxima and minima of current coincide with minima and maxima of e.m.f. and zero phase angles.

For such graphical constructions, polar coördinate paper and two angles α and δ are desirable, the angle α being the angle between current and change of e.m.f., $\tan \alpha = \dfrac{x}{r} = 4$, and the

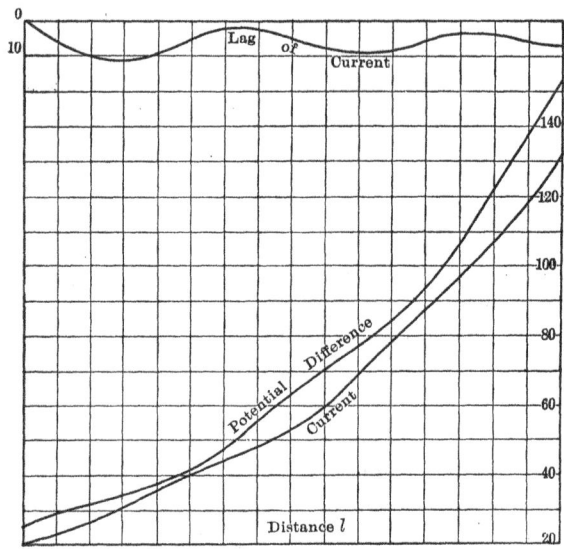

Fig. 80. Current, e.m.f. and space-phase angle between current and e.m.f. in a transmission line. Non-inductive load.

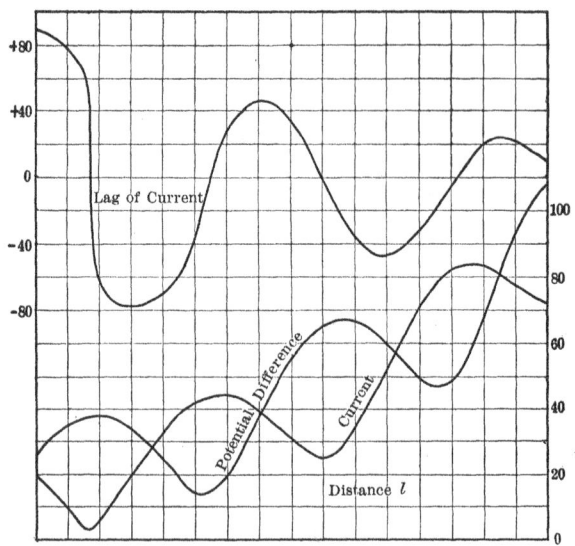

Fig. 81. Current, e.m.f. and space-phase angle between current and e.m.f. in a transmission line. Inductive load.

angle δ the angle between e.m.f. and change of current, $\tan \delta = \dfrac{b}{g} = 20$ in above instance.

With non-inductive load, Fig. 80, these oscillations of intensity have almost disappeared, and only traces of them are noticeable in the fluctuations of the space-phase angle and the relative values of current and e.m.f. along the line.

Towards the generator end of the line, that is, towards rising power, the curves can be extended indefinitely, approaching more and more the conditions of non-inductive circuit. Towards decreasing power, however, all curves ultimately reach the conditions of a wattless receiving circuit, as Figs. 81 and 82, at the point where the total energy input into the line has been consumed therein, and at this point the two curves for lead and for lag join each other as shown in Fig. 83, the one being a prolongation of the other, and the power in the line reverses. Thus in Fig. 83 energy flows from both sides of the line towards the point of zero power marked by 0, where the current and e.m.f. are in quadrature with each other, the current being leading with regard to the power from the left and lagging with regard to the power from the right side of the diagram.

16. It is of interest to investigate some special cases of such circuits of distributed constants.

(A) *Open circuit at the end of the line.*

Assuming a constant alternating e.m.f. E_1 impressed upon a circuit at one end while the other end of the circuit is open.

Counting the distance l from the open end of the line, and denoting the length of the line by l_0, for $l = 0$,

$$ I = I_0 = 0, $$

and for $l = l_0$,

$$ E = E_1; $$

hence, substituting in equations (17),

$$ 0 = A_1 - A_2, $$

$$ E_1 = \sqrt{\frac{Z}{Y}} \left\{ A_1 \varepsilon^{+al_0}(\cos \beta l_0 + j \sin \beta l_0) + A_2 \varepsilon^{-al_0}(\cos \beta l_0 - j \sin \beta l_0) \right\}; $$

hence,
$$ A_2 = A_1 = A $$

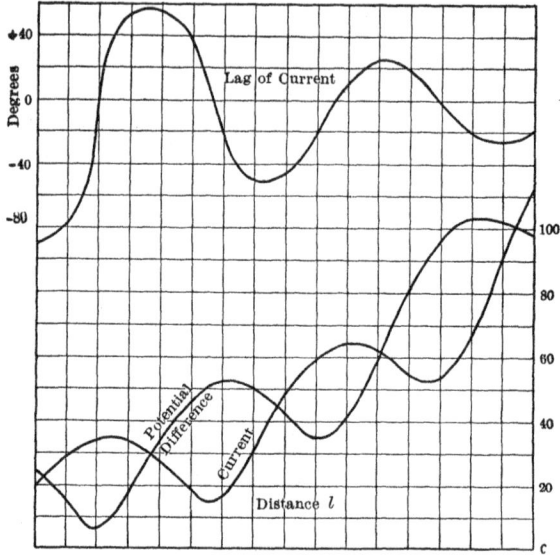

Fig. 82. Current, e.m.f. and space-phase angle between current and e.m.f. in a transmission line. Anti-inductive load.

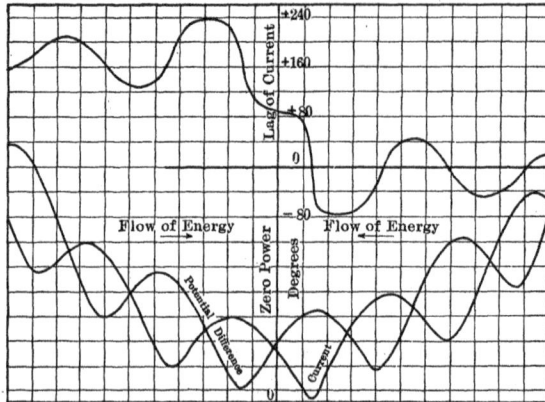

Fig. 83. Current, e.m.f. and space-phase angle between current and e.m.f. in a transmission line.

and

$$A = \frac{E_1\sqrt{\dfrac{Y}{Z}}}{\varepsilon^{+al_0}(\cos\beta l_0 + j\sin\beta l_0) + \varepsilon^{-al_0}(\cos\beta l_0 - j\sin\beta l_0)}$$

$$= \frac{E_1\sqrt{\dfrac{Y}{Z}}}{(\varepsilon^{+al_0} + \varepsilon^{-al_0})\cos\beta l_0 + j(\varepsilon^{+al_0} - \varepsilon^{-al_0})\sin\beta l_0};$$

hence, substituting in (17),

$$\left.\begin{aligned}
I &= E_1\sqrt{\frac{Y}{Z}}\,\frac{(\varepsilon^{+al} - \varepsilon^{-al})\cos\beta l + j(\varepsilon^{+al} + \varepsilon^{-al})\sin\beta l}{(\varepsilon^{+al_0} + \varepsilon^{-al_0})\cos\beta l_0 + j(\varepsilon^{+al_0} - \varepsilon^{-al_0})\sin\beta l_0} \\
\text{and} & \\
E &= E_1\,\frac{(\varepsilon^{+al} + \varepsilon^{-al})\cos\beta l + j(\varepsilon^{+al} - \varepsilon^{-al})\sin\beta l}{(\varepsilon^{+al_0} + \varepsilon^{-al_0})\cos\beta l_0 + j(\varepsilon^{+al_0} - \varepsilon^{-al_0})\sin\beta l_0}.
\end{aligned}\right\} \quad (38)$$

At $l = 0$, or the open end of the line, by equations (38),

$$I_0 = 0$$

and

$$E_0 = \frac{2E_1}{(\varepsilon^{+al_0} + \varepsilon^{-al_0})\cos\beta l_0 + j(\varepsilon^{+al_0} - \varepsilon^{-al_0})\sin\beta l_0}. \quad (39)$$

The absolute values of I and E follow from equations (38) and (39):

$$I = E_1\sqrt{\frac{y}{z}}\sqrt{\frac{(\varepsilon^{+al} - \varepsilon^{-al})^2\cos^2\beta l + (\varepsilon^{+al} + \varepsilon^{-al})^2\sin^2\beta l}{(\varepsilon^{+al_0} + \varepsilon^{-al_0})^2\cos^2\beta l_0 + (\varepsilon^{+al_0} - \varepsilon^{-al_0})^2\sin^2\beta l_0}},$$

which expanded gives

$$\left.\begin{aligned}
I &= E_1\sqrt{\frac{y}{z}\,\frac{\varepsilon^{+2al} + \varepsilon^{-2al} - 2\cos 2\beta l}{\varepsilon^{+2al_0} + \varepsilon^{-2al_0} + 2\cos 2\beta l_0}} \\
\text{and} & \\
E &= E_1\sqrt{\frac{\varepsilon^{+2al} + \varepsilon^{-2al} + 2\cos 2\beta l}{\varepsilon^{+2al_0} + \varepsilon^{-2al_0} + 2\cos\beta l_0}}
\end{aligned}\right\} \quad (40)$$

and

$$E_0 = \frac{2E_1}{\sqrt{\varepsilon^{+2al_0} + \varepsilon^{-2al_0} + 2\cos 2\beta l_0}}. \quad (41)$$

As function of l, the e.m.f. E or the current I is a maximum or minimum for

$$\frac{d}{dl}\left(\varepsilon^{+2al} + \varepsilon^{-2al} \pm 2\cos 2\,\beta l\right) = 0;$$

hence,

$$\alpha\left(\varepsilon^{+2al} - \varepsilon^{-2al}\right) = \pm\,2\,\beta\sin 2\,\beta l. \tag{42}$$

For $l = 0$, and since α is a small quantity, the left side of (42) also is small, and for values of $\sin 2\,\beta l$ approximating zero, that is, in the neighborhood of $l = \dfrac{n\pi}{2\,\beta}$, or where βl is a multiple of a quadrant, equation (42) becomes zero. At $\beta l \cong 2\,n\dfrac{\pi}{2}$, or the even quadrants, E is a maximum, I a minimum, at $\beta l \cong (2\,n - 1)\dfrac{\pi}{2}$, or the odd quadrants, E is a minimum, I a maximum.

The even quadrants, therefore, are nodes of current and wave crests of e.m.f., and the odd quadrants are nodes of e.m.f. and crests of current.

A maximum voltage point, or wave crest, occurs at the open end of the line at $l = 0$, and is given by equation (41). As function of the length l_0 of the line this is a maximum for

$$\frac{d}{dl_0}\left(\varepsilon^{+2al_0} + \varepsilon^{-2al_0} + 2\cos 2\,\beta l_0\right) = 0,$$

or

$$\alpha\left(\varepsilon^{+2al_0} - \varepsilon^{-2al_0}\right) = 2\,\beta\sin 2\,\beta l_0,$$

or approximately at

$$\left.\begin{array}{l} \beta l_0 = \dfrac{\pi}{2} \\[2mm] l_0 = n\,\dfrac{\pi}{2\,\beta}\,; \end{array}\right\} \tag{43}$$

that is, when the line is a quarter wave length or an odd multiple thereof.

Substituting in (41), $\beta l_0 = \dfrac{\pi}{2}$ gives $\qquad$ (44)

$$E_0 = \frac{2 E_1}{\sqrt{\varepsilon^{+2 \, al_0} + \varepsilon^{-2 \, al_0} - 2}}$$

$$= \frac{2 E_1}{\varepsilon^{+al_0} - \varepsilon^{-al_0}} \cdot \qquad (45)$$

Since

$$\varepsilon^{\pm \, al_0} = 1 \pm a l_0 + \tfrac{1}{2} a^2 l_0{}^2 \pm \frac{1}{2 \times 3} a^3 l_0{}^3 + \pm \cdots,$$

for small values of al_0 we have

$$\varepsilon^{+al_0} - \varepsilon^{-al_0} = 2 \, a l_0$$

and

$$E_0 = \frac{E_1}{al_0}, \qquad (46)$$

which is the maximum voltage that can occur at the open end of a line with voltage E_1 impressed upon it at the other end.

Since, approximately,

$$\beta = \sqrt{xb}$$
$$= 2 \pi f \sqrt{LC},$$

by (44) we have

$$f = \frac{1}{4 \, l_0 \, \sqrt{LC}}, \qquad (47)$$

the frequency which at the length of line l_0 produces maximum voltage at the open end.

For the constants in the example discussed in paragraph 11 we have $l_0 = 200$ miles, $r = 0.21$ ohm, $L = 1.2 \times 10^{-3}$ henry, $C = 0.03 \times 10^{-6}$ farad, $g = 0$, $f = 208$ cycles per sec., $x = 1.57$ ohms, $z = 1.58$ ohms, $b = 39 \times 10^{-6}$ mho, $a = 0.53 \times 10^{-3}$, and $E_0 = 9.3 \, E_1$.

(B) *Line grounded at the end.*

17. Let the circuit be grounded or connected to the return conductor at one end, $l = 0$, and supplied by a constant impressed e.m.f. E_1 at the other end, $l = l_0$.

Then for $l = 0$,

$$E = E_0 = 0$$

and for $l = l_0$,

$$E = E_1;$$

hence, substituting in (17),

$$0 = A_1 + A_2$$

and

$$E_1 = \sqrt{\frac{Z}{Y}} \left\{ A_1 \varepsilon^{+al_0}(\cos \beta l_0 + j \sin \beta l_0) + A_2 \varepsilon^{-al_0}(\cos \beta l_0 - j \sin \beta l_0) \right\};$$

hence,

$$A_1 = -A_2 = A,$$

and

$$A = \frac{E_1 \sqrt{\dfrac{Y}{Z}}}{(\varepsilon^{+al_0} - \varepsilon^{-al_0}) \cos \beta l_0 + j (\varepsilon^{+al_0} + \varepsilon^{-al_0}) \sin \beta l_0},$$

and, substituting in (17),

$$
\left.
\begin{aligned}
I &= E_1 \sqrt{\frac{Y}{Z}} \; \frac{(\varepsilon^{+al} + \varepsilon^{-al}) \cos \beta l + j (\varepsilon^{+al} - \varepsilon^{-al}) \sin \beta l}{(\varepsilon^{+al_0} - \varepsilon^{-al_0}) \cos \beta l_0 + j (\varepsilon^{+al_0} + \varepsilon^{-al_0}) \sin \beta l_0} \\
E &= E_1 \frac{(\varepsilon^{+al} - \varepsilon^{-al}) \cos \beta l + j (\varepsilon^{+al} + \varepsilon^{-al}) \sin \beta l}{(\varepsilon^{+al_0} - \varepsilon^{-al_0}) \cos \beta l_0 + j (\varepsilon^{+al_0} + \varepsilon^{-al_0}) \sin \beta l_0}.
\end{aligned}
\right\} \quad (48)
$$

At the grounded end, $l = 0$,

$$I_0 = \frac{2 E_1 \sqrt{\dfrac{Y}{Z}}}{(\varepsilon^{+al_0} - \varepsilon^{-al_0}) \cos \beta l_0 + j (\varepsilon^{+al_0} + \varepsilon^{-al_0}) \sin \beta l_0}. \quad (49)$$

Substituting (49) in (48) gives

$$
\left.
\begin{aligned}
I &= \tfrac{1}{2} I_0 \left\{ (\varepsilon^{+al} + \varepsilon^{-al}) \cos \beta l + j (\varepsilon^{+al} - \varepsilon^{-al}) \sin \beta l \right\} \\
E &= \tfrac{1}{2} I_0 \sqrt{\frac{Z}{Y}} \left\{ (\varepsilon^{+al} - \varepsilon^{-al}) \cos \beta l + j (\varepsilon^{+al} + \varepsilon^{-al}) \sin \beta l. \right\}
\end{aligned}
\right\} \quad (50)
$$

In this case nodes of voltage and crests of current appear at $l = 0$ and at the even quadrants, $\beta l = 2 n \dfrac{\pi}{2}$, and nodes of current and crests of voltage appear at the odd quadrants, $\beta l = (2n - 1) \dfrac{\pi}{2}$.

(C) *Infinitely long conductor.*

18. If an e.m.f. E_0 is impressed upon an infinitely long conductor, that is, a conductor of such length that of the power input no appreciable part reaches the end, we have, for $l = 0$,

$$E = E_0$$

and for $l = \infty$,

$$E = 0 \text{ and } I = 0;$$

hence, substituting in (23) gives

$$A_2 = 0$$

and

$$E_0 = \sqrt{\frac{Z}{Y}} A_1;$$

hence,

$$\left.\begin{aligned} I &= E_0 \sqrt{\frac{Y}{Z}} \varepsilon^{-al} (\cos \beta l - j \sin \beta l) \\[2mm] E &= E_0 \varepsilon^{-al} (\cos \beta l - j \sin \beta l). \end{aligned}\right\} \tag{51}$$

From (51) it follows that

$$\frac{E}{I} = \sqrt{\frac{Z}{Y}} ;$$

that is, an infinitely long conductor acts like an impedance,

$$Z_1 = \sqrt{\frac{Z}{Y}} = r_1 + jx_1,$$

and the current at every point of the conductor thus has the same space-phase angle to the voltage,

$$\tan \alpha_1 = \frac{x_1}{r_1} .$$

The equivalent impedance of the infinite conductor is

$$Z_1 = \sqrt{\frac{Z}{Y}} = \frac{V}{Y} = \frac{\alpha + j\beta}{g + jb}$$

$$= \frac{\alpha g + \beta b}{y^2} + j\frac{\beta g - \alpha b}{y^2} \tag{52}$$

and the space-phase angle is

$$\tan \alpha_1 = \frac{\beta g - \alpha b}{\alpha g + \beta b}. \tag{53}$$

If $g = 0$ and $x = 0$, we have

$$\alpha = \beta = \sqrt{\frac{br}{2}}$$

and

$$\tan \alpha_1 = 1,$$

or

$$\alpha_1 = 45°;$$

that is, current and e.m.f. differ by one-eighth period.

This is approximately the case in cables, in which the dielectric losses and the inductance are small.

An infinitely long conductor therefore shows the wave propagation in its simplest and most perspicuous form, since the reflected wave is absent.

(D) *Generator feeding into a closed circuit.*

19. Let $l = 0$ be the center of the circuit; then

$$E_l = -E_{-l} \text{ and } I_l = I_{-l};$$

hence, $\qquad E = 0 \text{ at } l = 0,$

and the equations are the same as those of a line grounded at the end $l = 0$, which have been discussed under (B).

(E) *Line of quarter wave length.*

20. Interesting is the case of a line of quarter wave length.

Let the length l_0 of the line be one quarter wave of the impressed e.m.f.

$$\beta l_0 = \frac{\pi}{2}. \tag{54}$$

To illustrate the general character of the phenomena, we may as first approximation neglect the energy losses in the circuit, that is, assume the resistance r and the conductance g as negligible compared with x and b,

$$r = 0 = g.$$

These values substituted in (14) give

$$\alpha = 0 \quad \text{and} \quad \beta = \sqrt{xb}. \tag{55}$$

Counting the distance l from the end of the line l_0 we have for $l = 0$,

$$E_0 = e_0 - je_0' \\ I_0 = i_0 - ji_0', \tag{56}$$

and

and at the beginning of the line for $l = l_0$,

$$E_1 = e_1 - je_1' \\ I_1 = i_1 - ji_1', \tag{57}$$

and

and by (54) and (55),

$$l_0 = \frac{\pi}{2\sqrt{xb}}. \tag{58}$$

Substituting (56), (57), and (54) in (17) gives

$$I_0 = A_1 - A_2$$

and

$$E_0 = \sqrt{\frac{Z}{Y}}\,(A_1 + A_2) = \sqrt{\frac{x}{b}}\,(A_1 + A_2),$$

or

$$I_1 = j\,(A_1 + A_2)$$

and

$$E_1 = -j\sqrt{\frac{Z}{Y}}\,(A_1 - A_2) = +j\sqrt{\frac{x}{b}}\,(A_1 - A_2);$$

hence, eliminating A_1 and A_2 gives the relations between the electric quantities at the generator end of the quarter-wave line, E_1, I_1, and at the receiving end, E_0, I_0:

$$E_0 = -j\sqrt{\frac{x}{b}}\,I_1$$

and

$$I_0 = -j\sqrt{\frac{b}{x_0}}\,E_1,$$

(59)

and the absolute values are

$$E_0 = \sqrt{\frac{x}{b}}\,I_1$$

and

$$I_0 = \sqrt{\frac{b}{x}}\,E_1;$$

(60)

which means that if the supply voltage E_1 is constant, the output current I_0 is constant and lags 90 space-degrees behind the input voltage; if the supply current I_1 is constant, the output voltage E_0 is constant, and lags 90 space-degrees, and inversely. A quarter-wave line of negligible losses thus converts from constant potential to constant current, or from constant current to constant voltage. (Constant-potential constant-current transformation.)

Multiplying (60) gives

$$E_0 I_0 = I_1 E_1,$$

or

$$E_0 = \frac{E_1 I_1}{I_0};$$

hence, if $I_0 = 0$, that is, the line is open at the end, $E_0 = \infty$, and with a finite voltage supply to a line of quarter-wave length, an infinite (extremely high) voltage is produced at the other end.

Such a circuit thus may be used to produce very high voltages.

Since $x_0 = l_0 x =$ total reactance and $b_0 = l_0 b =$ total susceptance of the circuit, by (58) we have

$$x_0 b_0 = \frac{\pi^2}{4},$$

(61)

or the condition of quarter-wave length.

Substituting $x_0 = 2\pi f L_0$ and $b_0 = 2\pi f C_0$, we have

$$L_0 C_0 = \frac{1}{16 f^2}, \qquad (62)$$

or
$$f = \frac{1}{4 \sqrt{L_0 C_0}}, \qquad (63)$$

the condition of quarter-wave transmission.

21. If the resistance, r, and the conductance, g, of a quarter-wave circuit are not negligible, substituting (56), (54) and (57) in (17) we have, for $l = 0$,

and
$$\left.\begin{array}{l} I_0 = A_1 - A_2 \\[2mm] E_0 = \sqrt{\dfrac{Z}{Y}}\left(A_1 + A_2\right), \end{array}\right\} \qquad (64)$$

and for $l = l_0$,

and
$$\left.\begin{array}{l} I_1 = j\left\{A_1 \varepsilon^{+al_0} + A_2 \varepsilon^{-al_0}\right\} \\[2mm] E_1 = j\sqrt{\dfrac{Z}{Y}}\left\{A_1 \varepsilon^{+al_0} - A_2 \varepsilon^{-a}\right\}. \end{array}\right\} \qquad (65)$$

From (64) it follows that

and
$$\left.\begin{array}{l} A_1 = \dfrac{1}{2}\left(I_0 + \sqrt{\dfrac{Y}{Z}} E_0\right) \\[3mm] A_2 = -\dfrac{1}{2}\left(I_0 - \sqrt{\dfrac{Y}{Z}} E_0\right), \end{array}\right\} \qquad (66)$$

and substituting in (65) and rearranging we have

and
$$\left.\begin{array}{l} I_1 = j\left\{E_0\sqrt{\dfrac{Y}{Z}}\,\dfrac{\varepsilon^{+al_0} + \varepsilon^{-al_0}}{2} + I_0\,\dfrac{\varepsilon^{+al_0} - \varepsilon^{-al_0}}{2}\right\} \\[3mm] E_1 = j\left\{I_0\sqrt{\dfrac{Z}{Y}}\,\dfrac{\varepsilon^{+al_0} + \varepsilon^{-al_0}}{2} + E_0\,\dfrac{\varepsilon^{+al_0} - \varepsilon^{-al_0}}{2}\right\}, \end{array}\right\} \qquad (67)$$

or,

$$
I_0 = -j \left\{ E_1 \sqrt{\frac{Y}{Z}} \frac{\varepsilon^{+al_0} + \varepsilon^{-al_0}}{\varepsilon^{+2al_0} + \varepsilon^{-2al_0}} - I_1 \frac{\varepsilon^{+al_0} - \varepsilon^{-al_0}}{\varepsilon^{+2al_0} + \varepsilon^{-2al_0}} \right\}
$$

and

$$
E_0 = -j \left\{ I_1 \sqrt{\frac{Z}{Y}} \frac{\varepsilon^{+al_0} + \varepsilon^{-al_0}}{\varepsilon^{+2al} + \varepsilon^{-2al_0}} - E_1 \frac{\varepsilon^{+al_0} - \varepsilon^{-al_0}}{\varepsilon^{+2al_0} + \varepsilon^{-2al_0}} \right\},
$$

(68)

or, analogous to equation (59),

$$
I_0 = -\sqrt{\frac{Y}{Z}} \left\{ \frac{2\,j\dot{E}_1}{\varepsilon^{+al_0} + \varepsilon^{-al_0}} + \frac{\dot{E}_0 \left(\varepsilon^{+al_0} - \varepsilon^{-al_0}\right)}{\varepsilon^{+al_0} + \varepsilon^{-al_0}} \right\}
$$

and

$$
E_0 = -\sqrt{\frac{Z}{Y}} \left\{ \frac{2\,j\dot{I}_1}{\varepsilon^{+al_0} + \varepsilon^{-al_0}} + \frac{\dot{I}_0 \left(\varepsilon^{+al_0} - \varepsilon^{-al_0}\right)}{\varepsilon^{+al_0} + \varepsilon^{-al_0}} \right\}.
$$

(69)

In these equations the second term is usually small, due to the factor $(\varepsilon^{+al_0} - \varepsilon^{-al_0})$, and the first term represents constant potential-constant current transformation.

22. In a quarter-wave line, at constant impressed e.m.f. E_1, the current output I_0 is approximately constant and lagging 90 degrees behind E_1; it falls off slightly, however, with increasing load, that is, increasing I_1, due to the second term in equation (68); the voltage at the end of the line, E_0, at constant impressed voltage, is approximately proportional to the load, but does not reach infinity at open circuit, but a finite, though high, limiting value.

Inversely, at constant current input the voltage output is approximately constant and the output current proportional to the load.

The deviation from constancy, at constant E_1, of I_0, or at constant I_1, of E_0, therefore, is due to the second term, with factor $(\varepsilon^{+al_0} - \varepsilon^{-al_0})$.

Substituting (54),

$$
al_0 = \frac{\alpha}{\beta} \frac{\pi}{2},
$$

hence, al_0 is usually a very small quantity, and $\varepsilon^{\pm al_0} = \varepsilon^{\pm \frac{\alpha}{\beta} \frac{\pi}{2}}$ thus can be represented by the first terms of the series:

$$\varepsilon^{\pm\frac{\alpha}{\beta}\frac{\pi}{2}} = 1 \pm \frac{\alpha}{\beta}\frac{\pi}{2} + \frac{1}{2}\left(\frac{\alpha}{\beta}\frac{\pi}{2}\right)^2 \pm \frac{1}{2\times3}\left(\frac{\alpha}{\beta}\frac{\pi}{2}\right)^3 + \pm\cdots$$

$$= 1 \pm \frac{\alpha}{\beta}\frac{\pi}{2};$$

hence,

$$\frac{\varepsilon^{+al_0} + \varepsilon^{-al_0}}{2} = 1,$$

and

$$\frac{\varepsilon^{+al_0} - \varepsilon^{-al_0}}{2} = \frac{\alpha}{\beta}\frac{\pi}{2},$$

and, by (69),

$$\left.\begin{array}{l} I_0 = -j\sqrt{\dfrac{Y}{Z}}\,E_1 - \dfrac{\alpha}{\beta}\dfrac{\pi}{2}\sqrt{\dfrac{Y}{Z}}\,E_0 \\[3mm] E_0 = -j\sqrt{\dfrac{Z}{Y}}\,I_1 - \dfrac{\alpha}{\beta}\dfrac{\pi}{2}\sqrt{\dfrac{Z}{Y}}\,I_0. \end{array}\right\} \tag{70}$$

and

If r and g are small compared with x and b,

$$\beta = \sqrt{\tfrac{1}{2}\,(zy - rg + xb)} = \sqrt{xb}$$

and

$$\alpha = \sqrt{\tfrac{1}{2}\,(zy + rg - xb)};$$

substituting, by the binomial theorem,

$$zy = \sqrt{(r^2 + x^2)\,(g^2 + b^2)} = xb\left\{\left[1 + \left(\frac{r}{x}\right)^2\right]\left[1 + \left(\frac{g}{b}\right)^2\right]\right\}^{\frac{1}{2}}$$

$$= xb\left\{1 + \frac{1}{2}\left(\frac{r}{x}\right)^2 + \frac{1}{2}\left(\frac{g}{b}\right)^2 + \cdots\right\}$$

gives

$$\alpha = \sqrt{\frac{1}{2}\left\{\frac{1}{2}\left(\frac{r}{x}\right)^2 + \frac{1}{2}\left(\frac{g}{b}\right)^2 + \frac{rg}{bx}\right\}bx}$$

$$= \frac{\sqrt{bx}}{2}\left(\frac{r}{x} + \frac{g}{b}\right),$$

and

$$\frac{\alpha}{\beta} = \frac{1}{2}\left(\frac{r}{x} + \frac{g}{b}\right).$$

The quantity

$$\frac{\alpha}{\beta} = \frac{1}{2}\left(\frac{r}{x} + \frac{g}{b}\right) = u \tag{71}$$

may be called the *time constant of the circuit*.

The equations of *quarter-wave transmission* thus are

$$I_0 = -\sqrt{\frac{Y}{Z}} \left\{ j\, E_1 + \frac{u\pi}{2}\, E_0 \right\}$$

and

$$E_0 = -\sqrt{\frac{Z}{Y}} \left\{ j\, I_1 + \frac{u\pi}{2}\, I_0 \right\}, \tag{72}$$

and the maximum voltage E_0, at the open end of the circuit, at constant impressed e.m.f. E_1, is

$$I_0 = 0$$

and

$$E_0 = -\frac{2\, j E_1}{u\pi}, \tag{73}$$

and the current input is

$$I_1 = j\sqrt{\frac{Y}{Z}}\, E_0$$

$$= \frac{2\, E_1}{u\pi}\sqrt{\frac{Y}{Z}}, \tag{74}$$

where, approximately,

$$\sqrt{\frac{Z}{Y}} = \sqrt{\frac{L}{C}}. \tag{75}$$

23. Consider as an example a high potential coil of a transformer with one of its terminals connected to a source of high potential, for testing its insulation to ground, while the other terminal is open.

Assume the following constants per unit length of circuit: $r = 0.1$ ohm, $L = 0.02$ h., $C = 0.01 \times 10^{-6}$ farad, and $g = 0$; then, with a length of circuit $l_0 = 100$, the quarter-wave frequency is, by (47),

$$f = \frac{1}{4\, l_0\sqrt{LC}} = 177 \text{ cycles per sec.,}$$

ɔr very close to the third harmonic of a 60-cycle impressed voltage.

If, therefore, the testing frequency is low, 59 cycles, the circuit is a quarter wave of the third harmonic.

Assuming an impressed e.m.f. of 50,000 volts and 59 cycles, containing a third harmonic of 10 per cent, or $E_1 = 5000$ volts at 177 cycles, for this harmonic, we have $x = 22.2$ ohms and $b = 11.1 \times 10^{-6}$ ohm; hence, $u = 0.00225$ and

$$E_0 = \frac{2}{u\pi} E_1 = 283\ E_1;$$

therefore at $E_1 = 5000$ volts, $E_0 = 1,415,000$ volts;
that is, infinity, as far as insulation strength is concerned.

Quarter-wave circuits thus may be used, and are used, to produce extremely high voltages, and if a sufficiently high frequency is used — 100,000 cycles and more, as in wireless telegraphy, etc. — the length of the circuit is moderate.

This method of producing high voltages has the disadvantage that it does not give constant potential, but the high voltage is due to the tendency of the circuit to regulate for constant current, which means infinite voltage at infinite resistance or open circuit, but as soon as current is taken off the high potential point the voltage falls. The great advantage of the quarter-wave method of producing high voltage is its simplicity and ease of insulation; as the voltage gradually builds up along the circuit, the high voltage point or end of circuit may be any distance away from the power supply, and thus can easily be made safe.

24. As a quarter-wave circuit converts from constant potential to constant current, it is not possible, with constant voltage impressed upon a circuit of approximately a quarter-wave length, to get constant voltage at the other or receiving end of the circuit. Long before the circuit approaches quarter-wave length, and as soon as it becomes an appreciable part of a quarter wave, this tendency to constant current regulation makes itself felt by greaᵗ variations of voltage with changes of load at the receiving end of the circuit, constant voltage being impressed upon the generator end; that is, with increasing length of transmission lines the voltage regulation at the receiving end becomes seriously impaired hereby, even if the line resistance is moderate, and the operation of apparatus which require approximate constancy of voltage but do not operate on constant current—as synchronous apparatus—becomes more difficult.

Hence, at the end of very long transmission lines the voltage regulation becomes poor, and synchronous machines tend to instability and have to be provided with powerful steadying devices, giving induction motor features, and with a line approaching quarter-wave length, voltage regulation at the receiving end ceases, unless very powerful voltage controlling devices are used, such as large synchronous condensers, that is, synchronous machines establishing a fixed voltage and controlling the line by automatically drawing leading or lagging currents, in correspondence with the line conditions.

In this case of a line of approximately quarter-wave length, the constant potential-constant current transformation may be used to produce constant or approximately constant voltage at the load, by supplying constant current to the line; that is, the transmission line is made a quarter-wave length by modifying its constants, or choosing the proper frequency, the generators are designed to regulate for constant current and thus give a voltage varying with the load, and are connected in series (with constant current generators series connection is stable, parallel connections unstable) and feed constant current, at variable voltage, into the quarter-wave line. At the receiving end of the line, constant voltage then exists with varying load, or rather a voltage. which slightly falls off with the load, due to the power loss in the line. To maintain constant receiver voltage at all loads, then, would require a slight increase of generator current with increase of load, that is, increase of generator voltage, which can be produced by compounding regulated by the voltage.

In such a quarter-wave transmission the voltage at the receiving end then remains constant, while the current output from the line increases from nothing at no load. At the generator end the current remains approximately constant, increasing from no load to full load by the amount required to take care of the line loss, while the voltage at the generators increases from nearly nothing at no load, with increasing load, approximately proportional thereto.

25. There is, however, a serious limitation imposed upon quarter-wave transmission by considerations of voltage; to use the transmission line economically the voltage throughout it should not differ much, since the insulation of the line depends on the maximum, the efficiency of transmission, however, on the average voltage, and a line in which the voltage at the two ends is very different is uneconomical.

To use line copper and line insulation economically, in a quarter-wave transmission, the voltages at the two ends should be approximately equal at maximum load. These voltages are related to each other and to the current by the line constants, by equations (72).

By these equations (neglecting the term with u), reduced to absolute values, we have approximately

$$i_0 = \sqrt{\frac{y}{z}}\, e_1$$

and

$$e_0 = \sqrt{\frac{z}{y}}\, i_1,$$

and if $e_1 = e_0$,

$$i_0 = \sqrt{\frac{y}{z}}\, e_0;$$

hence, the power is

$$p_0 = e_0 i_0 = \sqrt{\frac{y}{z}}\, e_0^2,$$

or

$$e_0^2 = p_0 \sqrt{\frac{z}{y}}; \tag{76}$$

hence, the voltage e_0 required to transmit the power p_0 without great potential differences in the line depends on the power p_0 and the line constants, and inversely.

26. As an example of a quarter-wave transmission may be considered the transmission of 60,000 kilowatts over a distance of 700 miles, for the supply of a general three-phase distribution system, of 95 per cent power factor, lag.

The design of the transmission line is based on a compromise between different and conflicting requirements: economy in first cost requires the highest possible voltage and smallest conductor section, or high power loss in the line; economy of operation requires high voltage and large conductor section, or low power loss; reliability of operation of the line requires lowest permissible voltage and therefore large conductor section or high power loss; reliability of operation of the receiving system requires good voltage regulation and thus low line resistance, etc., etc.

Assume that the maximum effective voltage between the line conductors is limited to 120,000, and that there are two separate pole lines, each carrying three wires of 500,000 circular mils cross section, placed 6 feet between wires, and provided with a grounded neutral.

If there were no energy losses in the line and no increase of capacity due to insulators, etc., the speed of propagation would be the velocity of light, $S = 188,000$ miles per second, and the quarter-wave frequency of a line of $l_0 = 700$ miles would be

$$f = \frac{S}{4\,l_0} = 67 \text{ cycles per sec.};$$

hence, fairly close to the standard frequency of 60 cycles.

The loss of power in the line, and thus the increase of inductance by the magnetic field inside of the conductor (which would not exist in a conductor of perfect conductivity or zero resistance loss), the increase of capacity by insulators, poles, etc., lowers the frequency below that corresponding to the velocity of light and brings it nearer to 60 cycles.

In a line as above assumed the constants per mile of double conductor are: $r = 0.055$ ohm; $L = 0.001$ henry, and $C = 0.032 \times 10^{-6}$ farad, and, neglecting the conductance, $g = 0$, the quarter-wave frequency is

$$f = \frac{1}{4\,l_0\,\sqrt{LC}} = 63 \text{ cycles per sec.}$$

Either then the frequency of 63 cycles per second, or slightly above standard, may be chosen, or the line inductance or line capacity increased, to bring the frequency down to 60 cycles.

Assuming the inductance increased to $L = 0.0011$ henry gives $f = 60$ cycles per second, and the line constants then are $l_0 = 700$ miles; $f = 60$ cycles per second; $r = 0.055$ ohm; $L = 0.0011$ henry; $C = 0.032 \times 10^{-6}$ farad, and $g = 0$; hence, $x = 0.415$ ohm; $z = 0.42$ ohm; $Z = 0.055 + 0.415\,j$ ohm;

$b = 12.1 \times 10^{-6}$ mho; $y = 12.1 \times 10^{-6}$ mho, and $Y = j \ 12.1 \times 10^{-6}$ mho, and

$$\sqrt{\frac{Z}{Y}} = 186 + 12 \ j,$$

$$\sqrt{\frac{z}{y}} = 186,$$

$$u = \frac{1}{2} \left(\frac{r}{x} + \frac{g}{b} \right) = 0.066,$$

$$\beta = 2.247 \times 10^{-3}$$

$$\alpha = u\beta = 0.148 \times 10^{-3}.$$

At 60,000 kilowatts total input, or 20,000 kilowatts per line, and 120,000 volts between lines, or $\dfrac{120,000}{\sqrt{3}} = 69,000$ volts per line, and about 95 per cent power factor, the current input at full load is 306 amp. per line (of two conductors in multiple).

To get at full load $p = 20 \times 10^6$ watts, approximately the same voltage at both ends of the line, by equation (76), we must have

$$e^2 = p \sqrt{\frac{z}{y}},$$

or $e = 61,000$ volts.

Assuming therefore at the receiving end the voltage of 110,000 between the lines, or, 63,500 volts per line, and choosing the output current as zero vector, and counting the distance from the receiving end towards the generator, we have for $l = 0$,

$$I = I_0 = i_0,$$

and the voltage, at 95 per cent power factor, or $\sqrt{1 - 0.95^2} = 0.312$ inductance factor, is

$$E = E_0 = e_0 (0.95 + 0.312 \ j)$$
$$= 60,300 + 19,800 \ j.$$

Substituting these values in equations (72) gives

$$i_0 = \frac{\{ - jE_1 - 0.104 \ (60,300 + 19,800 \ j) \}}{186 - 12 \ j},$$

$$60,300 + 19,800 \ j = - (186 - 12 \ j) \ \{ jI_1 + 0.104 \ i_0 \};$$

hence,　　$-jE_1 = (186 - 12\,j)\,i_0 + (6250 + 2060\,j)$

and　　　$-jI_1 = \dfrac{60{,}300 - 19{,}800\,j}{186 - 12\,j} + 0.104\,i_0$

$$= 317 + 128\,j + 0.104\,i_0,$$

and the absolute values are

$$e_1 = \sqrt{(186\,i_0 + 6250)^2 + (12\,i_0 - 2060)^2}$$

and　　　$i_1 = \sqrt{(317 + 0.104\,i_0)^2 + 128^2};$

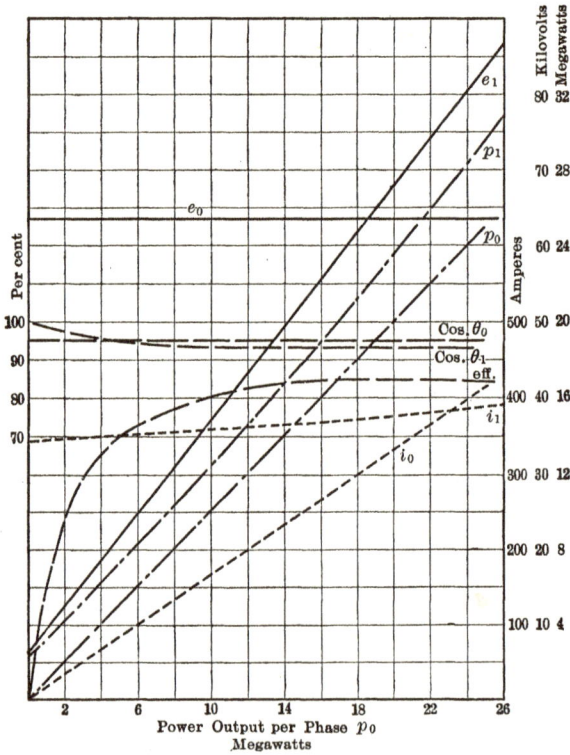

Fig. 84.　Long-distance quarter-wave transmission.

herefrom the power output and input, efficiency, power factor, etc., can be obtained.

In Fig. 84, with the power output per phase as abscissas, are shown the following quantities: voltage input e_1 and output e_0, in drawn lines; amperes input i_1 and output i_0, in dotted lines; power input p_1 and output p_0, in dash-dotted lines, and efficiency and power factor in dashed lines.

As seen, the power factor at the generator is above 93 per cent leading, and the efficiency reaches nearly 85 per cent.

At full load input of 20,000 kilowatts per phase, and 95 per cent power factor, lagging, of the output, the generator voltage is 58,500, or still 8 per cent below the output voltage of 63,500. The generator voltage equals the output voltage at 10 per cent overload, and exceeds it by 14 per cent at 25 per cent overload.

To maintain constant voltage at the output side of the line, the generator current has to be increased from 342 amperes at no load to 370 amperes at full load, or by 8.2 per cent, and inversely, at constant-current input, the output voltage would drop off, from no load to full load, by about 8 per cent. This, with a line of 15 per cent resistance drop, is a far closer voltage regulation than can be produced by constant potential supply, except by the use of synchronous machines for phase control.

CHAPTER III.

27. An interesting application of the equations of the long distance transmission line given in the preceding chapter can be made to the determination of the *natural period of a transmission line;* that is, the frequency at which such a line discharges an accumulated charge of atmospheric electricity (lightning), or oscillates because of a sudden change of load, as a break of circuit, or in general a change of circuit conditions, as closing the circuit, etc.

The discharge of a condenser through a circuit containing self-inductance and resistance is oscillating (provided the resistance does not exceed a certain critical value depending upon the capacity and the self-inductance); that is, the discharge current alternates with constantly decreasing intensity. The frequency of this oscillating discharge depends upon the capacity C and the self-inductance L of the circuit, and to a much lesser extent upon the resistance, so that, if the resistance of the circuit is not excessive, the frequency of oscillation can, by neglecting the resistance, be expressed with fair, or even close, approximation by the formula

$$f = \frac{1}{2\,\pi\,\sqrt{CL}}.$$

An electric transmission line represents a circuit having capacity as well as self-inductance; and thus when charged to a certain potential, for instance, by atmospheric electricity, as by induction from a thunder-cloud passing over or near the line, the transmission line discharges by an oscillating current.

Such a transmission line differs, however, from an ordinary condenser in that with the former the capacity and the self-inductance are distributed along the circuit.

In determining the frequency of the oscillating discharge of such a transmission line, a sufficiently close approximation is obtained by neglecting the resistance of the line, which, at the relatively high frequency of oscillating discharges, is small compared with the reactance. This assumption means that the dying out of the discharge current through the influence of the resistance of the circuit is neglected, and the current assumed as an alternating current of approximately the same frequency and the same intensity as the initial waves of the oscillating discharge current. By this means the problem is essentially simplified.

28. Let l_0 = total length of a transmission line; l = the distance from the beginning of the line; r = resistance per unit length; x = reactance per unit length = $2 \pi f L$, where $L \doteq$ inductance per unit length; g = conductance from line to return (leakage and discharge into the air) per unit length; b = capacity susceptance per unit length = $2 \pi f C$, where C = capacity per unit length.

Neglecting the line resistance and line conductance,

$$r = 0 \text{ and } g = 0,$$

the line constants α and β, by equations (14), Chapter II, then assume the form

$$\alpha = 0 \text{ and } \beta = \sqrt{xb}, \tag{1}$$

and the line equations (17) of Chapter II become

$$I = (A_1 - A_2) \cos \beta l + j (A_1 + A_2) \sin \beta l$$

and

$$E = \sqrt{\frac{Z}{Y}} \left\{ (A_1 + A_2) \cos \beta l + j (A_1 - A_2) \sin \beta l \right\},$$

or writing

$$A_1 - A_2 = C_1 \text{ and } A_1 + A_2 = C_2,$$

and substituting

$$\sqrt{\frac{Z}{Y}} = \sqrt{\frac{x}{b}} = \sqrt{\frac{L}{C}} \tag{2}$$

we have

and

$$\left. \begin{array}{l} I = C_1 \cos \beta l + j C_2 \sin \beta l \\[2mm] E = \sqrt{\frac{L}{C}} \left\{ C_2 \cos \beta l + j C_1 \sin \beta l \right\} . \end{array} \right\} \tag{3}$$

A free oscillation of a circuit implies that energy is neither supplied to the circuit nor abstracted from it. This means that at both ends of the circuit, $l = 0$ and $l = l_0$, the power equals zero.

If this is the case, the following conditions may exist:

(1) The current is zero at one end, the voltage zero at the other end.

(2) Either the current is zero at both ends or the voltage is zero at both ends.

(3) The circuit has no end but is closed upon itself.

(4) The current is in quadrature with the voltage. This case does not represent a free oscillation, since the frequency depends also on the connected circuit, but rather represents a line supplying a wattless or reactive load.

In free oscillation the circuit thus must be either open or grounded at its ends or closed upon itself.

(1) *Circuit open at one end, grounded at other end.*

29. Assuming the circuit grounded at $l = 0$, open at $l = l_0$, we have for $l = 0$,

$$E = E_0 = 0,$$

and for $l = l_0$,

$$I = I_1 = 0;$$

hence, substituting in equations (3), at $l = 0$,

$$C_2 = 0;$$

hence,

and

$$\left. \begin{array}{l} I = C_1 \cos \beta l \\[2mm] E = j\sqrt{\dfrac{L}{C}} C_1 \sin \beta l, \end{array} \right\} \tag{4}$$

and at $l = l_0$,

$$C_1 \cos \beta l_0 = 0,$$

and since C_1 cannot be zero without the oscillation disappearing altogether,

$$\cos \beta l_0 = 0; \tag{5}$$

hence,

$$\beta l_0 = (2n - 1)\frac{\pi}{2}, \tag{6}$$

where $n = 1, 2, 3 \ldots$ or any integer and

$$\beta l = (2\,n - 1)\,\frac{\pi}{2\,l_0}\,l. \tag{7}$$

Substituting (1) in (6) gives

$$\beta = \sqrt{xb} = \frac{(2\,n - 1)\,\pi}{2\,l_0}, \tag{8}$$

or substituting for x and b, $x = 2\,\pi f L$ and $b = 2\,\pi f C$, gives

$$2\,\pi f \sqrt{LC} = \frac{(2\,n - 1)\,\pi}{2\,l_0};$$

or

$$f = \frac{(2\,n - 1)}{4\,l_0\,\sqrt{LC}} \tag{9}$$

is the *frequency of oscillation* of the circuit.

The lowest frequency or *fundamental frequency of oscillation* is, for $n = 1$,

$$f_1 = \frac{1}{4\,l_0\,\sqrt{LC}}, \tag{10}$$

and besides this fundamental frequency, all its *odd multiples* or higher harmonics may exist in the oscillation

$$f = (2\,n - 1)\,f_1. \tag{11}$$

Writing $L_0 = l_0 L = $ total inductance, and $C_0 = l_0 C = $ total capacity of the circuit, equation (9) assumes the form

$$f_1 = \frac{1}{4\,\sqrt{L_0 C_0}}. \tag{12}$$

The fundamental frequency of oscillation of a transmission line open at one end and grounded at the other, and having a total inductance L_0 and a total capacity C_0, is, neglecting energy losses,

$$f_1 = \frac{1}{4\,\sqrt{L_0 C_0}},$$

while the frequency of oscillation of a localized inductance L_0 and localized capacity C_0, that is, the frequency of discharge of a condenser C_0 through an inductance L_0, is

$$f = \frac{1}{2\pi\sqrt{L_0 C_0}} \,. \tag{13}$$

The difference is due to the distributed character of L_0 and C_0 in the transmission line and the resultant phase displacement between the elements of the line, which causes the inductance and capacity of the line elements, in their effect on the frequency, not to add but to combine to a resultant, which is the projection of the elements of a quadrant, on the diameter, or $\dfrac{2}{\pi}$ times the sum, just as, for instance, the resultant m.m.f. of a distributed armature winding of n turns of i amperes is not ni but $\dfrac{2}{\pi}\,ni$.

Hence, the effective inductance of a transmission line in free oscillation is

$$L_0' = \frac{2}{\pi} l_0 L$$

and the effective capacity is

$$C_0' = \frac{2}{\pi} l_0 C, \tag{14}$$

and using the effective values L_0' and C_0', the fundamental frequency, equation (11), then appears in the form

$$f_1 = \frac{1}{2\pi\sqrt{L_0' C_0'}} \,; \tag{15}$$

that is, the same value as found for the condenser discharge.

In comparing with localized inductances and capacities, the distributed capacity and inductance, in free oscillation, thus are represented by their effective values (13) and (14).

30. Substituting in equations (4),

$$C_1 = c_1 + jc_2, \tag{16}$$

gives

$$I = (c_1 + jc_2) \cos \beta l$$

and

$$E = \sqrt{\frac{L}{C}} (c_2 - jc_1) \sin \beta l. \tag{17}$$

By the definition of the complex quantity as vector representation of an alternating wave the cosine component of the wave is represented by the real, the sine component by the imaginary term; that is, a wave of the form $c_1 \cos 2 \pi f t + c_2 \sin 2 \pi f t$ is represented by $c_1 + j c_2$, and inversely, the equations (17), in their analytic expression, are

$$i = (c_1 \cos 2 \pi f t + c_2 \sin 2 \pi f t) \cos \beta l$$

and

$$e = \sqrt{\frac{L}{C}} (c_2 \cos 2 \pi f t - c_1 \sin 2 \pi f t) \sin \beta l. \quad (18)$$

Substituting (7) and (11) in (18), and writing

$$\theta = 2 \pi f_1 t \text{ and } \tau = \frac{\pi l}{2 l_0} \quad (19)$$

gives

$$\begin{aligned}
i &= \{c_1 \cos (2n- 1)\theta + c_2 \sin (2n- 1)\theta\} \cos (2n -1) \tau \\
&= c \cos (2n - 1) (\theta - \gamma) \cos (2n - 1)\tau
\end{aligned}$$

and

$$\begin{aligned}
e &= \sqrt{\frac{L}{C}} \left\{ c_2 \cos (2n-1)\theta - c_1 \sin (2n-1)\theta \right\} \sin (2n-1)\tau \quad (20) \\
&= - \sqrt{\frac{L}{C}} c \sin (2n - 1) (\theta - \gamma) \sin (2n - 1)\tau,
\end{aligned}$$

where

$$\tan (2n - 1) \gamma = \frac{c_2}{c_1} \text{ and } c = \sqrt{c_1^2 + c_2^2}. \quad (21)$$

In the denotation (19), θ represents the *time angle*, with the complete cycle of the fundamental frequency of oscillation as one revolution or 360 degrees, and τ represents the *distance angle*, with the length of the line as a quadrant or 90 degrees. That is, distances are represented by angles, and the whole line is a quarter wave of the fundamental frequency of oscillation. This form of free oscillation may be called *quarter-wave oscillation*.

The fundamental or lowest discharge wave or oscillation of the circuit then is

$$i_1 = c \cos (\theta - \gamma_1) \cos \tau$$

and

$$e_1 = - \sqrt{\frac{L}{C}} c \sin (\theta - \gamma_1) \sin \tau. \quad (22)$$

With this wave the voltage is a maximum at the open end of the line, $l = l_0$, and gradually decreases to zero at the other end or beginning of the line, $l = 0$.

The current is zero at the open end of the line, and gradually increases to a maximum at $l = 0$, or the grounded end of the line.

Thus the relative intensities of current and potential along the line are as represented by Fig. 85, where the current is shown as I, the voltage as E.

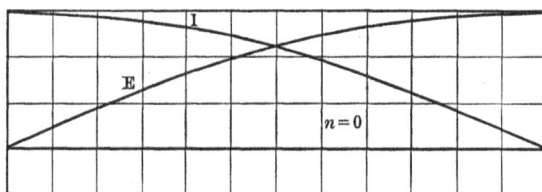

Fig. 85.   Discharge of current and e.m.f. along a transmission line open at one end.   Fundamental discharge frequency.

The next higher discharge frequency, for $n = 2$, gives

$$i_3 = c_3 \cos 3\ (\theta - \gamma_3) \cos 3\ \tau$$

and
$$e_3 = -\ c_3\ \sqrt{\frac{L}{C}}\ \sin 3\ (\theta - \gamma_3) \sin 3\ \tau. \tag{23}$$

Here the voltage is again a maximum at the open end of the line, $l = l_0$, or $\tau = \dfrac{\pi}{2} = 90°$, and gradually decreases, but reaches zero at two-thirds of the line, $l = \dfrac{2\,l_0}{3}$, or $\tau = \dfrac{\pi}{3} = 60°$, then increases again in the opposite direction, reaches a second but opposite maximum at one-third of the line, $l = \dfrac{l_0}{3}$, or $\tau = \dfrac{\pi}{6} = 30°$, and decreases to zero at the beginning of the line.   There is thus a node of voltage at a point situated at a distance of two-thirds of the length of the line.

The current is zero at the end of the line, $l = l_0$, rises to a maximum at a distance of two-thirds of the length of the line, decreases to zero at a distance of one-third of the length of the line, and rises again to a second but opposite maximum at the beginning of the line, $l = 0$. The current thus has a node at a point situated at a distance of one-third of the length of the line.

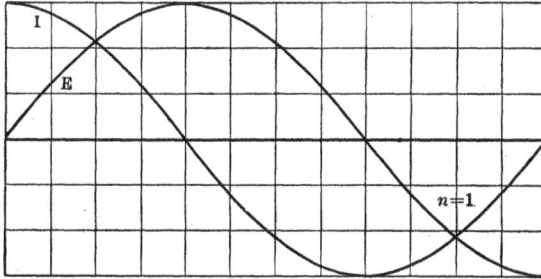

Fig. 86. Discharge of current and e.m.f. along a transmission
line open at one end.

The discharge waves, $n = 2$, are shown in Fig. 86, those with $n = 3$, with two nodal points, in Fig. 87.

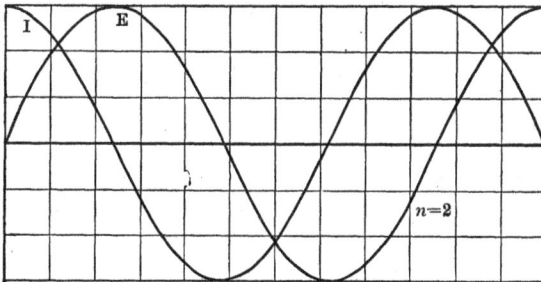

Fig. 87. Discharge of current and e.m.f. along a transmission
line open at one end.

31. In case of a lightning discharge the capacity C_0 is the capacity of the line against ground, and thus has no direct relation to the capacity of the line conductor against its return. The same applies to the inductance L_0.

If d = diameter of line conductor, l_h = height of conductor above ground, and l_0 = length of conductor, the capacity is

$$C_0 = \frac{1.11 \times 10^{-6}\, l_0}{2\, log\, \dfrac{4\, l_h}{d}}, \text{in mf.}$$

the self-inductance is

$$L_0 = 2 \times 10^{-6} l_0\, log\, \frac{4\, l_h}{d}, \text{ in mh.}$$

(24)

The fundamental frequency of oscillation, by substituting (24) in (10), is

$$f_1 = \frac{1}{4 \sqrt{C_0 L_0}} = \frac{7.5 \times 10^9}{l_0}; \tag{25}$$

that is, the frequency of oscillation of a line discharging to ground is independent of the size of line wire and its distance from the ground, and merely depends upon the length, l_0, of the line, being inversely proportional thereto.

We thus get the numerical values,
Length of line

$$= \begin{cases} 10 \quad 20 \quad 30 \quad 40 \quad 50 \quad 60 \quad\quad 80 \quad\quad 100 \text{ miles} \\ 1.6 \quad 3.2 \quad 4.8 \quad 6.4 \quad 8 \quad 9.6 \quad 12.8 \quad 16 \times 10^6 \text{ cm.} \end{cases}$$

hence frequency,

$$f_1 = \quad 4700 \; 2350 \; 1570 \; 1175 \; 940 \; 783 \quad 587 \quad 470 \text{ cycles per sec.}$$

As seen, these frequencies are comparatively low, and especially with very long lines almost approach alternator frequencies.

The higher harmonics of the oscillation are the odd multiples of these frequencies.

Obviously all these waves of different frequencies represented in equation (20) can occur simultaneously in the oscillating discharge of a transmission line, and, in general, the oscillating discharge of a transmission line is thus of the form

$$\left. \begin{aligned} i &= \sum_1^\infty n\, c_n \cos (2n-1)(\theta - \gamma_n) \cos (2n-1)\, \tau, \\ e &= -\sqrt{\frac{L}{C}} \sum_1^\infty n\, c_n \sin (2n-1)(\theta - \gamma_n) \sin (2n-1)\, \tau. \end{aligned} \right\} \tag{26}$$

A simple harmonic oscillation as a line discharge would require a sinoidal distribution of potential on the transmission line at the instant of discharge, which is not probable, so that probably all lightning discharges of transmission lines or oscillations produced by sudden changes of circuit conditions are complex waves of many harmonics, which in their relative magnitude depend upon the initial charge and its distribution — that is, in the case of the lightning discharge, upon the atmospheric electrostatic field of force.

The fundamental frequency of the oscillating discharge of a transmission line is relatively low, and of not much higher magnitude than frequencies in commercial use in alternating-current circuits. Obviously, the more nearly sinoidal the distribution of potential before the discharge, the more the low harmonics predominate, while a very unequal distribution of potential, that is a very rapid change along the line, causes the higher harmonics to predominate.

32. As an example the discharge of a transmission line may be investigated, the line having the following constants per mile: $r = 0.21$ ohm; $L = 1.2 \times 10^{-3}$ henry; $C = 0.03 \times 10^{-6}$ farad, and of the length $l_0 = 200$; hence, by equations (10), (19), $f_1 = 208$ cycles per sec.; $\theta = 1315\,t$, and $\tau = 0.00785\,l$, when charged to a uniform voltage of $e_0 = 60{,}000$ volts but with no current in the line before the discharge, and the line then grounded at one end, $l = 0$, while open at the other end, $l = l_0$.

Then, for $t = 0$ or $\theta = 0$, $i = 0$ for all values of τ except $\tau = 0$; hence, by (26),

$$\cos (2\,n - 1)\,\gamma_n = 0,$$

and thus

$$(2\,n - 1)\,\gamma_n = \frac{\pi}{2} \tag{27}$$

and

$$\cos (2\,n - 1)\,(\theta - \gamma_n) = \sin (2\,n - 1)\,\theta,$$
$$\sin (2\,n - 1)\,(\theta - \gamma_n) = -\cos (2\,n - 1)\,\theta;$$

hence,

$$\left.\begin{aligned}
i &= \sum_1^\infty n\, c_n \sin (2\,n - 1)\,\theta \cos (2\,n - 1)\,\tau \\[2mm]
e &= \sqrt{\frac{L}{C}} \sum_1^\infty n\, c_n \cos (2\,n - 1)\,\theta \sin (2\,n - 1)\,\tau.
\end{aligned}\right\} \tag{28}$$

Also for $t = 0$, or $\theta = 0$, $e = e_0$ for all values of τ except $\tau = 0$; hence, by (28),

$$e_0 = \sqrt{\frac{L}{C}} \sum_1^\infty n\, c_n \sin (2\,n - 1)\,\tau. \tag{29}$$

From equation (29), the coefficients c_n are determined in the usual manner of evaluating a Fourier series, that is, by multiplying with $\sin (2 m - 1) \tau$ (or $\cos (2 m - 1) \tau$) and integrating:

$$\int_0^\pi e_0 \sin (2 m - 1) \tau \, d\tau =$$

$$\sqrt{\frac{L}{C}} \sum_1^\infty {}^n c_n \int_0^\pi \sin (2 n - 1) \tau \sin (2 m - 1) \tau \, d\tau.$$

Since

$$\int_0^\pi \sin (2 n - 1) \tau \sin (2 m - 1) \tau \, d\tau$$

$$= \int_0^\pi \frac{\cos 2 (n - m) \tau - \cos 2 (n + m - 1) \tau}{2} \, d\tau,$$

which is zero for $n \neq m$, while for $m = n$ the term

$$\int_0^\pi \frac{\cos 2 (n - m) \tau}{2} \, d\tau = \int_0^\pi \frac{d\tau}{2} = \frac{\pi}{2}$$

and

$$\int_0^\pi e_0 \sin (2 n - 1) \tau \, d\tau = - e_0 \left[\frac{\cos (2 n - 1) \tau}{2 n - 1} \right]_0^\pi = + \frac{2 e_0}{2 n - 1},$$

we have

$$\frac{2 e_0}{2 n - 1} = c_n \frac{\pi}{2} \sqrt{\frac{L}{C}}$$

and

$$c_n = \frac{4 e_0}{(2 n - 1) \pi} \sqrt{\frac{C}{L}}; \qquad (30)$$

hence,

$$i = \frac{4}{\pi} e_0 \sqrt{\frac{C}{L}} \sum_1^\infty {}^n \frac{\sin (2 n - 1) \theta \cos (2 n - 1) \tau}{2 n - 1}$$

$$= \frac{4}{\pi} e_0 \sqrt{\frac{C}{L}} \left\{ \sin \theta \cos \tau + \frac{\sin 3 \theta \cos 3 \tau}{3} + \frac{\sin 5 \theta \cos 5 \tau}{5} + \cdots \right\}$$

$$\qquad (31)$$

$$= 382 \left\{ \sin \theta \cos \tau + \frac{\sin 3 \theta \cos 3 \tau}{3} + \frac{\sin 5 \theta \cos 5 \tau}{5} + \cdots \right\},$$

in amperes,

and

$$e = \frac{4}{\pi} e_0 \sum_1^\infty {}_n \frac{\cos (2n-1) \theta \sin (2n-1) \tau}{2n-1}$$

$$= \frac{4}{\pi} e_0 \left\{ \cos \theta \sin \tau + \frac{\cos 3\theta \sin 3\tau}{3} + \frac{\cos 5\theta \sin 5\tau}{5} + \cdots \right\}$$

$$\tag{32}$$

$$= 76{,}400 \left\{ \cos \theta \sin \tau + \frac{\cos 3\theta \sin 3\tau}{3} + \frac{\cos 5\theta \sin 5\tau}{5} + \cdots \right\},$$

in volts.

33. As further example, assume now that this line is short-circuited at one end, $l = 0$, while supplied with 25-cycle alternating power at the other end, $l = l_0$, and that the generator voltage drops, by the short circuit, to 30,000, and then the line cuts off from the generating system at about the maximum value of the short-circuit current, that is, at the moment of zero value of the impressed e.m.f.

At a frequency of $f_0 = 25$ cycles, the reactance per unit length of line or per mile is

$$x = 2\pi f_0 L = 0.188 \text{ ohm}$$

and the impedance is

$$z = \sqrt{r^2 + x^2} = 0.283 \text{ ohm},$$

or, for the total line,

$$z_0 = l_0 z = 56.6 \text{ ohms};$$

hence, the approximate short-circuit current

$$i = \frac{e}{z_0} = \frac{30{,}000}{56.6} = 530 \text{ amp.,}$$

and its maximum value is

$$i_0 = 530 \times \sqrt{2} = 750 \text{ amp.}$$

Therefore, in equations (26), at time $t = 0$, or $\theta = 0$, $e = 0$ for all values of τ except $\tau = \dfrac{\pi}{2}$; hence,

$$\sin (2n-1) \gamma_n = 0,$$

or, $$\gamma_n = 0,$$

and thus

$$i = \sum_{1}^{\infty} n\, c_n \cos (2\,n - 1)\, \theta \cos (2\,n - 1)\, \tau$$

and

$$e = -\sqrt{\frac{L}{C}} \sum_{1}^{\infty} n\, c_n \sin (2\,n - 1)\, \theta \sin (2\,n - 1)\, \tau.$$

$$(33)$$

However, at $t = 0$, or $\theta = 0$, for all values of τ except $\tau = \dfrac{\pi}{2}$,

$$i = i_0;$$

hence, substituting in (33),

$$i_0 = \sum_{1}^{\infty} n\, c_n \cos (2\,n - 1)\, \tau. \tag{34}$$

From equation (34), the coefficients c_n are determined in the same manner as in the preceding example, by multiplying with $\cos (2\,n - 1)\,\tau$ and integrating, as

$$c_n = - (-1)^n \frac{4\, i_0}{(2\,n - 1)\, \pi}; \tag{35}$$

hence,

$$\begin{aligned}
i &= -\frac{4\, i_0}{\pi} \sum_{1}^{\infty} n\, (-1)^n \frac{\cos (2\,n - 1)\, \theta \cos (2\,n - 1)\, \tau}{2\,n - 1} \\
&= \frac{4\, i_0}{\pi} \left\{ \cos \theta \cos \tau - \frac{\cos 3\, \theta \cos 3\, \tau}{3} + \frac{\cos 5\, \theta \cos 5\, \tau}{5} - + \cdots \right\}
\end{aligned}$$

$$(36)$$

$$= 956 \left\{ \cos \theta \cos \tau - \frac{\cos 3\, \theta \cos 3\, \tau}{3} + \frac{\cos 5\, \theta \cos 5\, \tau}{5} - + \cdots \right\},$$

in amperes,

and

$$\begin{aligned}
e &= -\frac{4\, i_0}{\pi} \sqrt{\frac{L}{C}} \sum_{1}^{\infty} n\, (-1)^n \frac{\sin (2\,n - 1)\, \theta \sin (2\,n - 1)\, \tau}{2\,n - 1} \\
&= \frac{4\, i_0}{\pi} \sqrt{\frac{L}{C}} \left\{ \sin \theta \sin \tau - \frac{\sin 3\, \theta \sin 3\, \tau}{3} + \frac{\sin 5\, \theta \sin 5\, \tau}{5} - + \cdots \right\}
\end{aligned}$$

$$(37)$$

$$= 191{,}200 \left\{ \sin \theta \sin \tau - \frac{\sin 3\, \theta \sin 3\, \tau}{3} + \frac{\sin 5\, \theta \sin 5\, \tau}{5} - + \cdots \right\},$$

in volts.

The maximum voltage is reached at time $\theta = \dfrac{\pi}{2}$, and is

$$e = \frac{4\,i_0}{\pi}\sqrt{\frac{L}{C}}\left\{\sin \tau + \frac{\sin 3\,\tau}{3} + \frac{\sin 5\,\tau}{5} + \cdots\right\},$$

and since the series

$$\sin \tau + \frac{\sin 3\,\tau}{3} + \frac{\sin 5\,\tau}{5} + \cdots = \frac{\pi}{2},$$

the maximum voltage is

$$e = i_0\sqrt{\frac{L}{C}} = 300{,}000 \text{ volts.}$$

As seen, very high voltages may be produced by the interruption of the short-circuit current.

(2a) *Circuit grounded at both ends.*

34. The method of investigation is the same as in paragraph 29; the terminal conditions are, for $l = 0$,

$$E = 0,$$

and for $l = l_0$,

$$E = 0.$$

Substituting $l = 0$ into equations (3) gives

$$C_2 = 0;$$

hence,

$$\left.\begin{aligned}
I &= C_1 \cos \beta l, \\
E &= jC_1\sqrt{\frac{L}{C}}\sin \beta l.
\end{aligned}\right\} \tag{38}$$

Substituting $l = l_0$ in (38) gives

$$E_1 = 0 = jC_1\sqrt{\frac{L}{C}}\sin \beta l_0;$$

hence,

$$\sin \beta l_0 = 0, \text{ or } \beta l = n\pi, \tag{39}$$

and, in the same manner as in (1),

$$\beta l = n \frac{\pi}{l_0} l = n\tau; \qquad (40)$$

that is, the length of the line, l_0, represents one half wave, or $\tau = \pi$, or a multiple thereof.

$$f = \frac{n}{2\,l_0\sqrt{LC}} = \frac{n}{2\,\sqrt{L_0C_0}}, \qquad (41)$$

and the *fundamental frequency of oscillation* is

$$f_1 = \frac{1}{2\,l_0\sqrt{LC}}, \qquad (42)$$

and

$$f = nf_1; \qquad (43)$$

that is, the line can oscillate at a fundamental frequency f_1, for which the length, l_0, of the line is a half wave, and at all multiples or higher harmonics thereof, the *even* ones as well as the *odd* ones.

This kind of oscillation may be called a *half-wave oscillation*.

35. Unlike the quarter-wave oscillation, which contains only the odd higher harmonics of the fundamental wave, the half-wave oscillation also contains the even harmonics of the fundamental frequency of oscillation.

Substituting $C_1 = c_1 + jc_2$ into (38) gives

and

$$\begin{aligned} I &= (c_1 + jc_2)\cos\beta l \\ E &= (c_2 - jc_1)\sqrt{\frac{L}{C}}\sin\beta l, \end{aligned} \right\} \qquad (44)$$

and replacing the complex imaginary by the analytic expression, that is, the real term by $\cos 2\,\pi ft$, the imaginary term by $\sin 2\,\pi ft$, gives

and

$$\begin{aligned} i &= \{c_1\cos 2\,\pi ft + c_2\sin 2\,\pi ft\}\cos\beta l \\ e &= \sqrt{\frac{L}{C}}\{c_2\cos 2\,\pi ft - c_1\sin 2\,\pi ft\}\sin\beta l, \end{aligned} \right\}$$

and substituting

$$2 \pi f_1 t = \theta, \\ 2 \pi f t = n\theta; \biggr\}$$

we have (45)

then (44) gives, by (40):

$$i = (c_1 \cos n\theta + c_2 \sin n\theta) \cos n\tau \\ e = \sqrt{\frac{L}{C}} (c_2 \cos n\theta - c_1 \sin n\theta) \sin n\tau; \biggr\}$$

and (46)

or writing

$$c_1 = c \cos n\gamma \\ c_2 = c \sin n\gamma \biggr\}$$

and (47)

gives

$$i = c \cos n (\theta - \gamma) \cos n\tau \\ e = - c \sqrt{\frac{L}{C}} \sin n (\theta - \gamma) \sin n\tau, \biggr\}$$

and (48)

and herefrom the *general equations of this half-wave oscillation* are

$$i = \sum_{1}^{\infty} {}^n c_n \cos n (\theta - \gamma_n) \cos n\tau \\ e = - \sqrt{\frac{L}{C}} \sum_{1}^{\infty} {}^n c_n \sin n (\theta - \gamma_n) \sin n\tau \biggr\}$$

and (49)

(2b) *Circuit open at both ends.*

36. For $l = 0$ we have

$$I = 0;$$

hence,

$$C_1 = 0$$

and

$$I = j C_2 \sin \beta l \\ E = \sqrt{\frac{L}{C}} C_2 \cos \beta l, \biggr\}$$

and (50)

while for $l = l_0$, $I = 0$;

hence,

$$\sin \beta l_0 = 0, \text{ or } \beta l_0 = n\pi; \tag{51}$$

that is, the circuit performs a *half-wave oscillation* of *fundamental frequency*,

$$f_1 = \frac{1}{2 \, l_0 \sqrt{LC}} \tag{52}$$

and all its higher harmonics, the even ones as well as the odd ones have a frequency

$$f = nf_1, \tag{53}$$

and the final equations are

$$i = -\sum_1^\infty n \, c_n \sin n \, (\theta - \gamma) \sin n\tau$$

and

$$\left.\begin{array}{l} \\ e = \sqrt{\dfrac{L}{C}} \sum_1^\infty n \, c_n \cos n \, (\theta - \gamma) \cos n\tau, \end{array}\right\} \tag{54}$$

where $\quad \theta = 2 \, \pi f_1 t \quad$ and $\quad \tau = \dfrac{\pi}{l_0} l.$ \hfill (55)

(3) *Circuit closed upon itself.*

37. If a circuit of length l_0 is closed upon itself, then the free oscillation of such a circuit is characterized by the condition that current and voltage at $l = l_0$ are the same as at $l = 0$, since $l = l_0$ and $l = 0$ are the same point of the circuit.

Substituting this condition in equations (3) gives

$$\left.\begin{array}{l} I = C_1 = C_1 \cos \beta l_0 + jC_2 \sin \beta l_0 \\[1mm] \sqrt{\dfrac{C}{L}} \, E = C_2 = C_2 \cos \beta l_0 + jC_1 \sin \beta l_0; \end{array}\right\} \tag{56}$$

and

herefrom follows

$$\left.\begin{array}{l} C_1 (1 - \cos \beta l_0) = + jC_2 \sin \beta l_0, \\[1mm] C_2 (1 - \cos \beta l_0) = + jC_1 \sin \beta l_0, \end{array}\right\} \tag{57}$$

hence, $\qquad (1 - \cos \beta l_0)^2 = - \quad \sin^2 \beta l_0$

or $\qquad\qquad \cos \beta l_0 = 1; \tag{58}$

hence,

$$\beta l_0 = 2\,n\pi; \tag{59}$$

that is, the circuit must be a complete wave or a multiple thereof.

The free oscillation of a circuit which is closed upon itself is a *full-wave oscillation*, containing a *fundamental wave* of frequency

$$f_1 = \frac{1}{l\sqrt{LC}}, \tag{60}$$

and all the higher harmonics thereof, the *even* ones as well as the *odd* ones,

$$f = nf_1. \tag{61}$$

Substituting in (3),

and

$$\left.\begin{array}{l} C_1 = c_1' + jc_1'' \\[2mm] C_2 = c_2' + jc_2'' \end{array}\right\}$$

gives

and

$$\left.\begin{array}{l} I = (c_1' + jc_1'')\cos\beta l + (c_2'' - jc_2')\sin\beta l \\[3mm] E = \sqrt{\dfrac{L}{C}}\{(c_2' + jc_2'')\cos\beta l + (c_1'' - jc_1')\sin\beta l\}. \end{array}\right\} \tag{62}$$

Substituting the analytic expression,

$$c_1' + jc_1'' = c_1'\cos 2\pi ft + c_1''\sin 2\pi ft, \text{ etc.,}$$

also

$$\left.\begin{array}{l} 2\pi f_1 t = \theta, \\[2mm] 2\pi ft = n\theta, \end{array}\right\} \tag{63}$$

and

$$\beta l = \frac{2\pi n}{l_0}l = n\tau,$$

where

$$\tau = \frac{2\pi}{l_0}l, \tag{64}$$

that is, the length of the circuit, $l = l_0$, is represented by the angle $\tau = 2\pi$, or a complete cycle, this gives

$$I = (c_1{}' \cos n\theta + c_1{}'' \sin n\theta) \cos n\tau$$
$$+ (c_2{}'' \cos n\theta - c_2{}' \sin n\theta) \sin n\tau$$

and

$$E = \sqrt{\frac{L}{C}} \{ (c_2{}' \cos n\theta + c_2{}'' \sin n\theta) \cos n\theta$$
$$+ (c_1{}'' \cos n\theta - c_1{}' \sin n\theta) \sin n\tau \},$$

(65)

or writing

$$c_1{}' = a \cos n\gamma$$
$$c_1{}'' = a \sin n\gamma$$
$$c_2{}' = b \cos n\chi$$
$$c_2{}'' = b \sin n\chi$$

gives

$$i = a \cos n\, (\theta - \gamma) \cos n\tau - b \sin n\, (\theta - \chi) \sin n\tau$$

and

$$e = \sqrt{\frac{L}{C}} \{ b \cos n\, (\theta - \chi) \cos n\tau - a \sin n\, (\theta - \gamma) \sin n\tau \}.$$

(66)

Thus in its most general form the full-wave oscillation gives the equations

$$i = \sum_1^\infty {}_n \left\{ a_n \cos n\, (\theta - \gamma_n) \cos n\tau - b_n \sin n\, (\theta - \chi_n) \sin n\tau \right\}$$

$$e = \sqrt{\frac{L}{C}} \sum_1^\infty {}_n \left\{ b_n \cos n\, (\theta - \chi_n) \cos n\tau - a_n \sin n\, (\theta - \gamma_n) \sin n\tau \right\},$$

(67)

where

$$\theta = 2\pi f_1 t,$$
$$f_1 = \frac{1}{l_0 \sqrt{LC}},$$
$$\tau = \frac{2\pi}{l_0} l,$$

(68)

and a_n, γ_n and b_n, χ_n are groups of four integration constants.

38. With a short circuit at the end of a transmission line, the drop of potential along the line varies fairly gradually and uniformly, and the instantaneous rupture of a short circuit — as by a short-circuiting arc blowing itself out explosively —

causes an oscillation in which the lower frequencies predominate, that is, a low-frequency high-power surge. A spark discharge from the line, a sudden high voltage charge entering the line locally, as directly by a lightning stroke, or indirectly by induction during a lightning discharge elsewhere, gives a distribution of potential which momentarily is very non-uniform, changes very abruptly along the line, and thus gives rise mainly to very high harmonics, but as a rule does not contain to any appreciable extent the lower frequencies; that is, it causes a high-frequency oscillation, more or less local in extent, and while of high voltage, of rather limited power, and therefore less destructive than a low-frequency surge.

At the frequencies of the high-frequency oscillation neither capacity nor inductance of the transmission line is perfectly constant: the inductance varies with the frequency, by the increasing screening effect or unequal current distribution in the conductor; the capacity increases by brush discharge over the insulator surface, by the increase of the effective conductor diameter due to corona effect, etc. The frequencies of the very high harmonics are therefore not definite but to some extent variable, and since they are close to each other they overlap; that is, at very high frequencies the transmission line has no definite frequency of oscillation, but can oscillate with any frequency.

A long-distance transmission line has a definite natural period of oscillation, of a relatively low fundamental frequency and its overtones, but can also oscillate with any frequency whatever, provided that this frequency is very high.

This is analogous to waves formed in a body of water of regular shape: large standing waves have a definite wave length, depending upon the dimensions of the body of water, but very short waves, ripples in the water, can have any wave length, and do not depend on the size of the body of water.

A further investigation of oscillations in conductors with distributed capacity, inductance, and resistance requires, however, the consideration of the resistance, and so leads to the investigation of phenomena transient in space as well as in time, which are discussed in Section IV.

39. In the equations discussed in the preceding, of the free oscillations of a circuit containing uniformly distributed resistance, inductance, capacity, and conductance, the energy losses in the circuit have been neglected, and voltage and current therefore appear alternating instead of oscillating. That is, these equations represent only the initial or maximum values of the phenomenon, but to represent it completely an exponential function of time enters as factor, which, as will be seen in Section IV, is of the form

$$\varepsilon^{-ut}, \tag{69}$$

where $u = \dfrac{1}{2}\left(\dfrac{r}{L} + \dfrac{g}{C}\right)$ may be called the "time constant" of the circuit.

While quarter-wave oscillations occasionally occur, and are of serious importance, the occurrence of half-wave oscillations and especially of full-wave oscillations of the character discussed before, that is, of a uniform circuit, is less frequent.

When in a circuit, as a transmission line, a disturbance or oscillation occurs while this circuit is connected to other circuits — as the generating system and the receiving apparatus — as is usually the case, the disturbance generally penetrates into the circuits connected to the circuit in which the disturbance originated, that is, the entire system oscillates, and this oscillation usually is a full-wave oscillation; that is, the oscillation of a circuit closed upon itself; occasionally a half-wave oscillation. For instance, if in a transmission system comprising generators, step-up transformers, high-potential lines, step-down transformers, and load, a short circuit occurs in the line, the circuit comprising the load, the step-down transformers, and the lines from the step-down transformers to the short circuit is left closed upon itself without power supply, and its stored energy is, therefore, dissipated as a full-wave oscillation. Or, if in this system an excessive load, as the dropping out of step of a synchronous converter, causes the circuit to open at the generating station, the dissipation of the stored energy — in this case that of the excessive current in the system — occurs as a full-wave oscillation, if the line cuts off from the generating station on the low-tension side of the step-up transformers, and the oscillating circuit comprises the high-tension coils of the step-up transformers, the transmission line, step-down transformers, and load. If the line disconnects from the generating system on the high-potential side of the step-up transformers, the oscillation is a half-wave oscillation, with the two ends of the oscillating circuit open.

Such oscillating circuits, however,— representing the most frequent and most important case of high-potential disturbances in transmission systems,—cannot be represented by the preceding equations since they are not circuits of uniformly distributed constants but compound circuits comprising several sections of different constants, and therefore of different ratios of energy consumption and energy storage, $\dfrac{r}{L}$ and $\dfrac{g}{C}$. During the free oscillation of such circuits an energy transfer takes place between the different sections of the circuit, and energy flows from those sections in which the energy consumption is small compared with the energy storage, as transformer coils and highly inductive loads, to those sections in which the energy consumption is large compared with the energy storage, as the more non-inductive parts of the system. This introduces into the equations exponential functions of the distance as well as the time, and requires a study of the phenomenon as one transient in distance as well as in time. The investigation of the oscillation of a compound circuit, comprising sections of different constants, is treated in Section IV.

CHAPTER IV.

DISTRIBUTED CAPACITY OF HIGH-POTENTIAL TRANSFORMERS.

40. In the high-potential coils of transformers designed for very high voltages phenomena resulting from distributed capacity occur.

In transformers for very high voltages — 100,000 volts and more, or even considerably less in small transformers — the high-potential coil contains a large number of turns, a great length of conductor, and therefore its electrostatic capacity is appreciable, and such a coil thus represents a circuit of distributed resistance, inductance, and capacity somewhat similar to a transmission line.

The same applies to reactive coils, etc., wound for very high voltages, and even in smaller reactive coils at very high frequency.

This capacity effect is more marked in smaller transformers, where the size of the iron core and therewith the voltage per turn is less, and therefore the number of turns greater than in very large transformers, and at the same time the exciting current and the full-load current are less; that is, the charging current of the conductor more comparable with the load current of the transformer or reactive coil.

It is, however, much more serious in large transformers, since in such the resistance is smaller compared to the inductance and capacity, and therefore the damping of any high frequency oscillation less, the possibility of the formation of sustained and cumulative oscillations greater.

However, even in large transformers and at moderately high voltages, capacity effects occur in transformers, if the frequency is sufficiently high, as is the case with the currents produced in overhead lines by lightning discharges, or by arcing grounds resulting from spark discharges between conductor and ground, or in starting or disconnecting the transformer. With such frequencies, of many thousand cycles, the internal capacity of the transformer becomes very marked in its effect on the distribution of voltage and current, and may produce dangerous high-voltage points in the transformer.

The distributed capacity of the transformer, however, is different from that of a transmission line.

348

In a transmission line the distributed capacity is shunted capacity, that is, can be represented diagrammatically by condensers shunted across the circuit from line to line, or, what amounts to the same thing, from line to ground and from ground to return line, as shown diagrammatically in Fig. 88.

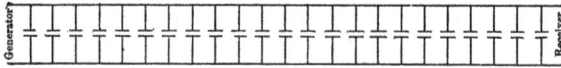

Fig. 88. Distributed capacity of a transmission line.

The high-potential coil of the transformer also contains shunted capacity, or capacity from the conductor to ground, and so each coil element consumes a charging current proportional to its potential difference against ground. Assuming the circuit as insulated, and the middle of the transformer coil at ground potential, the charge consumed by unit length of the coil increases from zero at the center to a maximum at the ends. If one terminal of the circuit is grounded, the charge consumed by the coil increases from zero at the grounded terminal to a maximum at the ungrounded terminal.

In addition thereto, however, the transformer coil also contains a capacity between successive turns and between successive layers. Starting from one point of the conductor, after a certain

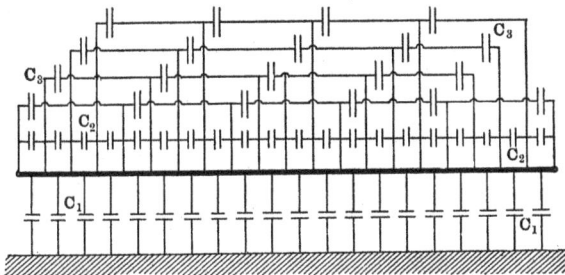

Fig. 89. Distributed capacity of a high-potential transformer coil.

length, the length of one turn, the conductor reapproaches the first point in the next adjacent turn. It again approaches the first point at a different and greater distance in the next adjacent layer.

first point at a different and greater distance in the next adjacent layer.

A transformer high-potential coil can be represented diagrammatically as a conductor, Fig. 89. C_1 represents the capacity against ground, C_2 represents the capacity between adjacent turns, and C_3 the capacity between adjacent layers of the coil.

The capacities C_2 and C_3 are not uniformly distributed but more or less irregularly, depending upon the number and arrangement of the transformer coils and the number and arrangement of turns in the coil. As approximation, however, the capacities C_2 and C_3 can be assumed as uniformly distributed capacity between successive conductor elements. If l = length of conductor, they may be assumed as a capacity between l and $l + dl$, or as a capacity across the conductor element dl.

This approximation is permissible in investigating the general effect of the distributed capacity, but omits the effect of the irregular distribution of C_2 and C_3, which leads to local oscillations of higher frequencies, extending over sections of the circuit, such as individual transformer coils, and may cause destructive voltage across individual transformer coils, without the appearance of excessive voltages across the main terminals of the transformer.

41. Let then, in the high-potential coil of a high-voltage transformer, e = the e.m.f. generated per unit length of conductor, as, for instance, per turn; $Z = r + jx$ = the impedance per unit length; $Y = g + jb$ = the capacity admittance against ground per unit length of conductor, and $Y' = pY$ = the capacity effective admittance representing the capacity between successive turns, successive layers, and successive coils, as represented by the condensers C_2 and C_3 in Fig. 89.

The charging current of a conductor element dl, due to the admittance Y', is made up of the charging currents against the next following and that against the preceding conductor element.

Let l_0 = length of conductor; l = distance along conductor; E = potential at point l, or conductor element dl, and I = current in conductor element dl; then

$$dE = \frac{d\dot{E}}{dl}\, dl = \text{the potential difference between successive}$$

conductor elements or turns.

$Y' \dfrac{d\dot{E}}{dl} dl$ = the charging current between one conductor element and the next conductor element or turn.

$- Y' \dfrac{d (\dot{E} - d\dot{E})}{dl} dl$ = the charging current between one conductor element and the preceding conductor element or turn, hence,

$Y' \dfrac{d^2\dot{E}}{dl^2} dl$ = the charging current of one conductor element due to capacity between adjacent conductors or turns.

If now the distance l is counted from the point of the conductor, which is at ground potential, $YEdl$ = the charging current of one conductor element against ground, and

$$dI = \left\{ YE + Y' \dfrac{d^2\dot{E}}{dl^2} \right\} dl = Y \left\{ E + p \dfrac{d^2\dot{E}}{dl^2} \right\} dl$$

is the total current consumed by a conductor element.

However, the e.m.f. consumed by impedance equals the e.m.f. consumed per conductor element; thus

$$dE_z = ZI\, dl.$$

This gives the two differential equations:

$$\dfrac{d\dot{I}}{dl} = Y \left\{ E + p \dfrac{d^2\dot{E}}{dl^2} \right\} \tag{1}$$

and

$$e - \dfrac{d\dot{E}}{dl} = ZI. \tag{2}$$

Differentiating (2) and substituting in (1) gives

$$\dfrac{d^2\dot{E}}{dl^2} = ZY \left\{ E + p \dfrac{d^2\dot{E}}{dl^2} \right\};$$

transposing,

$$\dfrac{d^2\dot{E}}{dl^2} = \dfrac{- \dot{E}}{p - \dfrac{1}{ZY}}, \tag{3}$$

or

$$\dfrac{d^2\dot{E}}{dl^2} = - a^2 E, \tag{4}$$

where
$$a^2 = \frac{1}{p - \dfrac{1}{ZY}} \cdot \tag{5}$$

If $\dfrac{1}{ZY}$ is small compared with p, we have, approximately,

$$a^2 = \frac{1}{p} \tag{6}$$

and
$$E = A \cos al + B \sin al, \tag{7}$$

and since, for $l = 0$, $E = 0$, if the distance l is counted from the point of zero potential, we have

$$E = B \sin al, \tag{8}$$

and the current is given by equation (2) as

$$I = \frac{1}{Z} \left\{ e - \frac{dE}{dl} \right\}; \tag{9}$$

substituting (8) in (9) gives

$$I = \frac{1}{Z} \left\{ e - aB \cos al \right\}. \tag{10}$$

If now $I_1 = $ the current at the transformer terminals, $l = l_0$, we have, from (10),

$$ZI_1 = e - aB \cos al_0$$

and
$$B = \frac{e - ZI_1}{a \cos al_0}; \tag{11}$$

substituting in (8) and (10),

$$\left. \begin{aligned} E &= (e - ZI_1) \frac{\sin al}{a \cos al_0} \\[2mm] \text{and} \qquad I &= \frac{1}{Z} \left\{ (e - ZI_1) \frac{\cos al}{\cos al_0} - e \right\}; \end{aligned} \right\} \tag{12}$$

for $I_1 = 0$, or open circuit of the transformer, this gives

$$\left. \begin{aligned} E &= e \frac{\sin al}{a \cos al_0} \\[2mm] \text{and} \qquad I &= \frac{e}{Z} \left(\frac{\cos al}{\cos al_0} - 1 \right). \end{aligned} \right\} \tag{13}$$

The e.m.f., E, thus is a maximum at the terminals,

$$E_1 = \frac{e}{a} \tan al_0,$$

the current a maximum at the zero point of potential, $l = 0$, where

$$I_0 = \frac{e}{Z}\left(\frac{1}{\cos al_0} - 1\right). \tag{14}$$

42. Of all industrial circuits containing distributed capacity and inductance, the high potential coils of large high voltage power transformers probably have the lowest attenuation constant of oscillations, that is the lowest ratio of r to $\sqrt{LC}$, and high frequency oscillations occurring in such circuits thus die out at a slower rate, hence are more dangerous than in most other industrial circuits. Nearest to them in this respect are the armature circuits of large high voltage generators, and similar considerations apply to them.

As the result hereof, the possibility of the formation of continual and cumulative oscillations, in case of the presence of a source of high frequency power, as an arc or a spark discharge in the system, is greater in high potential transformer coils than in most other circuits. Regarding such cumulative oscillations and their cause and origin, see the chapters on "Instability of Electric Circuits," in "Theory and Calculation of Electric Circuits."

The frequency of oscillation of the high potential circuit of large high voltage power transformers usually is of the magnitude of 10,000 to 30,000 cycles; the frequency of oscillation of individual transformer coils of this circuit is usually of the magnitude of 30,000 to 100,000 cycles. There then are the danger frequencies of large high voltage transformers.

CHAPTER V.

DISTRIBUTED SERIES CAPACITY.

43. The capacity of a transmission line, cable, or high-potential transformer coil is shunted capacity, that is, capacity from conductor to ground, or from conductor to return conductor, or shunting across a section of the conductor, as from turn to turn or layer to layer of a transformer coil.

In some circuits, in addition to this shunted capacity, distributed series capacity also exists, that is, the circuit is broken at frequent and regular intervals by gaps filled with a dielectric or insulator, as air, and the two faces of the conductor ends thus constitute a condenser in series with the circuit. Where the elements of the circuit are short enough so as to be represented, approximately, as conductor differentials, the circuit constitutes a circuit with distributed series capacity.

An illustration of such a circuit is afforded by the so-called "multi-gap lightning arrester," as shown diagrammatically in Fig. 90, which consists of a large number of metal cylinders p, q ..., with small spark gaps between the cylinders, connected between line L and ground G. This arrangement, Fig. 90, can be represented diagrammatically by Fig. 91. Each cylinder has a capacity C_0 against ground, a capacity C against the adjacent cylinder, a resistance r,— usually very small,— and an inductance L.

The series of insulator discs of a high voltage suspension — or strain — insulator also forms such a circuit.

If such a series of n equal capacities or spark gaps is connected across a constant supply voltage e_0, each gap has a voltage $e = \dfrac{e_0}{n}$.

If, however, the supply voltage is alternating, the voltage does not divide uniformly between the gaps, but the potential difference is the greater, that is, the potential gradient steeper the nearer the gap is to the line L, and this distribution of potential becomes the more non-uniform the higher the frequency; that is, the greater the charging current of the capacity of the cylinder against the ground. The charging currents against ground, of all the cylinders from q to the ground G, Figs. 90 and 91, must pass the gap between the adjacent cylinders p and q; that is, the charging current of the condenser represented by two adjacent

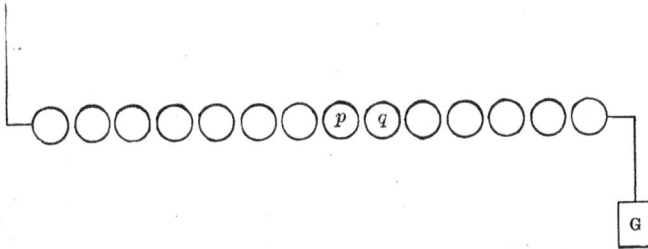

Fig. 90. Multi-gap lightning arrester.

cylinders p and q is the sum of all the charging currents from q to G; and as the potential difference between the two cylinders p and q is proportional to the charging current of the condenser

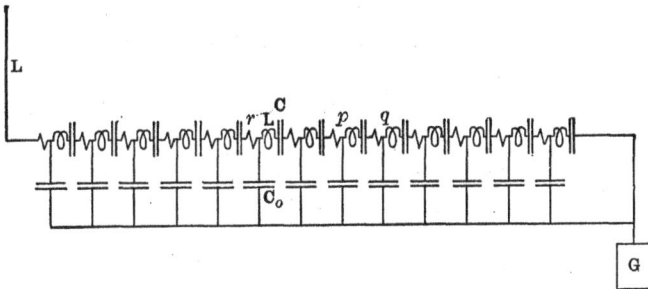

Fig. 91. Equivalent circuit of a multi-gap lightning arrester.

formed by these two cylinders, C, this potential difference increases towards L, being, at each point proportional to the vector sum of all the charging currents, against ground, of all the cylinders between this point and ground.

The higher the frequency, the more non-uniform is the potential gradient along the circuit and the lower is the total supply voltage required to bring the maximum potential gradient, near the line L, above the disruptive voltage, that is, to initiate the discharge. Thus such a multigap structure is discriminating regarding frequency; that is, the discharge voltage with increasing frequency, does not remain constant, but decreases with increase of frequency, when the frequency becomes sufficiently high to give appreciable charging currents. Hence high frequency oscillations discharge over such a structure at lower voltage than machine frequencies.

For a further discussion of the feature which makes such a multigap structure useful for lightning protection, see A. I. E. E. Transactions, 1906, pp. 431, 448, 1907, p. 425, etc.

44. Such circuits with distributed series capacity are of great interest in that it is probable that lightning flashes in the clouds are discharges in such circuits. From the distance traversed by lightning flashes in the clouds, their character, and the disruptive strength of air, it appears certain that no potential difference can exist in the clouds of such magnitude as to cause a disruptive discharge across a mile or more of space. It is probable that as the result of condensation of moisture, and the lack of uniformity of such condensation, due to the gusty nature of air currents, a non-uniform distribution of potential is produced between the rain drops in the cloud; and when the potential gradient somewhere in space exceeds the disruptive value, an oscillatory discharge starts between the rain drops, and gradually, in a number of successive discharges, traverses the cloud and equalizes the potential gradient. A study of circuits containing distributed series capacity thus leads to an understanding of the phenomena occurring in the thunder cloud during the lightning discharge.*

Only a general outline can be given in the following.

45. In a circuit containing distributed resistance, conductance, inductance, shunt, and series capacity, as the multigap lightning arrester, Fig. 90, represented electrically as a circuit in Fig. 91, let r = the effective resistance per unit length of circuit, or per circuit element, that is, per arrester cylinder; g = the shunt conductance per unit length, representing leakage, brush discharge, electrical radiation, etc.; L = the inductance per unit length of circuit; C = the series capacity per unit length of circuit, or circuit element, that is, capacity between adjacent arrester cylinders, and C_0 = the shunt capacity per unit length of circuit, or circuit element, that is, capacity between arrester cylinder and

---

* See paper, "Lightning and Lightning Protection," **N.E.L.A.**, 1907. Reprinted and enlarged in "General Lectures on Electrical Engineering," by Author.

ground. If then $f =$ the frequency of impressed e.m.f., the series impedance per unit length of circuit is

$$Z = r + j\,(x - x_c); \tag{1}$$

the shunt admittance per unit length of circuit is

$$Y = g + jb, \tag{2}$$

where

$$\left. \begin{array}{l} x = 2\,\pi f L, \\[2mm] x_c = \dfrac{1}{2\,\pi f C}, \\[2mm] b = 2\,\pi f C_0; \end{array} \right\} \tag{3}$$

or the absolute values are

$$\left. \begin{array}{l} z = \sqrt{r^2 + (x - x_c)^2} \\[2mm] y = \sqrt{g^2 + b^2}. \end{array} \right\} \tag{4}$$

and

If the distance along the circuit from line L towards ground G is denoted by l, the potential difference between point l and ground by E, and the current at point l by I, the differential equations of the circuit are *

$$\frac{d\dot{E}}{dl} = Z\dot{I} \tag{5}$$

and

$$\frac{d\dot{I}}{dl} = Y\dot{E}. \tag{6}$$

Differentiating (5) and substituting (6) therein gives

$$\frac{d^2\dot{E}}{dl^2} = YZ\dot{E}. \tag{7}$$

Equation (7) is integrated by

$$\dot{E} = A_1 \varepsilon^{-al} + A_2 \varepsilon^{+al}, \tag{8}$$

where

$$a = \sqrt{YZ} = \alpha + j\beta, \tag{9}$$

and

$$\left. \begin{array}{l} \alpha = \sqrt{\tfrac{1}{2}\{yz + gr - b\,(x - x_c)\}} \\[2mm] \beta = \sqrt{\tfrac{1}{2}\{yz - gr + b\,(x - x_c)\}}. \end{array} \right\} \tag{10}$$

* Section III, Chapter II, paragraph 7.

Substituting (10) in (8) and eliminating the imaginary exponents by the substitution of trigonometric functions,

$$E = A_1 \varepsilon^{-al} (\cos \beta l - j \sin \beta l) + A_2 \varepsilon^{+al} (\cos \beta l + j \sin \beta l). \quad (11)$$

46. However, if n = the total length of circuit from line L to ground G, or total number of arrester cylinders between line and ground, for $l = n$,

$$E = 0, \quad (12)$$

and for $l = 0$,

$$E = e_0 = \text{the impressed e.m.f.} \quad (13)$$

Substituting (12) and (13) into (11) gives

$$0 = A_1 \varepsilon^{-an} (\cos \beta n - j \sin \beta n) + A_2 \varepsilon^{+an} (\cos \beta n + j \sin \beta n)$$

and

$$e_0 = A_1 + A_2;$$

hence,

$$\left. \begin{array}{l} A_1 = \dfrac{e_0}{1 - \varepsilon^{-2an} (\cos 2 \beta n - j \sin 2 \beta n)}, \\[2mm] A_2 = - A_1 \varepsilon^{-2an} (\cos 2 \beta n - j \sin 2 \beta n), \end{array} \right\} \quad (14)$$

and the potential difference against ground is

$$E = e_0$$
$$\frac{\varepsilon^{-al}(\cos \beta l + j \sin \beta l) - \varepsilon^{-a(2n-l)} [\cos \beta (2n-l) - j \sin \beta(2n-l)]}{1 - \varepsilon^{-2an} (\cos 2 \beta n - j \sin 2 \beta n)}. \quad (15)$$

From equation (5), substituting (15) and (9), we have

$$I = - \sqrt{\frac{Y}{Z}} e_0$$
$$\frac{\varepsilon^{-al}(\cos \beta l - j \sin \beta l) + \varepsilon^{-a(2n-l)} [\cos \beta (2n-l) - j \sin \beta (2n-l)]}{1 - \varepsilon^{-2an}(\cos 2 \beta n - j \sin 2 \beta n)}. \quad (16)$$

Reduced to absolute terms this gives the potential difference against ground as

$$e = e_0 \sqrt{\frac{\varepsilon^{-2al} + \varepsilon^{-2a(2n-l)} - 2 \varepsilon^{-2an} \cos 2 \beta (n-l)}{1 + \varepsilon^{-4an} - 2 \varepsilon^{-2an} \cos 2 \beta n}}, \quad (17)$$

the current as

$$i = e_0 \sqrt{\frac{y}{z}} \sqrt{\frac{\varepsilon^{-2al} + \varepsilon^{-2a(2n-l)} + 2\varepsilon^{-2an} \cos 2\beta(n-l)}{1 + \varepsilon^{-4an} - 2\varepsilon^{-2an} \cos 2\beta n}}, \quad (18)$$

and the potential gradient, or potential difference between adjacent cylinders, is

$$e' = x_c i = e_0 x_c \sqrt{\frac{y}{z}} \sqrt{\frac{\varepsilon^{-2al} + \varepsilon^{-2a(2n-l)} + 2\varepsilon^{-2an} \cos 2\beta(n-l)}{1 + \varepsilon^{-4an} - 2\varepsilon^{-2an} \cos 2\beta n}}. \quad (19)$$

For an infinite length of line, $n = \infty$, that is, for a very large number of lightning arrester cylinders, where ε^{-2an} is negligible, as in the case where the discharge passes from the line into the arrester without reaching the ground, equations (17), (18), (19) simplify to

$$e = e_0 \varepsilon^{-al}, \quad (20)$$

$$i = e_0 \sqrt{\frac{y}{z}} \varepsilon^{-al}, \quad (21)$$

and

$$e' = e_0 x_c \sqrt{\frac{y}{z}} \varepsilon^{-al}; \quad (22)$$

that is, are simple exponential curves.

Substituting (4) and (3) in (21) and (22) gives

$$e' = e_0 \varepsilon^{-al} \sqrt[4]{\frac{C_0^2 + \left(\dfrac{g}{2\pi f}\right)^2}{C^2 \{[1 - (2\pi f)^2 CL]^2 + (2\pi f Cr)^2\}}} \quad (23)$$

and

$$i = 2\pi f C e'; \quad (24)$$

or, approximately, if r and g are negligible, we have

$$e' = e_0 \varepsilon^{-al} \sqrt{\frac{C_0}{C\{1 - (2\pi f)^2 CL\}}} \quad (25)$$

and

$$i = 2\pi f e_0 \varepsilon^{-al} \sqrt{\frac{CC_0}{1 - (2\pi f)^2 CL}}. \quad (26)$$

47. Assume, as example, a lightning arrester having the following constants: $L = 2 \times 10^{-8}$ henry; $C_0 = 10^{-13}$ farads;

$C = 4 \times 10^{-11}$ farads; $r = 1$ ohm; $g = 4 \times 10^{-6}$ mho; $f = 10^8 = 100$ million cycles per second; $n = 300$ cylinders, and $e_0 = 30,000$ volts; then from equation (3), $x = 12.6$ ohms, $x_c = 39.7$ ohms, and $b = 62.8 \times 10^{-6}$ mhos;

from equation (1),

$$Z = 1 - 27.1\,j \text{ ohms};$$

from equation (2),

$$Y = (4 + 62.8\,j)\,10^{-6} \text{ mho};$$

from equation (4),

$$z = 27.1 \text{ ohms and } y = 62.9 \times 10^{-6} \text{ mho};$$

from equation (10),

$$\alpha = 0.0021 \text{ and } \beta = 0.0412;$$

from equation (17),

$$e = 35,500 \sqrt{\varepsilon^{-0.0042\,l} + 0.08\,\varepsilon^{+0.0042\,l} - 0.568 \cos\,(24.72 - 0.0824\,l)};$$

from equation (18),

$$i = 54 \sqrt{\varepsilon^{-0.0042\,l} + 0.08\,\varepsilon^{+0.0042\,l} + 0.568 \cos\,(24.72 - 0.0824\,l)},$$

and from equation (19),

$$e' = 2140 \sqrt{\varepsilon^{-0.0042\,l} + 0.08\,\varepsilon^{+0.0042\,l} + 0.568 \cos\,(24.72 - 0.0824\,l)}.$$

Hence, at $l = 0$, $e = 30,000$ volts, $i = 64.6$ amperes, and $e' = 2560$ volts; and at $l = 300$, $e = 0$, $i = 57.5$ amperes, and $e' = 2280$ volts.

With voltages per gap varying from 2280 to 2560, 300 gaps would, by addition, give a total voltage of about 730,000, while the actual voltage is only about one-twenty-fourth thereof; that is, the sum of the voltages of many spark-gaps in series may be many times the resultant voltage, and a lightning flash may pass possibly for miles through clouds with a total potential of only a few hundred million volts. In the above example the 300 cylinders include 7.86 complete wave-lengths of the discharge.

CHAPTER VI.

ALTERNATING MAGNETIC FLUX DISTRIBUTION.

48. As carrier of magnetic flux iron is used, as far as possible, since it has the highest permeability or magnetic conductivity. If the magnetic flux is alternating or otherwise changing rapidly, an e.m.f. is generated by the change of magnetic flux in the iron, and to avoid energy losses and demagnetization by the currents produced by these e.m.fs. the iron has to be subdivided in the direction in which the currents would exist, that is, at right angles to the lines of magnetic force. Hence, alternating magnetic fields and magnetic structures desired to respond very quickly to changes of m.m.f. are built of thin wires or thin iron sheets, that is, are laminated.

Since the generated e.m.fs. are proportional to the frequency of the alternating magnetism, the laminations must be finer the higher the frequency.

To fully utilize the magnetic permeability of the iron, it therefore has to be laminated so as to give, at the impressed frequency, practically uniform magnetic induction throughout its section, that is, negligible secondary currents. This, however, is no longer the case, even with the thinnest possible laminations, at extremely high frequencies, as oscillating currents, lightning discharges, etc., and under these conditions the magnetic flux distribution in the iron is not uniform, but the magnetic flux density, $\mathfrak{B}$, decreases rapidly, and lags in phase, with increasing depth below the surface of the lamination, so that ultimately hardly any magnetic flux exists in the inside of the laminations, but practically only a surface layer carries magnetic flux. The apparent permeability of the iron thus decreases at very high frequency, and this has led to the opinion that at very high frequencies iron cannot follow a magnetic cycle. There is, however, no evidence of such a "viscous hysteresis." Magnetic investigations at 100,000 to 200,000 cycles per second have given the same magnetic cycles as at low frequencies. It therefore is probable that iron follows magnetically even at the highest frequencies, traversing practically the same hysteresis cycle irrespective of the frequency, if the true·m.m.f., that is, the resultant of the impressed m.m.f. and the m.m.f. of the secondary currents in the iron, is considered. Since with increasing frequency, at constant impressed m.m.f., the resultant m.m.f. decreases, due to the increase of the demagnetizing secondary currents, this simulates the effect of a viscous hysteresis.

Frequently also, for mechanical reasons, iron sheets of greater thickness than would give uniform flux density have to be used in an alternating field.

Since rapidly varying magnetic fields usually are alternating, and the subdivision of the iron is usually by lamination, it will be sufficient to consider as illustration of the method the distribution of alternating magnetic flux in iron laminations.

49. Let Fig. 92 represent the section of a lamination. The alternating magnetic flux is assumed to pass in a direction perpendicular to the plane of the paper.

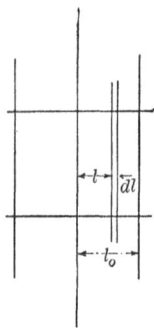

Let μ = the magnetic permeability, λ = the electric conductivity, l = the distance of a layer dl from the center line of the lamination, and $2\,l_0$ = the total thickness of the lamination. If then I = the current density in the layer dl, and E = the e.m.f. per unit length generated in the zone dl by the alternating magnetic flux, we have

$$I = \lambda E. \tag{1}$$

The magnetic flux density $\mathfrak{B}_1$ at the surface $l = l_0$ of the lamination corresponds to the impressed or external m.m.f. The density $\mathfrak{B}$ in the zone dl corresponds to the impressed m.m.f. plus the sum of all the m.m.fs. in the zones outside of dl, or from l to l_0.

Fig. 92. Alternating magnetic flux distribution in solid iron.

The current in the zone dl is

$$I\,dl = \lambda E\,dl \tag{2}$$

and produces the m.m.f.

$$\mathfrak{IC} = 0.4\,\pi\lambda E\,dl, \tag{3}$$

which in turn would produce the magnetic flux density

$$d\mathfrak{B} = 0.4\,\pi\lambda\mu E\,dl; \tag{4}$$

that is, the magnetic flux density $\mathfrak{B}$ at the two sides of the zone dl differs by the magnetic flux density $d\mathfrak{B}$ (equation (4)) produced by the m.m.f. in zone dl, and this gives the differential equation between $\mathfrak{B}$, E, and l,

$$\frac{d\dot{\mathfrak{B}}}{dl} = 0.4 \; \pi \lambda \mu \dot{E}. \tag{5}$$

The e.m.f. generated at distance l from the center of the lamination is due to the magnetic flux in the space from l to l_0. Thus the e.m.fs. at the two sides of the zone dl differ from each other by the e.m.f. generated by the magnetic flux $\dot{\mathfrak{B}}dl$ in this zone.

Considering now $\mathfrak{B}$, $\dot{E}$, and $\dot{I}$ as complex quantities, the e.m.f. $d\dot{E}$, that is, the difference between the e.m.fs. at the two sides of the zone dl, is in quadrature ahead of $\mathfrak{B}dl$, and thus denoted by

$$d\dot{E} = j \; 2 \; \pi f \mathfrak{B} \; 10^{-8} \; dl, \tag{6}$$

where $f =$ the frequency of alternating magnetism.

This gives the second differential equation

$$\frac{d\dot{E}}{dl} \; j \; 2 \; \pi f \mathfrak{B} \; 10^{-8}. \tag{7}$$

50. Differentiating (5) in respect to l, and substituting (7) therein, gives

$$\frac{d^2 \mathfrak{B}}{dl^2} = 0.8 \; j\pi^2 f \lambda \mu \; 10^{-8} \; \mathfrak{B}, \tag{8}$$

or, writing

$$c^2 = f a^2 = 0.4 \; \pi^2 f \lambda \mu \; 10^{-8}, \tag{9}$$

$$a^2 = 0.4 \; \pi^2 \lambda \mu \; 10^{-8}, \tag{10}$$

we have

$$\frac{d^2 \dot{\mathfrak{B}}}{dl^2} = 2 \; jc^2 \mathfrak{B}. \tag{11}$$

This differential equation is integrated by

$$\mathfrak{B} = A\varepsilon^{-vl}; \tag{12}$$

this equation substituted in (11) gives

$$v^2 = 2 \; jc^2; \tag{13}$$

hence,

$$v = \pm (1 + j) c \qquad (14)$$

and

$$\mathcal{B} = A_1 \varepsilon^{+(1+j)cl} + A_2 \varepsilon^{+(1-j)cl}.$$

Since $\mathcal{B}$ must have the same value for $-l$ as for $+l$, being symmetrical at both sides of the center line of the lamination,

$$A_1 = A_2 = A,$$

hence,

$$\mathcal{B} = A \{\varepsilon^{+(1+j)cl} + \varepsilon^{-(1+j)cl}\}; \qquad (15)$$

or, substituting

$$\varepsilon^{\pm jcl} = \cos cl \pm j \sin cl \qquad (16)$$

gives

$$\mathcal{B} = A \{(\varepsilon^{+cl} + \varepsilon^{-cl}) \cos cl + j (\varepsilon^{+cl} - \varepsilon^{-cl}) \sin cl\}. \qquad (17)$$

51. Denoting the flux density in the center of the lamination, for $l = 0$, by $\mathcal{B}_0$ from (17) we have

$$\mathcal{B}_0 = 2 A;$$

hence,

$$A = \tfrac{1}{2} \mathcal{B}_0 \qquad (18)$$

and

$$\mathcal{B} = \mathcal{B}_0 \left\{ \frac{\varepsilon^{+cl} + \varepsilon^{-cl}}{2} \cos cl + j \frac{\varepsilon^{+cl} - \varepsilon^{-cl}}{2} \sin cl \right\}. \qquad (19)$$

Denoting the flux density at the outside of the lamination, for $l = l_0$, that is, the density produced by the external m.m.f., by $\mathcal{B}_1$, substituted in (19), we have

$$\mathcal{B}_1 = \mathcal{B}_0 \left\{ \frac{\varepsilon^{+cl_0} + \varepsilon^{-cl_0}}{2} \cos cl_0 + j \frac{\varepsilon^{+cl_0} - \varepsilon^{-cl_0}}{2} \sin cl_0 \right\}, \qquad (20)$$

and substituting (20) in (19),

$$\mathcal{B} = \mathcal{B}_1 \frac{(\varepsilon^{+cl} + \varepsilon^{-cl}) \cos cl + j (\varepsilon^{+cl} - \varepsilon^{-cl}) \sin cl}{(\varepsilon^{+cl_0} + \varepsilon^{-cl_0}) \cos cl_0 + j (\varepsilon^{+cl_0} - \varepsilon^{-cl_0}) \sin cl_0}. \qquad (21)$$

The mean or apparent value of the flux density, i.e., the average throughout the lamination, is

$$\mathcal{B}_m = \frac{1}{l_1} \int_0^{l_0} \mathcal{B} \, dl. \qquad (22)$$

Using equation (15) as the more convenient for integration gives

$$\mathfrak{B}_m = \frac{A}{(1+j)\,cl_0}\left[\varepsilon^{+(1+j)\,cl} - \varepsilon^{-(1+j)\,cl}\right]_0^{l_0}$$

$$= \frac{A\left(\varepsilon^{+(1+j)\,cl_0} - \varepsilon^{-(1+j)\,cl_0}\right)}{(1+j)\,cl_0}, \tag{23}$$

and substituting herein (16), (18) and (20), gives

$$\left.\begin{aligned}
\mathfrak{B}_m &= \frac{\dot{\mathfrak{B}}_0}{(1+j)\,cl_0}\left\{\frac{\varepsilon^{+cl_0} - \varepsilon^{-cl_0}}{2}\cos cl_0 + j\,\frac{\varepsilon^{+cl_0} + \varepsilon^{-cl_0}}{2}\sin cl_0\right\} \\
&= \frac{\dot{\mathfrak{B}}_1}{(1+j)\,cl_0}\,\frac{(\varepsilon^{+cl_0} - \varepsilon^{-cl_0})\cos cl_0 + j\,(\varepsilon^{+cl_0} + \varepsilon^{-cl_0})\sin cl_0}{(\varepsilon^{+cl_0} + \varepsilon^{-cl_0})\cos cl_0 + j\,(\varepsilon^{+cl_0} - \varepsilon^{-cl_0})\sin cl_0}.
\end{aligned}\right\} \tag{24}$$

The absolute values of the flux densities are derived as square root of the sum of the squares of real and imaginary terms in equations (19), (20), (21), and (24), as

$$\mathfrak{B} = \frac{\mathfrak{B}_0}{2}\sqrt{\varepsilon^{+2cl} + \varepsilon^{-2cl} + 2\cos 2\,cl}, \tag{25}$$

$$\mathfrak{B}_1 = \frac{\mathfrak{B}_0}{2}\sqrt{\varepsilon^{+2cl_0} + \varepsilon^{-2cl_0} + 2\cos 2\,cl_0}, \tag{26}$$

$$\mathfrak{B} = \mathfrak{B}_1\sqrt{\frac{\varepsilon^{+2cl} + \varepsilon^{-2cl} + 2\cos 2cl}{\varepsilon^{+2cl_0} + \varepsilon^{-2cl_0} + 2\cos 2\,cl_0}}, \tag{27}$$

and

$$\left.\begin{aligned}
\mathfrak{B}_m &= \frac{\mathfrak{B}_0}{2\,cl_0\sqrt{2}}\sqrt{\varepsilon^{+2cl_0} + \varepsilon^{-2cl_0} - 2\cos 2\,cl_0} \\
&= \frac{\mathfrak{B}_1}{cl_0\sqrt{2}}\sqrt{\frac{\varepsilon^{+2cl_0} + \varepsilon^{-2cl_0} - 2\cos 2\,cl_0}{\varepsilon^{+2cl_0} + \varepsilon^{-2cl_0} + 2\cos 2\,cl_0}}.
\end{aligned}\right\} \tag{28}$$

52. Where the thickness of lamination, $2\,l_0$, or the frequency f, is so great as to give cl_0 a value sufficiently high to make ε^{-cl_0}, or the reflected wave, negligible compared with the main wave ε^{+cl_0}, the equations can be simplified by dropping ε^{-cl}. In this case the flux density, $\mathfrak{B}$, is very small or practically nothing in the interior, and reaches appreciable values only near the surface. It then is preferable to count the distance from the surface of the lamination into the interior, that is, substitute the independent variable

$$s = l_0 - l. \tag{29}$$

Dropping ε^{-cl} and ε^{-cl_0} in equation (21) gives

$$\mathcal{B} = \mathcal{B}_1 \frac{\varepsilon^{cl} \; (\cos cl + j \sin cl)}{\varepsilon^{cl_0} \; (\cos cl_0 + j \sin cl_0)}$$

$$= \mathcal{B}_1 \, \varepsilon^{-c \, (l_0 - l)} \{\cos c \; (l_0 - l) - j \sin c \; (l_0 - l)\} \, ;$$

hence,

$$\mathcal{B} = \mathcal{B}_1 \, \varepsilon^{-cs} \; (\cos cs - j \sin cs); \tag{30}$$

and the absolute value is

$$\mathcal{B} = \mathcal{B}_1 \, \varepsilon^{-cs}, \tag{31}$$

and at the center of the lamination,

$$\left.\begin{array}{l} \mathcal{B}_0 = \mathcal{B}_1 \, \varepsilon^{-cl_0} (\cos cl_0 + j \sin cl_0), \\[2mm] \mathcal{B}_0 = \mathcal{B}_1 \, \varepsilon^{-cl_0}. \end{array}\right\} \tag{32}$$

From equation (24) the mean value of flux density follows when dropping ε^{-cl_0} as negligible, thus:

$$\mathcal{B}_m = \frac{\mathcal{B}_1}{(1 + j) \, cl_0} = \frac{(1 - j) \, \mathcal{B}_1}{2 \, cl_0} \tag{33}$$

or the absolute value is

$$\mathcal{B}_m = \frac{\mathcal{B}}{cl_0 \sqrt{2}}. \tag{34}$$

53. As seen, the preceding equations of the distribution of alternating magnetic flux in a laminated conductor are of the same form as the equations of distribution of current and voltage in a transmission line, but more special in form, that is, the attenuation constant α and the wave length constant β have the same value, c. As result, the distribution of the alternating magnetic flux in the lamina depends upon one constant only, cl_0.

The wave length is given by

$$cl_w = 2 \pi;$$

hence
$$l_\omega = \frac{2\,\pi}{c}$$

and by (9)

$$l_w = \frac{2\,\pi}{a\sqrt{f}}$$

$$= \frac{10,000}{\sqrt{0.1\ \lambda \mu f}}, \tag{35}$$

and the attenuation during one wave length, or decrease of intensity of magnetism, per wave length, is

$$\varepsilon^{-2\pi} = 0.0019,$$

per half-wave length it is

$$\epsilon^{-\pi} = 0.043.$$

and per quarter-wave length

$$\epsilon^{-\frac{\pi}{2}} = 0.207.$$

At the depth $\dfrac{l_w}{4}$ below the surface, the magnetic flux lags 90 degrees and has decreased to about 20 per cent; at the depth $\dfrac{l_w}{2}$ it lags 180 degrees, that is, is opposite in direction to the flux at the surface of the lamination, but is very small, the intensity being less than 5 per cent of that at the surface, and at the depth l_w the flux is again in phase with the surface flux, but its intensity is practically nil, less than 0.2 per cent of the surface intensity; that is, the penetration of alternating flux into the laminated iron is inappreciable at the depth of one wave length.

By equations (33) and (34), the total magnetic flux per unit width of lamination is

$$2\,l_0\mathfrak{B}_m = \frac{2\mathfrak{B}_1}{(1+j)\,c} = \frac{(1-j)\,\mathfrak{B}_1}{c}$$

the absolute value is $\quad 2\,l_0\mathfrak{B}_m = \dfrac{2\,\mathfrak{B}_1}{c\sqrt{2}}\,;$

that is, the same as would be produced at uniform density in a thickness of lamination

$$2\,l_p = \frac{2}{(1+j)\,c} = \frac{(1-j)}{c}$$

or absolute value,

$$2\,l_p = \frac{\sqrt{2}}{c}\,;$$

which means that the resultant alternating magnetism in the lamination lags 45 degrees, or one-eighth wave behind the impressed m.m.f., and is equal to a uniform magnetic density penetrating to a depth

$$l_p = \frac{1}{c\sqrt{2}}. \tag{36}$$

l_p, therefore, can be called the depth of penetration of the alternating magnetism into the solid iron.

Since the only constant entering into the equation is cl_0, the distribution of alternating magnetism for all cases can be represented as function of cl_0.

If cl_0 is small, and therefore the density in the center of the lamination $\mathfrak{B}_0$ comparable with the density $\mathfrak{B}_1$ at the outside, the equations (19), (20), and (24) respectively (25), (26), and (28) have to be used; if cl_0 is large, and the flux density $\mathfrak{B}_0$ in the center of the lamination is negligible, the simpler equations (30) to (34) can be used.

54. As an example, let $\mu = 1000$ and $\lambda = 10^5$; then $a = 1.98$, and for $f = 60$ cycles per second, $c = a\sqrt{f} = 15.3$; hence, the thickness of effective layer of penetration is

$$l_p = \frac{1}{c\sqrt{2}} = 0.046 \ cm = 0.018 \ \text{inches}.$$

In Fig. 93 is shown, with cl as abscissas, the effective value of the magnetic flux, which from equation (25) is

$$\mathfrak{B} = \frac{\mathfrak{B}_0}{2} \sqrt{\varepsilon^{+2cl} + \varepsilon^{-2cl} + 2 \cos 2\ cl},$$

and also the space-phase angle between $\mathfrak{B}$ and $\mathfrak{B}_0$, which from equation (19) is

$$\tan \tau_0 = \frac{\varepsilon^{cl} - \varepsilon^{-cl}}{\varepsilon^{cl} + \varepsilon^{-cl}} \tan cl. \tag{37}$$

In Fig. 94 is shown, with cs as abscissas, the effective value of the magnetic flux, which from equation (31) is

$$\mathfrak{B} = \mathfrak{B}_1 \varepsilon^{-cs},$$

and also the space-phase angle between $\mathcal{B}$ and $\mathcal{B}_1$, which from equation (30) is

$$\tan \tau_1 = \tan cs. \tag{38}$$

The thickness of the equivalent layer is marked in Fig. 94.

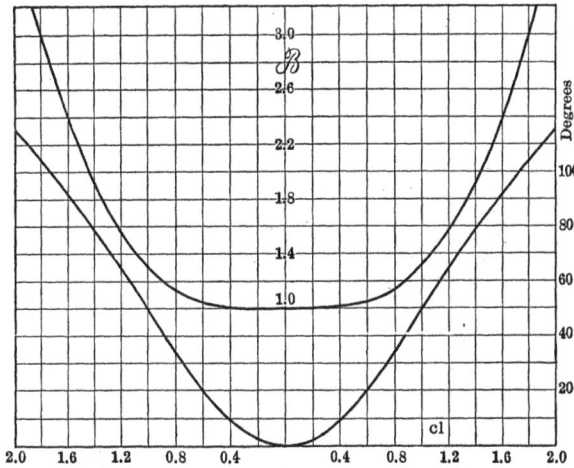

Fig. 93.   Alternating magnetic flux distribution in solid iron.

As further illustrations are shown in Fig. 95 the absolute values of magnetic flux density $\mathcal{B}$ throughout a layer of 14 mils thickness, that is, of $l_0 = 0.007$ inches $= 0.018$ cm. thickness.

For 60 cycles, by Curve I, $\qquad c = 15.3 \quad cl_0 = 0.275$

For 1000 cycles, by Curve II, $\quad c = 62.5 \quad cl_0 = 1.125$

For 10,000 cycles, by Curve III, $c = 198 \qquad cl_0 = 3.55$

It is seen that the density in Curve I is perfectly uniform, while in Curve III practically no flux penetrates to the center.

55. The effective penetration of the alternating magnetism into the iron, or the thickness l_p of surface layer which at constant induction $\mathcal{B}_1$ would give the same total magnetic flux as exists in the lamination, is

$$l_p = \frac{1}{(1+j)c} = \frac{1-j}{2\,c} \tag{39}$$

or the absolute value is

$$l_p = \frac{1}{c\sqrt{2}};$$

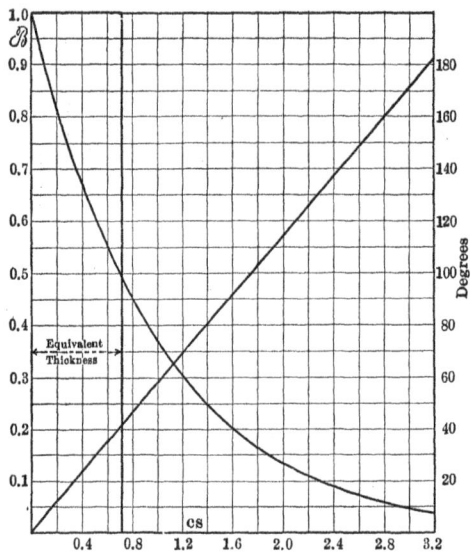

Fig. 94. Alternating magnetic flux distribution in solid iron.

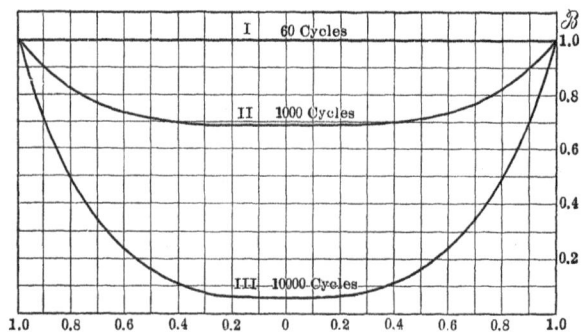

Fig. 95. Alternating magnetic flux distribution in solid iron.

hence, substituting for c from equation (9),

$$l_p = \frac{10^4}{\pi \sqrt{0.8 \, \lambda \mu f}} = \frac{3570}{\sqrt{\lambda \mu f}} \tag{40}$$

that is, the penetration of an alternating magnetic flux into a solid conductor is inversely proportional to the square root of the electric conductivity, the magnetic permeability, and the frequency.

The values of penetration, l_p, in centimeters for various materials and frequencies are given below.

Frequency.	25	60	1000	10,000	10^5	10^6
Soft iron, $\mu = 1000$, $\lambda = 10^5$........	0.0714	0.0460	0.0113	0.0036	0.00113	0.00036
Cast iron, $\mu = 200$, $\lambda = 10^4$........	0.504	0.325	0.080	0.0252	0.0080	0.0025
Copper, $\mu = 1$, $\lambda = 6 \times 10^5$........	0.922	0.595	0.144	0.0461	0.0144	0.0046
Resistance alloys, $\mu = 1$, $\lambda = 10^4$....	7.14	4.60	1.13	0.357	0.113	0.036

As seen, even at frequencies as low as 25 cycles alternating magnetism does not penetrate far into solid wrought iron, but penetrates to considerable depth into cast iron. It also is interesting to note that little difference exists in the penetration into copper and into cast iron, the high conductivity of the former compensating for the higher permeability of the latter.

56. The wave length, $l_w = \dfrac{2\pi}{c}$, substituting for c, from equation (9), is

$$l_w = \frac{31,600}{\sqrt{\lambda \mu f}}; \tag{41}$$

that is, the wave length of the oscillatory transmission of alternating magnetism in solid iron is inversely proportional to the square root of the electric conductivity, the magnetic permeability, and the frequency.

Comparing this equation (41) of the wave length l_w with equation (40) of the depth of penetration l_p, it follows that the depth of penetration is about one-ninth of the wave length, or 40 degrees, or, more accurately, since

$$l_p = \frac{1}{c\sqrt{2}} \text{ and } l_w = \frac{2\pi}{c},$$

we have

$$\frac{l_p}{l_w} = \frac{1}{2\sqrt{2}\pi} = \frac{1}{8.9}, \tag{42}$$

or 40.5 degrees.

The speed of propagation is

$$S = fl_w$$
$$= \frac{31,600\sqrt{f}}{\sqrt{\lambda\mu}} : \tag{43}$$

that is, the speed of propagation is inversely proportional to the square root of the electric conductivity and of the magnetic permeability, but directly proportional to the square root of the frequency. This gives a curious instance of a speed which increases with the frequency. Numerical values are given below.

Frequency.	25 Cycles.	10,000 Cycles.
Soft iron, $\mu=1000$, $\lambda=10^5$	$S=15.8$ cm.	316 cm.
Cast iron, $\mu=200$, $\lambda=10^4$	111 cm.	2230 cm.
Copper, $\mu=1$, $\lambda=6\times10^5$	204 cm.	4080 cm.

It is seen that these speeds are extremely low compared with the usual speeds of electromagnetic waves.

57. Since instead of $\mathfrak{B}_1$, corresponding to the impressed m.m.f. and permeability μ, the mean flux density in the lamina is $\mathfrak{B}_m$, the effect is the same as if the permeability of the material were changed from μ to

$$\mu' = \mu\frac{\dot{\mathfrak{B}}_m}{\mathfrak{B}_1}, \tag{44}$$

and μ' can be called the *effective permeability*, which is a function of the thickness of the lamination and of the frequency, that is, a function of cl_0; μ' appears in complex form thus,

$$\mu' = \mu_1' - j\mu_2';$$

that is, the permeability is reduced and also made lagging.

For high values of cl_0, that is, thin laminations or high frequencies, from (33), we have

$$\mu' = \frac{\mu}{(1+j)\cdot cl_0}$$
$$= \frac{\mu}{2\,cl_0} - j\,\frac{\mu}{2\,cl_0}. \tag{45}$$

or, absolute

$$\mu' = \frac{\mu}{cl_0\sqrt{2}}$$

58. As illustration, for iron of 14 mils thickness, or $l_0 = 0.018$ centimeters, and the constants $\mu = 1000$ and $\lambda = 10^6$, that is $a = 1.98$, the absolute value of the effective permeability is

and

$$\mu' = \frac{\mu}{cl_0 \sqrt{2}}$$

$$c = a \sqrt{f};$$

hence,

$$\mu' = \frac{\mu}{al_0 \sqrt{2f}}$$
$$= \frac{19,800}{\sqrt{f}} ; \tag{46}$$

that is, the effective or apparent permeability at very high frequencies decreases inversely proportional to the square root of the frequency. In the above instance the apparent permeability is:

> At low frequency, $\mu = 1000$;
> at 10,000 cycles, $\mu' = 198$;
> at 1,000,000 cycles, $\mu' = 19.8$;
> at 100 million cycles, $\mu' = 1.98$, and
> at 392 million cycles, $\mu' = 1$,

or the same as air, and at still higher frequencies the presence of iron reduces the magnetic flux.

It is interesting to note that with such a coarse lamination as a 14-mil sheet, even at the highest frequencies of millions of cycles, an appreciable apparent permeability is still left; that is, the magnetic flux is increased by the presence of iron; and the effect of iron in increasing the magnetic flux disappears only at 400 million cycles, and beyond this frequency iron lowers the magnetic flux. However, even at these frequencies, the presence of iron still exerts a great effect in the rapid damping of the oscillations by the lag of the mean magnetic flux by 45 degrees.

Obviously, in large solid pieces of iron, the permeability μ' falls below that of air even at far lower frequencies.

Where the penetration of the magnetic flux l_p is small compared with the dimensions of the iron, its shape becomes immaterial, since only the surface requires consideration, and so in this case any solid structure, no matter what shape, can be considered magnetically as its outer shell of thickness l_p when dealing with rapidly alternating magnetic fluxes.

At very high frequencies, when dealing with alternating magnetic circuits, the outer surface and not the section is, therefore, the dominating feature.

The lag of the apparent permeability represents an energy component of the e.m.f. of self-induction due to the magnetic flux, which increases with increasing frequency, and ultimately becomes equal to the reactive component.

CHAPTER VII.

DISTRIBUTION OF ALTERNATING-CURRENT DENSITY IN CONDUCTOR.

59. If the frequency of an alternating or oscillating current is high, or the section of the conductor which carries the current is very large, or its electric conductivity or its magnetic permeability high, the current density is not uniform throughout the conductor section, but decreases towards the interior of the conductor, due to the higher e.m.f. of self-inductance in the interior of the conductor, caused by the magnetic flux inside of the conductor. The phase of the current inside of the conductor also differs from that on the surface and lags behind it.

In consequence of this unequal current distribution in a large conductor traversed by alternating currents, the effective resistance of the conductor may be far higher than the ohmic resistance, and the conductor also contains internal inductance.

In the extreme case, where the current density in the interior of the conductor is very much lower than on the surface, or even negligible, due to this "screening effect," as it has been called, the current can be assumed to exist only in a thin surface layer of the conductor, of thickness l_p; that is, in this case the effective resistance of the conductor for alternating currents equals the ohmic resistance of a conductor section equal to the periphery of the conductor times the "depth of penetration."

Where this unequal current distribution throughout the conductor section is considerable, the conductor section is not fully utilized, but the material in the interior of the conductor is more or less wasted. It is of importance, therefore, in alternating-current circuits, especially in dealing with very large currents, or with high frequency, or materials of very high permeability, as iron, to investigate this phenomenon.

An approximate determination of this effect for the purpose of deciding whether the unequal current distribution is so small as to be negligible in its effect on the resistance of the conductor, or whether it is sufficiently large to require calculation and methods of avoiding it, is given in "Alternating-Current Phenomena," Chapter XIII, paragraph 113.

An appreciable increase of the effective resistance over the ohmic resistance may be expected in the following cases:

(1) In the low-tension distribution of heavy alternating currents by large conductors.

(2) When using iron as conductor, as for instance iron wires in high potential transmissions for branch lines of smaller power, or steel cables for long spans in transmission lines.

(3) In the rail return of single-phase railways.

(4) When carrying very high frequencies, such as lightning discharges, high frequency oscillations, wireless telegraph currents, etc.

In the last two cases, which probably are of the greatest importance, the unequal current distribution usually is such that practically no current exists at the conductor center, and the effective resistance of the track rail even for 25-cycle alternating current thus is several times greater than the ohmic resistance, and conductors of low ohmic resistance may offer a very high effective resistance to a lightning stroke.

By subdividing the conductor into a number of smaller conductors, separated by some distance from each other, or by the use of a hollow conductor, or a flat conductor, as a bar or ribbon, the effect is reduced, and for high-frequency discharges, as lightning arrester connections, flat copper ribbon offers a very much smaller effective resistance than a round wire. Stranding the conductor, however, has no direct effect on this phenomenon, since it is due to the magnetic action of the current, and the magnetic field in the stranded conductor is the same as in a solid conductor, other things being equal. That is, while eddy currents in the conductor, due to external magnetic fields, are eliminated by stranding the conductor, this is not the case with the increase of the effective resistance by unequal current distribution. Stranding the conductor, however, may reduce unequal current distribution indirectly, especially with iron as conductor material, by reducing the effective or mean permeability of the conductor, due to the break in the magnetic circuit between the iron strands, and also by the reduction of the mean conductivity of the conductor section. For instance, if in a stranded conductor 60 per cent of the conductor section is copper, 40 per cent space between the strands, the mean conductivity is 60 per cent of that of copper. If by the subdivision of an iron conductor into strands the reluctance of the magnetic circuit is increased tenfold, this represents a reduction of the mean permeability to one-tenth. Hence, if for the conductor material proper $\mu = 1000$, $\lambda = 10^5$, and the conductor section is reduced by stranding to 60 per cent, the permeability to one-tenth, the mean values would be

$$\mu_0 = 100 \quad \text{and} \quad \lambda_0 = 0.6 \times 10^5,$$

and the factor $\sqrt{\lambda\mu}$, in the equation of current distribution, is reduced from $\sqrt{\lambda\mu} = 10,000$ to $\sqrt{\lambda_0\mu_0} = 2450$, or to 24.5 per cent of its previous value. In this case, however, with iron as conductor material, an investigation must be made on the current distribution in each individual conductor strand.

Since the simplest way of reducing the effect of unequal current distribution is the use of flat conductors, the most important case is the investigation of the alternating-current distribution throughout the section of the flat conductor. This also gives the solution for conductors of any shape when the conductor section is so large that the current penetrates only the surface layer, as is the case with a steel rail of a single-phase railway. Where the alternating current penetrates a short distance only into the conductor, compared with the depth of penetration the curvature of the conductor surface can be neglected, that is, the conductor surface considered as a flat surface penetrated to the same depth all over. Actually on sharp convex surfaces the current penetrates somewhat deeper, somewhat less on sharp concave surfaces, so that the error is more or less compensated.

60. In a section of a flat conductor, as shown diagrammatically in Fig. 92, page 356, let λ = the electric conductivity of conductor material; μ = the magnetic permeability of conductor material; l = the distance counted from the center line of the conductor, and $2\,l_0$ = the thickness of conductor.

Furthermore, let E_0 = the impressed e.m.f. per unit length of conductor, that is, the voltage consumed per unit length in the conductor after subtracting the e.m.f. consumed by the self-inductance of the external magnetic field of the conductor; thus, if E_1 = the total supply voltage per unit length of conductor and E_2 = the external reactance voltage, or voltage consumed by the magnetic field outside of the conductor, between the conductors, we have

$$E_0 = E_1 - E_2.$$

Let

$I = i_1 - ji_2$ = current density in conductor element dl,
$\mathfrak{B} = b_1 - jb_2$ = magnetic density in conductor element dl,
E = e.m.f. consumed in the conductor element dl by the self-inductance due to the magnetic field inside of the conductor; then the current $I\,dl$ in the conductor element represents the m.m.f. or field intensity,

$$d\mathfrak{K} = 0.4\,\pi I\,dl, \tag{1}$$

which causes an increase of the magnetic density $\mathfrak{B}$ between the two sides of the conductor element dl by

$$d\mathfrak{B} = \mu d\mathfrak{K}$$
$$= 0.4\,\pi\mu I\,dl. \tag{2}$$

The e.m.f. consumed by self-inductance is proportional to the magnetic flux and to the frequency, and is 90 time-degrees ahead of the magnetic flux.

The increase of magnetic flux $\mathfrak{B}\,dl$, in the conductor element dl, therefore, causes an increase in the e.m.f. consumed by self-inductance between the two sides of the conductor element by

$$dE = -\,2\,j\pi f\mathfrak{B}\,10^{-8}\,dl, \tag{3}$$

where f = the frequency of the impressed e.m.f.

Since the impressed e.m.f. E_0 equals the sum of the e.m.f. consumed by self-inductance E and the e.m.f. consumed by the resistance of the conductor element,

$$E_0 = E + \frac{I}{\lambda}. \tag{4}$$

Differentiating (4) gives

$$dE = -\frac{1}{\lambda}dI, \tag{5}$$

and substituting (5) in (3) gives

$$dI = 2 j\pi f \lambda \mathfrak{B} \, 10^{-8} \, dl. \tag{6}$$

The two differential equations (6) and (2) contain $\mathfrak{B}$, I, and l, and by eliminating $\mathfrak{B}$, give the differential equation between I and l: differentiating (6) and substituting (2) therein gives

$$\frac{d^2 I}{dl^2} = 0.8 \, j\pi^2 f \, 10^{-8} \, \lambda \mu I; \tag{7}$$

or writing

$$c^2 = a^2 f = 0.4 \, \pi^2 \, 10^{-8} \, \lambda \mu f, \tag{8}$$

where

$$a^2 = 0.4 \, \pi^2 \, 10^{-8} \, \lambda \mu, \tag{9}$$

gives

$$\frac{d^2 I}{dl^2} = 2 \, jc^2 I. \, ^{*} \tag{10}$$

This differential equation (10) is integrated by

$$I = A \varepsilon^{-vl}, \tag{11}$$

and substituting (11) in (10) gives

$$v^2 = 2 \, jc^2,$$
$$v = \pm \, c \, (1 + j); \tag{12}$$

hence,

$$I = A_1 \varepsilon^{+c \, (1+j) \, l} + A_2 \varepsilon^{-c \, (1+j) \, l}. \tag{13}$$

Since I gives the same value for $+l$ and for $-l$,

$$A_1 = A_2 = A; \tag{14}$$

hence,

$$I = A \, \{ \varepsilon^{+c \, (1+j) \, l} + \varepsilon^{-c \, (1+j) \, l} \}. \tag{15}$$

Substituting

$$\varepsilon^{\pm jcl} = \cos cl \pm j \sin cl \tag{16}$$

gives

$$I = A \{ (\varepsilon^{+cl} + \varepsilon^{-cl}) \cos cl + j \, (\varepsilon^{+cl} - \varepsilon^{-cl}) \sin cl \}, \tag{17}$$

and for $l = l_0$, or at the conductor surface,

$$I_1 = A \{ (\varepsilon^{+cl_0} + \varepsilon^{-cl_0}) \cos cl_0 + j \, (\varepsilon^{+cl_0} - \varepsilon^{-cl_0}) \sin cl_0 \}. \tag{18}$$

* (10) is the same differential equation, and c has the same value as in the equation of alternating magnetic flux distribution (11) on page 363, and the alternating current distribution in a solid conductor thus is the same as the alternating magnetic flux distribution in the same conductor.

At the conductor surface, however, no e.m.f. of self-inductance due to the internal field exists, and

$$\dot{I}_1 = \lambda \dot{E}_0. \tag{19}$$

Substituting (19) in (18) gives the integration constant A, and this substituted in (17) gives the distribution of current density throughout the conductor section as

$$\dot{I} = \lambda \dot{E}_0 \frac{(\varepsilon^{+cl} + \varepsilon^{-cl}) \cos cl + j (\varepsilon^{+cl} - \varepsilon^{-cl}) \sin cl}{(\varepsilon^{+cl_0} + \varepsilon^{-cl_0}) \cos cl_0 + j (\varepsilon^{+cl_0} - \varepsilon^{-cl_0}) \sin cl_0}. \tag{20}$$

The absolute value is given as the square root of the sum of squares of real and imaginary terms,

$$I = \lambda E_0 \sqrt{\frac{\varepsilon^{+2cl} + \varepsilon^{-2cl} + 2 \cos 2 cl}{\varepsilon^{+2cl_0} + \varepsilon^{-2cl_0} + 2 \cos 2 cl_0}}. \tag{21}$$

The current density in the conductor center, $l = 0$, is

$$\dot{I}_0 = \frac{2 \lambda \dot{E}_0}{(\varepsilon^{+cl_0} + \varepsilon^{-cl_0}) \cos cl_0 + j (\varepsilon^{+cl_0} - \varepsilon^{-cl_0}) \sin cl_0}, \tag{22}$$

or the absolute value is

$$I_0 = \frac{2 \lambda E_0}{\sqrt{\varepsilon^{+2cl_0} + \varepsilon^{-2cl_0} + 2 \cos 2 cl_0}}. \tag{23}$$

61. It is seen that the distribution of alternating-current density throughout a solid flat conductor gives the same equation as the distribution of alternating magnetic density through an iron rail, equations of the same character as the equation of the long distance transmission line, but more special in form.

The mean value of current density throughout the conductor section,

$$\dot{I}_m = \frac{1}{l_0} \int_0^{l_0} \dot{I} \, dl, \tag{24}$$

which is derived in the same manner as in Chapter V, § 51, is

$$\dot{I}_m = \frac{\lambda E_0 \{(\varepsilon^{+cl_0} - \varepsilon^{-cl_0}) \cos cl_0 + j (\varepsilon^{+cl_0} + \varepsilon^{-cl_0}) \sin cl_0\}}{(1+j) cl_0 \{(\varepsilon^{+cl_0} + \varepsilon^{-cl_0}) \cos cl_0 + j (\varepsilon^{+cl_0} - \varepsilon^{-cl_0}) \sin cl_0\}}, \tag{25}$$

and the absolute value is

$$I_m = \frac{\lambda E_0}{cl_0 \sqrt{2}} \sqrt{\frac{\varepsilon^{+2cl_0} + \varepsilon^{-2cl_0} - 2\cos 2cl_0}{\varepsilon^{+2cl_0} + \varepsilon^{-2cl_0} + 2\cos 2cl_0}}. \tag{26}$$

Therefore, the increase of the effective resistance R of the conductor over the ohmic resistance R_0 is

$$\frac{R}{R_0} = \frac{\dot{I}_1}{I_m}, \tag{27}$$

and by (19):

$$\frac{R}{R_0} = (1 + j)\, cl_0 \frac{(\varepsilon^{+cl_0} + \varepsilon^{-cl_0})\cos cl_0 + j\,(\varepsilon^{+cl_0} - \varepsilon^{-cl_0})\sin cl_0}{(\varepsilon^{+cl_0} - \varepsilon^{-cl_0})\cos cl_0 + j\,(\varepsilon^{+cl_0} + \varepsilon^{-cl_0})\sin cl_0}. \tag{28}$$

or the absolute value is

$$\frac{R}{R_0} = cl_0 \sqrt{2} \sqrt{\frac{\varepsilon^{+2cl_0} + \varepsilon^{-2cl_0} + 2\cos 2cl_0}{\varepsilon^{+2cl_0} + \varepsilon^{-2cl_0} - 2\cos 2cl_0}}. \tag{29}$$

62. If cl_0 is so large that ε^{-cl_0} can be neglected compared with ε^{-cl_0}, then in the center of the conductor I is negligible, and for values of l near to l_0, or near the surface of the conductor, from equation (20) we have

$$I = \lambda E_0 \frac{\varepsilon^{+cl} (\cos cl + j \sin cl)}{\varepsilon^{+cl_0} (\cos cl_0 + j \sin cl_0)}$$

$$= \lambda E_0 \varepsilon^{c(l+l_0)} \{\cos c\,(l - l_0) + j \sin c\,(l - l_0)\}.$$

Substituting

$$s = l_0 - l, \tag{30}$$

where s is the depth below the conductor surface, we have

$$I = \lambda E_0 \varepsilon^{-cs} (\cos cs - j \sin cs), \tag{31}$$

and the absolute value is

$$I = \lambda E_0 \varepsilon^{-cs}; \tag{32}$$

the mean value of current density is, from (25)

$$I_m = \frac{\lambda \dot{E}_0}{(1 + j)\, cl_0} = \frac{(1 - j)\lambda \dot{E}_0}{2\, cl_0} \tag{33}$$

and the absolute value is

$$I_m = \frac{\lambda E_0}{cl_0 \sqrt{2}};$$ (34)

hence, the resistance ratio—or rather impedance ratio, as I_m lags behind E_0—is, since the current density at the surface, or density in the absence of a screening effect, is $I_1 = \lambda E_0$:

$$\frac{z}{r_0} = \frac{\dot{I}_1}{I_m} = (1 + j) cl_0$$

$$= cl_0 + jcl_0,$$ (35)

where r_0 = ohmic resistance, Z = effective impedance of the conductor, and the absolute value is

$$\frac{z}{r_0} = cl_0 \sqrt{2};$$ (36)

that is, the effective impedance Z of the conductor, as given by equation (28), and, for very thick conductors, from equation (35), appears in the form

$$Z = r_0(m_1 + jm_2),$$ (37)

which for very thick conductors gives for m_1 and m_2 the values

$$Z = r_0 (cl_0 + jcl_0).$$ (38)

63. As the result of the unequal current distribution in the conductor, the effective resistance is increased from the ohmic resistance r_0 to the value

$$r = r_0 m_1,$$
$$r = cl_0 r_0,$$

and in addition thereto an effective reactance

$$x = r_0 m_2,$$

or

$$x = cl_0 r_0,$$

is produced in the conductor.

In the extreme case, where the current does not penetrate much below the surface of the conductor, the effective resistance and the effective reactance of the conductor are equal and are

$$r = x = cl_0 r_0$$

where r_0 is the ohmic resistance of the conductor.

It follows herefrom that only $\dfrac{1}{cl_0}$ of the conductor section is effective; that is, the depth of the effective layer is

$$l_p = \frac{l_0}{cl_0} = \frac{1}{c};$$

or, in other words, the effective resistance of a large conductor carrying an alternating current is the resistance of a surface layer of the depth

$$l_p = \frac{1}{c}, \tag{39}$$

and in addition thereto an effective reactance equal to the effective resistance results from the internal magnetic field of the conductor.

Substituting (8) in (39) gives

or

$$\left. \begin{array}{c} l_p = \dfrac{10^4}{\pi \sqrt{0.4\, \lambda \mu f}}, \\[2ex] l_p = \dfrac{5030}{\sqrt{\lambda \mu f}}. \end{array} \right\} \tag{40}$$

It follows from the above equations that in such a conductor carrying an alternating current the thickness of the conducting layer, or the depth of penetration of the current into the conductor, is directly proportional, and the effective resistance and effective internal inductance inversely proportional, to the square root of the electric conductivity, of the magnetic permeability, and of the frequency.

From equation (40) it follows that with a change of conductivity λ of the material the apparent conductance, and therewith the apparent resistance of the conductor, varies proportionally to the square root of the true conductivity or resistivity.

Curves of distribution of current density throughout the section of the conductor are identical with the curves of distribution of magnetic flux, as shown by Figs. 93, 94, 95 of Chapter VI.

64. From the "depth of penetration," the actual or effective resistance of the conductor then is given by circumference of the conductor times depth of penetration. This method, of calculating the depth of penetration, has the advantage that it applies to all sizes and shapes of conductors, solid round conductors or hollow tubes, flat ribbon and even such complex shapes as the railway rail, provided only that the depth of penetration is materially less than the depth of the conductor material.

The effective resistance of the conductor then is, per unit length:

$$r = \frac{1}{\lambda l_p l_1} \qquad (41)$$

Thus, substituting (40):

$$\left. \begin{aligned} r &= \frac{\pi}{l_1} \sqrt{\frac{0.4\ \mu f}{\lambda}}\ 10^{-4} \\[2mm] &= \frac{\sqrt{\dfrac{\mu f}{\lambda}}}{5030\ l_1}\ \text{ohms per cm.} \end{aligned} \right\} \qquad (42)$$

where:

l_1 = circumference of conductor (actual, that is, following all indentations, etc.).

As the true ohmic resistance of the conductor is:

$$r_0 = \frac{1}{\lambda S}\ \text{ohms per cm.} \qquad (43)$$

where:

$$S = \text{conductor section,}$$

the "resistance ratio," or ratio of the effective resistance of unequal current distribution, to the true ohmic resistance, is:

$$\left. \begin{aligned} \frac{r}{r_0} &= \frac{S\pi}{l_1} \sqrt{0.4\ \lambda\mu f} \\[2mm] &= \frac{S}{5030\ l_1} \sqrt{\lambda\mu f} \end{aligned} \right\} \qquad (44)$$

The internal reactance of unequal current distribution equals the resistance, in the range considered:

$$x_1 = r \qquad (45)$$

while at low frequency, where the current distribution is still uniform, the internal reactance is, per unit length of conductor:

$$\left. \begin{aligned} x_{10} &= \pi f \mu\ 10^{-9}\ \text{henry per cm.} \\ &= \pi f\ 10^{-9}\ \text{for air}\ (\mu = 7) \end{aligned} \right\} \qquad (46)$$

65. It is interesting to calculate the depth of penetration of alternating current, for different frequencies, in different materials, to indicate what thickness of conductor may be employed.

Such values may be given for 25 cycles and 60 cycles as the machine frequencies, and for 10,000 cycles and 1,000,000 cycles as the limits of frequency, between which most high frequency oscillations, lightning discharges, etc., are found, and also for 1,000,000,000 cycles as about the highest frequencies which can be produced. The depth of penetration of alternating current in centimeters is given below.

Material	μ	λ	Penetration in cm. at				
			25 Cycles.	60 Cycles.	10,000 Cycles.	10^6 Cycles.	10^9 Cycles.
Very soft iron...	2000	1.1×10^5	0.068	0.044	3.4×10^{-3}	0.34×10^{-3}	0.011×10^{-3}
Steel rail........	1000	10^5	0.101	0.065	5.0×10^{-3}	0.5×10^{-3}	0.016×10^{-3}
Cast iron........	200	10^4	0.71	0.46	0.0355	3.55×10^{-3}	0.113×10^{-3}
Copper	1	6.2×10^5	1.28	0.82	0.064	6.4×10^{-3}	0.203×10^{-3}
Aluminum	1	3.7×10^5	1.65	1.07	0.082	8.2×10^{-3}	0.263×10^{-3}
German silver ..	1	0.33×10^5	5.53	3.57	0.276	27.6×10^{-3}	0.88×10^{-3}
Graphite	1	900	33 5	21.7	1 67	0.167	5.3×10^{-3}
Silicon	1	80	112.5	72.7	5.63	0.563	17.9×10^{-3}
Salt solu., conc.	1	0.2	2.25×10^3	1.45×10^3	112	11.2	0.36
Pure river water	1	10^{-4}	100.6×10^3	65×10^3	5030	503	16

It is interesting to note from this table that even at low machine frequencies the depth of penetration in iron is so little as to give a considerable increase of effective resistance, except when using thin iron sheets, while at lightning frequencies the depth of penetration into iron is far less than the thickness of sheets which can be mechanically produced. With copper and aluminum at machine frequencies this screening effect becomes noticeable only with larger conductors, approaching one inch in thickness, but with lightning frequencies the effect is such as to require the use of copper ribbons as conductor, and the thickness of the ribbon is immaterial; that is, increasing its thickness beyond that required for mechanical strength does not decrease the resistance, but merely wastes material. In general, all metallic conductors, at lightning frequencies give such small penetration as to give more or less increase of effective resistance, and their use for lightning protection therefore is less desirable, since they offer a greater resistance for higher frequencies, while the reverse is desirable.

Only pure river water does not show an appreciable increase of resistance even at the highest obtainable frequencies, and electrolytic conductors, as salt solution, give no screening effect within the range of lightning frequencies, while cast silicon can even at one million cycles be used in a thickness up to one-half inch without increase of effective resistance.

The maximum diameter of conductor which can be used with alternating currents without giving a serious increase of the effective resistance by unequal current distribution is given below.

At 25 cycles:

Steel wire........................ 0.30   cm. or 0.12 inch
Copper........................... 2.6   cm. or 1 inch
Aluminum .... 3.3   cm. or 1.3 inches

At 60 cycles:

Steel wire........................ 0.20   cm. or 0.08 inch
Copper........................... 1.6   cm. or 0.63 inch
Aluminum. .... 2.1   cm. or 0.83 inch

At lightning frequencies, up to one million cycles:

Copper........................... 0.013   cm. or 0.005 inch
Aluminum. .... 0.016   cm. or 0.0065 inch
German silver.................... 0.055   cm. or 0.022 inch
Cast silicon..................... 1.1   cm. or 0.44 inch
Salt solution.................... 22   cm. or 8.7 inches
River water......................   All sizes.

APPENDIX

Transient Unequal Current Distribution.

66. The distribution of a continuous current in a large conductor is uniform, as the magnetic field of the current inside of the conductor has no effect on the current distribution, being constant. In the moment of starting, stopping, or in any way changing a direct current in a solid conductor, the corresponding change of its internal magnetic field produces an unequal current distribution, which, however, is transient.

As in this case the distribution of current is transient in time as well as in space, the problem properly belongs in Section IV.

but may be discussed here, due to its close relation to the permanent alternating-current distribution in a solid conductor.

Choosing the same denotation as in the preceding paragraphs, but denoting current and e.m.f. by small letters as instantaneous values, equations (1), (2), and (4) of paragraph 61 remain the same:

$$d \, \mathfrak{K} = 0.4 \, \pi i \, dl, \tag{1}$$

$$d \, \mathfrak{B} = 0.4 \, \pi \mu i \, dl, \tag{2}$$

$$e_0 = e + \frac{i}{\lambda}, \tag{4}$$

where e_0 = voltage impressed upon the conductor (exclusive of its external magnetic field) per unit length, e = voltage consumed by the change of internal magnetic field, i = current density in conductor element dl at distance l from center line of flat conductor, μ = the magnetic permeability of the conductor, and λ = the electric conductivity of the conductor.

Equation (3), however,

$$dE = - \, 2 \, j\pi fB \, 10^{-8} \, dl,$$

changes to

$$de = - \frac{d\mathfrak{B}}{dt} 10^{-8} \, dl \tag{3}$$

when introducing the instantaneous values; that is, the integral or effective value of the e.m.f. E consumed by the magnetic flux density $\mathfrak{B}$ is proportional and lags 90 time-degrees behind $\mathfrak{B}$, while the instantaneous value i is proportional to the rate of change of $\mathfrak{B}$, that is, to its differential quotient.

Differentiating (3) with respect to dl gives

$$\frac{d^2e}{dl^2} = - \frac{d}{dt}\left(\frac{d\mathfrak{B}}{dl}\right) 10^{-8}, \tag{5}$$

and substituting herein equation (2) gives

$$\frac{d^2e}{dl^2} = - \, 0.4 \, \pi \mu \frac{di}{dt} 10^{-8}. \tag{6}$$

Differentiating (4) twice with respect to dl gives

$$0 = \frac{d^2e}{dl^2} + \frac{1}{\lambda}\frac{d^2i}{dl^2}, \tag{7}$$

and substituting (7) into (6) gives

$$\frac{d^2i}{dl^2} = + 0.4\,\pi\mu\lambda\,10^{-8}\frac{di}{dt} \qquad (8)$$

as the differential equation of the current density i in the conductor.

Substituting

$$c^2 = 0.4\,\pi\mu\lambda\,10^{-8} \qquad (9)$$

gives

$$\frac{d^2i}{dl^2} = c^2\frac{di}{dt}. \qquad (10)$$

This equation (10) is integrated by

$$i = A + B\varepsilon^{-a^2t-bl}, \qquad (11)$$

and substituting (11) in (10) gives the relation

$$b^2 = -c^2a^2;$$

hence,

$$b = \pm\,jca, \qquad (12)$$

and substituting (12) in (11), and introducing the trigonometric expressions for the exponential functions with complex imaginary exponents,

$$i = A + \varepsilon^{-a^2t}\,(C_1\cos cal + C_2\sin cal), \qquad (13)$$

where

$$C_1 = B_1 + B_2 \text{ and } C_2 = j\,(B_1 - B_2).$$

Assuming the current distribution as symmetrical with the axis of the conductor, that is, i the same for $+l$ and for $-l$, gives

$$C_2 = 0;$$

hence,

$$i = A + C\varepsilon^{-a^2t}\cos cal \qquad (13)$$

as the equation of the current distribution in the conductor.

It is, however, for $t = \infty$, or for uniform current distribution,

$$i = e_0\lambda;$$

hence, substituting in (13),

$$A = e_0 \lambda$$

and

$$i = e_0 \lambda + C\varepsilon^{-a^2 t} \cos cal. \tag{14}$$

At the surface of the conductor, or for $l = l_0$, no induction by the internal magnetic field exists, but the current has from the beginning the final value corresponding to the impressed e.m.f. e_0, that is, for $l = l_0$,

$$i = e_0 \lambda,$$

and substituting this value in (14) gives

$$e_0 \lambda = e_0 \lambda + C\varepsilon^{-a^2 t} \cos cal_0;$$

hence,

$$\cos cal_0 = 0$$

and

$$cal_0 = \frac{(2\kappa - 1)\pi}{2}, \tag{15}$$

or

$$a = \frac{(2\kappa - 1)\pi}{2 cl_0}, \tag{16}$$

where κ is any integer.

There exists thus an infinite series of transient terms, exponential in the time, t, and trigonometric in the distance, l, one of fundamental frequency, and with it all the odd harmonics, and the equation of current density, from (14), thus is

$$\left. \begin{array}{l} i = e_0 \lambda + \displaystyle\sum_1^\infty {}_\kappa C_\kappa \varepsilon^{-(2\kappa-1)^2 a_1^2 t} \cos (2\kappa - 1) ca_1 l \\[2mm] = e_0 \lambda + \displaystyle\sum_1^\infty {}_\kappa C_\kappa \varepsilon^{-(2\kappa-1)^2 a_1^2 t} \cos \dfrac{(2\kappa - 1)\pi l}{2 l_0}, \end{array} \right\} \tag{17}$$

where

$$a_1 = \frac{\pi}{2 cl_0}. \tag{18}$$

The values of the integration constants C_κ are determined by the terminal conditions, that is, by the distribution of current density at the moment of start of the transient phenomenon, or $t = 0$.

For $\qquad t = 0$,

$$i_0 = e_0\lambda + \sum_1^\infty {}_\kappa C_\kappa \cos \frac{(2\kappa - 1)\pi l}{2 l_0}. \qquad (19)$$

Assuming that the current density i_0 was uniform throughout the conductor section before the change of the circuit conditions which led to the transient phenomena — as would be expected in a direct-current circuit, — from (19) we have

$$\sum_1^\infty {}_\kappa C_\kappa \cos \frac{(2\kappa - 1)\pi l}{2 l_0} = - (e_0\lambda - i_0) = \text{constant}, \qquad (20)$$

and the coefficients C_κ of this Fourier series are derived in the usual manner of such series, thus:

$$C_\kappa = 2 \operatorname{avg}\left[- (e_0\lambda - i_0) \cos \frac{(2\kappa - 1)\pi l}{2 l_0} \right]_{l=0}^{l=l_0}$$

$$= (-1)^\kappa \frac{4}{\pi} (e_0\lambda - i_0), \qquad (21)$$

where $\operatorname{avg} [F(x)]_{x=x_1}^{x=x_2}$ denotes the average value of the function $F(x)$ between the limits $x = x_1$ and $x = x_2$

and equation (17) then assumes the form

$$i = e_0\lambda + \frac{4}{\pi}(e_0\lambda - i_0) \sum_1^\infty {}_\kappa \frac{(-1)^\kappa}{2\kappa - 1} \varepsilon^{-(2\kappa - 1)^2 a_1^2 t} \cos \frac{(2\kappa - 1)\pi l}{2 l_0}. \qquad (22)$$

This then is the final equation of the distribution of the current density in the conductor.

If now $l_1 = $ width of the conductor, then the total current in the conductor, of thickness $2 l_0$, is

$$I = l_1 \int_{-l_0}^{+l_0} i\, dl$$

$$= 2 e_0\lambda l_0 l_1 - \frac{16 l_0 l_1}{\pi^2}(e_0\lambda - i_0) \sum_1^\infty {}_\kappa \frac{\varepsilon^{-(2\kappa - 1)^2 a_1^2 t}}{(2\kappa - 1)^2},$$

or

$$I = 2 l_0 l_1 e_0\lambda \left\{ 1 - \frac{8}{\pi^2}\left(1 - \frac{i_0}{e_0\lambda}\right) \sum_1^\infty {}_\kappa \frac{1}{(2\kappa - 1)^2} \varepsilon^{-(2\kappa - 1)^2 a_1^2 t} \right\}. \qquad (23)$$

For the starting of current, that is, if the current is zero, $\iota_0 = 0$, in the conductor before the transient phenomenon, this gives

$$I = 2\,l_0 l_1 e_0 \lambda \left\{ 1 - \frac{8}{\pi^2} \sum_1^\infty {}^\kappa \frac{1}{(2\,\kappa - 1)^2}\, \varepsilon^{-(2\kappa-1)^2 a_1^2 t} \right\}. \qquad (24)$$

While the true ohmic resistance, r_0, per unit length of the conductor is

$$r_0 = \frac{1}{2\,l_0 l_1 \lambda}, \qquad (25)$$

the apparent or effective resistance per unit length of the conductor during the transient phenomenon is

$$r = \frac{e_0}{I} = \frac{r_0}{1 - \dfrac{8}{\pi^2} \displaystyle\sum_1^\infty {}^\kappa \frac{1}{(2\kappa - 1)^2}\, \varepsilon^{-(2\kappa-1)^2 a_1^2 t}}, \qquad (26)$$

and in the first moment, for $t = 0$, is

$$r = \infty,$$

since the sum is

$$\sum_1^\infty {}^\kappa \frac{1}{(2\kappa - 1)^2} = \frac{\pi^2}{8}.$$

The effective resistance of the conductor thus decreases from ∞ at the first moment, with very great rapidity — due to the rapid convergence of the series — to its normal value.

67. As an example may be considered the apparent resistance of the rail return of a direct-current railway during the passage of a car over the track.

Assume the car moving in the direction away from the station, and the current returning through the rail, then the part of the rail behind the car carries the full current, that ahead of the car carries no current, and at the moment where the car wheel touches the rail the transient phenomenon starts in this part of the rail. The successive rail sections from the wheel contact backwards thus represent all the successive stages of the transient phenomenon from its start at the wheel contact to permanent conditions some distance back from the car.

Assume the rail section as equivalent to a conductor of 8 cm. width and 8 cm. height, or $l_1 = 8$, $l_0 = 4$, and the car speed as 40 miles per hour, or 1800 cm. per second.

Assume a steel rail and let the permeability $\mu = 1000$ and the electric conductivity $\lambda = 10^5$.

Then
$$c = \sqrt{0.4 \, \pi \mu \lambda \, 10^{-8}} = \sqrt{1.2566} = 1.121,$$

$$a_1 = \frac{\pi}{2 \, c l_0} = 0.35,$$

$$a_1^2 = 0.122.$$

Since $i_0 = 0$, the current distribution in the conductor, by (22), is

$$i = e_0 \lambda \left\{ 1 + \frac{4}{\pi} \sum_1^\infty{}_\kappa \frac{(-1)^\kappa}{2\,\kappa - 1} \, \varepsilon^{-0.122\,(2\kappa-1)^2 t} \cos 0.393 \, (2 \, \kappa - 1) \, l \right\}$$

$$= e_0 \lambda \left\{ 1 - 1.27 \left[\varepsilon^{-0.122 t} \cos 0.393 \, l - \tfrac{1}{3} \varepsilon^{-1.10 t} \cos 1.18 \, l + \tfrac{1}{5} \varepsilon^{-3.05 t} \right. \right.$$
$$\left. \left. \cos 1.96 \, l - + \; \ldots \; \right] \right\},$$

the ohmic resistance per unit length of rail is

$$r_0 = \frac{1}{2 \, l_0 l_1 \lambda} = 0.156 \times 10^{-6} \text{ ohms per cm.}$$

and the effective resistance per unit length of rail, by (26), is

$$r = \frac{0.156 \times 10^{-6}}{1 - 0.81 \left[\varepsilon^{-0.122 t} + \tfrac{1}{9} \varepsilon^{-1.10 t} + \tfrac{1}{25} \varepsilon^{-3.05 t} + \tfrac{1}{49} \varepsilon^{-6.0 t} + \; \ldots \; \right]}.$$

At a velocity of 1800 cm. per second, the distance from the wheel contact to any point p of the rail, l', is given as function of the time t elapsed since the starting of the transient phenomenon at point p by the passage of the car wheel over it, by the expression $l' = 1800 \, t$, and substituting this in the equation of the effective resistance r gives this resistance as function of the distance from the car, after passage,

$$r = \frac{0.156 \times 10^{-6}}{1 - 0.81 \left[\varepsilon^{-68 \times 10^{-6} l'} + \tfrac{1}{9} \varepsilon^{-612 \times 10^{-6} l'} + \tfrac{1}{25} \varepsilon^{-1700 \times 10^{-6} l'} + \; \ldots \; \right]}$$
ohms per cm.

As illustration is plotted in Fig. 96 the ratio of the effective resistance of the rail to the true ohmic resistance, $\dfrac{r}{r_0}$, and with the distance from the car wheel, in meters, as abscissas, from the equation

$$\frac{r}{r_0} = \frac{1}{1 - 0.81\left[\varepsilon^{-0.0068\,l'} + \frac{1}{9}\varepsilon^{-0.0612\,l'} + \frac{1}{25}\varepsilon^{-0.17\,l'} + \frac{1}{49}\varepsilon^{-0.34\,l'} + \ldots\right]}.$$

As seen from the curve, Fig. 96, the effective resistance of the rail appreciably exceeds the true resistance even at a considerable distance behind the car wheel. Integrating the excess of the effective resistance over the ohmic resistance shows that

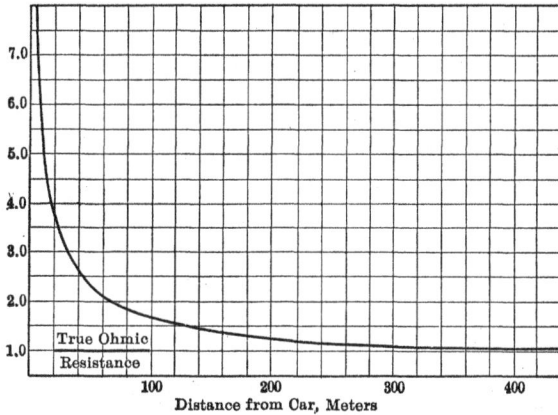

Fig. 96. Transient resistance of a direct-current railway rail return. Car speed 18 meters per second.

the excess of the effective or transient resistance over the ohmic resistance is equal to the resistance of a length of rail of about 300 meters, under the assumption made in this instance, and at a car speed of 40 miles per hour. This excess of the transient rail resistance is proportional to the car speed, thus less at lower speeds.

CHAPTER VIII.

VELOCITY OF PROPAGATION OF ELECTRIC FIELD.

68. In the theoretical investigation of electric circuits the velocity of propagation of the electric field through space is usually not considered, but the electric field assumed as instantaneous throughout space; that is, the electromagnetic component of the field is considered as in phase with the current, the electrostatic component as in phase with the voltage. In reality, however, the electric field starts at the conductor and propagates from there through space with a finite though very high velocity, the velocity of light; that is, at any point in space the electric field at any moment corresponds not to the condition of the electric energy flow at that moment but to that at a moment earlier by the time of propagation from the conductor to the point under consideration, or, in other words, the electric field lags the more, the greater the distance from the conductor.

Since the velocity of propagation is very high — about 3×10^{10} centimeters per second — the wave of an alternating or oscillating current even of high frequency is of considerable length; at 60 cycles the wave length is 0.5×10^9 centimeters, and even at a hundred thousand cycles the wave length is 3 kilometers, that is, very great compared with the distance to which electric fields usually extend.

The important part of the electric field of a conductor extends to the return conductor, which usually is only a few feet distant; beyond this, the field is the differential field of conductor and return conductor. Hence, the intensity of the electric field has usually already become inappreciable at a distance very small compared with the wave length, so that within the range in which an appreciable field exists this field is practically in phase with the flow of energy in the conductor, that is, the velocity of propagation has no appreciable effect, unless the return conductor is very far distant or entirely absent, or the frequency is so high, that the distance of the return conductor is an appreciable part of the wave length.

69. Consider, for instance, a circuit representing average transmission line conditions with 6 ft. = 182 cm., between conductors, traversed by a current of $f = 10^6$, or one million cycles, such as may be produced by a nearby lightning discharge. The wave length of this current and thus of its magnetic field then would be $\dfrac{S}{f} = \dfrac{3 \times 10^{10}}{10^6} = 30,000$ cm. and the distance of 182 cm. between the line conductors would be $\dfrac{182}{30,000}$ or $\dfrac{1}{165}$ of a wave length or $\dfrac{360}{165} = 2.2°$. That is, the magnetic field of the current, when it reaches the return conductor, would not be in phase with the current, but $\dfrac{1}{165}$ of a wave length or 2.2° behind the current. The voltage induced by the magnetic field would not be in quadrature with the current, or wattless, but lag $90 + 2.2 = 92.2°$ behind the current, thus have an energy component equal to $\cos 92.2° = 3.8$ per cent, giving rise to an effective resistance r_3, equal to 3.8 per cent of the reactance x. Even if at normal frequencies of 60 cycles the reactance is only equal to the ohmic resistance r_0—usually it is larger—the reactance x to 10^6 cycles would be $\dfrac{10^6}{60}, = 16,700$ times the ohmic resistance, and the effective resistance of magnetic radiation, r, being 3.8 per cent of this reactance, thus would be 630 times the ohmic resistance; $r_3 = 630 \, r_0$. Thus, the ohmic resistance would be entirely negligible compared with the effective resistance resulting from the finite velocity of the magnetic field.

Considering, however, a high frequency oscillation of 10^6 cycles, not between the line conductors, but between line conductor and ground, and assume, under average transmission line conditions, 30 ft. as the average height of the conductor above ground. The magnetic field of the conductor then can be represented as that between the conductor and its image conductor, 30 ft. below ground, and the distance between conductor and return conductor would be 2×30 ft. = 1820 cm. The lag of the magnetic field, due to the finite velocity of propagation, then becomes 22°, thus quite appreciable, and the energy component of the voltage induced by the magnetic field is $\cos (90 + 22°) = 37$ per cent.

This would give rise to an effective or radiation resistance $r_3 =$ 0.37 x. As in this case, the 60-cycle reactance usually is much larger than the ohmic resistance, assuming it as twice would make the radiation resistance $r_3 = 12,600\ r_0$, or more than ten thousand times the true ohmic resistance. It is true, that at these high frequencies, the ohmic resistance would be very greatly increased by unequal current distribution in the conductor. But the effective resistance of unequal current distribution increases only proportional to the square root of the frequency, and, assuming as instance conductor No. 00 B. & S. G., the effective resistance of unequal current distribution, r_1, would at 10^6 cycles be about 36 times the low frequency ohmic resistance r_0. Thus the effective resistance of magnetic radiation would still be $\dfrac{12,600}{36}$ = 350 times the effective resistance of unequal current distribution; $r_3 = 350\ r_1$, and the latter, therefore, be negligible compared with the former.

It is interesting to note, that the effective resistance of radiation, r_3, does not represent energy dissipation in the conductor by conversion into heat, but energy dissipation by radiation into space, and in distinction from the "radiation resistance," which dissipates energy into space, the effective resistance of unequal current distribution may be called a "thermal resistance," as it converts electric energy into heat.

Thus in this instance of a 10^6 cycle high frequency discharge between transmission line conductor and ground, the heating of the conductor would be increased 36 fold over that produced by a low frequency current of the same amperage, by the increase of resistance by unequal current distribution in the conductor; but the total energy dissipation by the conductor would be increased by magnetic radiation still 350 times more, so that the total energy dissipation by the 10^6 cycle current would be 12,600 times greater than it would be with a low frequency current of the same value, and the attenuation or rate of decay of the current thus would be increased many thousand times, over that calculated on the assumption of constant resistance at all frequencies.

70. The finite velocity of the electric field thus requires consideration, and may even become the dominating factor in the electrical phenomena:

(*a*) In the conduction of very high frequency currents, of hundred thousands of cycles.

(*b*) In the action, propagation and dissipation of high frequency disturbances in electric circuits.

(*c*) In flattening steep wave fronts and rounding the wave shape of complex waves and sudden impulses.

(*d*) In circuits having no return circuit or no well-defined return circuit, such as the lightning stroke, the discharge path and ground circuit of the lightning arrester, the wireless antenna, etc.

(*e*) Where the electric field at considerable distance from the conductor is of importance, as in radio-telegraphy and telephony.

As illustrations of the effect of the finite velocity of the electric field may be considered in the following:

(*A*) The inductance of a finite length of an infinitely long conductor without return conductor, self-inductance as well as mutual inductance.

Such circuit is more or less approximately represented by the lightning rod, by the ground circuit of the lightning arrester, a section of wireless antenna, etc.

(*B*) The inductance of a finite length of an infinitely long conductor with return conductor, self-inductance, as well as mutual inductance.

Such for instance is the circuit of a transmission line, or the circuit between transmission line and ground.

(*C*) The capacity of a finite section of an infinitely long conductor, without return conductor as well as with return conductor.

(*D*) The mutual inductance between two finite conductors at considerable distance from each other.

Sending and receiving antennæ of radio-communication represent such pair of conductors.

(*E*) The capacity of a sphere in space.

A. INDUCTANCE OF A LENGTH l_0 OF AN INFINITELY LONG CONDUCTOR WITHOUT RETURN CONDUCTOR.

71. Such for instance is represented by a section of a lightning stroke.

The inductance of the length l_0 of a straight conductor is usually given by the equation

$$L = 2l_0 \log \frac{l'}{l_r} \times 10^{-9}, \tag{1}$$

where l' = the distance of return conductor , l_r = the radius of the conductor, and the total length of the conductor is assumed as infinitely great compared with l_0 and l'. This is approximately the case with the conductors of a long distance transmission line.

For infinite distance l' of the return conductor, that is, a conductor without return conductor, equation (1) gives $L = \infty$; that is, a finite length of an infinitely long conductor without return conductor would have an infinite inductance L and inversely, zero capacity C.

In equation (1) the magnetic field is assumed as instantaneous, that is, the velocity of propagation of the magnetic field is neglected. Considering, however, the finite velocity of the magnetic field, the magnetic field at a distance l from the conductor and at time t corresponds to the current in the conductor at the time $t - t'$, where t' is the time required for the electric field to travel the distance l, that is, $t' = \dfrac{l}{S}$, where $S = 3 \times 10^{10}$ = the speed of light; or, the magnetic field at distance l and time t corresponds to the current in the conductor at the time $t - \dfrac{l}{S}$.

71. Representing the time t by angle $\theta = 2\pi ft$, where $f =$ the frequency of the alternating current in the conductor, and denoting

$$a = \frac{2\pi f}{S} = \frac{2\pi}{l_w}, \tag{2}$$

where

$$l_w = \frac{S}{f} = \text{the wave length of electric field,}$$

the field at distance l and time angle θ corresponds to time angle $\theta - al$, that is, lags in time behind the current in the conductor by the phase angle al.

Let

$$i = I \cos \theta = \text{current, absolute units.} \qquad (3)$$

The magnetic induction at distance l then is

$$\mathfrak{B} = \frac{2\,I}{l} \cos (\theta - al); \qquad (4)$$

hence, the total magnetic flux surrounding the conductor of length l_0, from distance l to infinity is

$$\Phi = l_0 \int_l^\infty \frac{2\,I}{l} \cos (\theta - al) \, dl$$

$$= 2\,Il_0 \left\{ \cos \theta \int_l^\infty \frac{\cos al}{l} \, dl + \sin \theta \int_l^\infty \frac{\sin al}{l} \, dl \right\}. \qquad (5)$$

$\int \dfrac{\cos al}{l} \, dl$ cannot be integrated in finite form, but represents a new function which in its properties is intermediate between the sine function

$$\int \cos al \, dl = \frac{1}{a} \sin al$$

and the logarithmic function

$$\int \frac{1}{l} dl = \log l$$

and thus may be represented by a new symbol,

$$\text{sine} - \text{logarithm} = \text{sil.}$$

In the same manner, $\int \dfrac{\sin al}{l} \, dl$ is related to $\dfrac{1}{a} \cos al$ and to $\log l$.

Introducing therefore for these two new functions the symbols

$$\text{sil } al = \int_l^\infty \frac{\cos al}{l} \, dl, \qquad (6)$$

$$\text{col } al = \int_l^\infty \frac{\sin al}{l} \, dl, \qquad (7)$$

gives

$$\Phi = 2\,Il_0\,\{\cos\theta\;\text{sil}\;al + \sin\theta\;\text{col}\;al\}. \tag{8}$$

The e.m.f. consumed by this magnetic flux, or e.m.f. of inductance, then is

$$e = \frac{d\Phi}{dt} = 2\,\pi f\,\frac{d\Phi}{d\theta}\,;$$

hence,

$$e = 4\,\pi f Il_0\,\{\cos\theta\;\text{col}\;al - \sin\theta\;\text{sil}\;al\}\,; \tag{9}$$

and since the current is

$$i = I\cos\theta,$$

the e.m.f. consumed by the magnetic field beyond distance l, or e.m.f. of inductance, contains a component in phase with the current, or *power component*,

$$e_1 = 4\,\pi f Il_0\;\text{col}\;al\,\cos\theta, \tag{10}$$

and a component in quadrature with the current, or *reactive component*,

$$e_2 = -\,4\,\pi f Il_0\;\text{sil}\;al\,\sin\theta, \tag{11}$$

which latter *leads* the current by a quarter period.

The reactive component e_2 is a true *self-induction*, that is, represents a surging of energy between the conductor and its electric field, but no power consumption. The effective component e_1, however, represents a power consumption

$$p = e_1 i$$
$$= 4\,\pi f I^2 l_0\;\text{col}\;al\,\cos^2\theta \tag{12}$$

by the magnetic field of the conductor, due to its finite velocity; that is, it represents the *power radiated into space by the conductor*.

The energy component e_1 gives rise to an *effective resistance*,

$$r = \frac{e_1}{i} = 4\,\pi f l_0\;\text{col}\;al, \tag{13}$$

and the reactive component gives rise to a *reactance*,

$$x = \left[\frac{e_2}{i}\right] = 4\,\pi f l_0\;\text{sil}\;al, \tag{14}$$

When considering the finite velocity of propagation of the electric field, self-inductance thus is not wattless, but contains an energy component, and so can be represented by an *impedance*,

$$Z = r + jx$$
$$= 4 \pi f l_0 (\text{col } al + j \text{ sil } al) 10^{-9} \text{ ohms.} \tag{15}$$

The *inductance* would be given by

$$L = - \frac{jZ}{2 \pi f}$$
$$= 2 l_0 \{\text{sil } al - j \text{ col } al\} 10^{-9} \text{ henrys,} \tag{16}$$

and the power radiated by the conductor is

$$p = i^2 r.$$

72. The functions

$$\text{sil } al = \int_l^\infty \frac{\cos al}{l} dl \tag{6}$$

and

$$\text{col } al = \int_l^\infty \frac{\sin al}{l} dl \tag{7}$$

can in general not be expressed in finite form, and so have to be recorded in tables.* Close approximations can, however, be derived for the two cases where l is very small and where l is very large compared with the wave length l_w of the electric field, and these two cases are of special interest, since the former represents the total magnetic field of the conductor, that is, its self-inductance for: $l = l_r$, and the latter the magnetic field interlinked with a distant conductor, such as the mutual inductance between sending and receiving conductor of radio telegraphy, or the induction of a lightning stroke on a transmission line.

It is

$$\left. \begin{array}{l} \text{sil } 0 = \infty, \\ \text{col } 0 = \dfrac{\pi}{2}, \\ \text{sil } \infty = 0, \\ \text{col } \infty = 0. \end{array} \right\} \tag{17}$$

* Tables of these and related functions are given in the appendix.

And it can be shown that for small values of al, that is, when l is only a small fraction of a wave length, the approximations hold:

$$\left.\begin{aligned}
\text{sil } al &= \log \frac{1}{al} - 0.5772 \\
&= \log \frac{0.5615}{al} \\
\text{col } al &= \frac{\pi}{2} - al
\end{aligned}\right\} \qquad (18)$$

while for large values of al, the approximations hold:

$$\left.\begin{aligned}
\text{sil } al &= -\frac{\sin al}{al} \\
\text{col } al &= \frac{\cos al}{al}
\end{aligned}\right\} \qquad (19)$$

As seen, for larger values of al, col al has the same sign as cos al; sil al has the opposite sign of sin al.

73. As Z in (15) is the impedance and L in (16) the inductance resulting from the magnetic field of the conductor l_0, beyond the distance l, Z in (15) represents the *mutually inductive impedance* and L in (16) the *mutual inductance* of a conductor l_0 without return conductor, upon a second conductor at distance l.

In this case, if l is large compared with the wave length, we get, by substituting (19) into (15) and (16):
the *mutually inductive impedance*:

$$\begin{aligned}
Z &= \frac{4\,\pi f l_0}{al}\left\{ \cos al + j \sin al \right\} 10^{-9} \\
&= \frac{2\,S l_0}{l}\left\{ \cos al - j \sin al \right\} 10^{-9} \text{ ohms}
\end{aligned}$$

or, absolute

$$\left.\begin{aligned}
Z &= \frac{2\,S l_0}{l} 10^{-9} \text{ ohms} \\
&= \frac{60\,l_0}{l} \text{ ohms}
\end{aligned}\right\} \qquad (20)$$

and, the *mutual inductance*:

$$\begin{aligned}
M &= \frac{2\,l_0}{al}\left\{ \sin al - j \cos al \right\} 10^{-9} \\
&= \frac{S l_0}{\pi f l}\left\{ \sin al - j \cos al \right\} 10^{-9} \text{ henrys}
\end{aligned}$$

or, absolute:

$$m = \frac{Sl_0}{\pi fl} 10^{-9} \text{ henrys} \left.\vphantom{\begin{array}{c} \\ \\ \\ \end{array}}\right\}$$
$$= \frac{30\, l_0}{\pi fl} \text{ henrys} \qquad (21)$$

Making, in (15) and (16):
$$l = l_r = \text{radius of the conductor,}$$
L becomes the self-inductance, and Z the self-inductive imped-ance of the conductor.

Since al_r always is a small quantity, equations (18) apply, and it is, substituting (2)

$$\text{col } al_r = \frac{\pi}{2} \left.\vphantom{\begin{array}{c} \\ \\ \\ \\ \\ \end{array}}\right\}$$
$$\text{sil } al_r = \log \frac{0.56}{al_r}$$
$$= \log \frac{0.56\, S}{2\, \pi fl_r} \qquad (21)$$
$$= \log \frac{0.56\, S}{fl_c}$$

where l_c is the circumference of the conductor.

Substituting (21) into (15), gives the *self-inductive radiation impedance* of the conductor without return conductor:

$$Z = r + jx$$
$$= 4\, \pi fl_0 \left\{ \frac{\pi}{2} + j \log \frac{0.56\, S}{fl_c} \right\} 10^{-9} \text{ ohms} \qquad (22)$$

comprising the *effective radiation resistance*:

$$r = 2\, \pi^2 fl_0\, 10^{-9} \text{ ohms} \qquad (23)$$

and the *effective radiation reactance*:

$$x = 4\, \pi fl_0 \log \frac{0.56\, S}{fl_c}\, 10^{-9} \text{ ohms} \qquad (24)$$

corresponding to the true *self-inductance* of the conductor without return conductor:

$$L = \frac{x}{2\, \pi f}$$
$$= 2\, l_0 \log \frac{0.56\, S}{fl_c}\, 10^{-9} \text{ henrys} \qquad (25)$$

As seen, the effective radiation resistance of a conductor without return conductor is proportional to the frequency f; the inductance L decreases with increasing frequency, but logarithmically, that is, much slower than the frequency, and the reactance x thus increases somewhat slower than the frequency.

Thus per meter of conductor, or $l_0 = 100$ cm., with a conductor of No. 00 B. & S. G., of $l_c = 2.93$ cm., it is, at

$$f = \quad 100{,}000 \text{ cycles} \qquad f = 1{,}000{,}000 \text{ cycles}$$
$$r = 0.195 \text{ ohms} \qquad r = 1.95 \text{ ohms}$$
$$L = 2.19 \times 10^{-6} \text{ henry} \qquad L = 1.73 \times 10^{-6} \text{ henry}$$
$$x = 1.38 \text{ ohms} \qquad x = 10.9 \text{ ohms}$$

For comparison, the true ohmic resistance of the conductor is $r_0 = 0.00026$ ohms, thus thousands of times less than the radiation resistance at these frequencies.

The power radiated by the conductor then is:

$$p = i^2 r$$
$$= 2\,\pi^2 f l_0 i^2\,10^{-9} \text{ watts} \tag{26}$$

hence, at 100 amperes, per meter length of conductor, at

$$f = \quad 100{,}000 \text{ cycles} : p = 1.95 \text{ kilowatts}$$
$$f = 1{,}000{,}000 \text{ cycles} : p = 19.5 \text{ kilowatts}$$

B. INDUCTANCE OF A LENGTH l_0 OF AN INFINITELY LONG CONDUCTOR, WITH RETURN CONDUCTOR AT DISTANCE l'.

74. Such for instance is represented by a section of a transmission line, etc. Let again

$$i = I \cos \theta = \text{current, absolute units.} \tag{27}$$

The magnetic induction, at distance l then is

$$\mathcal{B} = \frac{2I}{l} \cos (\theta - al) \tag{28}$$

hence, the total magnetic flux surrounding the conductor of length l_0, from its surface at distance l_r (the radius of the conductor) up to the return conductor at distance l', is

$$\Phi = l_0 \int_{l_r}^{l'} \frac{2\,I}{l} \cos (\theta - al)\, dl$$
$$= 2\,I l_0 \left\{ \cos \theta \int_{l_r}^{l'} \frac{\cos al}{l}\, dl + \sin \theta \int_{l_r}^{l'} \frac{\sin al}{l}\, dl \right\} \tag{29}$$

It is, however,

$$\left. \begin{array}{l} \int_{l_r}^{l'} \frac{\cos al}{l}\, dl = \int_{l_r}^{\infty} \frac{\cos al}{l}\, dl - \int_{l'}^{\infty} \frac{\cos al}{l}\, dl = \text{sil } al_r - \text{sil } al' \\[2mm] \int_{l_r}^{l'} \frac{\sin al}{l}\, dl = \text{col } al_r - \text{col } al' \end{array} \right\} \quad (30)$$

Hence:

$$\Phi = 2\, Il_0\{\cos\theta(\text{sil } al_r - \text{sil } al') + \sin\theta(\text{col } al_r - \text{col } al')\} \quad (31)$$

The voltage induced by this magnetic flux in the conductor is

$$e = \frac{d\Phi}{dt}\, 10^{-9} = 2\,\pi f\, \frac{d\Phi}{d\theta}\, 10^{-9} \text{ volts}$$

$$= 4\,\pi f Il_0\{\cos\theta\,(\text{col } al_r - \text{col } al') - \sin\theta\,(\sin al_r - \sin al')\}\, 10^{-9} \quad (32)$$

and the effective *self-inductive radiation impedance* of the conductor is

$$\frac{e}{I} = 4\,\pi f l_0\{\cos\theta\,(\text{col } al_r - \text{col } al') - \sin\theta\,(\text{sil } al_r - \text{sil } al')\}\, 10^{-9} \text{ ohms}$$

or, in symbolic expression,

$$Z = r + jx$$

$$= 4\,\pi f l_0\{(\text{col } al_r - \text{col } al') + j(\text{sil } al_r - \text{sil } al')\}\, 10^{-9} \text{ ohms} \quad (33)$$

and, the *effective radiation resistance:*

$$r = 4\,\pi f l_0\,(\text{col } al_r - \text{col } al')\, 10^{-9} \text{ ohms} \quad (34)$$

effective radiation self-inductance:

$$L = 2\, l_0\,(\text{sil } al_r - \text{sil } al') \text{ henrys} \quad (35)$$

Since even at a billion cycles, the wave length $l_w = 30$ cm. is large compared with the conductor radius l_r of a transmission line, al_r is very small, and we can therefore substitute (18), and get:

$$r = 4\,\pi f l_0 \left\{ \frac{\pi}{2} - \text{col } al' \right\} 10^{-9} \text{ ohms} \quad (36)$$

$$L = 2\, l_0 \left\{ \log \frac{0.56}{al_r} - \text{sil } al' \right\} 10^{-9} \text{ henrys} \quad (37)$$

75. Under transmission line conditions, for all except the highest frequencies al' is so small as to permit the approximations (18). This gives

$$r = 4\ \pi f l_0 \alpha l'\ 10^{-9} \text{ ohms} \tag{38}$$

$$L = 2\ l_0 \left\{ \log \frac{1}{al_r} - \log \frac{1}{al'} \right\} 10^{-9} \text{ henrys}$$

$$= 2\ l_0 \log \frac{l'}{l}\ 10^{-9} \text{ henrys} \tag{39}$$

Equation (39) is the usual inductance formula, derived without considering the finite velocity of the electric field.

Substituting (2) into (38) gives

$$r = \frac{8\ \pi^2 f^2 l' l_0}{S} 10^{-9} \text{ ohms}$$

$$= 2.63\ f^2 l' l_0\ 10^{-18} \text{ ohms} \tag{40}$$

The effective radiation resistance r thus is proportional to the square of the frequency, and proportional to the distance from the return conductor, for all frequencies of a wave length large compared with the distance from the return conductor. For such frequencies, the finite velocity of the field has no appreciable effect yet on the inductance of the conductor.

At $al = 0.1$ the error of approximations (18) is still less than 0.01 per cent.

and

$$al' < 0.1$$

gives, by (2):

$$f < \frac{S}{2\pi l'}$$

hence:

$$f < \frac{4770 \times 10^6}{l'}$$

At $l' = 6$ ft. $= 182$ cm., or 6 ft. distance between transmission line conductors, equations (39) and (40) thus are applicable up to frequencies of

$$f = \frac{4770}{l'} = 26.2 \text{ million cycles,}$$

thus for all frequencies which come into engineering consideration.

In a high-frequency oscillation between line and ground, as

due for instance to lightning, assuming the average height of the line as 30 ft., it is: $l' = 2 \times 30$ ft. $= 1820$ cm., hence,

$$f = 2.62 \text{ million cycles.}$$

However, even far beyond these frequencies, equations (39) and (40) are still approximately applicable.

76. As an instance may be calculated the effective radiation resistance per kilometer of long distance transmission conductor of No. 00 B. & S. G.,

   (a) Against the return conductor, at 6 ft. $= 182$ cm. distance;

   (b) Against the ground, at elevation of 30 ft. $= 910$ cm.

The true ohmic resistance is

$$r_0 = 0.257$$

The radiation resistance is, by (40),

$$r = 2.63 \ f^2 l' \ 10^{-18} \text{ ohms per cm.}$$
$$= 0.263 \ f^2 l' \ 10^{-12} \text{ ohms per km.}$$

hence for

$f =$	25	60	10^3
(a) $l' = 182;\ r =$ 3.0×10^{-6}		17.3×10^{-6}	48×10^{-6}
(b) $l' = 1820;\ r =$ 30×10^{-6}		173×10^{-6}	480×10^{-6}

10^4	10^5	10^6	10^7	10^8	cycles
0.0048	0.48	48	4800	225,000	ohms*
0.048	4.8	480	22,500*	194,000*	ohms.

By (40), the radiation resistance r is proportional to the square of the frequency, that is, $\dfrac{r_0}{f^2}$ is constant, up to those frequencies, where $al' = \dfrac{2\pi f l'}{S}$ becomes appreciable compared with the quarter wave length $\dfrac{\pi}{2}$, that is, as long as f is well below the value

$$f_0 = \frac{S}{4l'}$$

or, as long as the distance of the return conductor, l', is well below

$$l' = \frac{S}{4f}$$

that is, a quarter wave length.

   * In these values, the more complete equation (36) had to be used.

If the return conductor is at a distance equal to a quarter wave length,

$$l' = \frac{S}{4f}$$

then in equation (36), col $al' = 0$, and this equation becomes

$$r = 2\,\pi^2 f l_0\, 10^{-9}$$
$$= 19.7\,f l_0\, 10^{-9}\ \text{ohms}$$

which is the equation of the radiation resistance of a conductor without return conductor (23).   That is,

The radiation resistance of a conductor with return conductor at quarter wave length distance, is the same as that of a conductor without return conductor.

The radiation resistance of a conductor is a maximum with the return conductor at such distance l', which makes col al' in (36) a negative maximum.   This is for $al' = \pi$, hence,

$$l' = \frac{S}{2\pi}$$
$$r = 23.2\,f l_0\, 10^{-9}\ \text{ohms}$$

That is,

The radiation resistance of a conductor is a maximum with the return conductor at half a wave length distance.

C. CAPACITY OF A LENGTH l_0 OF AN INFINITELY LONG CONDUCTOR.

77. Such for instance is represented by a section of a transmission line, etc.

Let

$$e = E \cos \theta$$
$$= \text{voltage impressed upon conductor of radius } l_r. \quad (41)$$

Then

$$G = \frac{de}{dl}$$
$$= \text{dielectric gradient at distance } l \text{ from conductor} \quad (42)$$

$$K = \frac{G}{4\pi S^2}$$
$$= \frac{1}{4\pi S^2}\frac{de}{dl} \quad\quad\quad (43)$$

$$= \text{dielectric field intensity, and}$$

$$D = kK$$

$$= \frac{k}{4\pi S^2} \frac{de}{dl} \tag{44}$$

$$= \text{dielectric density}$$

$$\Psi = 2\,\pi l l_0 D$$

$$= \frac{k l l_0}{2 S^2} \frac{de}{dl} \tag{45}$$

$$= \text{dielectric flux, where}$$

$$2\,\pi l l_0 = \text{section traversed by flux.}$$

Thus,

$$\frac{de}{dl} = \frac{2 S^2}{k l l_0} \Psi \tag{46}$$

$$= \text{dielectric gradient, and}$$

$$e = \int_{l_r}^{l'} \frac{de}{dl}\, dl$$

$$= \frac{2 S^2}{k l_0} \int_{l_r}^{l'} \frac{\Psi}{l}\, dl \tag{47}$$

$$= \text{voltage impressed upon conductor, where:}$$

$$l' = \text{distance of return conductor.}$$

It is, however, by definition (see "Electric Discharges, Waves and Impulses," Chapter X),

$$\Psi = Ce \tag{48}$$

where

$$C = \text{capacity}$$

The dielectric flux Ψ, at distance l, lags behind the voltage impressed upon the conductor (41), by the phase angle al, thus is:

$$\Psi = CE \cos (\theta - al) \tag{49}$$

Substituting (41) and (49) into (47), and cancelling by E, give

$$\cos \theta = \frac{2 S^2 C}{k l_0} \int_{l_r}^{l'} \frac{\cos (\theta - al)}{l}\, dl$$

$$= \frac{2 S^2 C}{k l_0} \Big\{ \cos \theta \operatorname{sil} al + \sin \theta \operatorname{col} al \Big\}_{l_r}^{l'} \tag{50}$$

thus,

$$\frac{1}{C} = \frac{2 S^2}{k l_0} \Big\{ (\operatorname{sil} al_r - \operatorname{sil} al') + \tan \theta\, (\operatorname{col} al_r - \operatorname{col} al') \Big\} 10^{-9} \tag{51}$$

where the $\tan \theta$ represents the quadrature component of capacity, or effective dielectric radiation resistance, and the 10^{-9} reduces from absolute units to farads.

It then is, in complex expression:

$$\frac{1}{C} = \frac{2S^2}{kl_0}\left\{(\text{sil } al_r - \text{sil } al') + j(\text{col } al_r - \text{col } al')\right\}10^{-9} \quad (52)$$

78. The *dielectric radiation impedance* then is

$$Z = \frac{-j}{2\pi fC} = \frac{S^2}{\pi fkl_0}\left\{(\text{col } al_r - \text{col } al') - j(\text{sil } al_r - \text{sil } al')\right\}$$
$$10^{-9} \text{ ohms} \quad (53)$$

Since by (33) the *magnetic radiation impedance* is

$$Z' = 2\pi fLj = 4\pi fl_0\left\{(\text{col } al_r - \text{col } al') + j(\text{sil } al_r - \text{sil } al')\right\}$$
$$10^{-9} \text{ ohms} \quad (33)$$

it is, for the absolute values,

$$\frac{z'}{z} = 4\pi^2 f^2 LC = \frac{4\pi^2 f^2 kl_0^2}{S^2} \quad (54)$$

or, in air, for $k = 1$,

$$LC = \frac{l_0^2}{S^2} \quad (55)$$

and, per unit length, or $l_0 = 1$,

$$LC = \frac{1}{S^2} \quad (56)$$

where L and C are the radiation inductance and radiation capacity, including the wattless or reactive components, the true inductance and capacity, as well as the energy components, the effective magnetic and dielectric radiation resistances.

Equations (55) and (56) are the same equations which apply to the values of L and C calculated without considering the finite velocity of the field. Thus the capacity can be calculated from the inductance, and inversely, even in the general case.

It is, by (18),

$$\left.\begin{array}{l} \text{sil } al_r = \log\dfrac{0.56}{al_r} \\[3mm] \text{col } al_r = \dfrac{\pi}{2} \end{array}\right\} \quad (57)$$

since al_r is very small.

Substituting (57) into (53) gives, as the *dielectric radiation impedance*,

$$Z = \frac{S^2}{\pi f k l_0}\left\{\left(\frac{\pi}{2} - \text{col } al'\right) - j\left(\log\frac{0.56}{al_r} - \text{sil } al'\right)\right\} 10^{-9} \text{ ohms} \quad (58)$$

79. If l' is small compared with a quarter wave length, it is, by (18),

$$\left.\begin{array}{c} \text{sil } al' = \log\dfrac{0.56}{al'} \\[2mm] \text{col } al' = \dfrac{\pi}{2} - al' \end{array}\right\} \quad (59)$$

hence

$$\frac{1}{C} = \frac{2S^2}{kl_0}\left\{\log\frac{l'}{l_r} + jal'\right\}10^{-9}$$

and, substituting (2),

$$\frac{1}{C} = \frac{2S^2}{kl_0}\left\{\log\frac{l'}{l_r} + j\frac{2\pi f l'}{S}\right\}10^{-9} \quad (60)$$

$$= \left\{\frac{2S^2}{kl_0}\log\frac{l'}{l_r} + \frac{4\pi f l' S}{kl_0}\right\}10^{-9} \quad (61)$$

If

$$l' = \infty,$$

that is, *in a conductor without return conductor*, such as the vertical discharge path of a lightning arrester, it is, substituting (57) and (2) into (52):

$$\frac{1}{C} = \frac{2S^2}{kl_0}\left\{\log\frac{0.56}{2\pi f l_r} + j\frac{\pi}{2}\right\}10^{-9} \quad (62)$$

The capacity reactance, or *dielectric radiation impedance of the conductor without return conductor*, then is,

$$Z = \frac{-j}{2\pi f C}$$

$$Z = \frac{S^2}{\pi f k l_0}\left\{\frac{\pi}{2} - j\log\frac{0.56}{2\pi f l_r}\right\}10^{-9} \quad (63)$$

while the *dielectric radiation impedance of the conductor with return conductor at distance* l', is, by (60):

$$Z = \frac{S^2}{\pi f k l_0}\left\{\frac{2\pi f l'}{S} - j\log\frac{l'}{l_r}\right\}10^{-9}$$

$$= \left\{\frac{2Sl'}{kl_r} - j\frac{S^2}{\pi f k l_0}\log\frac{l'}{l_r}\right\}10^{-9} \quad (64)$$

Thus comprising an *effective dielectric radiation resistance*:

Conductor with return conductor,

$$r = \frac{2Sl'}{kl_0} 10^{-9} \text{ ohms} \qquad (65)$$

Conductor without return conductor,

$$r = \frac{S^2}{2fkl_0} 10^{-9} \text{ ohms} \qquad (66)$$

and an *effective dielectric radiation (capacity-) reactance*:
Conductor with return conductor,

$$x = \frac{S^2}{\pi fkl_0} \log \frac{l'}{l_r} 10^{-9} \text{ ohms} \qquad (67)$$

Conductor without return conductor,

$$x = \frac{S^2}{\pi fkl_0} \log \frac{0.56\,S}{2\,\pi fl_r} 10^{-9} \text{ ohms} \qquad (68)$$

An effective capacity:

$$C = \frac{1}{2\pi f x} \qquad (69)$$

Conductor with return conductor:

$$C = \frac{kl_0\,10^9}{2S^2 \log \dfrac{l'}{l_r}} \text{ farads} \qquad (70)$$

Conductor without return conductor:

$$C = \frac{kl_0\,10^9}{2S^2 \log \dfrac{0.56S}{2\pi fl_r}} \text{ farads} \qquad (71)$$

It is interesting to note that the values (70) and (67) of the capacity and the capacity reactance of a conductor with return conductor at distance l', is the same as derived without considering the velocity of propagation of the electric field. That is, the finite velocity of the electric field does not change the equation of the capacity (nor that of the inductance), as long as the return conductor is well within a quarter-wave distance.

As value of the *attenuation constant of dielectric radiation* then follows:

$$u_2 = \frac{g}{C} = \frac{2\pi fg}{b} = \frac{2\pi fr}{x}$$

and by (65) to (68) conductor with return conductor:

$$u_2 = \frac{4\pi^2 f^2 l' k}{S \log \dfrac{l'}{l_r}} \tag{72}$$

Conductor without return conductor:

$$u_2 = \frac{\pi^2 f}{\log \dfrac{0.56\ S}{2\ \pi f l_r}} \tag{73}$$

While the *attenuation constant of magnetic radiation* is, from (38) and (39), conductor with return conductor:

$$u_1 = \frac{r}{L} = \frac{4\ \pi^2 f^2 l'}{S \log \dfrac{l'}{l_r}} \tag{74}$$

and, from (23) and (25), conductor without return conductor:

$$u_1 = \frac{r}{L} = \frac{\pi^2 f}{\log \dfrac{0.56\ S}{2\ \pi f l_r}} \tag{75}$$

that is, the same values as the attenuation constant of dielectric radiation, as was to be expected.

It is interesting to note, that the effective dielectric capacity resistance, of the conductor with return conductor (65) is again proportional to the distance of the return conductor, l', like the effective resistance of magnetic radiation. As the dielectric capacity reactance (68) is inverse proportional to the frequency, the capacity current is (approximately) proportional to the frequency, and the power consumption by dielectric radiation, $i^2 r$, thus (approximately) proportional to the square of the frequency, just like the power consumption by magnetic radiation, in B.

Numerical values are given in the next chapter, and in section IV.

It is interesting to note, that the expressions of inductance L and capacity C, (39) and (70), are the same as the values of L and C, calculated without considering the finite velocity of the electric field, that is, are the "low frequency values" of external inductance and capacity. Thus,

As long as the distance of the return conductor is small compared with a quarter wave length, electric radiation due to the finite velocity of the electric field, does not affect or change the values of external inductance L and of capacity C, but causes energy dissipation by electromagnetic and by dielectric or electrostatic radiation, which is represented by effective resistances. The electromagnetic radiation resistance is represented by a series resistance, proportional to the square of the frequency; the electrostatic resistance by a shunted resistance in series with the capacity. It is independent of the frequency and the power consumed by either resistance thus is proportional to the square of the frequency.

D. MUTUAL INDUCTANCE OF TWO CONDUCTORS OF FINITE LENGTH AT CONSIDERABLE DISTANCE FROM EACH OTHER.

80. Such for instance are sending and receiving antennæ of radio-communication.   Or lightning stroke and transmission line.

The electric field of an infinitely long conductor without return conductor decreases inversely proportional to the distance l, and therefore is represented by

$$\psi = \frac{\Psi}{l} \tag{76}$$

where Φ is the intensity of the field at unit distance from the conductor.

The electric field of a conductor of finite length l_0 decreases inversely proportional to the distance l and also proportionally to the angle φ subtended by the conductor l_0 from the distance l,

$$\psi = \frac{\varphi}{\pi} \frac{\Psi}{l}. \tag{77}$$

Since this angle φ, for great distances l, is given by

$$\varphi = \frac{l_0}{l} \tag{78}$$

the electric field of a conductor without return conductor, of finite length l_0, at great distances l, is represented by

$$\psi = \frac{l_0 \Psi}{\pi l^2}. \tag{79}$$

Since the electric field of the return conductor is opposite to that of the conductor, it follows that the electric field of an infinitely long conductor, with the return conductor at distance l_1, is, by (76),

$$\psi = \frac{\Psi}{l - l'} - \frac{\Psi}{l + l'} \tag{80}$$

where $l' = l_1 \cos \tau$ is the projection of the distance l_1 between the conductors upon the direction l, that is, l' is the difference in the distance of the two conductors from the point l.

For large distances l, equation (80), becomes

$$\psi = \frac{l'\Psi}{l^2}. \tag{81}$$

In the same manner, from equation (79) it follows that the decrease of the electric field of the conductor of finite length l_0 with its return conductor at the distance l_1, that is, of a rectangular circuit of the dimensions of l_0 and l_1, is:

$$\psi = \frac{l_0\Psi}{\pi\left(l - \dfrac{l'}{2}\right)^2} - \frac{l_0\Psi}{\pi\left(l + \dfrac{l'}{2}\right)^2};$$

hence,

$$\psi = \frac{2\, l_0 l'\Psi}{\pi l^3} \tag{82}$$

81. Let l_1 and l_2 be the lengths of two conductors without return conductors, at distance l_0 from each other.

By equation (79), the electric field of a conductor of length l_1, without return conductor, at distance l, is given by:

$$\psi = \frac{l_1\Psi}{\pi l^2} \tag{79}$$

With the current

$$i = I \cos \theta \tag{83}$$

in the conductor l_1, the magnetic field at unit distance is

$$\Psi = 2\, i \tag{84}$$

and the electromagnetic component of the field at distance l thus is

$$\mathfrak{B} = \frac{2\, l_1 i}{\pi l^2}$$

$$= \frac{2\, l_1 I \cos\,(\theta - al)}{\pi l^2} \tag{85}$$

where the al represents the finite velocity of propagation from the conductor l_1 to the distance l.

The magnetic flux intercepted by the receiving conductor of length l_2, at distance l_0, then is:

$$\Phi = \int_{l_0}^{\infty} \frac{2\, l_1 l_2 I \cos(\theta - al)}{\pi l^2} dl$$

$$= \frac{2\, l_1 l_2 I}{\pi} \left\{ \cos\theta \int_{l_0}^{\infty} \frac{\cos al}{l^2}\, dl + \sin\theta \int_{l_0}^{\infty} \frac{\sin al}{l^2}\, dl \right\} \cdot \quad (86)$$

These integrals can not be integrated in finite form, but represent new functions, and as such may be denoted by

$$a \text{ coll } al = \int_{l}^{\infty} \frac{\cos al}{l^2}\, dl \qquad (87)$$

$$a \text{ sill } al = \int_{l}^{\infty} \frac{\sin al}{l^2}\, dl \qquad (88)$$

These functions are further discussed in the appendix.

82. Substituting (87) and (88) into (86), gives:

$$\Phi = \frac{2\, al_1 l_2 I}{\pi} \left\{ \cos\theta \text{ coll } al_0 + \sin\theta \text{ sill } al_0 \right\} \qquad (89)$$

and the *mutual inductance*,

$$M = \frac{\Phi}{I} \qquad (90)$$

thus is given, in symbolic representation:

$$M = \frac{2a}{\pi}\, l_1 l_2 \{ \text{coll } al_0 - j \text{ sill } al_0 \}\, 10^{-9} \text{ henry.} \qquad (91)$$

By partial integration, coll and sill can be reduced to sil and col, thusly:

$$a \text{ coll } al = \int_{l}^{\infty} \frac{\cos al}{l^2}\, dl = -\int_{l}^{\infty} \cos al\, d\left(\frac{1}{l}\right) = \frac{\cos al}{l} - a \text{ col } al \quad (92)$$

$$a \text{ sill } al = \int_{l}^{\infty} \frac{\sin al}{l^2}\, dl = -\int_{l}^{\infty} \sin al\, d\left(\frac{1}{l}\right) = \frac{\sin al}{l} + a \text{ sil } al. \quad (93)$$

For large values of l, by equations (20), col al approximates $\frac{\cos al}{al}$, and sil al approximates $-\frac{\sin al}{al}$, that is, coll al and sill al approximate zero, with increasing l, at a higher rate than do sil al and col al, as was to be expected.

Substituting (82) and (83) into (81) gives

$$M = \frac{2}{\pi} l_1 l_2 \left\{ \left(\frac{\cos a l_0}{l_0} - a \cot a l_0 \right) - j \left(\frac{\sin a l_0}{l_0} + a \sin a l_0 \right) \right\} 10^{-9}$$

henry   (94)

From the mutual inductance M follows the *mutual impedance*

$$Z = 4 \, j\pi f M$$

$$= 4 \, a f l_1 l_2 \{\text{sill } a l_0 + j \text{ coll } a l_0\} 10^{-9}$$

$$= 4 \, f l_1 l_2 \left\{ \left(\frac{\sin a l_0}{l_0} + a \sin a l_0 \right) + j \left(\frac{\cos a l_0}{l_0} - a \cot a l_0 \right) \right\} 10^{-9}$$

ohms   (95)

or, absolute,

$$z = 4 \, a f l_1 l_2 \sqrt{\text{sill}^2 \, a l_0 + \text{coll}^2 \, a l_0} \; 10^{-9}$$

$$= 4 \, f l_1 l_2 \sqrt{\left(\frac{\sin a l_0}{l_0} + a \sin a l_0 \right)^2 + \left(\frac{\cos a l_0}{l_0} - a \cot a l_0 \right)^2} \, 10^{-9}$$

ohms   (96)

For large values of $a l_0$, it is:

$$\text{sill } a l_0 = - \frac{\sin a l_0}{(a l_0)^2} \tag{97}$$

$$\text{coll } a l_0 = \frac{\cos a l_0}{(a l_0)^2}. \tag{98}$$

thus, substituted into (86):

$$z = 4 f \frac{l_1 l_2}{a l_0{}^2} 10^{-9}$$

$$= \frac{2 \, S l_1 l_2}{\pi l_0{}^2} 10^{-9} \text{ ohms.} \tag{99}$$

Thus, for instance:

$$l_1 = l_2 = 50 \text{ ft.} = 1515 \text{ cm.}$$
$$l_0 = 10 \text{ miles} = 1.61 \times 10^6 \text{ cm.}$$
$$f = 500{,}000 \text{ cycles}$$
$$z = 53 \times 10^{-6}$$

and, if
$$i_1 = 10 \text{ amps.}$$
$$e_2 = z \, i_1 = 0.53 \text{ millivolts.}$$

E. CAPACITY OF A SPHERE IN SPACE.

83. Let

l_0 = radius of the sphere.

$e_0 = E \cos \theta$ = voltage of the sphere.

e = voltage at distance l from centre of sphere.

The voltage gradient at distance l then is:

$$G = \frac{de}{dl} \qquad (100)$$

and the electrostatic field, in electromagnetic units:

$$K = \frac{G}{4 \pi S^2}$$

$$= \frac{1}{4 \pi S^2} \frac{de}{dl} \qquad (101)$$

thus the dielectric flux:

$$\Psi = 4 \pi l^2 K$$

$$= \frac{l^2}{S^2} \frac{de}{dl} \qquad (102)$$

Let C = capacity of the sphere. The dielectric flux then is

$$\Psi = Ce \qquad (103)$$

and lags by angle al behind the voltage, due to the finite velocity of the field (more correctly, by $a(l - l_0)$, but l_0 may be neglected against l).

Thus the flux at distance l is

$$\Psi = CE \cos (\theta - al) \qquad (104)$$

Substituting (94) into (92), and resolving, gives:

$$\frac{de}{dl} = \frac{CES^2 \cos (\theta - al)}{l^2} \qquad (105)$$

and, integrated from l_0 to ∞:

$$e = E \cos \theta = CES^2 \int_{l_0}^{\infty} \frac{\cos (\theta - al)}{l^2} dl \qquad (106)$$

hence,

$$\frac{1}{C} = S^2 \left\{ \int_{l_0}^{\infty} \frac{\cos al}{l^2} \, dl + \tan \theta \int_{l_0}^{\infty} \frac{\sin al}{l^2} \, dl \right\} 10^{-9} \qquad (107)$$

where the 10^{-9} reduces the farads.

The integrals are the same found in the case of mutual inductance, D, equations (87), (88), and reduce to sil and col by equations (92) and (93).

Writing them symbolically, (107) becomes:

$$\frac{1}{C} = S^2 \left\{ \left(\frac{\cos al_0}{l_0} - a \operatorname{col} al_0 \right) - j \left(\frac{\sin al_0}{l_0} + a \operatorname{sil} al_0 \right) \right\} 10^{-9} \quad (108)$$

If, as usual, l_0 is small compared with the wave length, it is:

$$
\left.
\begin{aligned}
\cos al_0 &= 1 \\
\operatorname{col} al_0 &= \frac{\pi}{2} \\
\sin al_0 &= 0 \\
\operatorname{sil} al_0 &= \log \frac{0.56}{al_0}
\end{aligned}
\right\} \qquad (109)
$$

thus,

$$\frac{1}{C} = S^2 \left\{ \left(\frac{1}{l_0} - \frac{a\pi}{2} \right) - ja \log \frac{0.56}{al_0} \right\} 10^{-9} \qquad (110)$$

and, substituting (2):

$$\frac{1}{C} = S^2 \left\{ \left(\frac{1}{l_0} - \frac{\pi^2 f}{S} \right) - j \frac{2\pi f}{S} \log \frac{0.56 \, S}{2\pi f l_0} \right\} 10^{-9} \qquad (111)$$

For $a = 0$, or infinite velocity of the electric field, (110) becomes

$$C = \frac{l_0}{S^2} \qquad (112)$$

which is the usual expression of the capacity of a sphere in space.

CHAPTER IX.

HIGH-FREQUENCY CONDUCTORS.

84. As the result of the phenomena discussed in the preceding chapters, conductors intended to convey currents of very high frequency, as lightning discharges, high frequency oscillations of transmission lines, the currents used in wireless telegraphy, etc., cannot be calculated by the use of the constants derived at low frequency, but effective resistance and inductance, and therewith the power consumed by the conductor, and the voltage drop, may be of an entirely different magnitude from the values which would be found by using the usual values of resistance and inductance. In conductors such as are used in the connections and the discharge path of lightning arresters and surge protectors, the unequal current distribution in the conductor (Chapter VII) and the power and voltage consumed by electric radiation, due to the finite velocity of the electric field (Chapter VIII), require consideration.

The true ohmic resistance in high frequency conductors is usually entirely negligible compared with the effective resistance resulting from the unequal current distribution, and still greater may be, at very high frequency, the effective resistance representing the power radiated into space by the conductor. The total effective resistance, or resistance representing the power consumed by the current in the conductor, thus comprises the true ohmic resistance, the effective resistance of unequal current distribution, and the effective resistance of radiation.

The power consumed by the effective resistance of unequal current distribution in the conductor is converted into heat in the conductor, and this resistance thus may be called the "thermal resistance" of the conductor, to distinguish it from the radiation resistance. The power consumed by the radiation resistance is not converted into heat in the conductor, but is dissipated in the space surrounding the conductor, or in any other conductor on which the electric wave impinges. That is, at very high frequency, the total power consumed by the effective resistance of the conductor does not appear as heating of the conductor, but a large part of it may be sent out into space as electric radiation, which accounts for the power exerted upon bodies near the path of a lightning stroke, as "side discharge."

It demonstrates that safety from lighting is not given by merely affording a discharge path, but while discharging through such path, most of the energy of the lightning may be communicated by radiation to other bodies.

The inductance is reduced by the unequal current distribution in the conductor, which, by deflecting most of the current into the outer layer of the conductor, reduces or practically eliminates the magnetic field inside of the conductor. The lag of the magnetic field in space, behind the current in the conductor, due to the finite velocity of radiation, also reduces the inductance to less than that from the conductor surface to a distance of one-half wave. An exact determination of the inductance is, however, not possible; the inductance is represented by the electro-magnetic field of the conductor, and this depends upon the presence and location of other conductors, etc., in space, on the length of the conductor, and the distance from the return conductor. Since very high frequency currents, as lightning discharges, frequently have no return conductor, but the capacity at the end of the discharge path returns the current as "displacement current," the extent and distribution of the magnetic field is indeterminate. If, however, the conductor under consideration is a small part of the total discharge -- as the ground connection of a lightning arrester, a small part of the discharge path from cloud to ground — and the frequency very high, so that the wave length is relatively short, and the space covered by the first half wave thus is known to be free of effective return conductors, the magnitude of the inductance can be calculated with fair approximation by assuming the conductor as a finite section of a conductor without return conductor.

85. In long distance transmission lines and other electric power circuits, disturbances leading to the appearance of high frequency currents may be either between the lines—such as caused by switching, sudden changes of load, spark discharges or short circuits between conductors, etc. Or they may be between line and ground, such as caused by lightning, by arcing grounds, short circuits to ground, with grounded neutral, etc.

In the former case, high frequency currents between the line conductors, the electric field is essentially contained between the line conductors, in a space which is usually practically free of other conductors. The effect of the finite velocity of the field on the inductance or rather the impedance of the conductor, the radiation resistance, etc., can be well approximated by the equations of a finite section of an infinitely long conductor, having its return conductor at a distance equal to the distance between the transmission conductors (Chapter VIII).

In the case of a high frequency disturbance between line and ground, all the line conductors may share in the conduction, that is carry current simultaneously in the same direction, as frequently the case with lightning discharges, etc. The impedance then is the joint impedance of all the line conductors (about one-third that of one conductor, with a three-wire line); the field is between the line and the ground, and usually fairly free of conducting bodies. Thus the radiation resistance, etc., can be calculated under the assumption of the image conductor as return conductor, that is a return conductor at a distance equal to twice the height of the line.

If the high frequency disturbance originates between one line conductor and the ground, as usual with arcing grounds, and occasionally with lightning discharges, etc., the high frequency field is between this conductor and the ground as return conductor, but the other line conductors (and other parallel circuits, as telephone lines, etc.) are within the high frequency field, and currents are induced in them by mutual induction. These currents, being essentially in reverse direction to the inducing current, act as partial return current, and the constants, as radiation resistance, etc., are intermediate between those of the two cases previously discussed.

With regards to unequal current distribution in the conductor, obviously the existence and location of the return conductor is of no moment.

In many cases therefore, for the two extremes—low frequency, where unequal current distribution and radiation are negligible, and very high frequency, where the current traverses only the outer layer and the total effect, contained within one wave length, is within a moderate distance of the conductor—the constants can be calculated; but for the intermediary case, of moderately high frequency, the conductor constants may be any-where between the two limits, i.e., the low frequency values and the values corresponding to an infinitely long conductor without return conductor.

Since, however, the magnitude of the conductor constants, as derived from the approximate equations of unequal current distribution and of radiation, are usually very different from the low frequency values, their determination is of interest even in the case of intermediate frequency, as indicating an upper limit of the conductor constants.

86. Using the following symbols, namely,

l = the length of conductor,

A = the sectional area,

l_1 = the circumference at conductor surface, that is, following all the indentations of the conductor,

l_2 = the shortest circumference of the conductor, that is, circumference without following its indentations,

l_r = the radius of the conductor,

l' = the distance from the return conductor,

λ = the conductivity of conductor material,

μ = the permeability of conductor material,

f = the frequency,

S = the speed of light = 3×10^{10} cm., and

$$a = \frac{2\pi f}{S} = 2.09f \ 10^{-10} = \text{the wave length constant,} \tag{1}$$

At low frequency, the current density throughout the conductor section is uniform, and its resistance is the true ohmic resistance:

$$r_0 = \frac{l}{\lambda A} \text{ ohms.} \tag{1}$$

The external reactance, that is, reactance due to the magnetic field outside of the conductor, is at low frequency, where the finite velocity of the magnetic field can be neglected, given by:

$$x_0 = 4\,\pi f l \log \frac{l'}{l_r} \ 10^{-9} \text{ ohms} \tag{2}$$

or, reducing to common logarithms, by multiplying with log 10 = 2.3026:

$$x_0 = 9.21 \, \pi f l \lg \frac{l'}{l_r} \, 10^{-9} \text{ ohms} \tag{3}$$

where lg may denote the common, log the natural logarithm.

In addition to the external reactance, there exists an internal reactance, due to the magnetic flux inside of the conductor. At low frequency, where the current density in the conductor is uniform, this is:

$$x_0' = \pi f \mu l \, 10^{-9} \text{ ohms} \tag{4}$$

and the total low frequency impedance thus is:

$$\begin{aligned}
Z_0 &= r_0 + j(x_0 + x_0') \\
&= l \left\{ \frac{1}{\lambda A} + j\pi f \left(4 \lg \frac{l'}{l_r} + \mu \right) 10^{-9} \right\} \\
&= l \left\{ \frac{1}{\lambda A} + j\pi f \left(9.21 \lg \frac{l'}{l_r} + \mu \right) 10^{-9} \right\} \text{ ohms}
\end{aligned} \right\} \tag{5}$$

and the low frequency inductance:

$$\begin{aligned}
L_0 &= \frac{x_0 + x_0'}{2 \pi f} \\
&= l \left(2 \log \frac{l'}{l_r} + \frac{\mu}{2} \right) 10^{-9} \\
&= l \left(4.6 \lg \frac{l'}{l_r} + \frac{\mu}{2} \right) 10^{-9} \text{ henry}
\end{aligned} \right\} \tag{6}$$

The magnetic field of the current surrounds this current and fills all the space outside thereof, up to the return current. Some of the magnetic field due to the current in the interior and the center of a conductor carrying current, thus is inside of the conductor, while all the magnetic field of the current in the outer layer of the conductor is outside of it. Therefore, more magnetic field surrounds the current in the interior of the conductor than the current in its outer layer, and the inductance therefore increases from the outer layer of the conductor towards its interior, by the "internal magnetic field." In the interior of the conductor, the reactance voltage thus is higher than on the outside. At low frequency, with moderate size of conductor, this difference is inappreciable in its effect. At higher frequencies, however, the higher reactance in the interior of the conductor, due to this internal magnetic field, causes the current density to decrease towards the interior of the conductor, and the current to lag, until finally the current flows practically only through a thin layer of the conductor surface.

As the result hereof, the effective resistance of the conductor is increased, due to the uneconomical use of the conductor material caused by the lower current density in the interior, and due to the phase displacement between the currents in the successive layers of the conductor, which results in the sum of the currents in the successive layers of the conductor being larger than the resultant current. Due to this unequal current distribution, the internal reactance of the conductor is decreased, as less current penetrates to the interior of the conductor, and thus produces less magnetic field inside of the conductor.

The equivalent *depth of penetration* of the current into the conductor, from Chapter VII, (40), is

$$l_p = \frac{10^4}{\pi \sqrt{0.4\,\lambda\mu f}} = \frac{5030}{\sqrt{\lambda\mu f}}; \tag{7}$$

hence, the effective *resistance of unequal current distribution*, or *thermal resistance* of the conductor, is, approximately,

$$r_1 = \frac{l}{\lambda l_p l_1} = \frac{l\pi}{l_1}\sqrt{\frac{0.4\,\mu f}{\lambda}}\,10^{-4} = \frac{1.98l}{l_1}\sqrt{\frac{\mu f}{\lambda}}\,10^{-4}\ \text{ohms}, \tag{8}$$

The *effective reactance of the internal flux*, at high frequency, approaches the value:

$$x_1 = r_1 = \frac{1.98\,l}{l_1}\sqrt{\frac{\mu f}{\lambda}}\,10^{-4}\ \text{ohms}. \tag{9}$$

87. The effective resistance resulting from the finite velocity of the electric field, or *radiation resistance*, by assuming the conductor as a section of an infinitely long *conductor without return conductor*, from Chapter VIII, (23), is

$$r_2 = 2\,l\pi^2 f\,10^{-9} = 1.97\,lf\,10^{-8}\ \text{ohms}, \tag{10}$$

and the *effective reactance of the external field* of a finite section of an infinitely long round conductor without return conductor, from Chapter VIII, (25), is

$$x_2 = 4\,\pi fl\left(\log\frac{1}{al_r} - 0.5772\right)10^{-9}\ \text{ohms}. \tag{11}$$

and, substituting (7),

$$x_2 = 4\pi fl \left(\log \frac{S}{2\pi fl_r} - 0.5772 \right) 10^{-9} \text{ ohms}$$
$$= 4\pi fl \, (21.72 - \log l_r f) \, 10^{-9} \text{ ohms} \tag{12}$$

or, substituting the common logarithm: $\log = 2.303 \lg$, gives:

$$x_2 = 2.89 \, fl(9.45 - \lg \, l_r f) \, 10^{-8} \text{ ohms} \tag{13}$$

Assuming now that the external magnetic field of a conductor of any shape is equal to that of a round conductor having the same minimum circumference, as is approximately the case, that is, substituting:

$$l_2 = 2 \pi l_r$$

in equations (12) and (13), gives

$$x_2 = 4\pi fl \left(\log \frac{S}{l_2 f} - 0.5772 \right) 10^{-9} \text{ ohms}$$
$$= 2.89 fl \, (10.25 - \lg \, l_2 f) \, 10^{-8} \text{ ohms} \tag{14}$$

While the case of a conductor without return conductor may be approximated under some conditions, such as lightning discharges, under other conditions, such as high frequency disturbances in transmission lines, the case of a conductor with return conductor at finite distance l' is more representative.

The effective radiation resistance and reactance of a section of an infinitely long *conductor with return conductor at distance* l', are, by Chapter VIII (42), and by (45) (44):

Radiation resistance,

$$r_3 = 4\pi fl \left(\frac{\pi}{2} - \text{col } al' \right) 10^{-9} \text{ ohms}$$

or, substituting (1),

$$r_3 = 4\pi fl \left(\frac{\pi}{2} - \text{col } \frac{2\pi fl'}{S} \right) 10^{-9} \text{ ohms} \tag{15}$$

and, if the distance of the return conductor, l', is small compared to the wave length, this becomes

$$r_3 = \frac{8\pi^2 f^2 l' l}{S} \, 10^{-9} \text{ ohms}$$
$$= 2.63 \, f^2 l' l \, 10^{-18} \text{ ohms} \tag{16}$$

Radiation reactance,

$$x_3 = 4\pi fl \left(\log \frac{1}{al_r} - 0.5772 - \text{sil } al' \right) 10^{-9} \text{ ohms} \tag{17}$$

or, substituting (7),

$$
\left.\begin{aligned}
x_3 &= 4\pi fl \left(\log \frac{S}{2\pi f l_r} - 0.5772 - \mathrm{sil}\, \frac{2\pi f l'}{S} \right) 10^{-9} \text{ ohms} \\
&= 4\pi fl \left(21.72 - \log l_r f - \mathrm{sil}\, \frac{2\pi f l'}{S} \right) 10^{-9} \text{ ohms}
\end{aligned}\right\} \quad (18)
$$

and, if the distance of the return conductor, l', is small compared to the wave length, (17) becomes the ordinary low frequency reactance formula:

$$
x_3 = 4\pi fl \log \frac{l'}{l_r} 10^{-9} \text{ ohms} = 4\pi fl \log \frac{2\pi l'}{l_2} 10^{-9} \text{ ohms} \quad (19)
$$

88. The total *impedance* of a conductor for high frequencies is therefore:

Conductor without return conductor:

$$
Z = (r_1 + r_2) + j(x_1 + x_2)
$$

$$
= l \left\{ \left[\frac{\pi}{l_1} \sqrt{\frac{0.4\mu f}{\lambda}} 10^{-4} + 2\pi^2 f\, 10^{-9} \right] + j \left[\frac{\pi}{l_1} \sqrt{\frac{0.4\mu f}{\lambda}} 10^{-4} + \right. \right.
$$

$$
\left. \left. 4\pi f \left(\log \frac{S}{l_2 f} - 0.5772 \right) 10^{-9} \right] \right\} \text{ ohms}
$$

$$
= l \left\{ \left[\frac{1.98}{l_1} \sqrt{\frac{\mu f}{\lambda}} 10^{-4} + 1.97 f\, 10^{-8} \right] + j \left[\frac{1.98}{l_1} \sqrt{\frac{\mu f}{\lambda}} 10^{-4} + \right. \right.
$$

$$
\left. \left. 2.89 f\, (10.25 - \lg l_2 f)\, 10^{-8} \right] \right\} \text{ ohms} \quad (20)
$$

Conductor with return conductor at distance l':

$$
Z = (r_1 + r_3) + j(x_1 + x_3)
$$

$$
= l \left\{ \left[\frac{\pi}{l_1} \sqrt{\frac{0.4\mu f}{\lambda}} 10^{-4} + \frac{8\pi^2 f^2 l'}{S} 10^{-9} \right] + j \left[\frac{\pi}{l_1} \sqrt{\frac{0.4\mu f}{\lambda}} 10^{-4} + \right. \right.
$$

$$
\left. \left. 4\pi f \log \frac{l'}{l_r} 10^{-9} \right] \right\}
$$

$$
= l \left\{ \left[\frac{1.98}{l_1} \sqrt{\frac{\mu f}{\lambda}} 10^{-4} + 2.63 f^2 l'\, 10^{-18} \right] + j \left[\frac{1.98}{l_1} \sqrt{\frac{\mu f}{\lambda}} 10^{-4} + \right. \right.
$$

$$
\left. \left. 2.89 f \lg \frac{l'}{l_r} 10^{-8} \right] \right\} \text{ ohms} \quad (21)
$$

or, if l' is of the same or higher magnitude as the wave length,

$$Z = (r_1 + r_3) + j(x_1 + x_3)$$

$$= l\left\{\left[\frac{\pi}{l_1}\sqrt{\frac{0.4\mu f}{\lambda}}\ 10^{-4} + 4\pi f\left(\frac{\pi}{2} - \text{col}\ \frac{2\pi f l'}{S}\right)10^{-9}\right] + \right.$$

$$\left. j\left[\frac{\pi}{l_1}\sqrt{\frac{0.4\mu f}{\lambda}}\ 10^{-4} + 4\pi f\left(\log\frac{S}{2\pi f l_r} - 0.5772 - \right.\right.\right.$$

$$\left.\left.\left. \text{sil}\ \frac{2\pi f l'}{S}\right)10^{-9}\right]\right\}$$

$$= l\left\{\left[\frac{1.98}{l_1}\sqrt{\frac{\mu f}{\lambda}}\ 10^{-4} + 4\pi f\left(\frac{\pi}{2} - \text{col}\ \frac{2\pi f l'}{S}\right)10^{-9}\right] + \right.$$

$$\left. j\left[\frac{1.98}{l_1}\sqrt{\frac{\mu f}{\lambda}}\ 10^{-4} + 4\pi f\left(21.72 - \log l_r f - \right.\right.\right.$$

$$\left.\left.\left. \text{sil}\ \frac{2\pi f l'}{S}\right)10^{-9}\right]\right\}\ \text{ohms}\quad (22)$$

The *inductance* $L = \dfrac{x}{2\pi f}$ is:

Conductor without return conductor,

$$L = l\left\{\frac{1}{l_1}\sqrt{\frac{0.1\mu}{\lambda f}}\ 10^{-4} + 2\left(\log\frac{S}{l_2 f} - 0.5772\right)10^{-9}\right\}\ \text{henry}\quad (23)$$

Conductor with return conductor at distance l',

$$L = l\left\{\frac{1}{l_1}\sqrt{\frac{0.1\mu}{\lambda f}}\ 10^{-4} + 2\log\frac{l'}{l_r}\ 10^{-9}\right\}\ \text{henry}\quad (24)$$

or

$$L = l\left\{\frac{1}{l_1}\sqrt{\frac{0.1\mu}{\lambda f}}\ 10^{-4} + 2\left(\log\frac{S}{2\pi f l_r} - 0.5772 - \right.\right.$$

$$\left.\left. \text{sil}\ \frac{2\pi f l'}{S}\right)10^{-9}\right\}\ \text{henry}$$

$$\left.\right\}\quad (25)$$

$$= l\left\{\frac{0.316}{l_1}\sqrt{\frac{\mu}{\lambda f}}\ 10^{-4} + 2\left(21.72 - \log l_r f - \right.\right.$$

$$\left.\left. \text{sil}\ \frac{2\pi f l'}{S}\right)10^{-9}\right\}\ \text{henry}$$

89. As an instance may be considered the high frequency impedance of a copper wire No. 00 B. & S. G., that is of the radius,

$$l_r = 0.1825\ \text{in.} = 0.463\ \text{cm.}$$

under the three conditions:

(a) Return conductor at $l' = 6$ ft. $= 182$ cm. distance, corresponding to transmission line conductors oscillating against each other.

(b) With the ground as return conductor, at 30 ft. distance, that is, $l' = 2 \times 30$ ft. $= 1820$ cm., corresponding to a transmission line conductor oscillating against ground.

(c) No return conductor, corresponding to the vertical discharge path of a lightning stroke.

With copper as conductor material, it is:

$$\lambda = 6.2 \times 10^5$$
$$\mu = 1$$

It is then, by the preceding equations, per meter length of conductor, or

$$l_0 = 100 \text{ cm.}$$

Low frequency values:
true ohmic resistance, (1),

$$r_0 = \frac{100}{\lambda \pi l_r^2} = 0.24 \times 10^{-3} \text{ ohms}$$

external reactance, (2):

$$x_0 = 0.4 \pi f \log \frac{l'}{l_r} 10^{-6} \text{ ohms}$$

hence,

(a) $l' = 182:$ $x_0 = \ \ 7.5 f \ 10^{-6}$ ohms
(b) $= 1820:$ $x_0 = 10.4 f \ 10^{-6}$ ohms
(c) $= \infty :$ $x_0 = \infty$

internal reactance, (4):

$$x_0' = 0.1 \pi f \ 10^{-6} = 0.314 f \ 10^{-6} \text{ ohms}$$

High frequency values:
thermal resistance, (8):

$$r_1 = \frac{100 \pi}{l_1} \sqrt{\frac{0.4 \ \mu f}{\lambda}} \ 10^{-4} = \frac{50}{l_r} \sqrt{\frac{0.4 \ \mu f}{\lambda}} \ 10^{-4} = 8.65 \sqrt{f} \ 10^{-6} \text{ ohms}$$

internal reactance, (9):

$$x_1 = r_1 = 8.65 \sqrt{f} \ 10^{-6} \text{ ohms}$$

radiation resistance,

(a) and (b) (16):

$$r_3 = \frac{0.8\ \pi^2 f^2 l'}{S}\ 10^{-6} = 263\ f^2 l'\ 10^{-18}\ \text{ohms}$$

(a) $l' = 182$: $r_3 = 0.048\ f^2\ 10^{-12}$ ohms
(b) $l' = 1820$: $r_3 = 0.48\ f^2\ 10^{-12}$ ohms
(c) $l' = \infty$ (10):

$$r_2 = 0.2\ \pi^2 f\ 10^{-6} = 1.97\ f\ 10^{-6}\ \text{ohms}$$

radiation reactance,

(a) and (b) (19):

$$x_3 = 0.4\ \pi f \log \frac{l'}{l_r}\ 10^{-6} = \mathfrak{X}_0$$

(a) $l' = 182$: $x_3 = 7.5\ f\ 10^{-6}$ ohms
(b) $l' = 1820$: $x_3 = 10.4\ f\ 10^{-6}$ ohms
(c) $l' = \infty$ (14):

$$x_2 = 0.4\ \pi f \left(\log \frac{S}{l_2 f} - 0.5772\right) 10^{-6} = f(28.5 - 2.89\ \lg f)10^{-6}\ \text{ohms}$$

For r_3 and x_3, for $f = 10^8$, $l' = 182$, and for $f = 10^7$ and 10^8, $l' = 1820$, the more complete equations (15) and (18) must be used, as l' exceeds a quarter wave length.

Table I gives numerical values, from 1 cycle to 10^8 cycles, of r, x, Z, cos a and the resistance ratios. These values are plotted. in Fig. 97, in logarithmic scale.

90. The low frequency values of resistance r_0 and external and internal reactance $x_0 + x_0'$, have no existence at the higher frequencies. But as they are the values calculated by the usual formulas, they are given in Table I for comparison with the true effective high frequency values. The values r_2 and x_2, though given for all frequencies, have a meaning only for the very high frequencies, 10^4 and higher, since at lower frequencies the condition of a conductor without return conductor can hardly be realized, as any conductor within a quarter wave length would act more or less as effective return conductor.

As seen from the equations, and illustrated in Table I, the thermal resistance of the conductor, that is, the resistance which converts electric energy into heat in the conductor, is the true ohmic resistance at low frequencies, but with increasing frequency

TABLE I.—CIRCUIT CONSTANTS OF COPPER WIRE NO. 00 B. & S. G. PER METER LENGTH OF CONDUCTOR.

(a) Return conductor at 6 ft. = 182 cm. distance.
(b) Return conductor at 2 × 30 ft. = 1820 cm. distance.
(c) No return conductor, conductivity: $\lambda = 6.2 \times 10^5$.

		$f=1$	10	10^2	10^3	10^4	10^5	10^6	10^7	10^8
Low Frequency										
$r_0=$		0.000240	0.000240	0.000240	0.000240	0.000240	0.000240	0.000240	0.000240	0.000240
$x_{10}=$		0.0000003	0.000003	0.00003	0.0003	0.003	0.03	0.3	3	30
$x_0=$	(a)	0.0000075	0.000075	0.00075	0.0075	0.075	0.75	7.5	75	750
	(b)	0.0000104	0.000104	0.00104	0.0104	0.104	1.04	10.4	104	1040
$z_0=$	(a)	0.000240	0.000252	0.000816	0.0078	0.078	0.78	7.8	78	780
	(b)	0.000240	0.000263	0.001097	0.0107	0.107	1.07	10.7	107	1070
$\cos \omega_0=$	(a)	1.000	0.952	0.293	0.0308	0.0031	0.000031	0.000031	0.000003	0.0000003
	(b)	1.000	0.912	2.218	0.225	0.0022	0.00022	0.000022	0.000002	0.0000002
High Frequency										
$r_1=$		0.000240	0.000240	0.000240	0.000273	0.000865	0.00273	0.00865	0.0273	0.0865
$x_1=$		0.0000003	0.000003	0.000030	0.000273	0.000865	0.00273	0.00865	0.0273	0.0865
$r_3=$	(a)	.....	.....	.....	.....	0.0000048	0.00048	0.048	4.8	225
$r_2=$	(b)	.....	.....	.....	.....	0.00048	0.0048	0.48	22.5	194
	(c)	0.000002	0.000020	0.000197	0.00197	0.0197	0.197	1.97	19.7	197
$x_3=$	(a)	0.0000075	0.000075	0.00075	0.0075	0.075	0.75	7.5	75	490
$x_2=$	(b)	0.0000104	0.000104	0.00104	0.0104	0.104	1.04	10.4	78	505
	(c)	0.0000285	0.000256	0.00227	0.0198	0.169	1.40	11.2	83	515
$z=$	(a)	0.000240	0.000252	0.000816	0.00778	0.0759	0.753	7.5	75.2	539
	(b)	0.000240	0.000263	0.001097	0.01068	0.1049	1.043	10.4	81.2	541
	(c)	0.000242	0.000367	0.00234	0.0201	0.170	1.42	11.4	85.3	551
$\cos \omega=$	(a)	1.000	0.952	0.293	0.0351	0.0115	0.0043	0.0076	0.0642	0.418
	(b)	1.000	0.912	0.218	0.0257	0.0087	0.0072	0.047	0.277	0.358
	(c)	1.000	0.710	0.186	0.112	0.121	0.141	0.173	0.232	0.358
$r_1+r_3=$ $\frac{r_0}{r_1+r_3}=$	(a)	1.000	1.000	1.000	1.14	3.62	13.4	236	20,100	937,000
	(b)	1.000	1.000	1.000	1.14	3.80	31.3	2030	94,000	808,000
	(c)	1.008	1.082	1.82	9.35	85.5	832	8250	82,000	820,000

begins to rise due to unequal current distribution in the conductor at about 1000 cycles in copper wire number 00 B. & S. G., and approaches proportionality with the square root of the frequency, hence reaches values many times the ohmic resistance, at very high frequencies.

The radiation resistance of the conductor without return conductor, r_2, is proportional to the frequency, but the radiation resistance of the conductor with return conductor, r_3, is proportional to the square of the frequency, hence very small until high frequencies are reached—10,000 to 100,000 cycles. The radiation resistance r_3 of the conductor with return conductor then increases very rapidly and reaches values many thousand times the ohmic resistance. At the very highest frequencies, many millions of cycles, its rate of increase becomes less again, and it approaches proportionality with the first power of the frequency, and approaches the value of radiation resistance r_2, of the conductor without return conductor, at frequencies of a wave length comparable with the distance of the return conductor. The radiation resistance r_3 of the conductor with return conductor is the larger, the greater the distance of the return conductor, and is proportional to this distance, within the range in which it is proportional to the square of the frequency. The radiation resistance of the conductor without return conductor, at the very highest frequencies, is the same as that of the conductor with return conductor, but, being proportional to the frequency, with decreasing frequency, it decreases at a lesser rate, and would even at commercial machine frequencies still be appreciable, if at such frequencies the conditions of a conductor without return conductor could be realized.

The total effective resistance of a conductor under transmission line conditions, that is, with return conductor at finite distance, is at low frequencies constant and is the true ohmic resistance. With increasing frequency, it begins to increase first slowly—at about 1000 cycles under transmission line conditions—and approaches proportionality to the square root of the frequency, as the result of the screening effect of the unequal current distribution in the conductor. Then the increase becomes more rapid, due to the appearance of the radiation resistance—at about 100,000 cycles under transmission line conditions—and reaches proportionality with the square of the frequency, at values many thousand times the ohmic resistance. Finally, at the very highest frequencies—10 million cycles—the rate of increase becomes less again, and approaches proportionality with the frequency.

91. It is interesting to note that the external reactance of the conductor with return conductor, or radiation reactance x_3, has up to very high frequencies, millions of cycles, the same value x_0 as calculated by the low frequency formula, that is, by neglecting the finite velocity of the field, hence is proportional to the frequency. The internal reactance x_1 is $= x_0'$, and proportional to the frequency at low frequencies, but drops behind x_0' due to unequal current distribution in the conductor, and approaches proportionality with the square root of the frequency. As, however, the internal reactance is a small part of the total reactance, it follows, that the total reactance of the conductor and thus also the absolute value of the impedance (for all higher frequencies, where the reactance preponderates) can be calculated by the usual low frequency reactance formula, which neglects the finite velocity of the field. Hence, the inductance L of the conductor can be assumed as constant for all frequencies up to millions of cycles; it decreases only very slowly by the decreasing internal reactance of unequal current distribution. Only at the very highest frequencies, where the wave length is comparable with the distance of the return conductor, the inductance L decreases, and the reactance $x_1 + x_3$ increases less then proportional to the frequency.

In a conductor without return conductor, the reactance at the very highest frequencies is approximately the same as in a conductor with return conductor. With decreasing frequency, however, $x_1 + x_2$ decreases less than proportional to the frequency, that is, the inductance L increases—and becomes infinity for zero frequency, if such were possible.

Without considering unequal current distribution in the conductor and the finite velocity of electric field, the power factor cos ω would steadily decrease, from unity at very low frequencies, to zero at infinite frequency. Due to the increase of the effective resistance, the power factor cos ω first decreases, from unity at low frequency, down to a minimum at some high frequency, and then increases again to high values at very high frequencies. The minimum value of the power factor is the lower and occurs at the higher frequencies, the shorter the distance of the return conductor is. Thus with the return conductor at 6 ft. distance, the power factor is 0.43 per cent at 100,000 cycles; with the return conductor at 60 ft. distance, it is 0.72 per cent; in the conductor without return conductor the power factor is 11.2 per cent at 1000 cycles.

92. It is of interest to determine the effect of size, shape and material on the high-frequency constants of a conductor.

These high-frequency constants are, per unit length of conductor:

Internal constants: Thermal resistance and internal reactance:

$$r_1 = x_1 = \frac{\pi}{l_1} \sqrt{\frac{0.4\,\mu f}{\lambda}}\ 10^{-4} \text{ ohms per cm.} \qquad (8)\ (9)$$

External constants: Radiation resistance:

$$r_3 = \frac{8\,\pi^2 f^2 l'}{\lambda}\ 10^{-9} \text{ ohms per cm.} \qquad (16)$$

External reactance:

$$x_3 = 4\,\pi f \log \frac{2\pi l'}{l_2}\ 10^{-9} \text{ ohms per cm.} \qquad (19)$$

These approximations hold for all but the very highest, and very low frequencies, that is, are correct within the frequency range with lower limit of about 1000 to 10,000 cycles, and upper limit of about 10 million cycles. Thus they apply for all those high frequencies which are of importance in the disturbances occurring in industrial circuits, with the exception of the lowest harmonics of low-frequency surges.

The constants of the conductor material enter the equations only as the ratio $\left(\dfrac{\mu}{\lambda}\right)$, permeability to conductivity, in the internal constants r_1 and x_1. Thus higher permeability has the same effect in increasing the thermal resistance as lower conductivity, and for instance, a cast silicon rod of permeability $\mu = 1$, and conductivity $\lambda = 55$, has the same high-frequency resistance and reactance, as a rod of the same size, of wrought iron, of permeability $\mu = 2000$ and conductivity $\lambda = 1.1 \times 10^5$, that is, of the same $\dfrac{\mu}{\lambda} = 0.0182$, though the latter has 2000 times the conductivity of the former.

Provided, however, that size of conductor and frequency are such as to fulfill the conditions under which equations (8) and (9) are applicable, which is, that the conductor is large compared with the depth of penetration of the current into the conductor:

$$l_p = \frac{10^4}{\pi\sqrt{0.4\ \lambda\mu f}}.$$

Thus an iron rod of 2 inches (5 cm.) diameter has at one million cycles the same thermal resistance as a silicon rod of the same size: 0.17 ohms per meter, since the depth of penetration is $l_p = 0.00034$ cm. for iron, 0.68 cm. for silicon, thus in either case small compared with the radius of the conductor $l_r = 2.5$ cm.

At 10,000 cycles, however, the iron rod has the thermal resistance and internal reactance $r_1 = x_1 = 0.017$ ohms per meter, the penetration being $l_p = 0.0034$ cm., thus small. For the silicon rod, however, at 10,000 cycles the penetration is $l_p = 6.8$ cm., thus at the radius $l_r = 2.5$ cm. formulas (8) and (9) do not apply any more, but it is approximately (that is neglecting unequal current distribution): $r_1 = 0.093$ ohms per meter, or 5.5 times the resistance of the iron rod, while the internal reactance is $x_1 = 0.031$ ohms per meter, hence 80 per cent. higher than that of the iron rod of the same size.

93. In the equations of the external constants, the radiation resistance and reactance, the material constants of the conductor do not enter, and the radiation resistance and the external reactance thus are independent of the conductor material.

The dimensional constants of the conductor, size and shape, enter the equation only as the circumference of the conductor l_1, l_2, that is, only the circumference of the conductor counts in high-frequency conduction, and all conductors of the same material, regardless of size and shape, have the same high-frequency resistances and reactances as long as they have the same conductor circumference. Thus a solid copper rod or a thin copper cylinder of the same outer diameter as the rod, or a flat copper ribbon of a circumference equal to that of the rod, are equally good high-frequency conductors, though the hollow cylinder or the ribbon may contain only a small part of the material contained in the solid copper rod. Provided, however, that the thickness or depth of the conductor (the thickness of wall of the hollow cylinder, half the thickness of the copper ribbon) is larger than the depth of penetration of current into the conductor, which is

$$l_p = \frac{10^4}{\pi\sqrt{0.4\,\lambda\mu f}}$$

In the expression of the radiation resistance, r_3, neither the material nor the dimensions of the conductor enter, that is, the radiation resistance of a conductor is independent of size, shape or material of the conductor and depends only on frequency and distance of the return conductor.

Thus a thin steel wire or a wet string have the same radiation resistance as a large copper bar. Obviously, the thermal resistance of the former is much larger and the total effective resistance thus would be larger except at those very high frequencies, at which the radiation resistance dominates.

94. As illustration may be calculated, for frequencies from 10 thousand to 10 million cycles and for 60 cycles, the resistances and reactances and thus the total impedance, the power factor, the voltage drop per meter at 100 amperes, of various conductors, for the three conditions:

(*a*) High frequency between conductors 6 ft. apart:

$$l' = 182 \text{ cm.}$$

(*b*) High frequency between conductor and ground 30 ft. below conductor:

$$l' = 1820 \text{ cm.}$$

(*c*) High-frequency discharge through vertical conductor without return conductor:

$$l' = \infty$$

For the conductors:

(1) Copper wire No. 00 B. & S. G.

$$l_r = 0.463 \text{ cm.} \quad \lambda = 6.2 \times 10^5$$

(2) Iron wire of the same size as (1): $\mu = 2000$

$$\lambda = 1.1 \times 10^5$$

(3) Copper ribbon of thickness equal to twice the depth of penetration at 10,000 cycles, and the same amount of material as (1)

(4) Iron ribbon of the same size as (3).

(5) Two inch iron pipe, $\frac{1}{8}$ inch thick.

This gives the depth of penetration at $f = 10^4$ cycles:

for copper, $\qquad\qquad l_p = \dfrac{6.4}{\sqrt{f}} = 0.064$ cm.

for iron, $\qquad\qquad l_p = \dfrac{0.34}{\sqrt{f}} = 0.0034$ cm.

It thus is:

(1) and (2) copper wire: $l_1 = l_2 = 2\pi l_r = 2.91$ cm.

(3) The area of the copper wire is $A = \pi l_r^2 = 0.672$ cm.^{2} Twice the depth of penetration is: $2\,l_p = 0.128$ cm., hence the thickness of the copper ribbon of equal weight with the wire is 0.128 cm. $= 0.05$ inches, the width is $l_3 = 5.25$ cm. or about two inches. The circumference then is:

$$l_1 = l_2 = 10.75 \text{ cm. or } 3.7 \text{ times that of the wire.}$$

(4) The same dimensions as (3).

(5) $l_r = 1$ inch $= 2.54$ cm.

$l_1 = l_2 = 16$ cm.

$A = 4.6$ cm.$^2 = 6.85$ that of (1) to (4).

Table II gives the values of r_0, r_1, r_3, r_2, x_0, x_1, x_3, x_2, $r_1 + r_3$, $r_1 + r_2$, $x_1 + x_3$, $x_1 + x_2$, z, $\cos\omega$, e and l_p, for $f = 10^6$ cycles.

Table III gives the values of $r_1 + r_3$ or $r_1 + r_2$, z, $\cos\omega$ and e, for $f = 60$, 10^4, 10^5, 10^6, 10^7 cycles for the five kinds of conductors.

95. It is interesting to compare in Table III, the constants of the first four conductors, as they have the same section, representing about average section of transmission conductors, but represent two shapes, round wire and thin flat ribbon, and two kinds of material, copper—high conductivity and non-magnetic —and iron—magnetic material of medium conductivity.

As seen, the effect of conductor shape and conductor material is very great at machine frequencies, 60 cycles, but becomes small and almost negligible at extremely high frequencies. This is rather against the usual assumption.

TABLE II.—CIRCUIT CONSTANTS FOR $f = 10^6$ = ONE MILLION CYCLES, PER METER LENGTH OF CONDUCTOR.

(a) Return conductor at 6 ft. = 182 cm. distance.
(b) Return conductor at 2 × 30 ft. = 1820 cm. distance.
(c) No return conductor.

		(1) Copper Wire.	(2) Iron Wire.	(3) Copper Ribbon.	(4) Iron Ribbon.	(5) Iron Pipe.	
$r_0 = \dfrac{100}{A\lambda} =$		0.00024	0.00135	0.00024	0.00135	0.000197	
$x_0 = 0.4\pi f\left(\log \dfrac{2\pi l'}{l_2} + \dfrac{l_0\,\mu}{l_2\,4}\right)10^{-6} =$	(a) (b)	7.8 / 10.8	635 / 638	3.05 / 5.95	173 / 176	81 / 84	Ohms / Ohms
$r_1 = \dfrac{100\pi}{l_1}\sqrt{\dfrac{0.4\,\mu f}{\lambda}}\,10^{-4} =$		0.00865	0.92	0.00233	0.248	0.167	Ohms
$r_3 = \dfrac{0.8\,\pi^2 l'}{S}\,10^{-6} =$	(a) (b)	0.048 / 0.48	0.048 / 0.48	0.048 / 0.48	0.048 / 0.48	0.048 / 0.48	Ohms / Ohms
$r_2 = 0.2\,\pi f\,10^{-6} =$	(c)	1.97	1.97	1.97	1.97	1.97	Ohms
$x_1 = \dfrac{100\pi}{l_1}\sqrt{\dfrac{0.4\,\mu f}{\lambda}}\,10^{-4} =$		0.00865	0.92	0.00233	0.248	0.167	Ohms
$x_3 = 0.4\,\pi f \log \dfrac{2\pi l'}{l_2}\,10^{-6} =$	(a) (b)	7.5 / 10.4	7.5 / 10.4	2.97 / 5.87	2.97 / 5.87	2.47 / 5.37	Ohms / Ohms
$x_2 = 0.4\,\pi f\left(\log \dfrac{S}{l_4 f} - 0.5772\right)10^{-6} =$	(c)	11.2	11.2	9.5	9.5	8.9	Ohms

$r = r_1 + r_3 =$ $r = r_1 + r_2 =$	(a) (b) (c)	0.0567 0.489 1.98	0.97 1.40 2.89	0.0503 0.482 1.97	0.296 0.728 2.22	0.215 0.647 2.14	Ohms Ohms Ohms
$x = x_1 + x_3 =$ $x = x_1 + x_2 =$	(a) (b) (c)	7.5 10.4 11.2	8.42 11.32 12.12	2.97 5.87 9.50	3.22 6.12 9.75	2.64 5.54 9.07	Ohms Ohms Ohms
$z = \sqrt{r^2 + x^2} =$	(a) (b) (c)	7.5 10.4 11.4	8.46 11.4 12.45	2.97 5.89 9.7	3.23 6.16 10.0	2.65 5.58 9.3	Ohms Ohms Ohms
$\cos \omega = \dfrac{r}{z} =$	(a) (b) (c)	0.76 4.7 17.3	11.5 12.3 23.3	1.7 8.2 20.3	9.2 11.8 22.2	8.1 11.6 23.0	Per cent. Per cent. Per cent.
$e = 100\,z =$	(a) (b) (c)	750 1040 1140	846 1140 1245	296 589 970	323 616 1000	265 558 930	Volts Volts Volts
$l_p = \dfrac{10^1}{\pi \sqrt{0.4\lambda\mu f}} =$		0.0064	0.00034	0.0064	0.00034	0.00034	Cm.

Table III.

	Conductors 6 Ft. apart.					Conductor 30 Ft. above Ground.						Conductor without Return Conductor.				
Frequency, cycles:	60	10^4	10^5	10^6	10^7	60	10^4	10^5	10^6	10^7	60	10^4	10^5	10^6	10^7
Resistance, r, in ohms per meter;																
(1) Copper wire	0.00024	0.00087	0.0052	0.0567	4.827	0.00024	0.00091	0.0075	0.489	22.5		0.0206	0.200	1.98	19.7
(2) Iron wire	0.0072	0.092	0.292	0.97	7.72	0.0072	0.092	0.297	1.40	25.4		0.112	0.489	2.89	20.0
(3) Copper ribbon	0.00024	0.00024	0.00122	0.0503	4.81	0.00024	0.00028	0.0055	0.482	22.5		0.0199	0.198	1.97	19.7
(4) Iron ribbon	0.0019	0.0248	0.079	0.296	5.58	0.0019	0.0248	0.083	0.728	23.3		0.0435	0.265	2.22	20.5
(5) Iron pipe	0.0013	0.0167	0.053	0.215	5.33	0.0013	0.0167	0.058	0.647	23.0		0.0364	0.250	2.14	20.2
Impedance, z, in ohms per meter.																
(1) Copper wire	0.00053	0.076	0.753	7.51	75	0.00069	0.105	1.043	10.41	81		0.115	1.14	11.4	85
(2) Iron wire	0.0105	0.191	1.083	8.46	78	0.0106	0.218	1.37	11.4	85		0.233	1.49	12.45	88
(3) Copper ribbon	0.00030	0.030	0.298	2.97	30	0.00043	0.059	0.588	5.89	49		0.097	0.97	9.7	72
(4) Iron ribbon	0.0028	0.060	0.315	3.23	31	0.0030	0.087	0.67	6.16	51		0.127	1.06	10.0	74
(5) Iron pipe	0.00195	0.0446	0.305	2.65	26	0.0021	0.072	0.592	5.58	47		0.112	0.97	9.3	70
Power factor, cos ω, in per cent.																
(1) Copper wire	45.3	1.14	0.69	0.76	6.4	34.8	0.87	0.72	4.7	27.8		17.9	17.5	17.3	23.2
(2) Iron wire	69.5	48.2	27.01	11.5	9.9	68.0	42.2	21.6	12.3	30.0		48.0	32.7	23.3	22.7
(3) Copper ribbon	80.0	0.8	0.4	1.7	16.0	53.5	0.48	0.93	8.2	46.0		20.5	20.4	20.3	27.3
(4) Iron ribbon	68.0	41.3	25.1	9.2	18.0	63.3	28.6	12.4	11.8	45.8		34.3	25.0	22.2	27.7
(5) Iron pipe	66.7	37.5	17.4	8.1	20.4	62.0	23.2	9.8	11.6	49.0		32.5	25.8	23.0	29.0
Voltage, e, in volts per meter;																
(1) Copper wire	0.053	7.6	75	751	7500	0.069	10.5	104	1041	8100		11.5	114	1140	8500
(2) Iron wire	1.05	19.1	108	846	7800	1.06	21.8	137	1140	8500		23.3	149	1245	8800
(3) Copper ribbon	0.03	3.0	30	297	3000	0.043	5.9	59	589	4900		9.7	97	970	7200
(4) Iron ribbon	0.28	6.0	36	323	3100	0.30	8.7	67	616	5100		12.7	106	1000	7400
(5) Iron pipe	0.195	4.5	30	265	2600	0.21	7.2	59	558	4700		11.2	97	930	7000

The reason is, at machine frequencies, the unequal current distribution, or the screening effect of the internal magnetic field, is still practically absent in copper conductors, even in round wires of medium size, and it is practically complete in iron conductors, even in ribbon of $\frac{1}{20}$ inch thickness, while at very high frequencies the effect of radiation preponderates, which is independent of the material, and the radiation resistance even independent of the shape of the conductor.

Thus under transmission line conditions, first and second of Table III, at 60 cycles the impedance, and hence the voltage drop in the iron conductor is from 7 to 20 times that of the copper conductor; at 10,000 cycles the voltage drop in the iron conductor is only from 1.5 to 2.5 times that in the copper conductor; the difference has decreased to from 14 per cent. to 44 per cent. at 100,000 cycles, 5 per cent. to 12.5 per cent. at one million cycles, while at 10 million cycles the voltage drop in the iron conductor is only 3 to 5 per cent. higher, thus practically the same and the only difference is that due to the conductor shape.

The effective resistance, and thus the power consumption in the iron conductor at 60 cycles is from 8 to 30 times that of the copper conductor, but with increasing frequency the difference in the effective resistance increases to from 88 to 106 times at 10,000 cycles, reaches a maximum and then decreases again, and is from 15 to 65 times that of the copper conductor at 100,000 cycles, only $1\frac{1}{2}$ to 17 times at a million cycles, while at 10 million cycles and above all the differences in the effective resistance practically disappear.

As the result, the power factor of the conductor—being the same, 100 per cent., at extremely low frequency and not much different, and fairly high at machine frequency—decreases with increasing frequency, reaches a minimum and then increases again to considerable values at extremely high frequency—where the high radiation resistance comes into play. The difference between iron and copper, however, is that the minimum value of the power factor, at medium high frequencies, is very low in copper, a fraction of 1 per cent., while in the iron conductor the power factor always retains considerable values, the minimum being 10 or more times that of the copper conductor. Thus an oscillation in an iron conductor must die out at a much faster rate than in a copper conductor and the liability of the formation of a continual or cumulative oscillation may exist in copper conductors but hardly in iron conductors.

The effect of the shape of the conductor on the impedance or voltage drop is fairly uniform throughout the entire frequency range, the voltage drop being the smaller, the larger the circumference.

With regard to the effective resistance, however, the effect of the conductor shape is considerable already at very low frequencies in iron conductors, but still absent with copper conductors, due to the absence of the screening effect in copper at low frequency. With increasing frequency, the difference appears in the effective resistance of the copper conductor also, with the appearance of unequal current distribution, and the ratio of the resistance of the round conductor to that of the flat conductor approaches the same value in copper as in iron. With the approach of very high frequency, however, the difference decreases again, with the appearance of radiation effect, and finally vanishes.

96. Thus to convey currents of extremely high frequency, an iron conductor is almost as good as a copper conductor of the same shape and cross section. As iron is very much cheaper than copper, it follows that in high-frequency conduction an iron conductor under the conditions of Table III, should be better than a copper conductor of the same cost and the same general shape, due to the larger size or rather circumference of the iron conductor.

There is, however, a material advantage at extremely high frequencies as well as at moderately high frequencies, in lower voltage drop at the same current resulting from such a shape conductor as gives maximum circumference, such as ribbon or hollow conductor. This advantage of ribbon or hollow tube, over the solid round conductor, exists also in the resistance and thus power consumption at medium high frequencies, but not at extremely high frequencies, but at the latter, in power consumption all conductors, regardless of size, shape or material, are practically equal.

With the thickness of ribbon conductor considered in Table III, of about $\frac{1}{20}$ inch, which is about the smallest mechanically permissible under usual conditions, the screening effect even in copper conductors is practically complete already at 10,000 cycles,

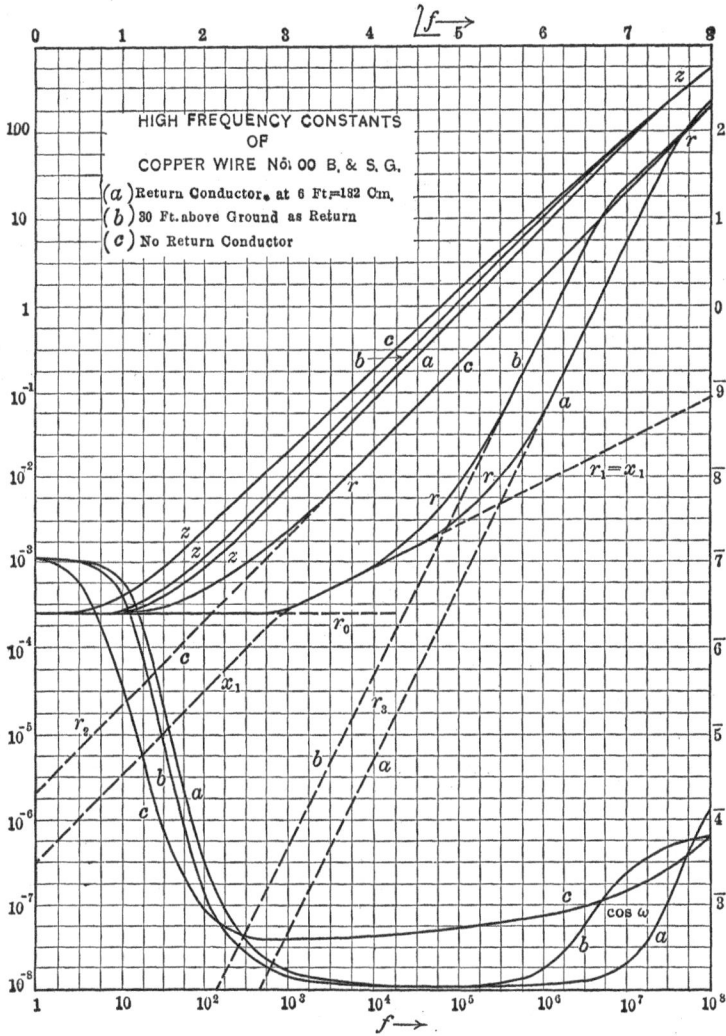

Fig. 97.

that is, the depth of penetration less than one-half the thickness of the conductor. Herefrom follows, that in the design of a high frequency conductor, the thickness of the ribbon or hollow cylinder is essentially determined by mechanical and not by electrical considerations; in other words, the thinnest mechanically permissible conductor usually is thicker than necessary for carrying the current. As iron usually cannot be employed in as thin ribbon as copper, due to its rusting, an iron conductor would have a larger section than a copper conductor of the same voltage drop and power consumption. Thereby a part of the advantage gained by the employment of the cheaper material would be lost.

97. The last section of Table III gives the constants of the conductor without return conductor, such as would be represented by the discharge circuit of a lightning arrester, by a wireless telegraph antenna, etc.; while the first two sections correspond to transmission line conditions, high-frequency currents between line conductors and between line and ground.

In the third section of Table III, the 60-cycle values are not given, and the values given for the lower high frequencies, 10^4 and even 10^5 cycles, usually have little meaning, are rarely realizable; they would correspond to a vertical conductor, as lightning arrester ground circuit, under conditions where no other conductor is within quarter-wave distance. Even at 10^5 cycles, however, the quarter-wave length is still 750 m. Thus there will practically always be other conductors within the field of the discharge conductor, acting as partial return conductor, and the actual values of impedance and resistance, that is, of voltage drop and power consumption in the conductor, thus will be intermediate between those given in the third section, for a conductor without return conductor, and those given in the first sections, for conductors with return conductors. Except at extremely high frequencies, at which the wave length gets so short that the condition of a conductor without return conductor becomes realizable. It is interesting to note, therefore, that at extremely high frequencies, the constants of the conductor without return conductor approach those of the conductor with return conductor. At lower high frequencies, impedance and resistance, and thus voltage drop and power consumption, of the return less conductor are much higher than those of the conductor with return, the more so, the lower the frequency. This is due to the considerable effect exerted already at low frequencies by electrical radiation.

However, while the case of the conductor without return conductor is not realizable at low and medium high frequencies in industrial circuits, it probably is more or less realized by the lightning discharge between ground and cloud, and the constants given in the third section of Table III would probably approximately represent the conditions met in the conductors of lightning rods such as used for the protection of buildings against lightning.

It is interesting to note that with such a conductor without return conductor, the power factor is always fairly high, even with copper as conductor material. The impedance and thus the voltage drop does not differ much from those of the conductor with return conductor. The resistance, however, and thus the power consumption are much higher, sometimes, in copper conductors, more than a hundred times as large, showing the large amount of energy radiated by the conductor—which reappears more or less destructively as "induced lightning stroke" in objects in the neighborhood of the lightning stroke.

SECTION IV

TRANSIENTS IN TIME AND SPACE

TRANSIENTS IN TIME AND SPACE

CHAPTER I.

GENERAL EQUATIONS.

1. Considering the flow of electric power in a circuit. Electric power p can be resolved in two components, one component, proportional to the magnetic effects, called the current i, and one component, proportional to the dielectric effects, called the voltage e:

$$p = ei$$

There may be energy dissipation, and energy storage in the electric circuit, and either may depend on the voltage, or on the current, giving four constants r, g, L and C, representing respectively the energy dissipation and the energy storage depending on current and on voltage respectively.

The rate of energy storage can not be proportional to the current i or voltage e, but only to their rate of change, $\frac{di}{dt}$ and $\frac{de}{dt}$: if the rate of energy storage depended on the current i itself, then at constant current i, energy storage would constantly take place, and the amount of stored energy continuously increase, at constant condition of the circuit, which obviously is impossible.

Energy dissipation however, in its simplest form, would be proportional to the current itself, respectively the voltage. Thus we have:

Energy dissipation: $\qquad ri$

$\qquad\qquad\qquad\qquad\quad ge$

Energy storage: $\qquad L\,\dfrac{di}{dt}$

$\qquad\qquad\qquad\qquad\quad C\,\dfrac{de}{dt}$

The energy relation of an electric circuit thus can be characterized by four constants, namely:

r = effective resistance, representing the power or rate of energy consumption depending upon the current, i^2r; or the power component of the e.m.f. consumed in the circuit, that is, with an alternating current, the voltage, ir, in phase with the current.

L = effective inductance, representing the energy storage depending upon the current, $\dfrac{i^2L}{2}$, as electromagnetic component of the electric field; or the voltage generated due to the change of the current, $L\dfrac{di}{dt}$, that is, with an alternating current, the reactive voltage consumed in the circuit jxi, where $x = 2\,\pi fL$ and f = frequency.

g = effective (shunted) conductance, representing the power or rate of energy consumption depending upon the voltage, e^2g; or the power component of the current consumed in the circuit, that is, with an alternating voltage, the current, eg, in phase with the voltage.

C = effective capacity, representing the energy storage depending upon the voltage, $\dfrac{e^2C}{2}$, as electrostatic component of the electric field; or the current consumed by a change of the voltage, $C\dfrac{de}{dt}$, that is, with an alternating voltage, the (leading) reactive current consumed in the circuit, jbe, where $b = 2\,\pi fC$ and f = frequency.

In the investigation of electric circuits, these four constants, r, L, g, C, usually are assumed as located separately from each other, or localized. Although this assumption can never be perfectly correct, — for instance, every resistor has some inductance and every reactor has some resistance, — nevertheless in most cases it is permissible and necessary, and only in some classes of phenomena, and in some kinds of circuits, such as high-frequency phenomena, voltage and current distribution in long-distance, high-potential circuits, cables, telephone circuits, etc., this assumption is not permissible, but r, L, g, C must be treated as distributed throughout the circuit.

In the case of a circuit with distributed resistance, inductance, conductance, and capacity, as r, L, g, C, are denoted the effective resistance, inductance, conductance, and capacity, respectively, per unit length of circuit. The unit of length of the circuit may be chosen as is convenient, thus: the centimeter in the high-frequency oscillation over the multigap lightning arrester circuit, or a mile or kilometer in a long-distance transmission circuit or high-potential cable, or the distance of the velocity of light, 300,000 km., as most convenient in studying the laws of electric waves, etc.

The permanent values of current and e.m.f. in such circuits of distributed constants have, for alternating-current circuits, been investigated in Section III, where it was shown that they can be treated as transient phenomena in space, of the complex variables, current I and e.m.f. E.

Transient phenomena in circuits with distributed constants, and, therefore, the general investigation of such circuits, leads to transient phenomena of two independent variables, time t and space or distance l; that is, these phenomena are transient in time and in space.

The difficulty met in studying such phenomena is that they are not alternating functions of time, and therefore can no longer be represented by the complex quantity.

It is possible, however, to derive from the constants of the circuit, r, L, g, C, and without any assumption whatever regarding current, voltage, etc., general equations of the electric circuits, and to derive some results and conclusions from such equations.

These general equations of the electric circuit are based on the single assumption that the constants r, L, g, C remain constant with the time t and distance l, that is, are the same for every unit length of the circuit or of the section of the circuit to which the equations apply. Where the circuit constants change, as where another circuit joins the circuit in question, the integration constants in the equations also change correspondingly.

Special cases of these general equations then are all the phenomena of direct currents, alternating currents, discharges of reactive coils, high-frequency oscillations, etc., and the difference between these different circuits is due merely to different values of the integration constants.

2. In a circuit or a section of a circuit containing distributed resistance, inductance, conductance, and capacity, as a transmission line, cable, high-potential coil of a transformer, telephone or telegraph circuit, etc., let r = the effective resistance per unit length of circuit; L = the effective inductance per unit length of circuit; g = the effective shunted conductance per unit length of circuit; C = the effective capacity per unit length of circuit; t = the time, l = the distance, from some starting point; e = the voltage, and i = the current at any point l and at any time t; then e and i are functions of the time t and the distance l.

In an element dl of the circuit, the voltage e changes, by de, by the voltage consumed by the resistance of the circuit element, $ri\,dl$, and by the voltage consumed by the inductance of the circuit element, $L\dfrac{di}{dt}dl$. Hence,

$$\frac{de}{dl} = ri + L\frac{di}{dt}\,. \tag{1}$$

In this circuit element dl the current i changes, by di, by the current consumed by the conductance of the circuit element, $ge\,dl$, and by the current consumed by the capacity of the circuit element, $C\dfrac{de}{dt}dl$. Hence,

$$\frac{di}{dl} = ge + C\frac{de}{dt}\,. \tag{2}$$

Differentiating (1) with respect to t and (2) with respect to l, and substituting then (1) into (2), gives

$$\frac{d^2i}{dl^2} = rgi + (rC + gL)\frac{di}{dt} + LC\frac{d^2i}{dt^2}, \tag{3}$$

and in the same manner,

$$\frac{d^2e}{dl^2} = rge + (rC + gL)\frac{de}{dt} + LC\frac{d^2e}{dt^2}\,. \tag{4}$$

These differential equations, of the second order, of current i and voltage e are identical; that is, in an electric circuit current and e.m.f. are represented by the same equations, which differ by the integration constants only, which are derived from the terminal conditions of the problem.

These differential equations are linear functions in the dependent variable and its derivates, and as the general exponential function is the only integral of such a differential equation, that is, is the only function linearly related to its derivates, these equations are integrated by the exponential function. That is:

Equation (3) is integrated by terms of the form

$$i = A\varepsilon^{-al-bt}. \tag{5}$$

Substituting (5) in (3) gives the identity

$$a^2 = rg - (rC + gL)\, b + LCb^2$$
$$= (bL - r)\,(bC - g). \tag{6}$$

In the terms of the form (5) the relation (6) thus must exist between the coefficients of l and t.

Substituting (5) into (1) gives

$$\frac{de}{dl} = (r - bL)\, A\varepsilon^{-al-bt}, \tag{7}$$

and, integrated,

$$e = \frac{bL - r}{a} A\varepsilon^{-al-bt}. \tag{8}$$

Or, substituting (5) in (8), and then substituting:

$$a = \sqrt{(bL - r)(bC - g)}$$

gives:
$$z = \frac{bL - r}{a} = \sqrt{\frac{bL - r}{bC - g}} \tag{9}$$

as what may be called the "surge impedance," or "natural impedance" of the circuit, and

$$e = \sqrt{\frac{bL - r}{bC - g}}\, A\varepsilon^{-al-bt} \tag{10}$$

or:
$$e = zi \tag{11}$$

The integration constant of (8) would be a function of t, and since it must fulfill equation (4), must also have the form (5) for the special value $a = 0$, hence, by (6), $b = \dfrac{r}{L}$ or $b = \dfrac{g}{C}$ and therefore can be dropped.

In their most general form the *equations of the electric circuit* are

$$i = \sum_n \left\{ A_n\varepsilon^{-a_nl - b_nt} \right\}, \tag{12}$$

$$e = \sum_n \left\{ \frac{b_nL - r}{a_n} A_n\varepsilon^{-a_nl - b_nt} \right\}, \tag{13}$$

$$a_n^2 - (b_nL - r)\,(b_nC - g) = 0, \tag{14}$$

where A_n and a_n and b_n are integration constants, the last two being related to each other by the equation (14).

3. These pairs of integration constants, A_n and (a_n, b_n), are determinated by the terminal conditions of the problem.

Some such terminal conditions, for instance, are:

Current i and voltage e given as a function of time at one point l_0 of the circuit — at the generating station feeding into the circuit or at the receiving end of the transmission line.

Current i given at one point, voltage e at another point — as voltage at the generator end, current at the receiving end of the line.

Voltage given at one point and the impedance, that is, the complex ratio $\dfrac{\text{volts}}{\text{amperes}}$, at another point, as voltage at the generator end, load at the receiving end of the circuit.

Current and voltage given at one time t_0 as function of the distance l, as distribution of voltage and current in the circuit at the starting moment of an oscillation, etc.

Other frequent terminal conditions are:

Current zero at all times at one point l_0, as the open end of the circuit.

Voltage zero at all times at one point l_0, as the grounded or the short-circuited end of the circuit.

Current and voltage at all times at one point l_0 of the circuit, equal to current and voltage at one point of another circuit, as the connecting point of one circuit with another one.

As illustration, some of these cases will be discussed **below.**

The quantities i and e must always be real; but since a_n and b_n appear in the exponent of the exponential function, a_n and b_n may be complex quantities, in which case the integration constants A_n must be such complex quantities that by combining the different exponential terms of the same index n, that is, corresponding to the different pairs of a and b derived from the same equation (13), the imaginary terms in A_n and

$$\frac{b_n L - r}{a_n} A_n \text{ cancel.}$$

In the exponential function

$$\varepsilon^{-al-bt},$$

writing

$$a = h + jk \quad \text{and} \quad b = p + jq, \qquad (15)$$

we have

$$\varepsilon^{-al-bt} = \varepsilon^{-hl-pt}\varepsilon^{-j(kl+qt)},$$

and the latter term resolves into trigonometric functions of the angle

$$kl + qt.$$

$$kl + qt = \text{constant} \qquad (16)$$

therefore gives the relation between l and t for constant phase of the oscillation or alternation of the current or voltage.

h and p thus are the coefficients of the transient, k and q the coefficients of the periodic term.

With change of time t the phase thus changes in position l in the circuit, that is, moves along the circuit.

Differentiating (16) with respect to t gives

$$k\frac{dl}{dt} + q = 0,$$

or

$$\frac{dl}{dt} = -\frac{q}{k}; \qquad (17)$$

that is, the phase of the oscillation or alternation moves along the circuit with the speed $-\frac{q}{k}$, or, in other words,

$$S = -\frac{q}{k} \qquad (18)$$

is the speed of *propagation of the electric phenomenon in the circuit,* and the phenomenon may be considered a wave motion.

(If no energy losses occur, $r = 0$, $g = 0$, in a straight conductor in a medium of unit magnetic and dielectric constant, that is, unit permeability and unit inductive capacity, S is the velocity of light.)

4. Since (14) is a quadratic equation, several pairs or corresponding values of a and b exist, which, in the most general case, are complex imaginary. The terms with conjugate complex imaginary values of a and b then have to be combined for the elimination of their imaginary form, and thereby trigonometric functions appear; that is, several terms in the equations (12) and

(13), which correspond to the same equation (14), and thus can be said to form a group, can be combined with each other.

Such a group of terms, of the same index n, is defined by the equation (14),

$$a_n{}^2 = (b_n L - r)(b_n C - g).$$

For convenience the index n may be dropped in the investigation of a group of terms of current and voltage, thus:

$$a^2 = (bL - r)(bC - g), \tag{19}$$

and the following substitutions may be made:

$$a = a_1 \sqrt{LC}, \tag{20}$$

$$\left.\begin{aligned} a &= h + jk, \\ a_1 &= h_1 + jk_1, \\ b &= p + jq, \end{aligned}\right\} \tag{21}$$

from which

$$\begin{aligned} h &= h_1 \sqrt{LC} \text{ and} \\ k &= k_1 \sqrt{LC}. \end{aligned} \tag{22}$$

Substituting (18) in (19),

$$(h_1 + jk_1)^2 = \left[(p + jq) - \frac{r}{L}\right]\left[(p + jq) - \frac{g}{C}\right]. \tag{23}$$

Carrying out and separating the real and the imaginary terms, equation (23) resolves into the two equations thus:

and
$$\left.\begin{aligned} h_1{}^2 - k_1{}^2 &= \left(p - \frac{r}{L}\right)\left(p - \frac{g}{C}\right) - q^2 \\ h_1 k_1 &= q\left(2\,p - \frac{r}{L} - \frac{g}{C}\right). \end{aligned}\right\} \tag{24}$$

Substituting

$$u = \frac{1}{2}\left(\frac{r}{L} + \frac{g}{C}\right) \tag{25}$$

$$m = \frac{1}{2}\left(\frac{r}{L} - \frac{g}{C}\right), \tag{26}$$

and $$p = s + u \tag{27}$$

into (24) gives

$$\left.\begin{aligned} h_1{}^2 - k_1{}^2 &= s^2 - q^2 - m^2, \\ h_1 k_1 &= sq, \end{aligned}\right\} \tag{28}$$

or

and

$$\left.\begin{aligned} s^2 - q^2 &= h_1{}^2 - k_1{}^2 + m^2, \\ sq &= h_1 k_1. \end{aligned}\right\} \tag{29}$$

or

Adding four times the square of the second equation to the square of the first equation of (28) and (29) respectively, gives

$$\left.\begin{aligned} h_1{}^2 + k_1{}^2 &= \sqrt{(s^2 - q^2 - m^2)^2 + 4\,s^2 q^2} \\ &= \sqrt{(s^2 + q^2 - m^2)^2 + 4\,q^2 m^2} \\ &= R_1{}^2 \end{aligned}\right\} \tag{30}$$

and

$$\left.\begin{aligned} s^2 + q^2 &= \sqrt{(h_1{}^2 - k_1{}^2 + m^2)^2 + 4\,h_1{}^2 k_1{}^2} \\ &= \sqrt{(h_1{}^2 + k_1{}^2 + m^2)^2 - 4\,k_1{}^2 m^2} \\ &= R_0{}^2, \end{aligned}\right\} \tag{31}$$

and substituting (22), gives, by (28), (29) and (30), (31)

$$\left.\begin{aligned} h &= \sqrt{LC}\,\sqrt{\tfrac{1}{2}\{R_1{}^2 + s^2 - q^2 - m^2\}}, \\ k &= \sqrt{LC}\,\sqrt{\tfrac{1}{2}\{R_1{}^2 - s^2 + q^2 + m^2\}}, \\ R_1{}^2 &= \sqrt{(s^2 + q^2 - m^2)^2 + 4\,q^2 m^2}, \end{aligned}\right\} \tag{32}$$

or

$$\left.\begin{aligned} s &= \frac{1}{\sqrt{LC}}\sqrt{\tfrac{1}{2}\{R_2{}^2 + h^2 - k^2 + LCm^2\}}, \\ q &= \frac{1}{\sqrt{LC}}\sqrt{\tfrac{1}{2}\{R_2{}^2 - h^2 + k^2 - LCm^2\}}, \\ R_2{}^2 &= \sqrt{(h^2 + k^2 + LCm^2)^2 - 4\,LCk^2 m^2}. \end{aligned}\right\} \tag{33}$$

If, however $(+ h + jk)$ and $(u + s + jq)$ satisfy equation (19), then any other one of the expressions

$$(\pm h \pm jk) \text{ and } (u \pm s \pm jq)$$

also satisfies equation (19), providing also the second equation of (28) or (29) is satisfied,

$$hk = LC \ sq; \tag{34}$$

that is, if s and q have the same sign, h and k must have the same sign, and inversely, if s and q have opposite signs, h and k must have opposite signs.

This then gives the corresponding values of a and b:

$$
\left.
\begin{array}{llll}
(1) & a = + h + jk & b = u - s - jq \\
    & + h - jk & u - s + jq \\
(2) & a = - h - jk & b = u - s - jq \\
    & - h + jk & u - s + jq \\
(3) & a = - h + jk & b = u + s - jq \\
    & - h - jk & u + s + jq \\
(4) & a = + h - jk & b = u + s - jq \\
    & + h + jk & u + s + jq
\end{array}
\right\} \tag{35}
$$

or eight pairs of corresponding values of a and b.

p is called the *attenuation constant*, since it represents the decrease of the electrical effect with the time.

u is called the *dissipation constant*, since it represents the dissipation of electrical energy in resistance and conductance.

m is called the *distortion constant*, since on it depends the distortion of the electric circuit, that is, the displacement between current and voltage, as will be seen hereafter.

s is called the *energy transfer constant*, since it represents the energy transfer, as will be seen hereafter.

5. Substituting the values (1) of (35) into one group of terms of equations (12) and (13),

and

$$
\left.
\begin{array}{l}
i = A\varepsilon^{-al-bt} \\[3mm]
e = \dfrac{bL - r}{a} A\varepsilon^{-al-bt},
\end{array}
\right\} \tag{36}
$$

gives

$$i_1 = A_1 \varepsilon^{-(h+jk)l-(u-s-jq)t} + A_1' \varepsilon^{-(h-jk)l-(u-s+jq)t}$$
$$= \varepsilon^{-hl-(u-s)t}\{A_1 \varepsilon^{-jkl+jqt} + A_1' \varepsilon^{+jkl-jqt}\},$$

and substituting for the exponential functions with imaginary exponents their trigonometric expressions by the equation

$$\varepsilon^{\pm jx} = \cos x \pm j \sin x$$

gives

$$i_1 = \varepsilon^{-hl-(u-s)t}\{A_1[\cos(qt-kl)+j\sin(qt-kl)]$$
$$+ A_1'[\cos(qt-kl)-j\sin(qt-kl)]\}$$
$$= \varepsilon^{-hl-(u-s)t}\{(A_1+A_1')\cos(qt-kl)+j(A_1-A_1')\sin(qt-kl)\};$$

hence, A_1 and A_1' must be conjugate complex imaginary quantities, and writing

$$C_1 = A_1 + A_1'$$

and

$$C_1' = j(A_1 - A_1') \qquad (37)$$

gives

$$i_1 = \varepsilon^{-hl-(u-s)t}\{C_1 \cos(qt-kl) + C_1' \sin(qt-kl)\}. \qquad (38)$$

Substituting in the same manner in the equation of e, in (36), gives

$$e_1 = \frac{(u-s-jq)L-r}{h+jk} A_1 \varepsilon^{-(h+jk)l-(u-s-jq)t}$$
$$+ \frac{(u-s+jq)L-r}{h-jk} A_1' \varepsilon^{-(h-jk)l-(u-s+jq)t}$$
$$= \varepsilon^{-hl-(u-s)t}\left\{\frac{[(u-s-jq)L-r](h-jk)}{h^2+k^2} A_1 \varepsilon^{+j(qt-kl)}\right.$$
$$\left.+ \frac{[(u-s+jq)L-r](h+jk)}{h^2+k^2} A_1' \varepsilon^{-j(qt-kl)}\right\};$$

hence expanding, and substituting the trigonometric expressions

$$e_1 = \varepsilon^{-hl-(u-s)t}\left\{\left(\frac{[(u-s)L-r]h-qkL}{h^2+k^2} - j\frac{[(u-s)L-r]k+qhL}{h^2+k^2}\right)\right.$$
$$A_1[\cos(qt-kl)+j\sin(qt-kl)] + \left(\frac{[(u-s)L-r]h-qkL}{h^2+k^2}\right.$$
$$\left.+ j\frac{[(u-s)L-r]k+qkL}{h^2+k^2}\right)A_1'[\cos(qt-kl)-j\sin(qt-kl)]\right\}, \qquad (39)$$

and introducing the denotations

$$c_1 = \frac{qkL + h\left[r - (u - s)\,L\right]}{h^2 + k^2} = \frac{qk + h\,(m + s)}{h^2 + k^2}\,L$$

$$c_1' = \frac{k\left[r - (u - s)\,L\right] - qhL}{h^2 + k^2} = \frac{k\,(m + s) - qh}{h^2 + k^2}\,L, \tag{40}$$

and substituting (40) in (39), gives

$$e_1 = \varepsilon^{-hl - (u-s)\,t}\{(-c_1 + jc_1')\,A_1\,[\cos\,(qt - kl) + j\,\sin\,(qt - kl)]$$
$$+ (-c_1 - jc_1')\,A_1'\,[\cos\,(qt - kl) - j\,\sin\,(qt - kl)]\}$$
$$= \varepsilon^{-hl - (u-s)\,t}\{[-c_1\,(A_1 + A_1') + jc_1'\,(A_1 - A_1')]\cos\,(qt - kl)$$
$$+ [-jc_1(A_1 - A_1') - c_1'(A_1 + A_1')]\sin\,(qt - kl)\}. \tag{41}$$

Substituting the denotations (37) into (41) gives

$$e_1 = \varepsilon^{-hl - (u-s)\,t}\{(c_1'C_1' - c_1C_1)\cos\,(qt - kl)$$
$$- (c_1C_1' + c_1'C_1)\sin\,(qt - kl)\}. \tag{42}$$

The second group of values of a and b in equation (35) differs from the first one merely by the reversal of the signs of h and k, and the values i_2 and e_2 thus are derived from those of i_1 and e_1 by reversing the signs of h and k.

Leaving then the same denotations c_1 and c_1' would reverse the sign of e_2, or, by reversing the sign of the integration constants C, that is, substituting

$$C_2 = - (A_2 + A_2')$$
and
$$C_2' = - j\,(A_2 - A_2'), \tag{43}$$

the sign of i_2 reverses; that is,

$$i_2 = - \varepsilon^{+hl - (u-s)\,t}\{C_2\cos\,(qt + kl) + C_2'\sin\,(qt + kl)\} \tag{44}$$

and

$$e_2 = \varepsilon^{+hl - (u-s)\,t}\{(c_1'C_2' - c_1C_2)\cos\,(qt + kl)$$
$$- (c_1C_2' + c_1'C_2)\sin\,(qt + kl)\}. \tag{45}$$

The third group in equation (35) differs from the first one by the reversal of the signs of h and s, and its values i_3 and e_3 therefore are derived from i_1 and e_1 by reversing the signs of h and s.

Introducing the denotations

$$c_2 = \frac{qk - h(m - s)}{h^2 + k^2}L,$$
$$c_2' = \frac{k(m - s) + qh}{h^2 + k^2}L \qquad (46)$$

and

$$C_3 = A_3 + A_3',$$
$$C_3' = j(A_3 - A_3'), \qquad (47)$$

gives

$$i_3 = \varepsilon^{+hl-(u+s)t}\{C_3 \cos(qt - kl) + C_3' \sin(qt - kl)\} \quad (48)$$

and

$$e_3 = \varepsilon^{+hl-(u+s)t}\{(c_2'C_3' - c_2 C_3) \cos(qt - kl)$$
$$- (c_2 C_3' + c_2' C_3) \sin(qt - kl)\}. \qquad (49)$$

The fourth group in (35) follows from the third group by the reversal of the signs h and k, and retaining the denotations c_2 and c_2', but introducing the integration constants,

$$C_4 = -(A_4 + A_4')$$
$$C_4' = -j(A_4 - A_4'), \qquad (50)$$

and

gives

$$i_4 = -\varepsilon^{-hl-(u+s)t}\{C_4 \cos(qt + kl) + C_4' \sin(qt + kl)\} \quad (51)$$

and

$$e_4 = \varepsilon^{-hl-(u+s)t}\{(c_2'C_4' - c_2 C_4) \cos(qt + kl)$$
$$- (c_2 C_4' + c_2' C_4) \sin(qt + kl)\}. \qquad (52)$$

6. This then gives as the general expression of the *equations of the electric circuit:*

$$i = \sum[\varepsilon^{-hl-(u-s)t}\{C_1 \cos(qt - kl) + C_1' \sin(qt - kl)\}(i_1)$$
$$- \varepsilon^{+hl-(u-s)t}\{C_2 \cos(qt + kl) + C_2' \sin(qt + kl)\}(i_2)$$
$$+ \varepsilon^{+hl-(u+s)t}\{C_3 \cos(qt - kl) + C_3' \sin(qt - kl)\}(i_3)$$
$$- \varepsilon^{-hl-(u+s)t}\{C_4 \cos(qt + kl) + C_4' \sin(qt + kl)\}](i_4) \qquad (53)$$

and

$$
\begin{aligned}
e = \sum \big[&\varepsilon^{-hl-(u-s)t} \big\{ (c_1'C_1' - c_1C_1) \cos (qt - kl) \\
&\quad - (c_1'C_1 + c_1C_1') \sin (qt - kl) \big\} \ (e_1) \\
+ &\varepsilon^{+hl-(u-s)t} \big\{ (c_1'C_2' - c_1C_2) \cos (qt + kl) \\
&\quad - (c_1'C_2 + c_1C_2') \sin (qt + kl) \big\} \ (e_2) \\
+ &\varepsilon^{+hl-(u+s)t} \big\{ (c_2'C_3' - c_2C_3) \cos (qt - kl) \\
&\quad - (c_2'C_3 + c_2C_3') \sin (qt - kl) \big\} \ (e_3) \\
+ &\varepsilon^{-hl-(u+s)t} \big\{ (c_2'C_4' - c_2C_4) \cos (qt + kl) \\
&\quad - (c_2'C_4 + c_2C_4') \sin (qt + kl) \big\} \big] \ (e_4),
\end{aligned}
\tag{54}
$$

where C_1, C_1', C_2, C_2', C_3, C_3', C_4, C_4' and two of the four values s, q, h, k are integration constants, depending on the terminal conditions, and

$$
\left.
\begin{aligned}
c_1 &= \frac{qk + h\,(m + s)}{h^2 + k^2}\,L, \\[1mm]
c_1' &= \frac{k\,(m + s) - qh}{h^2 + k^2}\,L, \\[1mm]
c_2 &= \frac{qk - h\,(m - s)}{h^2 + k^2}\,L, \\[1mm]
c_2' &= \frac{k\,(m - s) + qh}{h^2 + k^2}\,L,
\end{aligned}
\right\}
\tag{55}
$$

and

$$
\left.
\begin{aligned}
u &= \frac{1}{2}\Big(\frac{r}{L} + \frac{g}{C}\Big), \\[1mm]
m &= \frac{1}{2}\Big(\frac{r}{L} - \frac{g}{C}\Big),
\end{aligned}
\right\}
\tag{56}
$$

and h, k and s, q are related by the equations

$$
\left.
\begin{aligned}
h &= \sqrt{LC}\ \sqrt{\tfrac{1}{2}\{R_1^2 + s^2 - q^2 - m^2\}}, \\
k &= \sqrt{LC}\ \sqrt{\tfrac{1}{2}\{R_1^2 - s^2 + q^2 + m^2\}},
\end{aligned}
\right\}
\tag{57}
$$

and

$$
R_1^2 = \sqrt{(s^2 + q^2 - m^2)^2 + 4\,q^2 m^2};
$$

hence,
$$
h^2 + k^2 = LCR_1^2,
\tag{58}
$$

or

$$s = \frac{1}{\sqrt{LC}} \sqrt{\tfrac{1}{2} \left\{ R_2^{\;2} + h^2 - k^2 + LCm^2 \right\}},$$

$$q = \frac{1}{\sqrt{LC}} \sqrt{\tfrac{1}{2} \left\{ R_2^{\;2} - h^2 + k^2 - LCm^2 \right\}}, \qquad (59)$$

and

$$R_2^{\;2} = \sqrt{(h^2 + k^2 + LCm^2)^2 - 4\,LCk^2m^2};$$

hence,

$$s^2 + q^2 = \frac{R_2^{\;2}}{LC}. \qquad (60)$$

Writing

$$D\,(qt \pm kl) = C \cos (qt \pm kl) + C' \sin (qt \pm kl) \qquad (61)$$

and

$$\begin{aligned} H\,(qt \pm kl) &= (c'C' - cC) \cos (qt \pm kl) \\ &\quad - (c'C + cC') \sin (qt \pm kl), \end{aligned} \qquad (62)$$

equations (50) and (51) can be written thus:

$$\left. \begin{aligned} i = \sum \Big[&\varepsilon^{-hl-(u-s)t}\, D_1\,(qt-kl) - \varepsilon^{+hl-(u-s)t}\, D_2(qt+kl) \quad (i') \\ &+ \varepsilon^{+hl-(u+s)t}\, D_3\,(qt-kl) - \varepsilon^{-hl-(u+s)t}\, D_4(qt+kl) \Big] \quad (i'') \end{aligned} \right\} (63)$$

and

$$\left. \begin{aligned} e = \sum \Big[&\varepsilon^{-hl-(u-s)t}\, H_1\,(qt-kl) + \varepsilon^{+hl-(u-s)t}\, H_2(qt+kl) \quad (e') \\ &+ \varepsilon^{+hl-(u+s)t}\, H_3\,(qt-kl) + \varepsilon^{-hl-(u+s)t}\, H_4(qt+kl) \Big] \quad (e'') \end{aligned} \right\} (64)$$

CHAPTER II

DISCUSSION OF SPECIAL CASES

7. The general equations of the electric circuit, (12) and (13) of Chapter I, consist of groups of terms of the form:

$$i = A\epsilon^{-al\ -bt} \tag{1}$$

$$\left. \begin{aligned} e &= \frac{bL - r}{a} A\epsilon^{-al-bt} \\ &= \sqrt{\frac{bL - r}{bC - g}} A\epsilon^{-al-bt} \\ &= zi \end{aligned} \right\} \tag{2}$$

where

$$z = \sqrt{\frac{bL - r}{bC - g}} \tag{3}$$

is the "surge impedance," or "natural impedance" of the circuit and a and b are related by the equation (14) of Chapter I:

$$\left. \begin{aligned} a^2 &= b^2LC - b(gL + rC) + rg \\ &= (bL - r)(bC - g) \end{aligned} \right\} \tag{4}$$

These equations must represent every existing electric circuit and every circuit which can be imagined, from the lightning stroke to the house bell and from the underground cable or transmission line to the incandescent lamp, under the only condition, that r, L, g and C are constant, or can be assumed as such with sufficient approximation.

The difference between all existing circuits thus consists merely in the difference in value of the constants A, and the constants a and b, the latter being related to each other by the equation (4).

To illustrate this, some special cases may be considered.

In general, A, a and b are general numbers, that is, complex imaginary quantities, but as such must be of such form that in the final form of i and e the imaginary terms eliminate. Thus, whenever a term of the form $X + jY$ exists, another term must exist of the form $X - jY$.

464

I. Special Case $b = 0$: Permanents

8. $b = 0$ means, that the electrical phenomenon is not a function of time, that is, is not transient, periodic or varying, but is constant or *permanent*.

By (4) it is:

$$a = \pm \sqrt{rg} \tag{5}$$

That is, two values of a exist, either of which gives a term equation (1), (2), and any combination of these terms thus also satisfies the differential equations.

In this case, by (3):

$$z = \pm \sqrt{\frac{r}{g}} \tag{6}$$

hence, the general equation of a *permanent* is:

$$\left. \begin{aligned} i &= A_1\epsilon^{-l\sqrt{rg}} + A_2\epsilon^{+l\sqrt{rg}} \\ e &= \sqrt{\frac{r}{g}} \left\{ A_1\epsilon^{-l\sqrt{rg}} - A_2\epsilon^{+l\sqrt{rg}} \right\} \end{aligned} \right\} \tag{7}$$

These equations do not contain L and C, that is, inductance and capacity have no effect in a *permanent*, that is, on an electrical phenomenon, which does not vary in time.

Equations (7) are the equations of a direct current circuit having distributed leakage, such as a metallic conductor submerged in water, or the current flow in the armor of a cable laid in the ground, or the current flow in the rail return of a direct current railway, etc.

r is the series resistance per unit length, g the shunted or leakage conductance per unit length of circuit.

Where the leakage conductance is not uniformly distributed, but varies, the numerical values in (7) change wherever the circuit constants change, just as would be the case if the resistance r of the conductor changed. If the leakage conductance g is not uniformly distributed, but localized periodically in space—as at the ties of the railroad track—when dealing with a sufficient circuit length, the assumption of uniformity would be justified as approximation.

9. (*a*) If the conductor is of infinite length—that is, of such great length, that the current which reaches the end of the conductor, is negligible compared with the current entering the conductor—it is:

$$A_2 = 0$$

since otherwise the second term of equation (18) would become infinite for $l = \infty$.

This gives:

$$i = A\epsilon^{-l\sqrt{rg}}$$
$$e = A\sqrt{\frac{r}{g}}\,\epsilon^{-l\sqrt{rg}}$$

(8)

or,

$$e = \sqrt{\frac{r}{g}}\,i$$

(9)

that is, a conductor of infinite length (or very great length) of series resistance r and shunted conductance g, has the effective resistance $r_0 = \sqrt{\frac{r}{g}}$.

It is interesting to note, that at a change of r or of g the effective resistance r_0, and thus the current flowing into the conductor at constant impressed voltage, or the voltage consumed at constant current, changes much less than r or g.

(b) If the circuit is open at $l = l_0$, it is:

$$i = A_1\epsilon^{-l_0\sqrt{rg}} + A_2\epsilon^{+l_0\sqrt{rg}} = 0$$

hence, if

$$A = A_1\epsilon^{-l_0\sqrt{rg}} = -A_2\epsilon^{+l_0\sqrt{rg}}$$

it is

$$i = A\left\{\epsilon^{+(l_0-l)\sqrt{rg}} - \epsilon^{-(l_0-l)\sqrt{rg}}\right\}$$
$$e = A\sqrt{\frac{r}{g}}\left\{\epsilon^{+(l_0-l)\sqrt{rg}} + \epsilon^{-(l_0-l)\sqrt{rg}}\right\}$$

(10)

(c) If the circuit is closed upon itself at $l = l_0$, it is,

$$e = \sqrt{\frac{r}{g}}\left\{A_1\epsilon^{-l_0\sqrt{rg}} - A_2\epsilon^{+l_0\sqrt{rg}}\right\} = 0$$

hence, if

$$A = A_1\epsilon^{-l_0\sqrt{rg}} = A_2\epsilon^{+l_0\sqrt{rg}}$$

it is

$$i = A\left\{\epsilon^{+(l_0-l)\sqrt{rg}} + \epsilon^{-(l_0-l)\sqrt{rg}}\right\}$$
$$e = A\sqrt{\frac{r}{g}}\left\{\epsilon^{+(l_0-l)\sqrt{rg}} - \epsilon^{-(l_0-l)\sqrt{rg}}\right\}$$

(11)

If, in (11), $l_0 = 0$, that is, the circuit is closed at the starting point, it is

$$i = A\left\{\epsilon^{-l\sqrt{rg}} + \epsilon^{+l\sqrt{rg}}\right\}$$

$$e = A\sqrt{\frac{r}{g}}\left\{\epsilon^{-l\sqrt{rg}} - \epsilon^{+l\sqrt{rg}}\right\}$$

or, counting the distance in opposite direction, that is, changing the sign of l:

$$i = A\left\{\epsilon^{+l\sqrt{rg}} + \epsilon^{-l\sqrt{rg}}\right\}$$

$$e = \dot{A}\sqrt{\frac{r}{g}}\left\{\epsilon^{+l\sqrt{rg}} - \epsilon^{-l\sqrt{rg}}\right\} \tag{12}$$

Assuming now l to be infinitely small,

$$l \cong 0$$

we get, by

$$\epsilon^{\pm s} = 1 \pm s + \pm$$

$$i = 2A$$

$$e = 2A\sqrt{\frac{r}{g}}\,l\sqrt{rg} = 2Arl = rli$$

$rl = r_0$ is, however, the total resistance of the circuit, and the equations (23), for infinitely small l, thus assume the form:

$$e = r_0 i \tag{13}$$

that is, the equation of the direct current circuit with massed constants, which so appears as special case of the general direct current circuit.

10. (*d*) If the circuit, at $l = l_0$, is closed by a resistance r_0, it is:

$$\left.\left|\frac{e}{i}\right|\right|_{l = l_0} = r_0$$

hence,

$$\frac{A_1\epsilon^{-l_0\sqrt{rg}} - A_2\epsilon^{+l_0\sqrt{rg}}}{A_1\epsilon^{-l_0\sqrt{rg}} + A_2\epsilon^{+l_0\sqrt{rg}}} = \frac{r_0}{\sqrt{\dfrac{r}{g}}}$$

$$\frac{A_2\epsilon^{+l_0\sqrt{rg}}}{A_1\epsilon^{-l_0\sqrt{rg}}} = \frac{\sqrt{\dfrac{r}{g}} - r_0}{\sqrt{\dfrac{r}{g}} + r_0}$$

or,

$$A_2 = A_1 \epsilon^{-2l_0\sqrt{rg}} \frac{\sqrt{\frac{r}{g}} - r_0}{\sqrt{\frac{r}{g}} + r_0}$$

$$i = A \left\{ \epsilon^{-l\sqrt{rg}} - \frac{r_0 - \sqrt{\frac{r}{g}}}{r_0 + \sqrt{\frac{r}{g}}} \epsilon^{-(2l_0 - l)\sqrt{rg}} \right\}$$

$$e = A\sqrt{\frac{r}{g}} \left\{ \epsilon^{-l\sqrt{rg}} + \frac{r_0 - \sqrt{\frac{r}{g}}}{r_0 + \sqrt{\frac{r}{g}}} \epsilon^{-(2l_0 - l)\sqrt{rg}} \right\}$$

$$(14)$$

These equations (12) and (14) can be written in various different forms. They are interesting in showing in a direct current circuit features which usually are considered as characteristic of alternating currents, that is, of wave motion.

The first term of (12) or (14) is the outflowing or main current respectively voltage, the second term is the reflected one.

At the end of the circuit with distributed constants, reflection occurs at the resistance r_0.

If $r_0 > \sqrt{\frac{r}{g}}$, the coefficient of the second term is positive, and partial reflection of current occurs, while the return voltages add itself to the incoming voltage.

If $r_0 < \sqrt{\frac{r}{g}}$, reflection of voltage occurs, while the return current adds itself to the incoming current.

If $r_0 = \sqrt{\frac{r}{g}}$, the second term vanishes, and the equations (14) become those of (8), of an infinitely long conductor. That is:

A resistance r_0, equal to the effective resistance (surge impedance) $\sqrt{\frac{r}{g}}$ of a direct current circuit of distributed constants, passes current and voltage without reflection. A higher resistance r_0 partially reflects the voltage—completely so for $r_0 = \infty$, or open circuit. A lower resistance r_0 partially reflects the current—completely so for $r_0 = 0$, or short circuit.

$\sqrt{\dfrac{r}{g}}$ thus takes in direct current circuits the same position as the surge impedance in alternating current or transient circuits.

II. Special Case: $a = 0$

11. By equation (4), this gives two solutions:

$$b = \frac{r}{L} \text{ and } b = \frac{g}{C}$$

(a)
$$b = \frac{r}{L}$$

substituted in (1) and (2) gives:

$$\left. \begin{aligned} i &= A\epsilon^{-\frac{r}{L}t} \\ e &= 0 \end{aligned} \right\} \tag{15}$$

that is, the inductive discharge of a circuit closed upon itself.

(b) Substituting in equations (1) and (2):

$$B = \frac{bL - r}{a} A$$

hence,

$$i = B \frac{a}{bL - r} \epsilon^{-al-bt}$$
$$e = B\epsilon^{-al-bt}$$

and then substituting,

$$a = 0; b = \frac{g}{C}$$

gives
$$\left. \begin{aligned} i &= 0 \\ e &= B\epsilon^{-\frac{g}{C}t} \end{aligned} \right\} \tag{16}$$

that is, if the condenser C is shunted by the conductance g, at voltage e on the condenser and thus also on the conductance g, the current in the external or supply circuit is zero, that is, the current in the conductance g is equal and opposite to that in the condenser:

$$i_1 = ge = gB\epsilon^{-\frac{g}{C}t} \tag{17}$$

(16) and (17) thus are the equations of the condenser discharge through the conductance g.

III. Special Case: $l \cong 0$: Massed Constants

12. In most electrical circuits, the length of the circuit is so short, that at the rate of change of the electrical phenomena, no appreciable difference exists between the different parts of the circuit as the result of the finite velocity of propagation.   Or in other words, the current is the same throughout the entire circuit or circuit section.

In general, therefore, the electrical constants, resistance, inductance, capacity, conductance, can be assumed as massed together locally, and not as distributed along the circuit.

The case of distributed constants mainly requires consideration only in case of circuits of such length, that the length of the circuit is an appreciable part of the wave length of the current, or the length of the impulse, etc.   This is the case:

1. In circuits of great length, such as transmission lines.

2. When dealing with transients of very short duration, or very high frequency, such as high frequency oscillations, switching impulses, etc.

The equations of the circuit of massed constants—the usual alternating or direct current circuit—should thus be derived from the equations (1) and (2) by the condition.

$$l = 0$$

In any electric circuit, there must be a point of zero potential difference.   Before substituting $l = 0$, into equations (12) and (13), that is, considering only an infinitely short part of the circuit at $l = 0$, these equations must be so modified as to bring the zero point of potential difference within this part of the circuit.

Thus, we first have to substitute in (1) and (2):

$$e = 0 \text{ at } l = 0$$

by equations (4) it is:

$$a = \pm \sqrt{(bL - r)(bC - g)} \tag{18}$$

that is, to every value of b correspond two values of a, equal numerically, but of opposite signs.

Substituting these in equations (1) and (2), gives

$$i = A_1 \epsilon^{-al-bt} + A_2 \epsilon^{+al-bt}$$
$$e = \frac{bL - r}{a} \{ A_1 \epsilon^{-al-bt} - A_2 \epsilon^{+al-bt} \} \tag{19}$$

and for $e = 0$ at $l = 0$:

$$A_2 = A_1 = A$$

thus,

$$i = A \epsilon^{-bt} (\epsilon^{-al} + \epsilon^{+al})$$
$$e = \frac{bL - r}{a} A \epsilon^{-bt} (\epsilon^{-al} - \epsilon^{+al}) \tag{20}$$

substituting now, for infinitely small l,

$$\epsilon^{\pm al} = 1 \pm al$$

gives, as the *general equation of the circuit with massed constants*:

$$i = B\epsilon^{-bt}$$
$$e = (r_0 - bL_0)B\epsilon^{-bt} \tag{21}$$

where

$$B = 2A$$
$$r_0 = lr = \text{total resistance of circuit}$$
$$L_0 = bL = \text{total inductance of circuit.}$$

The equations of voltage, in (21), may also be written:

$$e = r_0 B \epsilon^{-bt} - L_0 b B \epsilon^{-bt}$$
$$= r_0 i - L_0 \frac{di}{dt} \tag{22}$$

which is the equation of the inductive circuit with massed constants.

Or

$$e = (r_0 - bL_0)i \tag{23}$$

(a) *D. C.: Direct Currents.*

13. $b = 0$, that is, the electric effect is a permanent, does not vary with the time. Substituted in equations (21) gives:

$$i = B$$
$$e = r_0 B \tag{24}$$

the equations of the direct current circuit.

(b) *I. C.: Impulse Currents.*

$$b = \text{real}$$

that is, as function of time, the electrical effect is not periodic, but is transient, or is an *impulse*.

Solving equation (4) for b, gives:

$$b = u \mp s \tag{25}$$

where

$$s = \sqrt{\frac{a^2}{LC} + m^2} \tag{26}$$

and

$$\left. \begin{aligned} u &= \frac{1}{2}\left(\frac{r}{L} + \frac{g}{C}\right) \\ m &= \frac{1}{2}\left(\frac{r}{L} - \frac{g}{C}\right) \end{aligned} \right\} \tag{27}$$

Substituting these values into the general equation of massed constants, (21), gives:

$$i = B_1 \epsilon^{-(u-s)t} + B_2 \epsilon^{-(u+s)t}$$

or:

$$i = \epsilon^{-ut}\{B_1 \epsilon^{+st} + B_2 \epsilon^{-st}\} \tag{28}$$

$$e = B_1[r - (u - s)L]\epsilon^{-(u-s)t} + B_2[r - (u + s)L]\epsilon^{-(u+s)t}$$

or:

$$e = \epsilon^{-ut}\{(r - uL)(B_1\epsilon^{+st} + B_2\epsilon^{-st}) + sL(B_1\epsilon^{+st} - B_2\epsilon^{-st})\} \tag{29}$$

Assuming $i = 0$ for $t = 0$, that is, counting the time from the zero value of current, gives:

$$B_1 = -B_2 = B$$

hence,

$$\left. \begin{aligned} i &= B\epsilon^{-ut}(\epsilon^{+st} - \epsilon^{-st}) \\ e &= B\epsilon^{-ut}\{(r - uL)(\epsilon^{+st} - \epsilon^{-st}) + sL(\epsilon^{+st} + \epsilon^{-st})\} \end{aligned} \right\} \tag{30}$$

In this case, by (25) and (26), it must be:

$$u^2 \geq s^2 \geq m^2 \tag{31}$$

14. (ba) In the Special Case, that

$$s = u$$

it is, by (28) and (29),

$$\left. \begin{aligned} i &= B_1 + B_2\epsilon^{-2ut} \\ e &= rB_1 + B_2(r - 2uL)\epsilon^{-2ut} \end{aligned} \right\} \tag{32}$$

or, substituting for u, gives:

$$
\left.
\begin{aligned}
i &= B_1 + B_2\epsilon^{-\frac{r}{L}t}\epsilon^{-\frac{g}{C}t} \\
e &= rB_1 + \frac{gL}{C}B_2\epsilon^{-\frac{r}{L}t}\epsilon^{-\frac{g}{C}t}
\end{aligned}
\right\}
\tag{33}
$$

This is the general equation of a direct current circuit having inductance, resistance, capacity and conductance, including permanent as well as transient term.

If $i = 0$ at $t = 0$, it is

$$
B_1 = -B_2 = B
$$

hence,

$$
\left.
\begin{aligned}
i &= B\left\{ 1 - \epsilon^{-\frac{r}{L}t}\epsilon^{-\frac{g}{C}t} \right\} \\
e &= rB\left\{ 1 - \frac{gL}{rC}\epsilon^{-\frac{r}{L}t}\epsilon^{-\frac{g}{C}t} \right\}
\end{aligned}
\right\}
\tag{34}
$$

These are the general equations of a direct current with starting transient.

For

$$
g = 0
$$

that is, no losses in the condenser circuit, equations (34) assume the usual form of the direct current starting transient:

$$
\left.
\begin{aligned}
i &= B\left(1 - \epsilon^{-\frac{r}{L}t} \right) \\
e &= rB
\end{aligned}
\right\}
\tag{35}
$$

(c) *A. C.*: *Alternating Currents.*

$$
b = \pm jq
$$

15. That is, as function of time, the electrical effect is periodic (since the exponential function with imaginary exponent is the trigonometric function), or alternating.

Substituted in equations (21), this gives:

$$
\left.
\begin{aligned}
i &= B_1\epsilon^{+jqt} + B_2\epsilon^{-jqt} \\
e &= (r + jqL)\,B_1\epsilon^{+jqt} + (r - jqL)\,B_2\epsilon^{-jqt}
\end{aligned}
\right\}
\tag{36}
$$

and, substituting,

$$
\epsilon^{\pm jqt} = \cos qt \pm j \sin qt
$$

into (36), gives

$$i = A_1 \cos qt + A_2 \sin qt \\ e = (rA_1 + qLA_2) \cos qt + (rA_2 - qLA_1) \sin qt \quad\quad (37)$$

where

$$A_1 = B_1 + B_2 \\ A_2 = j (B_1 - B_2) \quad\quad (38)$$

substituting,

$$qt = \phi \\ q = 2\pi f \\ 2\pi fL = x \quad\quad (39)$$

gives

$$i = A_1 \cos \phi + A_2 \sin \phi \\ e = (rA_1 + xA_2) \cos \phi + (rA_2 - xA_1) \sin \phi \\ = r (A_1 \cos \phi + A_2 \sin \phi) + x (A_2 \cos \phi - A_1 \sin \phi) \quad\quad (40)$$

(d) O. C.: Oscillating Currents.

16. $$b = p \pm jq$$

In the general case of the circuit of massed constants, where b is a general number or complex quantity, it is, substituting in (21)

$$i = \epsilon^{-pt}\{B_1\epsilon^{+jqt} + B_2\epsilon^{-jqt}\} \\ e = \epsilon^{-pt}\{r - pL + jqL)B_1\epsilon^{+jqt} + (r - pL - jqL)B_2\epsilon^{-jqt}\} \quad (41)$$

and, substituting again the trigonometric function for the imaginary exponential function in (41), gives, in the same manner as in (c)

$$i = \epsilon^{-pt}\{A_1 \cos qt + A_2 \sin qt\} \\ e = \epsilon^{-pt}\{[(r - pL)A_1 + qLA_2] \cos qt + [(r - pL)A_2 \\ - qLA_1] \sin qt\} \\ = \epsilon^{-pt}\{(r - pL)(A_1 \cos qt + A_2 \sin qt) + \\ qL (A_2 \cos qt - A_1 \sin qt)\} \quad (42)$$

These are the equations of the oscillating currents and voltages, in a circuit of massed constants, consisting of the transient term ϵ^{-pt}, and the alternating or periodic term. The latter is the same as with the alternating currents, equations (37) or (40), except that in (42), in the equations of voltage $(r - pL)$ takes the place of the resistance r in (37). That is, the effective resistance, with oscillating currents, is lowered by the negative resistance of energy return, pL.

Substituting $p = 0$ in (42) gives (37).

IV. Special Case: Impulse Currents.

$$b = \text{real.}$$

17. By the equation (4)

$$a^2 = b^2LC - b\,(rC + gL) + rg,$$

to every value of b, correspond two values of a, equal numerically but opposite in sign:

$$\pm\,a$$

To every value of a correspond two values of b:

$$b = u \pm s \tag{43}$$

where

$$\left.\begin{aligned}
s &= \sqrt{m^2 + \frac{a^2}{LC}} \\
u &= \frac{1}{2}\left(\frac{r}{L} + \frac{g}{C}\right) \\
m &= \frac{1}{2}\left(\frac{r}{L} - \frac{g}{C}\right)
\end{aligned}\right\} \tag{44}$$

As b is assumed to be real, it must be positive, as otherwise the time exponential ϵ^{bt} would be increasing indefinitely. Thus it is, from (43):

$$s^2 \leqq u^2$$

Since u and b are real, s must be real. That is, by (44)

$$m^2 + \frac{a^2}{LC}$$

must be real and positive.

As m^2 is real, a^2 must be real, and must either be positive, or, if negative, $-\dfrac{a^2}{LC}$ must be less than m^2.

However, a can never be complex imaginary, without also making b complex imaginary, and impulse currents thus are characterized by the condition, that a is either real, or purely imaginary.

We thus get two cases of impulse currents:

$$b = u \mp s$$

(1) a^2 positive
  $a = \pm h$

$$s = \sqrt{m^2 + \frac{h^2}{LC}}$$

$$h = \sqrt{LC(s^2 - m^2)}$$

$$u^2 \geq s^2 \geq m^2$$

(45)

(2) a^2 negative
  $a = \pm jk$

$$s = \sqrt{m^2 - \frac{k^2}{LC}}$$

$$k = \sqrt{LC(m^2 - s^2)}$$

$$s^2 \leq m^2$$

$$k^2 \leq m^2 LC$$

(46)

non-periodic in space.

a.	b:	
$+ h$	$u - s$	
$- h$	$u - s$	(47)
$- h$	$u + s$	
$+ h$	$u + s$	

periodic in space.

a:	b:	
$+ jk$	$u - s$	
$- jk$	$u - s$	(48)
$+ jk$	$u + s$	
$- jk$	$u + s$	

(1) Non-periodic Impulse Currents.

18. Substituting (47) into (1) and (2) gives

$$i = \epsilon^{-ut}\{ A_1 \epsilon^{-hl+st} + A_2 \epsilon^{+hl+st} + A_3 \epsilon^{+hl-st} + A_4 \epsilon^{+hl-st}\} \quad (49)$$

$$e = \epsilon^{-ut}\left\{ A_1 \frac{(u-s)L - r}{h}\epsilon^{-hl+st} - A_2 \frac{(u-s)L - r}{h}\epsilon^{+hl+st} - \right.$$
$$\left. A_3 \frac{(u+s)L - r}{h}\epsilon^{+hl-st} + A_4 \frac{(u+s)L - r}{h}\epsilon^{-hl-sl}\right\} \quad (50)$$

It is, however,

$$\frac{(u \mp s)L - r}{h} = \frac{uL - r \mp sL}{h}$$

and, substituting for h from (45), and substituting for u

$$\frac{(u \mp s)L - r}{h} = \frac{\frac{1}{2}\left(\frac{r}{L} + \frac{g}{C}\right)L - r \mp sL}{\sqrt{LC}(s^2 - m^2)} = \frac{-mL \mp sL}{\sqrt{LC}(s^2 - m^2)},$$

$$= \mp \sqrt{\frac{L}{C}} \sqrt{\frac{s \pm m}{s \mp m}}$$

or, substituting,

$$\sqrt{\frac{s + m}{s - m}} = c \qquad (51)$$

it is

$$\frac{(u - s) L - r}{h} = - c \sqrt{\frac{L}{C}}$$

$$\frac{(u + s) L - r}{h} = + \frac{1}{c} \sqrt{\frac{L}{C}}$$

and, substituting these values into (50), gives

$$e = - \sqrt{\frac{L}{C}} \epsilon^{-ut} \left\{ c A_1 \epsilon^{-hl+st} - c A_2 \epsilon^{+hl+st} + \frac{1}{c} A_3 \epsilon^{+hl-st} \right.$$

$$\left. - \frac{1}{c} A_4 \epsilon^{-hl-st} \right\} \qquad (52)$$

The equations (49) and (52), of current and voltage respectively, of the non-periodic impulse, are the same as derived as equation (9) in Chapter III, as special case of the general circuit equation.

The constants A_1, A_2, A_3, A_4 of (60) are denoted by C_1, $- C_2$, C_3, $- C_4$ in (9) of Chapter III, and the further discussion of the equations given there.

(2) Periodic Impulse Currents.

19. Substituting (48) into (1) gives:

$$i = \epsilon^{-ut} \{ \epsilon^{+st} (A_1 \epsilon^{-jkl} + A_2 \epsilon^{+jkl}) + \epsilon^{-st} (A_3 \epsilon^{-jkl} + A_4 \epsilon^{+jkl}) \}$$

and, substituting the trigonometric expressions, this gives

$$i = \epsilon^{-ut} \{ \epsilon^{+st} (D_1 \cos kl - D_2 \sin kl) + \epsilon^{-st} (D_3 \cos kl - D_4 \sin kl) \} \qquad (53)$$

where

$$\left. \begin{array}{l} D_1 = A_1 + A_2 \\ D_2 = j (A_1 - A_2) \\ D_3 = A_3 + A_4 \\ D_4 = j (A_3 - A_4) \end{array} \right\} \qquad (54)$$

The expression

$$\frac{bL - r}{a}$$

in equation (2) assumes the form, by substituting (46) and (48):

$$\frac{bL - r}{a} = \frac{(u \mp s) L - r}{\pm jk} = \frac{-(m \pm s) L}{\pm j\sqrt{LC} (m^2 - s^2)}$$

$$= \pm j\sqrt{\frac{L}{C}} \sqrt{\frac{m \pm s}{m \mp s}}$$

and, substituting,

$$c = \sqrt{\frac{m + s}{m - s}} \qquad (55)$$

gives

$$\frac{bL - r}{a} = \pm jc\sqrt{\frac{L}{C}} \text{ respectively } \pm \frac{j}{c}\sqrt{\frac{L}{C}} \qquad (56)$$

Substituting now (48) and (56) into (2) gives

$$e = \sqrt{\frac{L}{C}} \epsilon^{-ut} \left\{ jc\epsilon^{+st}(A_1\epsilon^{-jkl} - A_2\epsilon^{-jkl}) + \frac{j}{c} \epsilon^{-st}(A_3\epsilon^{-jkl} - A_4\epsilon^{+jkl}) \right\}$$

and, substituting the trigonometric expression:

$$e = \sqrt{\frac{L}{C}} \epsilon^{-ut} \left\{ c\epsilon^{+st}(D_2 \cos kl + D_1 \sin kl) + \right.$$

$$\left. \frac{1}{c} \epsilon^{-st} (D_4 \cos kl + D_3 \sin kl) \right\} \qquad (57)$$

The equations (53) and (57), of current and voltage repectively, of the periodic impulse, are the equations (24) of Chapter III, derived by specialization from the general circuit equations, and their further discussion is given there.

V. Special Case: Alternating Currents

$$b = \text{imaginary.}$$

20. If

$$b = \pm jq$$

by equation (4), it is

$$a = \pm\sqrt{(r \mp jqL) (g \mp jqC)} \qquad (58)$$

$$= \pm (h \pm jk)$$

That is, a, as square root of a complex quantity, also becomes complex imaginary.

For

$$b = + jq$$

it is

$$a^2 = (r - jqL)(g - jqC)$$
$$= (rg - q^2L) - jq\,(rC + gL)$$

that is, the sign of the imaginary term of a^2 is negative, and the sign of the imaginary term of a must be negative, that is, $a = \pm (h - jk)$, and inversely.

The corresponding values of b and a thus are:

b:	a:
$+ jq$	$+ h - jk$
$- jq$	$+ h + jk$
$- jq$	$- h - jk$
$+ jq$	$- h + jk$

Substituting these values into (1) gives:

$$i = \epsilon^{-hl}\{A_1\epsilon^{+j(kl-qt)} + A_2\epsilon^{-j(kl-qt)}\} + \epsilon^{+hl}\{A_3\epsilon^{+j(kl+qt)} + A_4\epsilon^{-j(kl+qt)}\}$$

Thus, in trigonometric expression:

$$i = \epsilon^{-hl}\{B_1 \cos (kl - qt) + B_2 \sin (kl - qt)\}$$
$$+ \epsilon^{+hl}B_3 \cos (kl + qt) + B_4 \sin (kl + qt)\} \quad (59)$$

where

$$B_1 = A_1 + A_2 \qquad\qquad B_3 = A_3 + A_4$$
$$B_2 = j(A_1 - A_2) \qquad\qquad B_4 = j(A_3 - A_4)$$

Resolving in equation (59) the trigonometric function, and re-arranging, gives:

$$i = \epsilon^{-hl}\{(B_1 \cos qt - B_2 \sin qt) \cos kl + (B_2 \cos qt + B_1 \sin qt) \sin kl\}$$
$$+ \epsilon^{+hl}\{(B_3 \cos qt + B_4 \sin qt) \cos kl + (B_4 \cos qt - B_3 \sin qt) \sin kl\} \quad (60)$$

21. The equations (60) can be written in symbolic expression, by representing the terms with $\cos qt$ by the real, the terms with $\sin qt$ by the imaginary vector, that is, substituting,

$$B_1 \cos qt - B_2 \sin qt = B_1 + jB_2 = A_1$$

hence

$$B_2 \cos qt + B_1 \sin qt = B_2 - jB_1 = - jA_1$$
$$B_3 \cos qt + B_4 \sin qt = B_3 - jB_4 = - A_2$$
$$B_4 \cos qt - B_3 \sin qt = B_4 + jB_3 = - jA_2$$

hence

$$I = A_1\epsilon^{-hl} (\cos kl - j \sin kl) - A_2\epsilon^{+hl} (\cos kl + j \sin kl) \quad (61)$$

This is the equation (23) on page 295 in Section III, derived there as the current in a general alternating-current circuit of distributed constants.

Analogous, by substituting into equation (2), the equation of the voltage is derived:

$$E = \sqrt{\frac{Z}{Y}} \{ A_1 \epsilon^{-hl} (\cos kl - j \sin kl) + A_2 \epsilon^{+hl} (\cos kl + j \sin kl) \} \quad (62)$$

Where

$$Z = r + jqL$$
$$Y = g + jqC$$

and

$$q = 2\pi f$$

Alternating currents, representing the special case, where b is purely imaginary, and impulse currents, representing the case, where b is real, thus represent two analogous special cases of the general circuit equations.

These two classes are industrially the most important types of current, though their relation to the operation of electric systems is materially different:

The alternating currents are the useful currents in our large electric power systems.

The impulse currents are the harmful currents in our large electric power systems.

The alternating currents have been extensively studied, and the most of the preceding Section III is devoted to the general alternating-current circuit.

Very little work has been done in the study of impulse current, and the next Chapter thus shall be devoted to an outline of their theory.

CHAPTER III

IMPULSE CURRENTS

22. The terms of the general equations of the electric circuit (53) and (54) contain the constants:

The eight values, C and C'.

The four values, c and c'.

And the exponential constants u, m, h, k, s, q.

Of these, the values c and c' are expressions of L and of the exponential constants, by equation (55), thus are not independent constants.

u and m are circuit constants, and as such are the same for all terms of the general equation.

Of the four terms h, k, s, q, two are dependent upon the other two by the equations (57) and (59).

Thus there remains ten independent constants, which are to be determined by the terminal conditions:

Two of the four exponential terms h, k, s, q, and the eight coefficients C and C'.

h and s are attenuation constants, in length or space and in time respectively, and k and q are wave constants, in space and in time respectively.

In a non-periodic electrical effect, k and q thus would be zero, while $k = 0$ but $q \neq 0$ gives a phenomenon periodic in time, but non-periodic in space, and inversely, $q = 0$, but $k \neq 0$ give a phenomenon, periodic in space, but non-periodic in time.

If of the four constants: s, q, h, k, one equals zero, another one also must be zero; if

$$s = 0$$

it is, by equations (58) and (57):

$$R_1{}^2 = q^2 + m^2$$

$$h = 0$$

$$k = \sqrt{LC\,(q^2 + m^2)}$$

if

$$q = 0$$

it is, either

$$R_1{}^2 = s^2 - m^2$$

$$h = \sqrt{LC\,(s^2 - m^2)}$$

$$k = 0$$

or,

$$R_1{}^2 = m^2 - s^2$$

$$h = 0$$

$$k = \sqrt{LC\,(m^2 - s^2)}$$

if,

$$h = 0$$

it is, either

$$R_2{}^2 = k^2 - LCm^2$$

$$s = 0$$

$$q = \sqrt{\frac{k^2}{LC} - m^2}$$

or

$$R_2{}^2 = LCm^2 - k^2$$

$$s = \sqrt{m^2 - \frac{k^2}{LC}}$$

$$q = 0$$

if

$$k = 0$$

it is

$$R_2{}^2 = h^2 + LCm^2$$

$$s = \sqrt{\frac{h^2}{LC} + m^2}$$

$$q = 0$$

and this gives the three sets of values:

s	q	h	k
0	$\sqrt{\dfrac{k^2}{LC} - m^2}$	0	$\sqrt{LC\,(q^2 + m^2)}$
$\sqrt{m^2 - \dfrac{k^2}{LC}}$	0	0	$\sqrt{LC\,(m^2 - s^2)}$
$\sqrt{\dfrac{h^2}{LC} + m^2}$	0	$\sqrt{LC\,(s^2 - m^2)}$	0

$s = 0$ and $h = 0$ gives an electrical effect which is periodic in time and in space, and is transient in time. This case leads to the equation of the stationary oscillation of the circuit of distributed resistance, inductance, capacity and conductance, such as the transmission line, more fully discussed in Section III, and in Chapters V and VII of Section IV.

$q = 0$ gives a non-periodic transient.

Such a non-periodic transient is called an *impulse*, and currents, voltages and power of this character then are denoted as *impulse currents, impulse voltages,* and *impulse power.*

Impulse currents thus are non-periodic transients, while alternating currents are periodic non-transients or permanents.

23. An impulse thus is characterized by the condition:

$$q = 0.$$

Substituting this in equation (57) gives:

$$\left.\begin{array}{llll} R_1 = s^2 - m^2 & \text{or} & R_1 = m^2 - s^2 \\ h = \sqrt{LC\,(s^2 - m^2)} & \text{or} & h = 0 \\ k = 0 & \text{or} & k = \sqrt{LC\,(m^2 - s^2)} \end{array}\right\} \quad (2)$$

and inversely, by equation (59):

$$\left.\begin{array}{llll} R_2 = h^2 + LCm^2 & \text{or} & R_2 = LCm^2 - k^2 \\ s = \sqrt{\dfrac{h^2}{LC} + m^2} & \text{or} & s = \sqrt{m^2 - \dfrac{k^2}{LC}} \\ q = 0 & \text{or} & q = 0 \end{array}\right\} \quad (3)$$

and

$$m^2 - s^2 + \frac{h^2}{LC} = 0 \qquad m^2 - s^2 - \frac{k^2}{LC} = 0 \qquad (4)$$

That is, either $k = 0$, that is, the impulse is non-periodic in space also, or $h = 0$, that is, the impulse is periodic in space, has no attenuation with the distance.

By the character of the space distribution, we thus distinguish non-periodic and periodic impulses. One of the two constants, h or k, must always be zero, if $q = 0$, that is, the space distribution of an impulse can not be oscillatory, but is either exponential or trigonometric.

If

$s^2 > m^2$, the impulse is non-periodic; and $h^2 > LCs^2$

if

$s^2 < m^2$, the impulse is periodic, and $k^2 < LCm^2$

We thus distinguish two classes of impulses:

Non-periodic impulses and periodic impulses. The "periodic" here refers to the distribution in space, as in time the impulse always must be exponential or transient.

A. NON-PERIODIC IMPULSES.

24. A non-periodic impulse is an electrical effect, in which the electrical distribution is non-periodic, or exponential, in time as well as in space. Its condition is:

$$\left.\begin{array}{l} q = 0 \\ s^2 > m^2 \end{array}\right.$$

hence:

$$\left.\begin{array}{l} k = 0 \\ h = \sqrt{LC\,(s^2 - m^2)} \end{array}\right\} \tag{5}$$

or

$$s = \sqrt{\dfrac{h^2}{LC} + m^2}$$

Substituting these values in equations (55) gives

$$\left.\begin{array}{l} c_1 = L\,\dfrac{s + m}{h} = \sqrt{\dfrac{L}{C}}\,\sqrt{\dfrac{s + m}{s - m}} = c\sqrt{\dfrac{L}{C}} \\[2mm] c_2 = L\,\dfrac{s - m}{h} = \sqrt{\dfrac{L}{C}}\,\sqrt{\dfrac{s - m}{s + m}} = \dfrac{1}{c}\sqrt{\dfrac{L}{C}} \\[2mm] c_1{}' = 0 \\[1mm] c_2{}' = 0 \end{array}\right\} \tag{6}$$

where

$$c = \sqrt{\dfrac{s + m}{s - m}} \tag{7}$$

Substituting (5), (6) and (7) into equations (53) and (54), the sin terms vanish, and with them the integration constants C', and only the integration constants C remain; the cos terms become unity, and these equations assume the form:

$$i = \Sigma\{C_1\epsilon^{-hl-(u-s)t} - C_2\epsilon^{+hl-(u-s)t} + C_3\epsilon^{+hl-(u+s)t} - $$
$$C_4\epsilon^{-hl-(u+s)t}\}$$

$$e = -\Sigma\sqrt{\dfrac{L}{C}}\Big\{cC_1\epsilon^{-hl-(u-s)t} + cC_2\epsilon^{+hl-(u-s)t} + \dfrac{1}{c}C_3\epsilon^{+hl-\,u+s)t} $$
$$+ \dfrac{1}{c}C_4\epsilon^{-hl-\,u+s)t}\Big\} \tag{8}$$

or, considering only one group of the series, and separating ϵ^{-ut}:

$$i = \epsilon^{-ut}\{C_1\epsilon^{-hl+st} - C_2\epsilon^{+hl+st} + C_3\epsilon^{+hl-st} - C_4\epsilon^{-hl-st}\}$$

$$e = -\sqrt{\frac{L}{C}}\,\epsilon^{-ut}\left\{cC_1\epsilon^{-hl+st} + cC_2\epsilon^{+hl+st} + \frac{1}{c}C_3\epsilon^{+hl-st} + \frac{1}{c}C_4\epsilon^{-hl-st}\right\} \tag{9}$$

25. The equations (9) can be simplified by shifting the zero points of time and distance, by the substitution:

$$
\begin{aligned}
C_1 &= D_1\epsilon^{+hl_1-st_1} & C_2 &= D_2\epsilon^{-hl_1-st_1}\\
C_3 &= \pm\,D_1\epsilon^{-hl_1+st_1} & C_4 &= \pm\,D_2\epsilon^{+hl_1+st_1}
\end{aligned}\tag{10}
$$

hence,

$$D_1 = \sqrt{\pm C_1C_3} \qquad\qquad D_2 = \sqrt{\pm C_2C_4} \tag{11}$$

$$\epsilon^{+2hl_1-2st_1} = \pm\frac{C_1}{C_3} \qquad\qquad \epsilon^{+2hl_1+2st_1} = \pm\frac{C_4}{C_2}$$

$$\epsilon^{+4hl_1} = \pm\frac{C_1C_4}{C_2C_3}$$

$$\epsilon^{+4st_1} = \pm\frac{C_3C_4}{C_1C_2} \tag{12}$$

or

$$
\begin{aligned}
l_1 &= \frac{1}{4h}\log\left(\pm\frac{C_1C_4}{C_2C_3}\right)\\
t_1 &= \frac{1}{4s}\log\left(\pm\frac{C_3C_4}{C_1C_2}\right)
\end{aligned}\tag{13}
$$

where the $\pm$ denotes the positive value of the C product.

Making also the further substitution:

$$c = \epsilon^{+st_0}; \qquad \frac{1}{c} = \epsilon^{-st_0} \tag{14}$$

hence,

$$t_0 = \frac{\log c}{s} = \frac{1}{2s}\log\frac{s+m}{s-m} \tag{15}$$

(10) and (14) substituted in (9) gives:

$$i = \epsilon^{-ut}\{D_1(\epsilon^{-hl'+s(t'-t_0)} \pm \epsilon^{+hl'-s(t'-t_0)}) - D_2(\epsilon^{+hl'+s(t'-t_0)} \pm \epsilon^{-hl'-s(t'-t_0)})\}$$

$$e = -\sqrt{\frac{L}{C}}\,\epsilon^{-ut}\{D_1(\epsilon^{-hl'+st'} \pm \epsilon^{+hl'-st'}) + D_2(\epsilon^{+hl'+st'} \pm \epsilon^{-hl'-st'})\} \tag{16}$$

where

$$t' = t - t_1 + t_0$$
$$l_1 = l - l_1$$

(17)

By the substitution:

$$\epsilon^{+x} + \epsilon^{-x} = 2 \cosh x$$
$$\epsilon^{+x} - \epsilon^{-x} = 2 \sinh x$$

Equations (16) can be written in hyperbolic form thus:

$$i = \epsilon^{-ut}\{D_1' \cosh [hl' - s(t' - t_0)] - D_2' \cosh [hl' + s(t' - t_0)]\}$$
$$e = - \sqrt{\frac{L}{C}} \epsilon^{-ut}\{D_1' \cosh [hl' - st'] + D_2' \cosh [hl' + st']\}$$

(17b)

Or the corresponding sinh function, in case of the minus sign in equations (16).

26. The impulse thus is the combination of two single impulses of the form

$$\epsilon^{-ut}(\epsilon^{-hl+st} \pm \epsilon^{+hl-st})$$

which move in opposite direction, the D_1 impulse toward rising l: $\frac{dl}{dt} > 0$, and the D_2 impulse toward decreasing l: $\frac{dl}{dt} < 0$

The voltage impulse differs from the current impulse by the factor $\sqrt{\frac{L}{C}}$ (the "surge impedance"), and by a *time displacement* t_0. That is, in the general impulse, voltage e and current i are displaced in time.

t_0 thus may be called the time displacement, or time lag of the current impulse behind the voltage impulse.

t_0 is positive, that is, the current *lags* behind the voltage impulse, if in equation (15) the log is positive, that is, m is positive, or: $\frac{r}{L} > \frac{g}{C}$, that is, the resistance-inductance term preponderates.

Inversely, t_0 is negative, and the current *leads*, or the voltage impulse lags behind the current impulse, if m is negative, that is, $\frac{r}{L} < \frac{g}{C}$, or the capacity term preponderates.

If $g = 0$, that is, no shunted conductance, the current impulse always lags behind the voltage impulse.

If $m = 0$, that is, $\dfrac{r}{L} = \dfrac{g}{C}$, or $\dfrac{r}{g} = \dfrac{L}{C}$, $t_0 = 0$, that is, the voltage impulse and the current impulse are in phase with each other, that is, there is no time displacement, and current and voltage impulses have at any time or at any space the same shape: *distortionless circuit.* m therefore is called the *distortion constant* of the circuit.

27. If

$$s = m \tag{18}$$

it is

$$h = 0 \tag{19}$$

hence, substituting into equations (16)

$$\left. \begin{aligned} i &= A\epsilon^{-ut}(\epsilon^{+m(t'-t_0)} \pm \epsilon^{-m(t'-t_0)}) \\ e &= B\sqrt{\frac{L}{C}}\,\epsilon^{-ut}(\epsilon^{+mt'} \pm \epsilon^{-mt'}) \end{aligned} \right\} \tag{20}$$

where

$$\left. \begin{aligned} A &= D_1 - D_2 \\ B &= D_1 + D_2 \end{aligned} \right\}$$

that is, a simple impulse.

Or, substituting for u and m their values

$$\left. \begin{aligned} i &= A_1\epsilon^{-\frac{g}{C}t} \pm A_2\epsilon^{-\frac{r}{L}t} \\ e &= \sqrt{\frac{L}{C}}\left\{ B_1\epsilon^{-\frac{g}{C}t} \pm B_2\epsilon^{-\frac{r}{L}t} \right\} \end{aligned} \right\} \tag{21}$$

Thus, the capacity effect, as first term, and the inductance effect, as second term, appear separate.

28. In the individual impulse:

$$\epsilon^{-ut}(\epsilon^{-hl+st} \pm \epsilon^{+hl-st}) = \epsilon^{-hl}\epsilon^{-(u-s)t} \pm \epsilon^{+hl}\epsilon^{-(u+s)t} \tag{18}$$

the term ϵ^{-ut} is the attenuation of the impulse by the energy dissipation in the circuit, that is, represents the rate at which the impulse would die out by its energy dissipation.

The first term, $\epsilon^{-(u-s)t}$, dies out at a slower rate than given by the energy dissipation, that is, in this term, at any point l, energy is supplied, is left behind by the passing impulse, and as the result, this term decreases with increasing distance l, by the factor ϵ^{-hl}; inversely, the second term, $\epsilon^{-(u+s)t}$, dies out more rapidly with the time, than corresponds to the energy losses, that is, at any point l, this term abstracts energy and shifts it along the circuit, and thereby gives an increase of energy in the direction of propagation, by ϵ^{+hl}. In other words, of the two terms of the impulse, the one drops energy while moving along the line, and the other picks it up and carries it along.

The terms $\epsilon^{\pm st}$ thus represent the dropping and picking up of energy with the time, the terms $\epsilon^{\pm hl}$ the dropping and picking up of energy in space along the line. In distinction to u, which may be called the *energy dissipation constant*, s (and its corresponding h) thus may be called the *energy transfer constant* of the impulse. The higher s is, the greater then is the rate of energy transfer, that is, the steeper the wave front, and s thus may also be called the *wave front constant* of the impulse.

The transformation from the four constants C to the two constants D obviously merely represents the shifting of the zero value of time and distance to the center of the impulse.

Also, in equations (16), in the energy dissipation term ϵ^{-ut}, obviously also the new time t' may be used: $\epsilon^{-ut'} = \epsilon^{-u(t-t_1)}$, and this would merely mean a change of the constants D_1 and D_2 by the constant factor ϵ^{-ut_1}.

Substituting in equations (16): $l = 0$, gives the equation of the impulse in a circuit with massed constants

$$\left.\begin{aligned}
i &= A\epsilon^{-ut}(\epsilon^{+st} \pm \epsilon^{-st}) \\
e &= B\sqrt{\frac{L}{C}}\,\epsilon^{-ut}(\epsilon^{+(s)t-t_0} \pm \epsilon^{-s(t-t_0)})
\end{aligned}\right\}$$

where

$$\left.\begin{aligned}
A &= D_1 - D_2 \\
B &= D_1 + D_2
\end{aligned}\right\}$$

29. The equations (9) can be brought into a different form by the substitution,

$$\left.\begin{aligned}
C_1 &= B\epsilon^{+hl_1}\epsilon^{-st_1} \\
C_2 &= \pm\, B\epsilon^{-h(l_1+l_0)}\epsilon^{-st_1} \\
C_3 &= \pm\, B\epsilon^{-h(l_1+l_0)}\epsilon^{+st_1} \\
C_4 &= \pm\, B\epsilon^{+hl_1}\epsilon^{-st_1}
\end{aligned}\right\} \tag{22}$$

where the $+$ sign applies, if C_2, etc., is of the same sign with C_1, the $-$ sign, if it is of opposite sign.

Substituting also equations (14), this transforms equations (9) into the form

$$i = B\epsilon^{-ut}\{(\epsilon^{-hl'+s(t'-t_0)} \mp \epsilon^{-hl'-s(t'-t_0)}) - (\epsilon^{-h(l_0-l')+s(t'-t_0)} \mp \epsilon^{-h(l_0-l')-s(t'-t_0)})\}$$

$$e = -\sqrt{\frac{L}{C}}\,B\epsilon^{-ut}\{(\epsilon^{-hl'+st'} \pm \epsilon^{-hl'-st'}) + (\epsilon^{-h(l_0-l')+st'} \pm \epsilon^{-h(l_0-l')-st'})\} \tag{23}$$

where

$$\left.\begin{array}{c} l' = l - l_1 \\ t' = t - t_1 + t_0 \end{array}\right\} \tag{24}$$

or, re-arranging equations (23)

$$i = B\epsilon^{-ut}\{\epsilon^{-hl'} - \epsilon^{-h(l_0-l')}\}\{\epsilon^{+s(t'-t_0)} \pm \epsilon^{-s(t'-t_0)}\}$$

$$e = -\sqrt{\frac{L}{C}}\,B\,\epsilon^{-ut}\{\epsilon^{-hl'} + \epsilon^{-h(l_0-l')}\}\{\epsilon^{+st'} \pm \epsilon^{-st'}\} \tag{25}$$

Or substituting

$$\left.\begin{array}{c} B = A\,\epsilon^{+\frac{hl_0}{2}} \\[4pt] l = l' - \dfrac{l_0}{2} \end{array}\right\} \tag{26}$$

and

gives

$$i = A\epsilon^{-ut}\{\epsilon^{-hl} - \epsilon^{+hl}\}\{\epsilon^{+st('-t_0(} \mp \epsilon^{-s(t'-t_0)}\}$$

$$e = \sqrt{\frac{L}{C}}\,A\epsilon^{-ut}\{\epsilon^{-hl} + \epsilon^{+hl}\}\{\epsilon^{+st'} \pm \epsilon^{-st'}\} \tag{27}$$

30. These equations show the non-periodic impulse as consisting of the product of a time impulse

$$\epsilon^{+st} \pm \epsilon^{-st}$$

and a space impulse

$$\epsilon^{-hl} \pm \epsilon^{+hl}$$

Or, the impulse of current and of voltage consists of a main impulse, decaying along the line by the factor

$$\epsilon^{-hl'}$$

and a reflected impulse, or impulse traveling in the opposite direction, from the reflection point $\frac{l_0}{2}$, and dying out at the rate

$$\epsilon^{-h(l_0-l')}$$

and, by shifting the starting point of distance l to the reflection point $\frac{l_0}{2}$, the simplest form of impulse equations (27) are derived.

As seen from equations (27), and also from (25):

In the voltage impulse, the main and the reflected impulse add, in the current impulse they subtract.

The current impulse lags behind the voltage impulse by time t_0.

t_0 may be positive, or negative, that is, the current lag or lead, depending whether the resistance—inductance term, or the capacity term preponderates in the line constants, as discussed before.

Equations (27) may also be written in the form of hyperbolic function, as

$$i = A_0 \epsilon^{-ut} \sinh hl \sinh s(t' - t_0)$$

$$e = \sqrt{\frac{L}{C}} A_0 \epsilon^{-ut} \cosh hl \cosh st'$$

or
$$\tag{28}$$

$$i = A_0 \epsilon^{-ut} \sinh hl \cosh s(t' - t_0)$$

$$e = \sqrt{\frac{L}{C}} A_0 \epsilon^{-ut} \cosh hl \sinh st'$$

31. Still another form of equations (9) is given by the substitution,

$$C_1 = B\epsilon^{+hl_1}\epsilon^{-st_1}$$
$$C_2 = B\epsilon^{-hl_1}\epsilon^{-st_1}$$
$$C_3 = B\epsilon^{-h(l_1+l_0)}\epsilon^{+st_1}$$
$$C_4 = B\epsilon^{+h(l_1+l_0)}\epsilon^{+st_1}$$
$$c = \epsilon^{+st_0}$$
$$\tag{29}$$

this gives

$$i = B\epsilon^{-ut}\{\epsilon^{+s(t'-t_0)}(\epsilon^{+hl'} \mp \epsilon^{-hl'}) - \epsilon^{-s(t'-t_0)}(\epsilon^{+h(l'-l_0)} \pm \epsilon^{-h(l'-l_0)})\}$$

$$e = -\sqrt{\frac{L}{C}} B\epsilon^{-ut}\{\epsilon^{+st'}(\epsilon^{+hl'} \pm \epsilon^{-hl'}) + \epsilon^{-st'}(\epsilon^{+h(l'-l_0)} \pm \epsilon^{-h(l'-l_0)})\}$$
$$\tag{30}$$

where

$$l' = l - l_1$$
$$t' = t - t_1 + t_0$$
$$\tag{31}$$

Equations (30) can be written in hyperbolic functions in the form

$$i \doteq B\epsilon^{-ut}\{\epsilon^{+s(t'-t_0)} \cosh hl' - \epsilon^{-s(t'-t_0)} \cosh h\,(l' - l_0)\}$$

$$e = -B\sqrt{\frac{L}{C}}\epsilon^{-ut}\{\epsilon^{+st'} \sinh hl' + \epsilon^{-st'} \sinh h\,(l' - l_0)\}$$

or

$$i = B\epsilon^{-ut}\{\epsilon^{+s\,(t'-t_0)} \sinh hl' - \epsilon^{-s(t'-t_0)} \sinh h\,(l' - l_0)\}$$

$$e = -B\sqrt{\frac{L}{C}}\,\epsilon^{-ut}\{\epsilon^{+st'} \cosh hl' + \epsilon^{-st'} \cosh h\,(l' - l_0)\}$$

$$(32)$$

B. Periodic Impulses

32. A periodic impulse is an electrical effect, in which the electrical distribution is periodic in space, or can be represented by a periodic function, such as a trigonometric series. In time, however, it is non-periodic.

As any function can be represented within limits by a trigonometric series, it follows, that any electrical distribution can be represented by a trigonometric series with the complete circuit as one fundamental wave, or a fraction thereof, and thus every electrical impulse can be considered as periodic within the limits of its circuit.

As seen in the preceding (page 483), the periodic impulse is characterized by

$$
\begin{aligned}
s^2 &< m^2 \\
q &= 0 \\
h &= 0 \\
k &= \sqrt{LC\,(m^2 - s^2)}
\end{aligned}
$$

or

$$s = \sqrt{m^2 - \frac{k^2}{LC}}$$

$$(33)$$

Substituting these values in equations (55) gives

$$
\begin{aligned}
c_1 &= 0 \\
c_2 &= 0 \\
c_1' &= L\frac{m+s}{k} = \sqrt{\frac{L}{C}}\sqrt{\frac{m+s}{m-s}} = c\sqrt{\frac{L}{C}} \\
c_2' &= L\frac{m-s}{k} = \sqrt{\frac{L}{C}}\sqrt{\frac{m-s}{m+s}} = \frac{1}{c}\sqrt{\frac{L}{C}}
\end{aligned}
$$

$$(34)$$

where

$$c = \sqrt{\frac{m + s}{m - s}} \tag{35}$$

Substituting (33), (34) and (35) into equations (53) and (54), and substituting,

$$\left.\begin{array}{c} C_1 - C_2 = D_1 \\ C_1' + C_2' = D_2 \\ C_3 - C_4 = D_3 \\ C_3' + C_4' = D_4 \end{array}\right\} \tag{36}$$

gives

$$\left.\begin{array}{l} i = \Sigma\{\epsilon^{-(u-s)t}(D_1 \cos kl - D_2 \sin kl) + \\ \qquad\qquad \epsilon^{-(u+s)t}(D_3 \cos kl - D_4 \sin kl)\} \\ e = \Sigma\sqrt{\dfrac{L}{C}} \left\{ c\epsilon^{-(u-s)t}(D_2 \cos kl + D_1 \sin kl) + \\ \qquad\qquad \dfrac{1}{c}\epsilon^{-(u+s)t}(D_4 \cos kl + D_3 \sin kl)\right\} \end{array}\right\} \tag{37}$$

or, considering only one group of the series, separating ϵ^{-ut}, and substituting,

$$c = \epsilon^{+st_0} \tag{38}$$

hence,

$$t_0 = + \frac{\log c}{s} = \frac{1}{2s} \log \frac{m + s}{m - s} \tag{39}$$

gives

$$\left.\begin{array}{l} i = \epsilon^{-ut}\{\epsilon^{+st}(D_1 \cos kl - D_2 \sin kl) + \epsilon^{-st}(D_3 \cos kl - \\ \qquad\qquad\qquad\qquad D_4 \sin kl)\} \\ e = \sqrt{\dfrac{L}{C}}\,\epsilon^{-ut}\{\epsilon^{+s(t+t_0)}(D_2 \cos kl + D_1 \sin kl) + \\ \qquad\qquad \epsilon^{-s(t+t_0)}(D_4 \cos kl + D_3 \sin kl)\} \end{array}\right\} \tag{40}$$

33. Substituting,

$$\left.\begin{array}{ll} D_1 = A_1 \cos kl_1 & D_3 = A_2 \cos k\,(l_1 - l_0) \\ D_2 = A_1 \sin kl_1 & D_4 = A_2 \sin k\,(l_1 - l_0) \end{array}\right\} \tag{41}$$

and

$$\left.\begin{array}{l} A_1 = A\epsilon^{-st_1} \\ A_2 = \pm A\epsilon^{+st_1} \end{array}\right\} \tag{42}$$

gives

$$\left.\begin{array}{l} i = A\epsilon^{-ut}\{\epsilon^{+s(t-t_1)} \cos k\,(l + l_1) \pm \epsilon^{-s(t-t_1)} \cos k\,(l + l_1 - l_0)\} \\ e = A\sqrt{\dfrac{L}{C}}\,\epsilon^{-ut}\{\epsilon^{+s)t-t_1+t_0)} \sin kl\,(l + l_1) \pm \epsilon^{-s(t-t_1+t_0)} \\ \qquad\qquad\qquad\qquad\qquad\qquad \sin k(l + l_1 - l_0)\} \end{array}\right\} \tag{43}$$

substituting the new coördinates of time and distance, that is, changing the zero point of time and of distance, by the expression:

$$\left. \begin{array}{l} t' = t - t_1 + t_0 \\ l' = l + l_1 \end{array} \right\} \tag{44}$$

gives

$$\left. \begin{array}{l} i = A\epsilon^{-ut}\{\epsilon^{+s(t'-t_0)} \cos kl' \pm \epsilon^{-s(t'-t_0)} \cos k \, (l' - l_0)\} \\[2mm] e = A\sqrt{\dfrac{L}{C}}\, \epsilon^{-ut}\{\epsilon^{+st'} \sin kl' \pm \epsilon^{-st'} \sin k \, (l' - l_0)\} \end{array} \right\} \tag{45}$$

or, substituting,

$$A = B \, \epsilon^{+u(t_1-t_0)} \tag{46}$$

gives

$$\left. \begin{array}{l} i = B\epsilon^{-ut'}\{\epsilon^{+s(t'-t_0)} \cos kl' \pm \epsilon^{-s(t'-t_0')} \cos k \, (l' - l_0)\} \\[2mm] e = B\sqrt{\dfrac{L}{C}}\, \epsilon^{-ut'}\{\epsilon^{+st'} \sin kl' \pm \epsilon^{-st'} \sin k \, (l' - l_0)\} \end{array} \right\} \tag{47}$$

where the $+$ sign applies, if A_1 and A_2 are of the same, the $-$ sign, if A_1 and A_2 are of different sign.

Substituting, instead of the equations (41), the equations

$$\left. \begin{array}{ll} D_1 = A_1 \sin kl_1 & D_3 = A_2 \sin k \, (l_1 - l_0) \\[2mm] D_2 = A_1 \cos kl_1 & D_4 = A_2 \cos k \, (l_1 - l_0) \end{array} \right\} \tag{48}$$

gives

$$\left. \begin{array}{l} i = B\epsilon^{-ut'}\{\epsilon^{+s(t'-t_0)} \sin kl' \pm \epsilon^{-s(t'-t_0)} \sin k \, (l' - l_0)\} \\[2mm] e = - B\sqrt{\dfrac{L}{C}}\, \epsilon^{-ut'}\{\epsilon^{+st'} \cos kl' \pm \epsilon^{-st'} \cos k \, (l' - l_0)\} \end{array} \right\} \tag{49}$$

The equations (47) and (49) of the periodic impulse, are of the same form as the equations (32) of the non-periodic impulse, except that in the latter the hyperbolic functions take the place of the trigonometric functions in (47) and (49).

34. From these equations (37), (40), (43), (45), (47) and (49) it follows:

In the periodic impulse, (37), current and voltage each consist of two components, which are periodic functions of distance, but exponential functions of time. The second component dies out at a faster rate than the first component. Since the attenuation due to the energy dissipation by resistance and by conductance follows the exponential term ϵ^{-ut}, it follows, that the first component of current and voltage respectively dies out with the time at a slower rate, the second component at a faster rate than corresponds to the energy dissipation in the circuit. That is, the first component receives energy, the second gives off energy, and the second component thus continuously transfers energy to the first component.

The coefficient u thus may be called the *energy dissipation constant*, the coefficient s the *energy transfer constant*, while $u - s$ and $u + s$ respectively are the *attenuation constants* of the two respective components.

By energy transfer, the first component thus increases in energy by ϵ^{+st}, the second decreases by ϵ^{-st}, while both simultaneously decrease by ϵ^{-ut}, by energy dissipation, as seen from (40).

From (43), (45) and (47) it follows, that current and voltage are in quadrature with each other in their distribution in space, in either of the two components of the periodic impulse. That is, in each of the two components maximum current coincides with zero voltage, and inversely.

From (45) and (47) it follows, that the two components of the periodic impulse differ in the phase of their space distribution by the distance l_0, the second component lagging behind the first component by the distance l_0.

In each of the two components of the periodic impulse, the current lags behind the voltage by the time t_0.

Current and voltage thus are in quadrature with each other in space, and displaced from each other in time, by the "time displacement" t_0.

From equations (39) it follows, that t_0 is positive, that is, the current lags behind the voltage by time t_0, if m is positive, and t_0 is negative, that is, the current leads the voltage, if m is negative.

Since $m = \frac{1}{2}\left(\frac{r}{L} - \frac{g}{C}\right)$, it follows:

The current lags behind the voltage, if $\frac{r}{L} > \frac{g}{C}$, that is, if the inductance effect preponderates, and

The current leads the voltage, if $\frac{g}{C} > \frac{r}{L}$, that is, if the capacity effect preponderates.

35. From (47) it follows:

The voltage equals the current times the surge impedance $z = \sqrt{\dfrac{L}{C}}$, but is in quadrature with it in space, and the current is lagging by t_0 in time.

By the conditions of existence of the periodic impulse, s must numerically be smaller than m.

$$s = 0 \text{ gives}$$

By (39) $\qquad\qquad st_0 = 0$

By (33) $\qquad\qquad k = m\sqrt{LC}$

and by (49)

$$i = B\epsilon^{-ut}\{ \cos kl' \pm \cos k\,(l' - l_0)\}$$
$$e = B\sqrt{\frac{L}{C}}\,\epsilon^{-ut'}\{\sin kl' \pm \sin k\,(l' - l_0)\} \qquad (50)$$

hence, current and voltage are in phase in time, but in quadrature in space.

$$s = m \text{ gives}$$
$$k = 0$$

Hence, from (40)

$$i = \epsilon^{-ut}(D_1\epsilon^{+mt} + D_3\epsilon^{-mt})$$
$$e = \sqrt{\frac{L}{C}}\,\epsilon^{-ut}(D_2\epsilon^{+m(t+t_0)} + D_4\epsilon^{-m(t+t_0)})$$

hence, substituting for u and m,

$$i = D_1\epsilon^{-\frac{g}{C}t} + D_3\epsilon^{-\frac{r}{L}t}$$
$$e = \sqrt{\frac{L}{C}}\,\{D_2'\epsilon^{-\frac{g}{C}(t+t_0)} + D_4'\epsilon^{-\frac{r}{L}(t+t_0)}\} \qquad (51)$$

where

$$D_2' = D_2\epsilon^{+ut_0}$$
$$D_4' = D_4\epsilon^{+ut_0} \qquad (52)$$

In this impulse, the capacity terms and the inductance terms are separate, and current and voltage are uniform throughout the entire circuit.

36. The constants D or A or B are determined, as integration constants, by the terminal conditions of the problem.

For instance, if at the starting moment of the impulse, that is, at time $t = 0$, the distribution of current and of voltage throughout the circuit are given, it is, by (37), for $t = 0$,

$$i = \Sigma\{(D_1 + D_3)\cos kl - (D_2 + D_4)\sin kl\}$$

$$e = \sqrt{\frac{L}{C}} \Sigma \left\{\left(cD_2 + \frac{D_4}{c}\right)\cos kl + \left(cD_1 + \frac{D_3}{c}\right)\sin kl\right\} \qquad (53)$$

The development of the given distribution of current and voltage into a Fourier series thus gives in the coefficients of this series the equations determining the constants D_1, D_2, D_3, D_4.

CHAPTER IV.

DISCUSSION OF GENERAL EQUATIONS.

37. In Chapter I the general equations of current and voltage were derived for a circuit or section of a circuit having uniformly distributed and constant values of r, L, g, C. These equations appear as a sum of groups of four terms each, characterized by the feature that the four terms of each group have the same values of s, q, h, k.

Of the four terms of each group, i_1, i_2, i_3, i_4 or e_1, e_2, e_3, e_4 respectively (equations (53) and (54) of Chapter I), two contain the angles $(qt - kl)$: i_1, e_1 and i_3, e_3; and two contain the angles $(qt + kl)$: i_2, e_2 and i_4, e_4.

In the terms i_1, e_1 and i_3, e_3, the speed of propagation of the phenomena follows from the equation

$$qt - kl = \text{constant},$$

thus:

$$\frac{dl}{dt} = +\frac{q}{k},$$

hence is positive, that is, the propagation is from lower to higher values of l, or towards increasing l.

In the terms i_2, e_2 and i_4, e_4, the speed of propagation from

$$qt + kl = \text{constant}$$

is

$$\frac{dl}{dt} = -\frac{q}{k}$$

hence is negative, that is, the propagation is from higher to lower values of l, or towards decreasing l.

Considering therefore i_1, e_1 and i_3, e_3 as direct or main waves, i_2, e_2 and i_4, e_4 are their return waves, or reflected waves, and i_2, e_2 is the reflected wave of i_1, e_1; i_4, e_4 is the reflected wave of i_3, e_3.

497

Obviously, i_2, e_2 and i_4, e_4 may be considered as main waves, and then i_1, e_1 and i_3, e_3 are reflected waves. Substituting $(-l)$ for $(+l)$ in equations (53) and (54) of Chapter I, that is, looking at the circuit in the opposite direction, terms i_2, e_2 and i_1, e_1 and terms i_4, e_4 and i_3, e_3 merely change places, but otherwise the equations remain the same, except that the sign of i is reversed, that is, the current is now considered in the opposite direction.

Each group thus consists of two waves and their reflected waves: $i_1 - i_2$ and $e_1 + e_2$ is the first wave and its reflected wave, and $i_3 - i_4$ and $e_3 + e_4$ is the second wave and its reflected wave.

In general, each wave and its reflected wave may be considered as one unit, that is, we can say: $i' = i_1 - i_2$ and $e' = e_1 + e_2$ is the first wave, and $i'' = i_3 - i_4$ and $e'' = e_3 + e_4$ is the second wave.

In the first wave, i', e', the amplitude decreases in the direction of propagation, ε^{-hl} for rising, ε^{+hl} for decreasing l, and the wave dies out with increasing time t by $\varepsilon^{-(u-s)t} = \varepsilon^{-ut}\varepsilon^{+st}$.

In the second wave, i'', e'', the amplitude increases in the direction of propagation, ε^{+hl} for rising, ε^{-hl} for decreasing l, but the wave dies out with the increasing time t by $\varepsilon^{-(u+s)t} = \varepsilon^{-ut}\varepsilon^{-st}$, that is, faster than the first wave.

If the amplitude of the wave remained constant throughout the circuit — as would be the case in a free oscillation of the circuit, in which the stored energy of the circuit is dissipated, but no power supplied one way or the other — that is, if $h = 0$, from equation (59) of Chapter I, $s = 0$; that is, both waves coincide and form one, which dies out with the time by the decrement ε^{-ut}.

It thus follows: In general, two waves, with their reflected waves, traverse the circuit, of which the one, i'', e'', increases in amplitude in the direction of propagation, but dies out correspondingly more rapidly in time, that is, faster than a wave of constant amplitude, while the other, i', e', decreases in amplitude but lasts a longer time, that is, dies out slower than a wave of constant amplitude. In the one wave, i'', e'', an increase of amplitude takes place at a sacrifice of duration in time, while in the other wave, i', e', a slower dying out of the wave with the time is produced at the expense of a decrease of amplitude during its propagation, or, in i'', e'' duration in time is sacrificed to duration in distance, and inversely in i', e'.

It is interesting to note that in a circuit having resistance, inductance, and capacity, the mathematical expressions of the two cases of energy flow; that is, the gradual or exponential and the oscillatory or trigonometric, are both special cases of the equations (63) and (64) of Chapter I, corresponding respectively to $q = 0$, $k = 0$ and to $h = 0$, $s = 0$.

38. In the equations (53) and (54) of Chapter I

$$qt = 2\,\pi$$

gives the time of a complete cycle, that is, the *period of the wave*,

$$t_0 = \frac{2\,\pi}{q},$$

and *the frequency of the wave* is

$$f = \frac{q}{2\,\pi}.$$

$$kl = 2\,\pi$$

gives the distance of a complete cycle, that is, the *wave length*,

$$l_w = \frac{2\,\pi}{k},$$

$$(u - s)\,t = 1 \quad \text{and} \quad (u + s)\,t = 1$$

give the time,

$$t_1' = \frac{1}{u - s} \quad \text{and} \quad t_1'' = \frac{1}{u + s},$$

during which the wave decreases to $\dfrac{1}{\varepsilon} = 0.3679$ of its value, and

$$hl = 1$$

gives the distance,

$$l_1 = \frac{1}{h},$$

over which the wave decreases to $\dfrac{1}{\varepsilon} = 0.3679$ of its value;

that is, q is the *frequency constant* of the wave,

$$f = \frac{q}{2\,\pi}, \qquad t_0 = \frac{2\,\pi}{q}; \tag{1}$$

k is the *wave length constant*,

$$l_w = \frac{2\,\pi}{k}\,; \tag{2}$$

$(u - s)$ and $(u + s)$ are the *time attenuation constants* ot the wave,

$$\left.\begin{aligned} t_1' &= \frac{1}{u - s}, \\[2mm] t_1'' &= \frac{1}{u + s}, \end{aligned}\right\} \tag{3}$$

and h is the *distance attenuation constant* of the wave,

$$l_1 = \frac{1}{h}. \tag{4}$$

39. If the frequency of the current and e.m.f. is very high, thousands of cycles and more, as with traveling waves, lightning disturbances, high-frequency oscillations, etc., q is a very large quantity compared with s, u, m, h, k, and k is a large quantity compared with h, then by dropping in equations (53) to (64) of Chapter I the terms of secondary order the equations can be simplified.

From (57) of Chapter I,

$$\begin{aligned} R_1{}^2 &= \sqrt{(s^2 + q^2 - m^2)^2 + 4\,q^2 m^2} = \sqrt{(q^2 + m^2)^2 + 2\,s^2\,(q^2 - m^2)} \\[1mm] &= (q^2 + m^2)\left\{1 + \frac{2\,s^2\,(q^2 - m^2)}{(q^2 + m^2)^2}\right\}^{\frac{1}{2}} \\[1mm] &= q^2 + m^2 + s^2\,\frac{q^2 - m^2}{q^2 + m^2} \\[1mm] &= q^2 + m^2 + s^2 \\[1mm] &= q^2, \end{aligned}$$

$$\left.\begin{aligned} h &= \sqrt{LC}\,\sqrt{\tfrac{1}{2}\left\{R_1{}^2 + s^2 - q^2 - m^2\right\}} = s\,\sqrt{LC}, \\[1mm] k &= \sqrt{LC}\,\sqrt{\tfrac{1}{2}\left\{R_1{}^2 - s^2 + q^2 + m^2\right\}} = \sqrt{LC\,(q^2 + m^2)} = q\,\sqrt{LC}, \end{aligned}\right\} \tag{5}$$

$$h^2 + k^2 = (s^2 + q^2)\,LC = q^2 LC,$$

and

$$c_1 = \frac{qk + h\,(m + s)}{h^2 + k^2}\,L = \frac{qL}{k} = \sqrt{\frac{L}{C}},$$

$$c_2 = \frac{qk - h(m - s)}{h^2 + k^2} L = \frac{qL}{k} = \sqrt{\frac{L}{C}},$$

$$c_1' = \frac{k(m + s) - qh}{h^2 + k^2} L = \frac{q\sqrt{LC}(m + s) - qs\sqrt{LC}}{q^2 LC} L = \frac{m}{q}\sqrt{\frac{L}{C}},$$

$$c_2' = \frac{k(m - s) + qh}{h^2 + k^2} L = \frac{q\sqrt{LC}(m - s) + qs\sqrt{LC}}{q^2 LC} L = \frac{m}{q}\sqrt{\frac{L}{C}};$$

that is,

$$\left. \begin{aligned} c_1 = c_2 &= \sqrt{\frac{L}{C}} \\ c_1' = c_2' &= \frac{m}{q}\sqrt{\frac{L}{C}}. \end{aligned} \right\} \tag{9}$$

and

Writing

$$\left. \begin{aligned} \sigma &= \sqrt{LC}, \\ c &= \sqrt{\frac{L}{C}}, \end{aligned} \right\} \tag{7}$$

where σ is the reciprocal of the speed of propagation (velocity of light), we have

$$\left. \begin{aligned} h &= \sigma s, \\ k &= \sigma q, \end{aligned} \right\} \tag{8}$$

and

$$\left. \begin{aligned} c_1 = c_2 &= c \\ c_1' = c_2' &= \frac{m}{q}c \end{aligned} \right\} \tag{9}$$

and introducing the new independent variable, as distance,

$$\lambda = \sigma l, \tag{10}$$

we have

and

$$\left. \begin{aligned} kl &= q\lambda \\ hl &= s\lambda; \end{aligned} \right\} \tag{11}$$

hence, the wave length is given by

$$q\lambda = 2\pi$$

as

$$\lambda_0 = \frac{2\pi}{q}, \tag{12}$$

and since the period is

$$t_0 = \frac{2\pi}{q},$$

it follows that *by the introduction of the denotation* (10) *distances are measured with the velocity of propagation as unit length,* and wave length l_w and period t_0 thus have the same numerical values.

The use of the velocity of propagation as unit of length of electric circuits such as transmission lines offers many advantages in dealing with transients, and therefore is generally advisable.

Substituting now in equations (63) and (64) of Chapter I gives

$$i = \varepsilon^{-ut}\sum\{\varepsilon^{+s(t-\lambda)}\,D_1\,[q\,(t-\lambda)]-\varepsilon^{+s(t+\lambda)}\,D_2\,[q\,(t+\lambda)]\;(i')$$

$$+ \varepsilon^{-s(t-\lambda)}\,D_3\,[q(t-\lambda)]-\varepsilon^{-s(t+\lambda)}\,D_4\,[q\,(t+\lambda)]\}\;(i'')\;(13)$$

and

$$e = \varepsilon^{-ut}\sum\{\varepsilon^{+s(t-\lambda)}\,H_1\,[q\,(t-\lambda)]+\varepsilon^{+s(t+\lambda)}\,H_2[q\,(t+\lambda)]\;(e')$$

$$+ \varepsilon^{-s(t-\lambda)}\,H_3\,[q\,(t-\lambda)]+\varepsilon^{-s(t+\lambda)}\,H_4[q\,(t+\lambda)]\},\,(e'')\;(14)$$

where

$$D\,[q\,(t\pm\lambda)] = C\cos q\,(t\pm\lambda) + C'\sin q\,(t\pm\lambda)$$

and

$$H\,[q\,(t\pm\lambda)] = \sqrt{\frac{L}{C}}\left\{\left(\frac{m}{q}C' - C\right)\cos q\,(t\pm\lambda)\right.$$

$$\left.-\left(\frac{m}{q}C + C'\right)\sin q\,(t\pm\lambda)\right\}. \tag{15}$$

40. As seen from equations (13) and (14), the waves are products of ε^{-ut} and a function of $(t-\lambda)$ for the main wave,

$(t + \lambda)$ for the reflected wave, thus:

$$i_1 + i_3 = \varepsilon^{-ut} f_1 (t - \lambda) \left.\vphantom{\begin{array}{c}a\\b\end{array}}\right\}$$

and

$$i_2 + i_4 = \varepsilon^{-ut} f_2 (t + \lambda); \qquad (16)$$

hence, for constant $(t - \lambda)$ on the main waves, and for constant $(t + \lambda)$ on the reflected waves, we have

$$i_1 + i_3 = B\varepsilon^{-ut} \left.\vphantom{\begin{array}{c}a\\b\end{array}}\right\}$$

and

$$i_2 + i_4 = B'\varepsilon^{-ut}; \qquad (17)$$

that is, during its passage along the circuit the wave decreases by the decrement ε^{-ut}, or at a constant rate, independent of frequency, wave length, etc., and depending merely on the circuit constants r, L, g, C. The decrement of the traveling wave in the direction of its motion is

$$\varepsilon^{-ut} = \varepsilon^{-\frac{1}{2}\left(\frac{r}{L} + \frac{g}{C}\right)t},$$

and therefore is independent of the character of the wave, for instance its frequency, etc.

41. The physical meaning of the two waves i' and e' can best be appreciated by observing the effect of the wave when traversing a fixed point λ of the circuit.

Consider as example the main wave only, $i' = i_1 + i_3$, and neglect the reflected waves, for which the same applies.

From equation (74),

$$i = \varepsilon^{-s\lambda - (u-s)t} D_1 [q (t - \lambda)] + \varepsilon^{+s\lambda - (u+s)t} D_3 [q (t - \lambda)]; \quad (18)$$

or the absolute value is

$$I = D_1 \varepsilon^{-s\lambda - (u-s)t} + D_3 \varepsilon^{+s\lambda - (u+s)t}, \qquad (19)$$

where D_1 and D_3 have to be combined vectorially.

Assuming then that at the time $t = 0$, $I = 0$, for constant λ we have

$$I = D (\varepsilon^{-(u-s)t} - \varepsilon^{-(u+s)t}), \qquad (20)$$

the amplitude of I at point λ.

Since (81) is the difference of two exponential functions of different decrement, it follows that as function of the time t, I

rises from 0 to a maximum and then decreases again to zero, as shown in Fig. 98, where

$$I_1 = D\varepsilon^{-(u-s)\,t},$$

$$I_3 = D\varepsilon^{-(u+s)\,t},$$

$$I = I_1 - I_3,$$

and the actual current i is the oscillatory wave with I as envelope.

The combination of two waves thus represents the passage of a wave across a given point, the amplitude rising during the arrival and decreasing again after the passage of the wave.

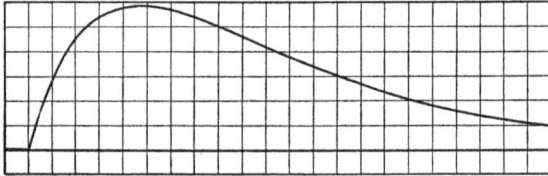

Fig. 98. Amplitude of electric traveling wave.

42. If h and so also s equal zero, i', e' and i'', e'' coincide in equations (13) and (14), and C_1 and C_3 thus can be combined into one constant B_1, C_2 and C_4 into one constant B_2, thus:

$$\left.\begin{aligned}
C_1 + C_3 &= B_1, \\
C_2 + C_4 &= B_2, \\
C_1' + C_3' &= B_1', \\
C_2' + C_4' &= B_2',
\end{aligned}\right\} \qquad (21)$$

and (13), (14) then assume the form

$$i = \varepsilon^{-ut} \sum \{[B_1 \cos q\,(t-\lambda) + B_1' \sin q\,(t-\lambda)]$$
$$- [B_2 \cos q\,(t+\lambda) + B_2' \sin q\,(t+\lambda)]\}, \qquad (22)$$

$$e = \sqrt{\frac{L}{C}}\,\varepsilon^{-ut} \sum \left\{ \left[\left(\frac{m}{q}B_1' - B_1\right)\cos q\,(t-\lambda) - \left(\frac{m}{q}B_1 + B_1'\right) \right.\right.$$
$$\left.\sin q\,(t-\lambda)\right] + \left[\left(\frac{m}{q}B_2' - B_2\right)\cos q\,(t+\lambda) - \left(\frac{m}{q}B_2 + B_2'\right)\sin q\,(t+\lambda)\right]\right\}.$$
$$(23)$$

These equations contain the distance λ only in the trigonometric but not in the exponential function; that is, i and e vary in phase throughout the circuit, but not in amplitude; or, in other words, the oscillation is of uniform intensity throughout the circuit, dying out uniformly with the time from an initial maximum value; however, the wave does not travel along the circuit, but is a stationary or standing wave. It is an oscillatory discharge of a circuit containing a distributed r, L, g, C, and therefore is analogous to the oscillating condenser discharge through an inductive circuit, except that, due to the distributed capacity, the phase changes along the circuit. The free oscillations of a circuit such as a transmission line are of this character.

For $\lambda = 0$, that is, assuming the wave length of the oscillation as so great, hence the circuit as such a small fraction of the wave length, that the phase of i and e can be assumed as uniform throughout the circuit, the equations (22) and (23) assume the form

$$i = \varepsilon^{-ut}\{B_0 \cos qt + B_0' \sin qt\}$$

and (24)

$$e = \sqrt{\frac{L}{C}}\,\varepsilon^{-ut}\left\{\left(\frac{m}{q}B' - B\right)\cos qt - \left(\frac{m}{q}B + B'\right)\sin qt\right\};$$

these are the usual equations of the condenser discharge through an inductive circuit, which here appear as a special case of a special case of the general circuit equations.

If q equals zero, the functions D and H in equations (13) and (14) become constant, and these equations so assume the form

$$i = \varepsilon^{-ut}\sum\left\{[C_1\varepsilon^{+s(t-\lambda)} + C_3\varepsilon^{-s(t-\lambda)}]\right.$$
$$\left. - [C_2\varepsilon^{+s(t+\lambda)} + C_4\varepsilon^{-s(t+\lambda)}]\right\}$$

and (25)

$$e = \sqrt{\frac{L}{C}}\,\varepsilon^{-ut}\left\{[B_1\varepsilon^{+s(t-\lambda)} + B_3\varepsilon^{-s(t-\lambda)}]\right.$$
$$\left. + [B_2\varepsilon^{+s(t+\lambda)} + B_4\varepsilon^{-s(t+\lambda)}]\right\},$$

where

$$B = \frac{m}{q} C' - C. \tag{26}$$

This gives expressions of current and e.m.f. which are no longer oscillatory but exponential, thus representing a gradual change of i and e as functions of time and distance, corresponding to the gradual or logarithmic condenser discharge. For $\lambda = 0$, these equations change to the equations of the logarithmic condenser discharge.

These equations (25) are only approximate, however, since in them the quantities s, u, h have been neglected compared with q, assuming the latter as very large, while now it is assumed as zero.

43. If, however,

$$m = \frac{1}{2} \left(\frac{r}{L} - \frac{g}{C} \right) = 0, \tag{27}$$

that is,

$$\frac{r}{L} = \frac{g}{C}, \tag{28}$$

or

$$r \div g = L \div C,$$

or, in words, the power coefficients of the circuit are proportional to the energy storage coefficients, or the time constant of the electromagnetic field of the circuit, $\frac{r}{L}$, equals the time constant of the electrostatic field of the circuit, $\frac{g}{C}$, then

$$u = \frac{r}{L} = \frac{g}{C} = \text{time constant of the circuit}, \tag{29}$$

and from equation (57) of Chapter I

$$\left. \begin{array}{l} R_1{}^2 = s^2 + q^2, \\ h = s\sqrt{LC} = \sigma s, \\ k = q\sqrt{LC} = \sigma q, \end{array} \right\} \tag{30}$$

and from equation (55) of Chapter I

$$
\left.\begin{aligned}
c_1 &= \frac{L}{\sigma} = \sqrt{\frac{L}{C}} = c, \\[4pt]
c_1' &= 0, \\[4pt]
c_2 &= \frac{L}{\sigma} = \sqrt{\frac{L}{C}} = c, \\[4pt]
c_2' &= 0;
\end{aligned}\right\} \tag{31}
$$

and

hence, substituting in equations (53) and (54) of Chapter I,

$$
i = \varepsilon^{-\frac{r}{L}t}\left\{ \varepsilon^{+s(t-\lambda)} D_1[q(t-\lambda)] - \varepsilon^{+s(t+\lambda)} D_2[q(t+\lambda)] \right.
$$
$$
\left. + \varepsilon^{-s(t-\lambda)} D_3[q(t-\lambda)] - \varepsilon^{-s(t+\lambda)} D_4[q(t+\lambda)] \right\} \tag{32}
$$

and

$$
e = -\sqrt{\frac{L}{C}}\,\varepsilon^{-\frac{r}{L}t}\left\{ \varepsilon^{+s(t-\lambda)} D_1[q(t-\lambda)] + \varepsilon^{+s(t+\lambda)} D_2[q(t+\lambda)] \right.
$$
$$
\left. + \varepsilon^{-s(t-\lambda)} D_3[q(t-\lambda)] + \varepsilon^{-s(t+\lambda)} D_4[q(t+\lambda)] \right\}. \tag{33}
$$

These equations are similar to (13) and (14), but are derived here for the case $m = 0$, without assumptions regarding the relative magnitude of q and the other quantities: "distortionless circuit."

These equations (32) and (33) therefore also apply for $q = 0$, and then assume the form

$$
i = \varepsilon^{-\frac{r}{L}t}\left\{ [C_1\varepsilon^{+s(t-\lambda)} + C_3\varepsilon^{-s(t-\lambda)}] - [C_2\varepsilon^{+s(t+\lambda)} + C_4\varepsilon^{-s(t+\lambda)}] \right\}, \tag{34}
$$

$$
e = -\sqrt{\frac{L}{C}}\,\varepsilon^{-\frac{r}{L}t}\left\{ [C_1\varepsilon^{+s(t-\lambda)} + C_3\varepsilon^{-s(t-\lambda)}] \right.
$$
$$
\left. + [C_2\varepsilon^{+s(t+\lambda)} + C_4\varepsilon^{-s(t+\lambda)}] \right\}. \tag{35}
$$

These equations (34) and (35) are the same as (25), but in the present case, where $m = 0$, apply irrespective of the relative values of the quantities s, etc.

Therefore in a circuit in which $m = 0$ a transient term may appear which is not oscillatory in time nor in space, but changing gradually.

If the constant h in equations (53) and (54) differs from zero, the oscillation (using the term *oscillation* here in the most general sense, that is, including also *alternation*, as an oscillation of zero attenuation) travels along the circuit, but it becomes stationary, as a standing wave, for $h = 0$; that is, the distance attenuation constant h may also be called the *propagation constant* of the wave.

$h = 0$ thus represents a wave which does not propagate or move along the circuit, but stands still, that is, a stationary or standing wave.

If the constant h in equations (53) and (54) of Chapter I differs from zero, the oscillation (using the term *oscillation* here in the most general sense, that is, including also *alternation*, as an oscillation of zero attenuation) travels along the circuit, but it becomes stationary, as a standing wave, for $h = 0$; that is, the distance attenuation constant h may also be called the *propagation constant* of the wave.

$h = 0$ thus represents a wave which does not propagate or move along the circuit, but stands still, that is, a stationary or standing wave.

CHAPTER V.

STANDING WAVES.

44. If the propagation constant of the wave vanishes,

$$h = 0,$$

the wave becomes a stationary or standing wave, and the equations of the standing wave are thus derived from the general equations (53) to (64) of Chapter I, by substituting therein $h = 0$, which gives

$$R_2^2 = \sqrt{(k^2 - LCm^2)^2}; \tag{1}$$

hence, if

$$k^2 > LCm^2,$$
$$R_2^2 = k^2 - LCm^2;$$

and if

$$k^2 < LCm^2,$$
$$R_2^2 = LCm^2 - k^2.$$

Therefore, two different cases exist, depending upon the relative values of k^2 and LCm^2, and in addition thereto the intermediary or critical case, in which $k^2 = LCm^2$.

These three cases require separate consideration.

$$LCm^2 = LC\left\{\frac{1}{2}\left(\frac{r}{L} - \frac{g}{C}\right)\right\}^2 = \left\{\frac{1}{2}\left(r\sqrt{\frac{C}{L}} - g\sqrt{\frac{L}{C}}\right)\right\}^2 \tag{2}$$

is a circuit constant, while k is the wave length constant, that is, the higher k the shorter the wave length.

A. Short waves,

$$k^2 > LCm^2, \tag{3}$$

hence,

$$R_2^2 = k^2 - LCm^2 \tag{4}$$

and

$$s = 0,$$

$$q = \sqrt{\frac{k^2}{LC} - m^2}, \tag{5}$$

509

or approximately, for very large k,

$$q = \frac{k}{\sqrt{LC}}.$$ (6)

Herefrom then follows

$$
\left.
\begin{aligned}
c_1 &= \frac{qL}{k} = c, \\[2mm]
c_1' &= \frac{mL}{k} = c', \\[2mm]
c_2 &= \frac{qL}{k} = c, \\[2mm]
c_2' &= \frac{mL}{k} = c'.
\end{aligned}
\right\}
$$ (7)

and

Substituting now $h = 0$ and (5), (6) in equations (53), (54), of Chapter I, the two waves i', e' and i'', e'' coincide, and all the exponential terms reduce to ϵ^{-ut}; hence, substituting

$$
\left.
\begin{aligned}
B_1 &= C_1 + C_3, \\
B_2 &= C_2 + C_4, \\
B_1' &= C_1' + C_3', \\
B_2' &= C_2' + C_4',
\end{aligned}
\right|
$$ (8)

and

gives

$$
i = \epsilon^{-ut} \{[B_1 \cos(qt - kl) + B_1' \sin(qt - kl)] \\
- [B_2 \cos(qt + kl) + B_2' \sin(qt + kl)]\}
$$ (9)

and

$$
e = \frac{L}{k} \epsilon^{-ut} \{[(mB_1' - qB_1) \cos(qt - kl) - (mB_1 + qB_1') \sin(qt - kl)] \\
+ [(mB_2' - qB_2) \cos(qt + kl) - (mB_2 + qB_2') \sin(qt + kl)]\}.
$$ (10)

Equations (9) and (10) represent a *stationary electrical oscillation* or standing wave on the circuit.

B. Long waves,

$$k^2 < LCm^2;$$ (11)

hence,

$$R_2{}^2 = LCm^2 - k^2 \tag{12}$$

and

$$s = \sqrt{m^2 - \frac{k^2}{LC}}, \\ q = 0, \tag{13}$$

or approximately, for very small values of k,

$$s = m = \frac{1}{2}\left(\frac{r}{L} - \frac{g}{C}\right); \tag{14}$$

herefrom then follows

$$c_1 = c_2 = 0, \\ c_1{}' = \frac{(m + s)\,L}{k}$$

and

$$c_2{}' = \frac{(m - s)\,L}{k}\;. \tag{15}$$

Substituting now $h = 0$ and (13), (15) into (53) and (54) of Chapter I, the two waves i', e' and i'', e'' remain separate, having different exponential terms, $\epsilon^{-(u-s)t}$ and $\epsilon^{-(u+s)t}$, but in each of the two waves the main wave and the reflected wave coincide, due to the vanishing of q.

Substituting then

$$B_1 = C_1 - C_2, \\ B_1{}' = C_1{}' + C_2{}', \\ B_2 = C_3 - C_4, \\ B_2{}' = C_3{}' + C_4{}', \tag{16}$$

and

gives

$$i = \epsilon^{-ut}\{(B_1\,\epsilon^{+st} + B_2\epsilon^{-st})\cos kl - (B_1{}'\epsilon^{+st} + B_2{}'\,\epsilon^{-st})\sin kl\} \tag{17}$$

and

$$e = \frac{L}{k} \varepsilon^{-ut} \{ [(m + s) B_1' \varepsilon^{+st} + (m - s) B_2' \varepsilon^{-st}] \cos kl$$

$$+ [(m + s) B_1 \varepsilon^{+st} + (m - s) B_2 \varepsilon^{-st}] \sin kl \}$$

$$= \frac{L}{k} \varepsilon^{-ut} \{ m[(B_1' \varepsilon^{+st} + B_2' \varepsilon^{-st}) \cos kl$$

$$+ (B_1 \varepsilon^{+st} + B_2 \varepsilon^{-st}) \sin kl]$$

$$+ s[(B_1' \varepsilon^{+st} - B_2' \varepsilon^{-st}) \cos kl$$

$$+ (B_1 \varepsilon^{+st} - B_2 \varepsilon^{-st}) \sin kl] \} \tag{18}$$

Equations (17) and (18) represent a *gradual or exponential circuit discharge*, and the distribution still is a trigonometric function of the distance, that is, a wave distribution, but dies out gradually with the time, without oscillation.

C. Critical case,

$$k^2 = LCm^2; \tag{19}$$

hence,

$$\left. \begin{array}{c} R_2^{\,2} = 0, \\ s = 0, \\ q = 0, \end{array} \right\} \tag{20}$$

and

$$\left. \begin{array}{c} c_1 = c_2 = 0, \\ c_1' = c_2' = \dfrac{mL}{k} = \sqrt{\dfrac{L}{C}} = c', \end{array} \right\} \tag{21}$$

and all the main waves and their reflected waves coincide when substituting $h = 0$, (20), (21) in (53) and (54) of Chapter I

Hence, writing

$$\left. \begin{array}{c} B = C_1 - C_2 + C_3 - C_4 \\ B' = C_1' + C_2' + C_3' + C_4' \end{array} \right\} \tag{22}$$

and

gives

$$i = \varepsilon^{-ut} \{ B \cos kl - B' \sin kl \} \tag{23}$$

and

$$e = \sqrt{\frac{L}{C}} \, \epsilon^{-ut} \left\{ B' \cos kl + B \sin kl \right\}. \tag{24}$$

In the critical case (23) and (24), the wave is distributed as a trigonometric function of the distance, but dies out as a simple exponential function of the time.

45. An electrical standing wave thus can have two different forms: it can be either oscillatory in time or exponential in time, that is, gradually changing. It is interesting to investigate the conditions under which these two different cases occur.

The transition from gradual to oscillatory takes place at

$$k_0^2 = m^2 LC; \tag{25}$$

for larger values of k the phenomenon is oscillatory; for smaller, exponential or gradual.

Since k is the wave length constant, the wave length, at which the phenomenon ceases to be oscillatory in time and becomes a gradual dying out, is given by (2) of Chapter **IV** as

$$\left. \begin{array}{l} l_{w0} = \dfrac{2\pi}{k_0} \\[2ex] \phantom{l_{w0}} = \dfrac{2\pi}{m\sqrt{LC}} . \end{array} \right\} \tag{26}$$

In an undamped wave, that is, in a circuit of zero r and zero g, in which no energy losses occur, the speed of propagation is

$$S = \frac{q}{k} = \frac{1}{\sqrt{LC}}, \tag{27}$$

and if the medium has unit permeability and unit inductivity, it is the speed of light,

$$S_0 = 3 \times 10^{10}. \tag{28}$$

In an undamped circuit, this wave length l_{w0} would correspond to the frequency,

$$q_0 = \frac{k_0}{\sqrt{LC}} = m;$$

hence, from (1) of Chapter IV,

$$f_0 = \frac{q_0}{2\pi} = \frac{m}{2\pi}.$$ (29)

The frequency at the wave length l_{w_0} is zero, since at this wave length the phenomenon ceases to be oscillatory; that is, due to the energy losses in the circuit, by the effective resistance r and effective conductance g, the frequency f of the wave is reduced below the value corresponding to the wave length l_w, the more, the greater the wave length, until at the wave length l_{w_0} the frequency becomes zero and the phenomenon thereby non-oscillatory. This means that with increasing wave length the velocity of propagation of the phenomenon decreases, and becomes zero at wave length l_{w_0}.

If $m^2 LC = 0,$

$k_0 = 0$ and $l_{w_0} = \infty$;

that is, the standing wave is always oscillatory.

If $m^2 LC = \infty,$

$k_0 = \infty$ and $l_{w_0} = 0;$

that is, the standing wave is always non-oscillatory, or gradually dying out.

In the former case, $m^2 LC = 0$, or oscillatory phenomenon, substituting for m^2, we have

$$\frac{1}{4}\left(\frac{r}{L} - \frac{g}{C}\right)^2 LC = 0,$$

$$r\sqrt{\frac{C}{L}} - g\sqrt{\frac{L}{C}} = 0,$$

and

$$\frac{r}{g} = \frac{L}{C};$$

or

$$rC - gL = 0 \quad \text{(distortionless circuit)}.$$

In the latter case, $m^2 LC = \infty$, or non-oscillatory or exponential standing wave, we have

$$r\sqrt{\frac{C}{L}} - g\sqrt{\frac{L}{C}} = \infty,$$

and since neither r, g, L, nor C can be equal infinity it follows that either $L = 0$ or $C = 0$.

Therefore, the standing wave in a circuit is always oscillatory, regardless of its wave length, if

$$rC - gL = 0, \tag{30}$$

or

$$\frac{r}{g} = \frac{L}{C}; \tag{31}$$

that is, the ratio of the energy coefficients is equal to the ratio of the reactive coefficients of the circuit.

The standing wave can never be oscillatory, but is always exponential, or gradually dying out, if either the inductance L or the capacity C vanishes; that is, the circuit contains no capacity or contains no inductance.

In all other cases the standing wave is oscillatory for waves shorter than the critical value $l_{w_0} = \dfrac{2\,\pi}{k_0}$, where

$$k_0{}^2 = m^2 LC = \frac{1}{4} \left\{ r \sqrt{\frac{C}{L}} - g \sqrt{\frac{L}{C}} \right\}^2, \tag{32}$$

and is exponential or gradual for standing waves longer than the critical wave length l_{w_0}; or for $k < k_0$ the standing wave is exponential, for $k > k_0$ it is oscillatory.

The value $k_0 = m \sqrt{LC}$ thus takes a similar part in the theory of standing waves as the value $r_0{}^2 = 4\,L_0 C_0$ in the condenser discharge through an inductive circuit; that is, it separates the exponential or gradual from trigonometric or oscillatory conditions.

The difference is that the condenser discharge through an inductive circuit is gradual, or oscillatory, depending on the circuit constants, while in a general circuit, with the same circuit constants, usually gradual as well as oscillatory standing waves exist, the former with greater wave length, or

$$m \sqrt{LC} > k, \tag{33}$$

the latter with shorter wave length, or

$$m \sqrt{LC} < k. \tag{34}$$

An idea of the quantity k_0, and therewith the wave length l_{w_0}, at which the frequency of the standing wave becomes zero, or the wave non-oscillatory, and of the frequency f_0, which, in an undamped circuit, will correspond to this critical wave length l_{w_0}, can best be derived by considering some representative numerical examples.

As such may be considered:

(1) A high-power high-potential overhead transmission line.
(2) A high-potential underground power cable.
(3) A submarine telegraph cable.
(4) A long-distance overhead telephone circuit.

(1) *High-power high-potential overhead transmission line.*

46. Assume energy to be transmitted 120 miles, at 40,000 volts between line and ground, by a three-phase system with grounded neutral. The line consists of copper conductors, wire No. 00 B. and S. gage, with 5 feet between conductors.

Choosing the mile as unit length,

$$r = 0.41 \text{ ohm per mile.}$$

The inductance of a conductor is given by

$$L = l \left(2 \log_\epsilon \frac{l_d}{l_r} + \frac{\mu}{2} \right) 10^{-9}, \text{ in henrys,} \qquad (35)$$

where $l =$ the length of conductor, in cm.; $l_r =$ the radius of conductor; $l_d =$ the distance from return conductor, and $\mu =$ the permeability of conductor material. For copper, $\mu = 1$.

As one mile equals 1.61×10^5 cm., substituting this, and reducing the natural logarithm to the common logarithm, by the factor 2.3026, gives

$$L = \left(0.7415 \log \frac{l_d}{l_r} + 0.0805 \right), \text{ in mh. per mile.} \qquad (36)$$

For $\qquad l_r = 0.1825$ inch and $l_d = 60$ inches,
$$L = 1.95 \text{ mh. per mile.}$$

The capacity of a conductor is given by

$$C = l \left(\frac{1 + \delta}{2 \, S_0^2 \, \log_\epsilon \dfrac{l_d}{l_r}} \right) 10^9, \text{ in farads.} \qquad (37)$$

where $S_0 = 3 \times 10^{10} =$ the speed of light, and $\delta =$ the allowance for capacity of insulation, tie wires, supports, etc., assumed as 5 per cent.

Substituting S_0, and reducing to one mile and common logarithm, gives

$$C = \frac{0.0408}{log \dfrac{l_d}{l_r}}, \text{in mf.;} \tag{38}$$

hence, in this instance,

$$C = 0.0162 \text{ mf.}$$

Estimating the loss in the static field of the line as 400 watts per mile of conductor gives an effective conductance,

$$g = \frac{400}{40,000^2} = 0.25 \times 10^{-6} \text{ mho,}$$

which gives the line constants per mile as $r = 0.41$ ohm; $L = 1.95 \times 10^{-3}$ henry; $g = 0.25 \times 10^{-6}$ mho, and $C = 0.0162 \times 10^{-6}$ farad.

Herefrom then follows

$$u = \frac{1}{2}\left(\frac{r}{L} + \frac{g}{C}\right) = \frac{1}{2}(210 + 15.5) = 113,$$

$$m = \frac{1}{2}\left(\frac{r}{L} - \frac{g}{C}\right) = 97,$$

$$\sigma = \sqrt{LC} = \sqrt{31.6} \times 10^{-6} = 5.62 \times 10^{-6},$$

$$k_0 = m\sqrt{LC} = 545 \times 10^{-6};$$

hence, the critical wave length is

$$l_{w_0} = \frac{2\pi}{k_0} = 11,500 \text{ miles,}$$

and in an undamped circuit this wave length would correspond to the frequency of oscillation,

$$f_0 = \frac{m}{2\pi} = 15.7 \text{ cycles per sec.}$$

Since the shortest wave at which the phenomenon ceases to be oscillatory is 11,500 miles in length, and the longest wave which can originate in the circuit is four times the length of the circuit, or 480 miles, it follows that whatever waves may originate in this circuit are by necessity oscillatory, and non-oscillatory currents or voltages can exist in this circuit only when impressed upon it by some outside source, and then are of such great wave length that the circuit is only an insignificant fraction of the wave, and great differences of voltage and current of non-oscillatory nature cannot exist, as *standing waves.*

Since the difference in length between the shortest non-oscillatory wave and the longest wave which can originate in the circuit is so very great, it follows that in high-potential long-distance transmission circuits all phenomena which may result in considerable potential differences and differences of current throughout the circuit are oscillatory in nature, and the solution case (*A*) is the one the study of which is of the greatest importance in long-distance transmissions.

With a length of circuit of 120 miles, the longest standing wave which can originate in the circuit has the wave length

$$l_w = 480 \text{ miles,}$$

and herefrom follows

$$k = \frac{2\pi}{l_w} = 0.0134$$

and

$$\frac{k^2}{LC} = \frac{0.0134^2}{31.6 \times 10^{-12}} = 5.7 \times 10^6;$$

hence, in the expression of q in equation (101),

$$q = \sqrt{\frac{k^2}{LC} - m^2}$$

$$= \sqrt{5.7 \times 10^6 - 0.00941 \times 10^6},$$

m^2 is negligible compared with $\dfrac{k^2}{LC}$; that is,

$$q = \frac{k}{\sqrt{LC}} = \frac{0.0134}{5.62 \times 10^{-6}} = 2380; \qquad (39)$$

or

$$f = \frac{q}{2\pi} = 380 \text{ cycles per sec.}$$

Hence, even for the longest standing wave which may originate in this transmission line, $q = 2380$ is such a large quantity compared with $m = 97$ that m can be neglected compared with q, and for shorter waves, the overtones of the fundamental wave, this is still more the case; that is, in equation (9) and (10) the terms with m may be dropped. In equation (10) $\frac{Lq}{k}$ thus become common factors, and since from equation (39)

$$\frac{Lq}{k} = \sqrt{\frac{L}{C}}, \tag{40}$$

by substituting $m = 0$ and (40) in (9) and (10) we get the *general equations of standing waves in long-distance transmission lines*, thus:

$$i = \varepsilon^{-ut}\{[B_1 \cos(qt - kl) + B_1{'} \sin(qt - kl)] \\ - [B_2 \cos(qt + kl) + B_2{'} \sin(qt + kl)]\}, \tag{41}$$

$$e = -\sqrt{\frac{L}{C}}\,\varepsilon^{-ut}\{[B_1 \cos(qt - kl) + B_1{'} \sin(qt - kl)] \\ + [B_2 \cos(qt + kl) + B_2{'} \sin(qt+kl)]\}, \tag{42}$$

or

$$e = \varepsilon^{-ut}\{[A_1 \cos(qt + kl) + A_1{'} \sin(qt + kl)] \\ + [A_2 \cos(qt - kl) + A_2{'} \sin(qt - kl)]\}, \tag{43}$$

$$i = \sqrt{\frac{C}{L}}\,\varepsilon^{-ut}\{[A_1 \cos(qt + kl) + A_1{'} \sin(qt + kl)] \\ - [A_2 \cos(qt - kl) + A_2{'} \sin(qt - kl)]\}, \tag{44}$$

where

$$A_1 = -\sqrt{\frac{L}{C}}\,B_2, \qquad A_1{'} = -\sqrt{\frac{L}{C}}\,B_2{'}, \text{ etc.}$$

(2) *High-potential underground power cable.*

47. Choose as example an underground power cable of 20 miles length, transmitting energy at 7000 volts between conductor and ground or cable armor, that is, a three-phase three-conductor 12,000-volt cable.

Assume the conductor as stranded and of a section equivalent to No. 00 B. and S. G.

the expression for the capacity, equation (23), multiplies with the expression for the capacity, equation (119), multiplies with the dielectric constant or specific capacity of the cable insulation, and that $\frac{l_d}{l_r}$ is very small, about three or less; or taking the values of the circuit constants from tests of the cable, we get values of the magnitude, per mile of single conductor, $r = 0.41$ ohm; $L = 0.4 \times 10^{-3}$ henry; $g = 10^{-6}$ mho, corresponding to a power factor of the cable-charging current, at 25 cycles, of 1 per cent; $C = .6 \times 10^{-6}$ farad.

Herefrom the following values are obtained: $u = 513, m = 512$, $\sigma = \sqrt{LC} = 15.5 \times 10^{-6}$, $k_0 = m\sqrt{LC} = 7.95 \times 10^{-3}$, and the critical wave length is $l_{w_0} = 790$ miles, and the frequency of an undamped oscillation, corresponding to l_{w_0}, is $f_0 = 81.5$ cycles per second.

As seen, in an underground high-potential cable the critical wave length is very much shorter than in the overhead long-distance transmission line. At the same time, however, the length of an underground cable circuit is very much shorter than that of a long-distance transmission line, so that the critical wave length still is very large compared with the greatest wave length of an oscillation originating in the cable, at least ten times as great. Which means that the discussion of the possible phenomena in any overhead line, under (1), applies also to the underground high-potential cable circuit.

In the present example the longest standing wave which may originate in the cable has the wave length

$$l_w = 80 \text{ miles,}$$

which gives

$$k = 0.0785$$

and

$$\frac{k}{\sqrt{LC}} = 5070,$$

or about ten times as large as m, so that m can still be neglected in equation (26) of Chapter IV, and we have

$$q = \frac{k}{\sqrt{LC}} = 5070,$$

or $\qquad f = 810$ cycles per second,

and the general equations of the phenomenon in long-distance transmission lines, (27) to (29), also apply as the general equations of standing waves in high-potential underground cable circuits.

(3) *Submarine telegraph cable.*

48. Choosing the following values: length of cable, single stranded-conductor, ground return, $= 4000$ miles; constants per mile of conductor: $r = 3$ ohms, $L = 10^{-3}$ henry, $g = 10^{-6}$ mho, and $C = 0.1 \times 10^{-6}$ farad, we get $u = 1500$; $m = 1500$; $\sigma = \sqrt{LC} = 10 \times 10^{-6}$, and $k_0 = m \sqrt{LC} = 15 \times 10^{-3}$, from which the critical wave length is $l_{w_0} = 418$ miles, and the corresponding frequency $f_0 = 239$ cycles per second.

From the above it is seen that in a submarine cable the critical wave length l_{w_0} is relatively short, so that in long submarine cables standing waves may appear which are not oscillatory in time but die out gradually, that is, are shown by the equation of case B. In such cables, due to their relatively high resistance, the damping effect is very great; $u = 1500$, and standing waves, therefore, rapidly die out.

In the investigation of the submarine cable, the complete equations must therefore be used, and q cannot always be assumed as large compared with m and u, except when dealing with local oscillations.

(4) *Long-distance overhead telephone circuit.*

49. Consider a telephone circuit of 1000 miles length, metallic return, consisting of two wires No. 4 B. and S. G., 24 inches distant from each other.

Calculating in the same way as discussed under (1), the following constants per mile of conductor are obtained: $r = 1.31$ ohms, $L = 1.84 \times 10^{-3}$ henry, and $C = .0172 \times 10^{-6}$ farad.

As conductance, g, we may assume

(a) $g = 0$; that is, very perfect insulation, as in dry weather.

(b) $g = 2.5 \times 10^{-6}$; that is, slightly leaky line.

(c) $g = 12 \times 10^{-6}$; that is, poor insulation, or a leaky line.

(d) $g = 40 \times 10^{-6}$; that is, extremely poor insulation, as during heavy rain.

The condition may also be investigated where the line is loaded with inductance coils spaced so close together that in their effect we can consider this additional inductance as uniformly distributed. Let the total inductance per unit length be increased by the loading coils to

$$L_1 = 9 \times 10^{-3} \, h,$$

or about five times the normal value.

Denoting then the constants of the loaded line by the index 1, we have:

Quantity	(a)	(b)	(c)	(d)
$u =$	356	429	706	1,518
$u_1 =$	73	146	423	1,236
$m =$	356	283	6	$- 806$
$m_1 =$	73	0	$- 277$	$- 1,090$
$\sigma = \sqrt{LC} =$		5.63×10^{-6}		
$\sigma_1 = \sqrt{L_1 C} =$		12.45×10^{-6}		
$k_0 = m \sqrt{LC} =$	2×10^{-3}	1.6×10^{-3}	33.7×10^{-6}	4.56×10^{-3}
$k_{01} = m_1 \sqrt{L_1 C} =$	910×10^{-6}	0	3.45×10^{-3}	13.5×10^{-3}
$l_{w_0} = \dfrac{2\pi}{k_0} =$	3,140	3,920	187,000	1,380
$l_{w_{01}} = \dfrac{2\pi}{k_{01}} =$	6,900	∞	1,820	464
$f_0 = \dfrac{m}{2\pi} =$	55.6	45	0.96	128
$f_{01} = \dfrac{m_1}{2\pi} =$	11.6	0	44	173

In a long-distance telephone line, distributed leakage up to a certain amount increases the critical wave length and thus makes even the long wave oscillatory. Beyond this amount leakage again decreases the wave length. Distributed inductance, as by loading the line, increases the critical wave length if the leakage is small, but in a very leaky line it decreases the critical wave length, and the amount of leakage up to which an increase of the critical wave length occurs is less in a loaded line, that is, in a line of higher inductance.

In other words, a moderate amount of distributed leakage improves a long-distance telephone line, an excessive amount of leakage spoils it. An increase of inductance, by loading the line, improves the line if the leakage is small, but may spoil the line if the leakage is considerable. The amount of leakage up to which improvement in the telephone line occurs is less in a loaded than in an unloaded line; that is, a loaded telephone line requires a far better insulation than an unloaded line.

CHAPTER VI.

TRAVELING WAVES.

50. As seen in Chapter V, especially in electric power circuits, overhead or underground, the longest existing standing wave has a wave length which is so small compared with the critical wave length — where the frequency becomes zero — that the effect of the damping constant on the frequency and the wave length is negligible. The same obviously applies also to traveling waves, generally to a still greater extent, since the lengths of traveling waves are commonly only a small part of the length of the circuit. Usually, therefore, in the discussion of traveling waves, the effect of the damping constants on the frequency constant q and the wave length constant k can be neglected, that is, frequency and wave length assumed as independent of the energy loss in the circuit.

Usually, therefore, the equations (13) and (14) of Chapter IV can be applied in dealing with the traveling wave.

In these equations the distance traveled by the wave per second is used as unit length by the substitution

$$\lambda = \sigma l,$$

where
$$\sigma = \sqrt{LC},$$

as this brings t and λ into direct comparison and eliminates h and k from the equations by the equation (11) of Chapter IV.

With this unit length the critical value of k, $k_0 = m\sqrt{LC}$, by substituting (8) and (7) of Chapter IV, gives $q_0 = m$, and the condition of the applicability of equations (13) and (14) of Chapter IV, therefore, is that q be a large quantity compared with $q_0 = m$.

In this case $\dfrac{m}{q}$ is a small quantity, and thus can usually be neglected in equations (15) and (14) of Chapter IV, except when C and C' are very different in magnitude.

524

This gives, under the limiting conditions discussed above, the *general equations of the traveling wave,* thus:

$$i = \varepsilon^{-ut}\{\varepsilon^{+s(t-\lambda)}[C_1 \cos q\,(t-\lambda)+ C_1{}' \sin q\,(t-\lambda)]$$
$$- \varepsilon^{+s(t+\lambda)}[C_2 \cos q\,(t+\lambda)+ C_2{}' \sin q\,(t+\lambda)]$$
$$+ \varepsilon^{-s(t-\lambda)}[C_3 \cos q\,(t-\lambda)+ C_3{}' \sin q\,(t-\lambda)]$$
$$- \varepsilon^{-s(t+\lambda)}[C_4 \cos q\,(t+\lambda)+ C_4{}' \sin q\,(t+\lambda)]\} \tag{1}$$

and

$$e = -\sqrt{\frac{L}{C}}\,\varepsilon^{-ut}\{\varepsilon^{+s(t-\lambda)}[C_1 \cos q\,(t-\lambda)+ C_1{}' \sin q\,(t-\lambda)]$$
$$+ \varepsilon^{+s(t+\lambda)}[C_2 \cos q\,(t+\lambda)+ C_2{}' \sin q\,(t+\lambda)]$$
$$+ \varepsilon^{-s(t-\lambda)}[C_3 \cos q\,(t-\lambda)+ C_3{}' \sin q\,(t-\lambda)]$$
$$+ \varepsilon^{-s(t+\lambda)}[C_4 \cos q\,(t+\lambda)+ C_4{}' \sin q\,(t+\lambda)]\}, \tag{2}$$

or

$$e = \varepsilon^{-ut)}\{\varepsilon^{+s(t+\lambda)}[A_1 \cos q\,(t+\lambda)+ A_1{}' \sin q\,(t+\lambda)]$$
$$+ \varepsilon^{+s(t-\lambda)}[A_2 \cos q\,(t-\lambda)+ A_2{}' \sin q\,(t-\lambda)]$$
$$+ \varepsilon^{-s(t+\lambda)}[A_3 \cos q\,(t+\lambda)+ A_3{}' \sin q\,(t+\lambda)]$$
$$+ \varepsilon^{-s(t-\lambda)}[A_4 \cos q\,(t-\lambda)+ A_4{}' \sin q\,(t-\lambda)]\} \tag{3}$$

and

$$i = \sqrt{\frac{C}{L}}\,\varepsilon^{-ut}\{\varepsilon^{+s(t+\lambda)}[A_1 \cos q\,(t+\lambda)+ A_1{}' \sin q\,(t+\lambda)]$$
$$- \varepsilon^{+s(t-\lambda)}[A_2 \cos q\,(t-\lambda)+ A_2{}' \sin q\,(t-\lambda)]$$
$$+ \varepsilon^{-s(t+\lambda)}[A_3 \cos q\,(t+\lambda)+ A_3{}' \sin q\,(t+\lambda)]$$
$$- \varepsilon^{-s(t-\lambda)}[A_4 \cos q\,(t-\lambda)+ A_4{}' \sin q\,(t-\lambda)]\}, \tag{4}$$

where

$$u = \frac{1}{2}\left(\frac{r}{L}+\frac{g}{C}\right),$$

$$\lambda = \sigma l, \tag{5}$$

and

$$\sigma = \sqrt{LC}.$$

In these equations (1) to (4) the sign of λ may be reversed, which merely means counting the distance in opposite direction.

This gives the following equations:

$$i = \varepsilon^{-ut}\{\varepsilon^{+s(t-\lambda)}\,[B_1\cos q\,(t-\lambda)+B_1{}'\sin q\,(t-\lambda)]$$
$$-\,\varepsilon^{+s(t+\lambda)}\,[B_2\cos q\,(t+\lambda)+B_2{}'\sin q\,(t+\lambda)]$$
$$+\,\varepsilon^{-s(t-\lambda)}\,[B_3\cos q\,(t-\lambda)+B_3{}'\sin q\,(t-\lambda)]$$
$$-\,\varepsilon^{-s(t+\lambda)}\,[B_4\cos q\,(t+\lambda)+B_4{}'\sin q\,(t+\lambda)]\},\qquad (6)$$

and

$$e = \sqrt{\frac{L}{C}}\,\varepsilon^{-ut}\{\varepsilon^{+s(t-\lambda)}\,[B_1\cos q\,(t-\lambda)+B_1{}'\sin q\,(t-\lambda)]$$
$$+\,\varepsilon^{+s(t+\lambda)}\,[B_2\cos q\,(t+\lambda)+B_2{}'\sin q\,(t+\lambda)]$$
$$+\,\varepsilon^{-s(t-\lambda)}\,[B_3\cos q\,(t-\lambda)+B_3{}'\sin q\,(t-\lambda)]$$
$$+\,\varepsilon^{-s(t+\lambda)}\,[B_4\cos q\,(t+\lambda)+B_4{}'\sin q\,(t+\lambda)]\},$$
$$(7)$$

or

$$e = \varepsilon^{-ut}\{\varepsilon^{+s(t-\lambda)}\,[A_1\cos q\,(t-\lambda)+A_1{}'\sin q\,(t-\lambda)]$$
$$+\,\varepsilon^{+s(t+\lambda)}\,[A_2\cos q\,(t+\lambda)+A_2{}'\sin q\,(t+\lambda)]$$
$$+\,\varepsilon^{-s(t-\lambda)}\,[A_3\cos q\,(t-\lambda)+A_3{}'\sin q\,(t-\lambda)]$$
$$+\,\varepsilon^{-s(t+\lambda)}\,[A_4\cos q\,(t+\lambda)+A_4{}'\sin q\,(t+\lambda)]\}\qquad (8)$$

and

$$i = \sqrt{\frac{C}{L}}\,\varepsilon^{-ut}\{\varepsilon^{+s(t-\lambda)}\,[A_1\cos q\,(t-\lambda)+A_1{}'\sin q\,(t-\lambda)]$$
$$-\,\varepsilon^{+s(t+\lambda)}\,[A_2\cos q\,(t+\lambda)+A_2{}'\sin q\,(t+\lambda)]$$
$$+\,\varepsilon^{-s(t-\lambda)}\,[A_3\cos q\,(t-\lambda)+A_3{}'\sin q\,(t-\lambda)]$$
$$-\,\varepsilon^{+s(t+\lambda)}\,[A_4\cos q\,(t+\lambda)+A_4{}'\sin q\,(t+\lambda)]\}.$$
$$(9)$$

In these equations (1) to (9) the values A, B, C, etc., are integration constants, which are determined by the terminal conditions of the problem.

The terms with $(t-\lambda)$ may be considered as the main wave, the terms with $(t+\lambda)$ as the reflected wave, or inversely, depending on the direction of propagation of the wave.

51. As the traveling wave, equations (1) to (9), consists of a main wave with variable $(t-\lambda)$ and a reflected wave of the same character but moving in opposite direction, thus with the variable $(t+\lambda)$, these waves may be studied separately, and afterwards the effect of their combination investigated.

Thus, considering at first one of the waves only, that with the variable $(t - \lambda)$, from equations (8) and (9) we have

$$
\begin{aligned}
e &= \varepsilon^{-ut}\{\varepsilon^{+s\,(t-\lambda)}[A_1 \cos q\,(t - \lambda) + A_1{}' \sin q\,(t - \lambda)] \\
&\quad + \varepsilon^{-s\,(t-\lambda)}[A_3 \cos q\,(t - \lambda) + A_3{}' \sin q\,(t - \lambda)]\} \\
&= \varepsilon^{-ut}\{(A_1\varepsilon^{+s\,(t-\lambda)} + A_3\varepsilon^{-s\,(t-\lambda)}) \cos q\,(t - \lambda) \\
&\quad + (A_1{}'\varepsilon^{+s\,(t-\lambda)} + A_3{}'\varepsilon^{-s\,(t-\lambda)}) \sin q\,(t - \lambda)\}
\end{aligned}
$$

$$(10)$$

and

$$
i = \sqrt{\frac{C}{L}}\, e;
\tag{11}
$$

that is, in a single traveling wave current and voltage are in phase with each other, and proportional to each other with an *effective impedance*, the *surge impedance* or *natural impedance* of the circuit

$$
z = \frac{e}{i} = \sqrt{\frac{L}{C}}.
\tag{12}
$$

This proportionality between e and i and coincidence of phase obviously no longer exist in the combination of main waves and reflected waves, since in reflection the current reverses with the reversal of the direction of propagation, while the voltage remains in the same direction, as seen by (8) and (9).

In equation (10) the time t appears only in the term $(t - \lambda)$ except in the factor ε^{-ut}, while the distance λ appears only in the term $(t - \lambda)$. Substituting therefore

$$
t_l = t - \lambda,
$$

hence

$$
t = t_l + \lambda;
$$

that is, counting the time differently at any point λ, and counting it at every point of the circuit from the same point in the phase of the wave from which the time t is counted at the starting point of the wave, $\lambda = 0$, or, in other words, shifting the starting point of the counting of time with the distance λ, and substituting in (150), we have

$$
\begin{aligned}
e &= \varepsilon^{-ut} \left\{ \varepsilon^{+st_l} (A_1 \cos qt_l + A_1' \sin qt_l) \right. \\
&\quad \left. + \varepsilon^{-st_l} (A_3 \cos qt_l + A_3' \sin qt_l) \right\} \\
&= \varepsilon^{-u\lambda} \varepsilon^{-ut_l} \left\{ \varepsilon^{+st_l} (A_1 \cos qt_l + A_1' \sin qt_l) \right. \\
&\quad \left. + \varepsilon^{-st_l} (A_3 \cos qt_l + A_3 \sin qt_l) \right\} \\
&= \varepsilon^{-u\lambda} \varepsilon^{-ut_l} \left\{ (A_1 \varepsilon^{+st_l} + A_3 \varepsilon^{-st_l}) \cos qt_l \right. \\
&\quad \left. + (A_1' \varepsilon^{+st_l} + A_3' \varepsilon^{-st_l}) \sin qt_l \right\}.
\end{aligned}
\tag{13}
$$

The latter form of the equation is best suited to represent the variation of the wave, at a fixed point λ in space, as function of the local time t_l.

Thus the wave is the product of a term $\varepsilon^{-u\lambda}$ which decreases with increasing distance λ, and a term

$$
\begin{aligned}
e_0 &= \varepsilon^{-ut_l} \left\{ \varepsilon^{+st_l} (A_1 \cos qt_l + A_1' \sin qt_l) \right. \\
&\quad \left. + \varepsilon^{-st_l} (A_3 \cos qt_l + A_3' \sin qt_l) \right\} \\
&= \varepsilon^{-ut_l} \left\{ (A_1 \varepsilon^{+st_l} + A_3 \varepsilon^{-st_l}) \cos qt_l \right. \\
&\quad \left. + (A_1' \varepsilon^{+st_l} + A_3' \varepsilon^{-st_l}) \sin qt_l \right\},
\end{aligned}
\tag{14}
$$

which latter term is independent of the distance, but merely a function of the time t_l when counting the time at any point of the line from the moment of the passage of the same phase of the wave.

Since the coefficient in the exponent of the distance decrement $\varepsilon^{-u\lambda}$ contains only the circuit constant,

$$
u = \frac{1}{2}\left(\frac{r}{L} + \frac{g}{C}\right),
$$

but does not contain s and q or the other integration constants, resubstituting from equations (10) to (7) of **Chapter IV**,

$$
\lambda = \sigma l = l \sqrt{LC},
$$

we have

$$
\begin{aligned}
u\lambda &= u \sqrt{LC}\, l \\
&= \frac{1}{2}\left\{ r \sqrt{\frac{C}{L}} + g \sqrt{\frac{L}{C}} \right\} l,
\end{aligned}
$$

where l is measured in any desired length.

Therefore the attenuation constant of a traveling wave is

$$u_0 = u\sqrt{LC} = \frac{1}{2}\left\{r\sqrt{\frac{C}{L}} + g\sqrt{\frac{L}{C}}\right\}, \tag{15}$$

and hence the distance decrement of the wave,

$$\varepsilon^{-u\lambda} = \varepsilon^{-ul\sqrt{LC}},$$

depends upon the circuit constants r, L, g, C only, but does not depend upon the wave length, frequency, voltage, or current; hence, all traveling waves in the same circuit die out at the same rate, regardless of their frequency and therefore of their wave shape, or, in other words, a complex traveling wave retains its wave shape when traversing a circuit, and merely decreases in amplitude by the distance decrement $\varepsilon^{-u\lambda}$. The wave attenuation thus is a constant of the circuit.

The above statement obviously applies only for waves of constant velocity, that is, such waves in which q is large compared with s, u, and m, and therefore does not strictly apply to extremely long waves, as discussed in 13.

52. By changing the line constants, as by inserting inductance L in such a manner as to give the effect of uniform distribution (loading the line), the attenuation of the wave can be reduced, that is, the wave caused to travel a greater distance l with the same decrease of amplitude.

As function of the inductance L, the attenuation constant (155) is a minimum for

$$\frac{du_0}{dL} = 0;$$

hence,

or

$$\left.\begin{array}{l} rC - gL = 0, \\[2mm] \dfrac{L}{C} = \dfrac{r}{g}, \end{array}\right\} \tag{16}$$

and if the conductance $g = 0$ we have $L = \infty$; hence, in a perfectly insulated circuit, or rather a circuit having no energy losses depending on the voltage, the attenuation decreases with increase of the inductance, that is, by "loading the line," and the more inductance is inserted the better the telephonic transmission.

In a leaky telephone line increase of inductance decreases the attenuation, and thus improves the telephonic transmission, up to the value of inductance,

$$L = \frac{rC}{g},\tag{17}$$

and beyond this value inductance is harmful by again increasing the attenuation.

For instance, if a long-distance telephone circuit has the following constants per mile: $r = 1.31$ ohms, $L = 1.84 \times 10^{-3}$ henry, $g = 1.0 \times 10^{-6}$ mho, and $C = 0.0172 \times 10^{-6}$ farad, the attenuation of a traveling wave or impulse is

$$u_0 = 0.00217;$$

hence, for a distance or length of line of $l_0 = 2000$ miles,

$$\varepsilon^{-u_0 l_0} = \varepsilon^{-4.34} = 0.0129;$$

that is, the wave is reduced to 1.29 per cent of its original value.

The best value of inductance, according to (17), is

$$L = \frac{r}{g} C = 0.0225 \text{ henry},$$

and in this case the attenuation constant becomes

$$u_0 = 0.00114,$$

and thus

$$\varepsilon^{-u_0 l_0} = \varepsilon^{-2.24} = 0.1055,$$

or 10.55 per cent of the original value of the wave; which means that in this telephone circuit, by adding an additional inductance of $22.5 - 1.84 = 20.7$ mh. per mile, the intensity of the arriving wave is increased from 1.29 per cent to 10.55 per cent, or more than eight times.

If, however, in wet weather the leakage increases to the value $g = 5 \times 10^{-6}$, we have in the unloaded line

$$u_0 = 0.00282 \text{ and } \varepsilon^{-u_0 l} = 0.0035,$$

while in the loaded line we have

$$u_0 = 0.00341 \text{ and } \varepsilon^{-u_0 l} = 0.0011,$$

and while with the unloaded line the arriving wave is still 0.35 per cent of the outgoing wave, in the loaded line it is only 0.11 per cent; that is, in this case, loading the line with inductance has badly spoiled telephonic communication, increasing the decay of the wave more than threefold. A loaded telephone line, therefore, is much more sensitive to changes of leakage g, that is, to meteorological conditions, than an unloaded line.

53. The equation of the traveling wave (13),

$$e = \varepsilon^{-u\lambda}\, \varepsilon^{-ut_l} \{ \varepsilon^{+st_l}\, (A_1 \cos qt_l + A_1{}' \sin qt_l) \\ + \varepsilon^{-st_l}(A_3 \cos qt_l + A_3{}' \sin qt_l) \},$$

can be reduced to the form

$$e = \varepsilon^{-u\lambda} \{ E_1 \varepsilon^{-ut_{l_1}} (\varepsilon^{+st_{l_1}} - \varepsilon^{-st_{l_1}}) \sin qt_{l_1} \\ + E_2 \varepsilon^{-ut_{l_2}} (\varepsilon^{+st_{l_2}} - \varepsilon^{-st_{l_2}}) \cos qt_{l_2} \}; \qquad (18)$$

where

and
$$\left. \begin{array}{l} t_{l_1} = t_l - \gamma_1 = t - \lambda - \gamma_1 \\[2mm] t_{l_2} = t_l - \gamma_2 = t - \lambda - \gamma_2. \end{array} \right\} \qquad (19)$$

By substituting (19) in (18), expanding, and equating (18) with (13), we get the identities

$$\left. \begin{array}{l} E_1 \varepsilon^{-s\gamma_1} \cos q\gamma_1 - E_2 \varepsilon^{-s\gamma_2} \sin q\gamma_2 = A_1, \\[2mm] E_1 \varepsilon^{-s\gamma_1} \sin q\gamma_1 + E_2 \varepsilon^{-s\gamma_2} \cos q\gamma_2 = A_1{}', \\[2mm] E_1 \varepsilon^{+s\gamma_1} \cos q\gamma_1 - E_2 \varepsilon^{+s\gamma_2} \sin q\gamma_2 = -A_3, \\[2mm] E_1 \varepsilon^{+s\gamma_1} \sin q\gamma_1 + E_2 \varepsilon^{+s\gamma_2} \cos q\gamma_2 = -A_3{}', \end{array} \right\} \qquad (20)$$

and these four equations determine the four constants E_1, E_2, γ_1, γ_2.

Any traveling wave can be resolved into, and considered as consisting of, a combination of two waves:
the *traveling sine wave*,

$$e_1 = E_1 \varepsilon^{-u\lambda}\, \varepsilon^{-ut_{l_1}} (\varepsilon^{+st_{l_1}} - \varepsilon^{-st_{l_1}}) \sin qt_{l_1}, \qquad (21)$$

and the *traveling cosine wave*,

$$e_2 = E_2 \varepsilon^{-u\lambda}\, \varepsilon^{-ut_{l_2}} (\varepsilon^{+st_{l_2}} - \varepsilon^{-st_{l_2}}) \cos qt_{l_2}. \qquad (22)$$

Since q is a large quantity compared with u and s, the two component traveling waves, (21) and (22), differ appreciably from each other in appearance only for very small values of t_l, that is, near $t_{l_1} = 0$ and $t_{l_2} = 0$. The traveling sine wave rises in the first half cycle very slightly, while the traveling cosine wave rises rapidly; that is, the tangent of the angle which the wave makes with the horizontal, or $\dfrac{de}{dt}$, equals 0 with the sine wave and has a definite value with the cosine wave.

All traveling waves in an electric circuit can be resolved into constituent elements, traveling sine waves and traveling cosine waves, and the general traveling wave consists of four component waves, a sine wave, its reflected wave, a cosine wave and its reflected wave.

The elements of the traveling wave, the traveling sine wave e_1, and the traveling cosine wave e_2 contain four constants: the intensity constant, E; the attenuation constant, u, and u_0 respectively; the frequency constant, q, and the constant, s.

The wave starts from zero, builds up to a maximum, and then gradually dies out to zero at infinite time.

The absolute term of the wave, that is, the term which represents the values between which the wave oscillates, is

$$e_0 = E\varepsilon^{-u\lambda}\varepsilon^{-ut_l} \left(\varepsilon^{+st_l} - \varepsilon^{-st_l}\right). \tag{23}$$

The term e_0 may be called the *amplitude* of the wave. It is a maximum for the value of t_l, given by

$$\frac{de_0}{dt_l} = 0,$$

which gives

$$-(u - s)\,\varepsilon^{-(u-s)\,t_0} + (u + s)\,\varepsilon^{-(u+s)\,t_0} = 0;$$

hence,

$$\varepsilon^{+2st_0} = \frac{u + s}{u - s}$$

and

$$t_{l_0} = \frac{1}{2\,s} \log \frac{u + s}{u - s}, \tag{24}$$

and substituting this value into the equation of the absolute term of the wave, (163), gives

$$e_m = E\varepsilon^{-u\lambda} \frac{2\,s}{\sqrt{u^2 - s^2}} \left(\frac{u + s}{u - s}\right)^{-\frac{v}{2s}} \qquad ((25)$$

The rate of building up of the wave, or the steepness of the wave front, is given by

$$G_0 = \left[\frac{de_0}{dt_l}\right]_{t_l = 0}$$

as

$$G_0 = E\varepsilon^{-u\lambda} \left[- (u - s)\, \varepsilon^{-(u-s)\, t_l} + (u + s)\, \varepsilon^{-(u+s)\, t_l}\right]_0$$
$$= 2\, sE\varepsilon^{-u\lambda}; \qquad ((26)$$

that is, the constant s, which above had no interpretation, represents the rapidity of the rise of the wave.

Referring, however, the rise of the wave to the maximum value e_m of the wave, and combining (165) with (166), we have

$$G_0 = e_m \frac{(u + s)^{\frac{u+s}{2s}}}{(u - s)^{\frac{u-s}{2s}}}. \qquad ((27)$$

The rapidity of the rise of the wave is a maximum, that is, t_l a minimum, for the value of s, in equation (164), given by

$$\frac{dt_{l_0}}{ds} = 0,$$

which gives

$$\log \frac{u + s}{u - s} = \frac{2\,us}{u^2 - s^2};$$

hence, $s = 0$, or the standing wave, which rises infinitely fast, that is, appears instantly.

The smaller therefore s is, the more rapidly is the rise of the traveling wave, and therefore s may be called the *acceleration constant* of the traveling wave.

54. In the components of the traveling wave, equations (21) and (22), the traveling sine wave,

$$e_1 = E\epsilon^{-u\lambda}\epsilon^{-ut_l} \left(\epsilon^{+st_l} - \epsilon^{-st_l}\right) \sin qt_l \quad (21),$$

and the traveling cosine wave (22),

$$e_2 = E\varepsilon^{-u\lambda}\varepsilon^{-ut_l}(\varepsilon^{+st_l} - \varepsilon^{-st_l})\cos qt_l$$

with the amplitude,

$$e_0 = E\varepsilon^{-u\lambda}\varepsilon^{-ut_l}(\varepsilon^{+st_l} - \varepsilon^{-st_l}), \tag{28}$$

we have

and

$$\left.\begin{array}{l} e_1 = e_0 \sin qt_l \\[2ex] e_2 = e_0 \cos qt_l. \end{array}\right\} \tag{29}$$

If $t_l = 0$, $e_0 = 0$; that is, t_l is the time counted from the beginning of the wave.

It is

$$t_l = t - \lambda - \gamma,$$

or, if we change the zero point of distance, that is, count the distance λ from that point of the line at which the wave starts at time $t = 0$, or, in other words, count time t and distance λ from the origin of the wave,

$$t_l = t - \lambda,$$

and the traveling wave thus may be represented by the amplitude,

$$e_0 = E\varepsilon^{-ut}(\varepsilon^{+st_l} - \varepsilon^{-st_l});$$

the sine wave,

$$e_1 = E\varepsilon^{-ut}(\varepsilon^{+st_l} - \varepsilon^{-st_l})\sin qt_l = e_0 \sin qt_l;$$

the cosine wave,

$$e_2 = E\varepsilon^{-ut}(\varepsilon^{+st_l} - \varepsilon^{-st_l})\cos qt_l = e_0 \cos qt_l; \tag{30}$$

and $t_l = t - \lambda$ can be considered as the distance, counting backwards from the wave front, or the temporary distance; that is, distance counted with the point λ, which the wave has just reached, as zero point, and in opposite direction to λ.

Equation (30) represents the distribution of the wave along the line at the moment t.

As seen, the wave maintains its shape, but progresses along the line, and at the same time dies out, by the time decrement ε^{-ut}.

Resubstituting,

$$t_l = t - \lambda,$$

the equation of the amplitude of the wave is

$$e_0 = E\varepsilon^{-ut} \left(\varepsilon^{+s(t-\lambda)} - \varepsilon^{-s(t-\lambda)} \right). \tag{31}$$

As function of the distance λ, the amplitude of the traveling wave, (171), is a maximum for

$$\frac{de_0}{d\lambda} = 0,$$

which gives

$$\lambda = 0;$$

that is, the amplitude of the traveling wave is a maximum at all times at its origin, and from there decreases with the distance.

This obviously applies only to the single wave, but not to a combination of several waves, as a complex traveling wave.

For $$\lambda = 0,$$

$$e_0 = E\varepsilon^{-ut} \left(\varepsilon^{+st} - \varepsilon^{-st} \right),$$

and as function of the time t this amplitude is a maximum, according to equations (23) to (25), at

and is

$$\left. \begin{aligned} t_0 &= \frac{1}{2\,s} \log \frac{u+s}{u-s}, \\[2mm] E_0 &= E \frac{2\,s}{\sqrt{u^2 - s^2}} \left(\frac{u+s}{u-s} \right)^{-\frac{u}{2\,s}}. \end{aligned} \right\} \tag{32}$$

At any other point λ of the circuit, the amplitude therefore is a maximum, according to equation (24), at the time

and is

$$\left. \begin{aligned} t_m &= t_0 + \lambda, \\[2mm] e_m &= E \frac{2\,s\varepsilon^{-u\lambda}}{\sqrt{u^2 - s^2}} \left(\frac{u+s}{u-s} \right)^{-\frac{u}{2\,s}}. \end{aligned} \right\} \tag{33}$$

55. As an example may be considered a traveling wave having the constants $u = 115$, $s = 45$, $q = 2620$, and $E = 100$, hence,

$$e_0 = 100\, \varepsilon^{-115t} \left(\varepsilon^{+45\,t_l} - \varepsilon^{-45\,t_l} \right),$$

$$= 100\, \varepsilon^{-115\lambda} \left(\varepsilon^{-70\,t_l} - \varepsilon^{-160\,t_l} \right),$$

where $t_l = t - \lambda$.

In Fig. 99 is shown the amplitude e_0 as function of the distance λ, for the different values of time,

$$t = 2,\ 4,\ 8,\ 12,\ 16,\ 20,\ 24,\ \text{and}\ 32 \times 10^{-3},$$

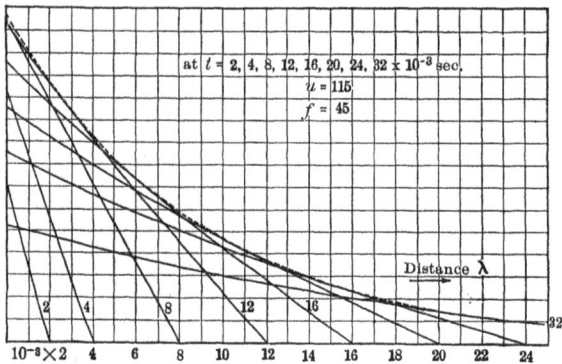

Fig. 99. Spread of amplitude of electric traveling wave.

with the maximum amplitude e_m, in dotted line, as envelope of the curve of e_0.

As seen, the amplitude of the wave gradually rises, and at the same time spreads over the line, reaching the greatest value at the starting point $\lambda = 0$ at the time $t_0 = 9.2 \times 10^{-3}$ sec., and then decreases again while continuing to spread over the line, until it gradually dies out.

It is interesting to note that the distribution curves of the amplitude are nearly straight lines, but also that in the present instance even in the longest power transmission line the wave has reached the end of the line, and reflection occurs before the maximum of the curve is reached. The unit of length λ is the distance traveled by the wave per second, or 188,000 miles, and during the rise of the wave, at the origin, from its start to the maximum, or 9.2×10^{-3} sec., the wave thus has traveled 1760 miles, and the reflected wave would have returned to the origin before the maximum of the wave is reached, providing the circuit is shorter than 880 miles.

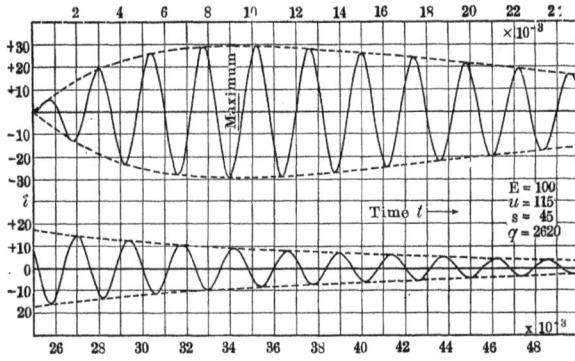

Fig. 100. Passage of traveling wave at a given point of a transmission line

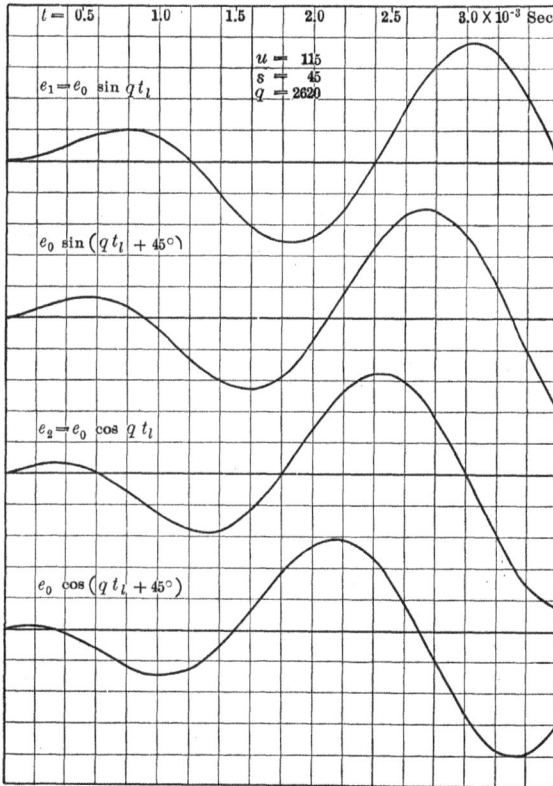

Fig. 101. Beginning of electric traveling waves.

With $s = 1$ it would be $t_0 = 8.7 \times 10^{-3}$ sec., or nearly the same, and with $s = 0.01$ it would be $t_0 = 3.75 \times 10^{-3}$ sec., or, in other words, the rapidity of the rise of the wave increases very little with a very great decrease of s.

Fig. 100 shows the passage of the traveling wave, $e_1 = e_0 \sin qt_l$, across a point λ of the line, with the local time t_l as abscissas and the instantaneous values of e_1 as ordinates. The values are given for $\lambda = 0$, where $t_l = t$; for any other point of the line λ the wave shape is the same, but all the ordinates reduced by the factor $\varepsilon^{-115\lambda}$ in the proportion as shown in the dotted curve in Fig. 99.

Fig. 101 shows the beginning of the passage of the traveling wave across a point $\lambda = 0$ of the line, that is, the starting of a wave, or its first one and one-half cycles, for the trigonometric functions differing successively by 45 degrees, that is,

$$e_1 = e_0 \sin qt_l,$$

$$\frac{e_1 + e_2}{\sqrt{2}} = e_0 \sin \left(qt_l + \frac{\pi}{4}\right),$$

$$e_2 = e_0 \cos qt_l = e_0 \sin \left(qt_l + \frac{\pi}{2}\right),$$

$$\frac{e_2 - e_1}{\sqrt{2}} = e_0 \cos \left(qt_l + \frac{\pi}{4}\right) = e_0 \sin \left(qt_l + \frac{3\pi}{4}\right).$$

The first curve of Fig. 101 therefore is the beginning of Fig. 100.

In waves traveling over a water surface shapes like Fig. 101 can be observed.

For the purpose of illustration, however, in Figs. 100 and 101 the oscillations are shown far longer than they usually occur; the value $q = 2620$ corresponds to a frequency $f = 418$ cycles, while traveling waves of frequencies of 100 to 10,000 times as high are more common.

Fig. 102 shows the beginning of a wave having ten times the attenuation of that of Fig. 101, that is, a wave of such rapid decay that only a few half waves are appreciable, for values of the phase differing by 30 degrees.

56. A specially interesting traveling wave is the wave in which

$$s = u, \tag{34}$$

since in this wave the time decrement of the first main wave and
its reflected wave vanishes,

$$\varepsilon^{-(u-s)t} = 1;$$ (35)

that is, the first main wave and its reflected wave are not transient but permanent or alternating waves, and the equations of

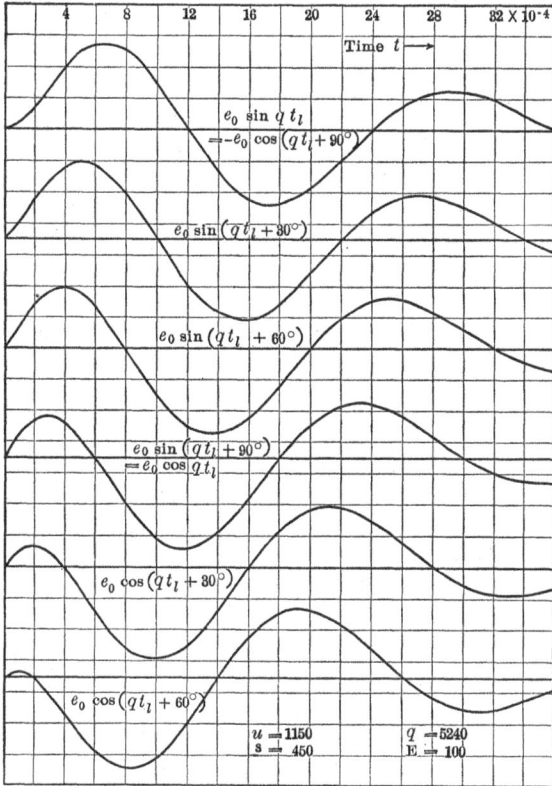

Fig. 102. Passage of a traveling wave at a given point of a line.

the first main wave give the equations of the alternating-current
circuit with distributed r, L, g, C, which thus appear as a special
case of a traveling wave.

Since in this case the frequency, and therewith the value of q,
are low and comparable with u and s, the approximations made

in the previous discussion of the traveling wave are not permissible, but the general equations (53) and (54) of Chapter I have to be used.

Substituting therefore in (53) and (54) of Chapter I,

$$s = u,$$

gives

$$
\begin{aligned}
i = \; & [\varepsilon^{-hl}\,\{C_1 \cos (qt - kl) + C_1{}' \sin (qt - kl)\} \\
& - \varepsilon^{+hl}\,\{C_2 \cos (qt + kl) + C_2{}' \sin (qt + kl)\}] \\
& - \varepsilon^{-2ut}\,[\varepsilon^{-hl}\,\{C_4 \cos (qt + kl) + C_4{}' \sin (qt + kl)\} \\
& - \varepsilon^{+hl}\,\{C_3 \cos (qt - kl) + C_3{}' \sin (qt - kl)\}]
\end{aligned}
\tag{36}
$$

and

$$
\begin{aligned}
e = \; & [\varepsilon^{-hl}\,\{(c_1{}'C_1{}' - c_1 C_1)\ \cos (qt - kl) \\
& \qquad\quad - (c_1{}'C_1 + c_1 C_1{}')\ \sin (qt - kl)\} \\
& + \varepsilon^{+hl}\,\{(c_1{}'C_2{}' - c_1 C_2)\ \cos (qt + kl) \\
& \qquad\quad - (c_1{}'C_2 + c_1 C_2{}')\ \sin (qt + kl)\}] \\
& + \varepsilon^{-2ut}\,[\varepsilon^{-hl}\,\{(c_2{}'C_4 - c_2 C_4)\ \cos (qt + kl) \\
& \qquad\quad - (c_2{}'C_4 + c_2 C_4{}')\ \sin (qt + kl)\} \\
& + \varepsilon^{+hl}\,\{(c_2{}'C_3{}' - c_2 C_3)\ \cos (qt - kl) \\
& \qquad\quad - (c_2{}'C_3 + c_2 C_3{}')\ \sin (qt - kl)\}].
\end{aligned}
\tag{37}
$$

In these equations of current i and e.m.f. e the first term represents the usual equations of the distribution of alternating current and voltage in a long-distance transmission line, and can by the substitution of complex quantities be reduced to a form given in Section III.

The second term is a transient term of the same frequency; that is, in a long-distance transmission line or other circuit of distributed r, L, g, C, when carrying alternating current under an alternating impressed e.m.f., at a change of circuit conditions, a transient term of fundamental frequency may appear which has the time decrement, that is, dies out at the rate

$$\varepsilon^{-2ut} = \varepsilon^{-\left(\frac{r}{L} + \frac{g}{C}\right) t} .$$

In this decrement the factor

$$\delta_1 = \varepsilon^{-\frac{r}{L} t}$$

is the usual decrement of a circuit of resistance r and inductance L, while the other factor,

$$\delta_2 = \varepsilon^{-\frac{g}{C}t}$$

may be attributed to the conductance and capacity of the circuit, and the total decrement is the product,

$$\delta = \delta_1 \delta_2.$$

A further discussion of the equations (36) and (37) and the meaning of their transient term requires the consideration of the terminal conditions of the circuit.

57. The alternating components of (36) and (37),

$$i_0 = \varepsilon^{-hl}\{C_1 \cos(qt - kl) + C_1' \sin(qt - kl)\}$$
$$- \varepsilon^{+hl}\{C_2 \cos(qt + kl) + C_2' \sin(qt + kl)\} \qquad (38)$$

and

$$e_0 = \varepsilon^{-hl}\left\{(c_1'C_1' - c_1C_1)\cos(qt - kl) - (c_1'C_1 + c_1C_1')\sin(qt - kl)\right\}$$
$$+ \varepsilon^{+hl}\left\{(c_1'C_2' - c_1C_2)\cos(qt + kl) - (c_1'C_2 + c_1C_2')\sin(qt + kl)\right\}, \qquad (39)$$

are reduced to their usual form in complex quantities by resolving the trigonometric function into functions of single angles, qt and kl, then dropping $\cos qt$, and replacing $\sin qt$ by the imaginary unit j. This gives

$$i_0 = \varepsilon^{-hl}\left\{(C_1 \cos kl - C_1' \sin kl)\cos qt \right.$$
$$+ (C_1' \cos kl + C_1 \sin kl)\cdot\sin qt\}$$
$$- \varepsilon^{+hl}\left\{(C_2 \cos kl + C_2' \sin kl)\cos qt \right.$$
$$+ (C_2' \cos kl - C_2 \sin kl)\sin qt\};$$

hence, in complex expression,

$$I = \varepsilon^{-hl}\left\{(C_1 - jC_1')\cos kl - (C_1' + jC_1)\sin kl\right\}$$
$$- \varepsilon^{+hl}\left\{(C_2 - jC_2')\cos kl + (C_2' + jC_2)\sin kl\right\}, \qquad (40)$$

and in the same manner,

$$E = \varepsilon^{-hl}\left\{[c_1'(C_1' + jC_1) - c_1(C_1 - jC_1')]\cos kl \right.$$
$$+ [c_1'(C_1 - jC_1') + c_1(C_1' + jC_1)]\sin kl\}$$
$$+ \varepsilon^{+hl}\left\{[c_1'(C_2' + jC_2) - c_1(C_2 - jC_2')]\cos kl \right.$$
$$- [c_1'(C_2 - jC_2') + c_1(C_2' + jC_2)]\sin kl\}. \qquad (41)$$

However, from equation (55) of Chapter I,

$$c_1 = \frac{qk + h\,(m + s)}{h^2 + k^2}\,L \quad \text{and} \quad c_1' = \frac{k\,(m + s) - qh}{h^2 + k^2}\,L,$$

since

$$
\left.
\begin{aligned}
q &= 2\,\pi f, \\[4pt]
s &= u = \frac{1}{2}\left(\frac{r}{L} + \frac{g}{C}\right), \\[4pt]
m &= \frac{1}{2}\left(\frac{r}{L} - \frac{g}{C}\right), \\[4pt]
s + m &= \frac{r}{L},
\end{aligned}
\right\} \tag{42}
$$

and

we have

$$
\left.
\begin{aligned}
c_1 &= \frac{2\,\pi f L k + r h}{h^2 + k^2} = \frac{xk + rh}{h^2 + k^2} \\[6pt]
c_1' &= \frac{rk - 2\,\pi f L h}{h^2 + k^2} = \frac{rk - xh}{h^2 + k^2},
\end{aligned}
\right\} \tag{43}
$$

and

where $x = 2\,\pi f L =$ reactance per unit length. $\qquad$ (44)

From equation (54),

$$R_1{}^2 = \sqrt{(s^2 + q^2 - m^2)^2 + 4\,q^2 m^2};$$

hence, substituting (42) and (44) and also

$$b = 2\,\pi f C, \tag{45}$$

we have

$$R_1{}^2 = \frac{1}{LC}\sqrt{(r^2 + x^2)\,(g^2 + b^2)}$$

$$= \frac{zy}{LC},$$

where

$$
\left.
\begin{aligned}
z &= \sqrt{r^2 + x^2} = \text{impedance per unit length} \\[6pt]
y &= \sqrt{g^2 + b^2} = \text{admittance per unit length.}
\end{aligned}
\right\} \tag{46}
$$

and

From the above it follows that

and

$$h = \sqrt{LC}\,\sqrt{\tfrac{1}{2}\,\{R_1^{\,2} + s^2 - q^2 - m^2\}}$$
$$= \sqrt{\tfrac{1}{2}\,(zy + rg - xb)}$$
$$k = \sqrt{\tfrac{1}{2}(zy - rg + xb)}. \qquad\qquad ((47)$$

If we now substitute

and

$$C_1 - jC_1' = B_2\sqrt{Y}$$
$$C_2 - jC_2' = -B_1\sqrt{Y}, \qquad\qquad ((48)$$

or

and

$$C_1' + jC_1 = +jB_2\sqrt{Y}$$
$$C_2' + jC_2 = -jB_1\sqrt{Y}, \qquad\qquad ((49)$$

where

and

$$Z = r + jx$$
$$Y = g + jb; \qquad\qquad ((50)$$

in (40) and (41) we have

$$I = \sqrt{Y}\,\{B_1\varepsilon^{+hl}(\cos kl + j\sin kl) + B_2\varepsilon^{-hl}(\cos kl - j\sin kl)\} \quad ((51)$$

and

$$E = (c_1 - jc_1')\,\sqrt{Y}\,\{B_1\varepsilon^{+hl}(\cos kl + j\sin kl)$$
$$- B_2\varepsilon^{-hl}(\cos kl - j\sin kl)\}, \quad ((52)$$

and substituting (43) gives

$$c_1 - jc_1' = \frac{h\,(r - jx) - k\,(x - jr)}{h^2 + k^2} = \frac{(r + jx)\,(h - jk)}{h^2 + k^2} = \frac{r + jx}{h + jk}. \quad ((53)$$

However,

$$h + jk = \sqrt{(h + jk)^2} = \sqrt{(h^2 - k^2) + 2\,jhk}$$
$$= \sqrt{(rg - xb) + j\sqrt{(zy)^2 - (rg - xb)^2}}$$
$$= \sqrt{(rg - xb) + j\,\sqrt{(r^2 + x^2)\,(g^2 + b^2) - (rg - xb)^2}}$$
$$= \sqrt{(rg - xb) + j\,\sqrt{r^2b^2 + x^2g^2 + 2\,rgxb}}$$
$$= \sqrt{(rg - xb) + j\,(rb + xg)}$$
$$= \sqrt{(r + jx)\,(g + jb)};$$

or $\qquad \boldsymbol{h + jk = \sqrt{ZY}} \qquad\qquad (194)$

substituted in (33) gives

$$c_1 - jc_1' = \sqrt{\frac{Z}{Y}}, \tag{55}$$

and (55) substituted in (52) gives

$$E = \sqrt{Z} \left\{ B_1 \epsilon^{+hl} (\cos kl + j \sin kl) - B_2 \epsilon^{-hl} (\cos kl - j \sin kl) \right\}, \tag{56}$$

where B_1 and B_2 are the complex imaginary integration constants.

Writing

$$h = \alpha \text{ and } k = \beta,$$
$$B_1 = D_1 \text{ and } B_2 = - D_2$$

the equations (51) and (56) become identical with the equations of the long-distance transmission line derived in Section III, equations (22) of paragraph 8.

It is interesting to note that here the general equations of alternating-current long-distance transmission appear as a special case of the equations of the traveling wave, and indeed can be considered as a section of a traveling wave, in which the acceleration constant s equals the exponential decrement u.

CHAPTER VII.

FREE OSCILLATIONS

58. The general equations of the electric circuit, (53) and (54) of Chapter I, contain eight terms: four waves: two main waves and their reflected waves, and each wave consists of a sine term and a cosine term.

The equations contain five constants, namely: the frequency constant, q; the wave length constant, k; the time attenuation constant, u; the distance attenuation constant, h, and the time acceleration constant, s; among these, the time attenuation, u, is a constant of the circuit, independent of the character of the wave.

By the value of the acceleration constant, s, waves may be subdivided into three classes, namely: $s = 0$, standing waves, as discussed in Chapter V; $u > s > 0$, traveling waves, as discussed in Chapter VI; $s = u$, alternating-current and e.m.f. waves, as discussed in Section III.

The general equations contain eight integration constants C and C', which have to be determined by the terminal conditions of the problem.

Upon the values of these integration constants C and C' largely depends the difference between the phenomena occurring in electric circuits, as those due to direct currents or pulsating currents, alternating currents, oscillating currents, inductive discharges, etc., and the study of the *terminal conditions* thus is of the foremost importance.

59. By *free oscillations* are understood the transient phenomena occurring in an electric circuit or part of the circuit to which neither electric energy is supplied by some outside source nor from which electric energy is abstracted.

Free oscillations thus are the transient phenomena resulting from the dissipation of the energy stored in the electric field of the circuit, or inversely, the accumulation of the energy of the electric field; and their appearance therefore presupposes the possibility of energy storage in more than one form so as to allow an interchange or surge of energy between its different forms, electromagnetic and electrostatic energy. Free oscillations occur only in circuits containing both capacity C and inductance, L.

The absence of energy supply or abstraction defines the free oscillations by the condition that the power $p = ei$ at the two ends of the circuit or section of the circuit must be zero at all times, or the circuit must be closed upon itself.

The latter condition, of a circuit closed upon itself, leads to a full-wave oscillation, that is, an oscillation in which the length of the circuit is a complete wave or a multiple thereof. With a circuit of uniform constants as discussed here such a full-wave oscillation is hardly of any industrial importance. While the most important and serious case of an oscillation is that of a closed circuit, such a closed circuit never consists of a uniform conductor, but comprises sections of different constants; generating system, transmission line and load, thus is a complex circuit comprising transition points between the sections, at which partial reflection occurs.

The full-wave oscillation thus is that of a complex circuit, which will be discussed in the following chapters.

Considering then the free oscillations of a circuit having two ends at which the power is zero, and representing the two ends of the electric circuit by $l = 0$ and $l = l_0$, that is, counting the distance from one end of the circuit, the conditions of a free oscillation are

$$l = 0, \qquad p = 0.$$
$$l = l_0, \qquad p = 0.$$

Since $p = ei$, this means that at $l = 0$ and $l = l_0$ either e or i must be zero, which gives four sets of terminal conditions:

$$
\left.
\begin{array}{ll}
(1)\ e = 0 \text{ at } l = 0; & i = 0 \text{ at } l = l_0. \\
(2)\ i = 0 \text{ at } l = 0; & e = 0 \text{ at } l = l_0. \\
(3)\ e = 0 \text{ at } l = 0; & e = 0 \text{ at } l = l_0. \\
(4)\ i = 0 \text{ at } l = 0; & i = 0 \text{ at } l = l_0.
\end{array}
\right\} \quad (1)
$$

Case (2) represents the same conditions as (1), merely with the distance l counting from the other end of the circuit — a line open at one end and grounded at the other end. Case (3) represents a circuit grounded at both ends, and case (4) a circuit open at both ends.

60. In either of the different cases, at the end of the circuit $l = 0$, either $e = 0$, or $i = 0$.

Substituting $l = 0$ into the equations (53) and (54) of Chapter I gives

$$
\begin{aligned}
e_0 = \varepsilon^{-(u-s)t} &\{[c_1' \, (C_1' + C_2') - c_1 \, (C_1 + C_2)] \cos qt \\
&- [c_1' \, (C_1 + C_2) + c_1 \, (C_1' + C_2')] \sin qt\} \\
+ \varepsilon^{-(u+s)t} &\{[c_2' \, (C_3' + C_4') - c_2 \, (C_3 + C_4)] \cos qt \\
&- [c_2' \, (C_3 + C_4) + c_2 \, (C_3' + C_4')] \sin qt\}
\end{aligned} \tag{2}
$$

and

$$
\begin{aligned}
i_0 = \varepsilon^{-(u-s)t} &\{(C_1 - C_2) \cos qt + (C_1' - C_2') \sin qt\} \\
+ \varepsilon^{-(u+s)t} &\{(C_3 - C_4) \cos qt + (C_3' - C_4') \sin qt\}.
\end{aligned} \tag{3}
$$

If neither q nor s equals zero,

for $e_0 = 0$,

$$
\left. \begin{aligned}
c_1' \, (C_1' + C_2') - c_1 \, (C_1 + C_2) = 0 \\
c_1' \, (C_1 + C_2) + c_1 \, (C_1' + C_2') = 0;
\end{aligned} \right\}
$$

and

hence,

$$
\left. \begin{aligned}
C_2 &= -C_1, & C_4 &= -C_3, \\
C_2' &= -C_1', & C_4' &= -C_3',
\end{aligned} \right\} \tag{4}
$$

and for $i_0 = 0$,

$$
\left. \begin{aligned}
C_2 &= C_1, & C_4 &= C_3, \\
C_2' &= C_1', & C_4' &= C_3'.
\end{aligned} \right\} \tag{5}
$$

Substituting in (53) and (54) of Chapter I,

$$
\begin{aligned}
i = \varepsilon^{-(u-s)t} &\{C_1 \, [\varepsilon^{-hl} \cos (qt - kl) \pm \varepsilon^{+hl} \cos (qt + kl)] \\
&+ C_1' \, [\varepsilon^{-hl} \sin (qt - kl) \pm \varepsilon^{+hl} \sin (qt + kl)]\} \\
+ \varepsilon^{-(u+s)t} &\{C_3 \, [\varepsilon^{+hl} \cos (qt - kl) \pm \varepsilon^{-hl} \cos (qt + kl)] \\
&+ C_3' \, [\varepsilon^{+hl} \sin (qt - kl) \pm \varepsilon^{-hl} \sin (qt + kl)]\}
\end{aligned} \tag{6}
$$

and

$$e = \varepsilon^{-(u-s)t} \{ (c_1'C_1' - c_1C_1) [\varepsilon^{-hl} \cos(qt - kl)$$
$$\mp \varepsilon^{+hl} \cos(qt + kl)]$$
$$- (c_1'C_1 + c_1C_1') [\varepsilon^{-hl} \sin(qt - kl)$$
$$\mp \varepsilon^{+hl} \sin(qt + kl)] \}$$
$$+ \varepsilon^{-(u+s)t} \{ (c_2'C_3' - c_2C_3) [\varepsilon^{+hl} \cos(qt - kl)$$
$$\mp \varepsilon^{-hl} \cos(qt + kl)]$$
$$- (C_2'C_3 + c_2C_3') [\varepsilon^{+hl} \sin(qt - kl)$$
$$\mp \varepsilon^{-hl} \sin(qt + kl)] \}, \qquad (7)$$

where the upper sign refers to $e = 0$, the lower sign to $i = 0$ for $l = 0$.

61. In a free oscillation, either e or i must be zero at the other end of the oscillating circuit, or at $l = l_0$.

Substituting, therefore, $l = l_0$ in equations (6) and (7), and resolving and arranging the terms by functions of t, the respective coefficients of

$$\varepsilon^{-(u-s)t} \cos qt, \quad \varepsilon^{-(u-s)t} \sin qt, \quad \varepsilon^{-(u+s)t} \cos qt, \text{ and } \varepsilon^{-(u+s)t} \sin qt$$

must equal zero, either in equation (6), if $i = 0$ at $l = l_0$, or in equation (7), if $e = 0$ at $l = l_0$, provided that, as assumed above, neither s nor q vanishes.

This gives, for $i = 0$ at $l = l_0$, from equation (202),

$$\left.\begin{array}{l} C_1 (\varepsilon^{-hl_0} \pm \varepsilon^{+hl_0}) \cos kl_0 - C_1'(\varepsilon^{-hl_0} \mp \varepsilon^{+hl_0}) \sin kl_0 = 0, \\ C_1 (\varepsilon^{-hl_0} \mp \varepsilon^{+hl_0}) \sin kl_0 + C_1'(\varepsilon^{-hl_0} \pm \varepsilon^{-hl_0}) \cos kl_0 = 0, \end{array}\right\} \quad (8)$$

and analogously for C_3 and C_3'.

In equations (8), either C_1, C_1', C_3, C_3' vanish, and then the whole oscillation vanishes, or, by eliminating C_1 and C_1' from equations (8), we get

$$(\varepsilon^{-hl_0} \pm \varepsilon^{+hl_0})^2 \cos^2 kl_0 + (\varepsilon^{-hl_0} \mp \varepsilon^{+hl_0})^2 \sin^2 kl_0 = 0; \qquad (9)$$

hence,

$$\left.\begin{array}{l} (\varepsilon^{-hl_0} \pm \varepsilon^{+hl_0}) \cos kl_0 = 0 \\ (\varepsilon^{-hl_0} \mp \varepsilon^{+hl_0}) \sin kl_0 = 0; \end{array}\right\}$$

and

hence, for the upper sign, or if $e = 0$ for $l = 0$,

thus:
$$h = 0 \text{ and } \cos kl_0 = 0,$$
$$kl_0 = \frac{(2n + 1)\pi}{2}, \tag{10}$$

and for the lower sign, or if $i = 0$ for $l = 0$,

thus:
$$h = 0 \text{ and } \sin kl_0 = 0,$$
$$kl_0 = n\pi. \tag{11}$$

In the same manner it follows, for $e = 0$ at $l = l_0$, from equation (7), if $e = 0$ for $l = 0$, (6),

thus:
$$h = 0, \quad \sin kl_0 = 0$$
$$kl = n\pi, \tag{12}$$

and if $i = 0$ for $l = 0$,

thus:
$$h = 0 \text{ and } \cos kl_0 = 0,$$
$$kl_0 = \frac{(2n + 1)\pi}{2}. \tag{13}$$

From equations (10) to (13) it thus follows that $h = 0$, that is, *the free oscillation of a uniform circuit is a standing wave.*
Also
$$kl_0 = \frac{(2n + 1)\pi}{2}, \tag{14}$$

if $e = 0$ at one, $i = 0$ at the other end of the circuit, and
$$kl_0 = n\pi \tag{15}$$

if either $e = 0$ at both ends of the circuit or $i = 0$ at both ends of the circuit.

62. From (14) it follows that
$$kl_0 = \frac{\pi}{2},$$

or an *odd* multiple thereof; that is, the longest wave which can exist in the circuit is that which makes the circuit a quarter-wave length. Besides this fundamental wave, all its odd multiples can exist. Such an oscillation may be called a *quarter-wave oscillation*.

The oscillation of a circuit which is open at one end, grounded at the other end, is a quarter-wave oscillation, which can contain only the odd harmonics of the fundamental wave of oscillation.

From (15) it follows that

$$kl_0 = \pi,$$

or a multiple thereof; that is, the longest wave which can exist in such a circuit is that wave which makes the circuit a half-wave length. Besides this fundamental wave, all its multiples, odd as well as even, can exist. Such an oscillation may be called a *half-wave oscillation*.

The oscillation of a circuit which is open at both ends, or grounded at both ends, is a half-wave oscillation, and a half-wave oscillation can also contain the even harmonics of the fundamental wave of oscillation, and therefore also a constant term for $n = 0$ in (15).

It is interesting to note that in the half-wave oscillation of a circuit we have a case of a circuit in which higher *even* harmonics exist, and the e.m.f. and current wave, therefore, are not symmetrical.

From $h = 0$ follows, by equation (59) of Chapter **I**,

$$\left.\begin{array}{l} s = 0, \quad \text{if} \quad k^2 > LCm^2, \\[2mm] q = 0, \quad \text{if} \quad k^2 < LCm^2. \end{array}\right\} \tag{16}$$

and

The smallest value of k which can exist from equation (14) is

$$k = \frac{\pi}{2\,l_0},$$

and, as discussed in paragraph 15, this value in high-potential high-power circuits usually is very much larger than LCm^2, so that the case $q = 0$ is realized only in extremely long circuits, as long-distance telephone or submarine cable, but not in transmission lines, and the first case, $s = 0$, therefore, is of most importance.

Substituting, therefore, $h = 0$ and $s = 0$ into the equation (55) of Chapter I, gives

and

$$c_1 = c_2 = \frac{q}{k}L = c \left.\begin{array}{c}\\\\\\\\\\\end{array}\right\}$$

$$c_1' = c_2' = \frac{m}{k}L = c', \qquad (17)$$

and substituting into equations (6) and (7) of the free oscillation gives

$$i = \varepsilon^{-ut}\{A_1[\cos(qt - kl) \pm \cos(qt + kl)]$$
$$+ A_2[\sin(qt - kl) \pm \sin(qt + kl)]\} \qquad (18)$$

and

$$e = \frac{L}{k}\varepsilon^{-ut}\{(mA_2 - qA_1)[\cos(qt - kl) \mp \cos(qt + kl)]$$
$$- (mA_1 + qA_2)[\sin(qt - kl) \mp \sin(qt + kl)]\}. \qquad (19)$$

where: $A_1 = C_1 + C_3$ and $A_2 = C_1' + C_3'$.

Since k and therefore q are large quantities, m can be neglected compared with q, and

$$k = \sqrt{LC}q;$$

hence

$$\frac{Lq}{k} = \sqrt{\frac{L}{C}}$$

and the equation (19) assumes, with sufficient approximation, the form

$$e = -\sqrt{\frac{L}{C}}\varepsilon^{-ut}\{A_1[\cos(qt - kl) \mp \cos(qt + kl)]$$
$$+ A_2[\sin(qt - kl) \mp \sin(qt + kl)]\}, \qquad (20)$$

where the upper sign in (18) and (20) corresponds to $e = 0$ at $l = 0$, the lower sign to $i = 0$ at $l = 0$, as is obvious from the equations.

Substituting

$$A_1 = A \cos \gamma \text{ and } A_2 = A \sin \gamma \qquad (21)$$

into (18) and (20) gives the equations of the free oscillation, thus:

$$i = A\varepsilon^{-ut} \{\cos (qt - kl - \gamma) \mp \cos (qt + kl - \gamma)\}$$

and

$$e = - A \sqrt{\frac{L}{C}} \varepsilon^{-ut} \{\cos (qt - kl - \gamma) \mp \cos (qt + kl - \gamma)\}. \qquad (22)$$

With the upper sign, or for $e = 0$ at $l = 0$, this gives

$$i = 2 A\varepsilon^{-ut} \cos kl \cos (qt - \gamma)$$

and

$$e = - 2 A \sqrt{\frac{L}{C}} \varepsilon^{-ut} \sin kl \sin (qt - \gamma). \qquad (23)$$

With the lower sign, or for $i = 0$ at $l = 0$, this gives

$$i = 2 A\varepsilon^{-ut} \sin kl \sin (qt - \gamma)$$

and

$$e = - 2 A \sqrt{\frac{L}{C}} \varepsilon^{-ut} \cos kl \cos (qt - \gamma). \qquad (24)$$

63. While the free oscillation of a circuit is a standing wave, the general standing wave, as represented by equations (43) and (44) of Chapter V with four integration constants A_1, A_1', A_2, A_2', is not necessarily a free oscillation.

To be a free oscillation, the power ei, that is, either e or i, must be zero at two points of the circuit, the ends of the circuit or section of circuit which oscillates.

At a point l_1 of the circuit at which $e = 0$, the coefficients of $\cos qt$ and $\sin qt$ in equation (43) of Chapter V must vanish. This gives

$$(A_1 + A_2) \cos kl_1 + (A_1' - A_2') \sin kl_1 = 0$$

and

$$- (A_1 - A_2) \sin kl_1 + (A_1' + A_2') \cos kl_1 = 0. \qquad (25)$$

Eliminating $\sin kl_1$ and $\cos kl_1$ from these two equations gives

$$(A_1^2 - A_2^2) + (A_1'^2 - A_2'^2) = 0,$$

or

$$A_1^2 + A_1'^2 = A_2^2 + A_2'^2, \qquad (26)$$

as the condition which must be fulfilled between the integration constants.

The value l_1 then follows from (25) as

$$\tan kl_1 = \frac{A_1' + A_2'}{A_1 - A_2} = -\frac{A_1 + A_2}{A_1' - A_2'}. \tag{27}$$

At a point l_2 of the circuit at which $i = 0$ the coefficients of $\cos qt$ and $\sin qt$ in equation (44) of Chapter V must vanish. This gives, in the same manner as above,

$$(A_1^2 - A_2^2) + (A_1'^2 - A_2'^2) = 0,$$

that is, the same conditions as (221), and gives for l_2 the value

$$\tan kl_2 = \frac{A_1' - A_2'}{A_1 + A_2} = -\frac{A_1 - A_2}{A_1' + A_2'}. \tag{28}$$

From (223) and (224) it follows that

$$\tan kl_2 = -\frac{1}{\tan kl_1}; \tag{29}$$

That is, the angles kl_1 and kl_2 differ by one quarter-wave length or an odd multiple thereof.

Herefrom it then follows that if the integration constants of a standing wave fulfill the conditions

$$A_1^2 + A_1'^2 = A_2^2 + A_2'^2 = B^2, \tag{30}$$

the circuit of this wave contains points l_1, distant from each other by a half-wave length, at which $e = 0$, and points l_2, distant from each other by a half-wave length, at which $i = 0$, and the points l_2 are intermediate between the points l_1, that is, distant therefrom by one quarter-wave length. Any section of the circuit, from a point l_1 or l_2 to any other point l_1 or l_2, then is a freely oscillating circuit.

In the free oscillation of the circuit the circuit is bounded by one point l_1 and one point l_2; that is, the e.m.f. is zero at one end and the current zero at the other end of the circuit, case (1) or (2) of equation (1), and the circuit is then a quarter-wave or an odd multiple thereof, or the circuit is bounded by two points l_1 or by two points l_2, and then the voltage is zero at both ends of the circuit in the former case, number (3) in equation (1), or the current is zero at both ends of the circuit in the latter case, number (4) in equation (1), and in either case the circuit is one half-wave or a multiple thereof.

Choosing one of the points l_1 or l_2 as starting point of the distance, that is, substituting $l - l_1$ or $l - l_2$ respectively, instead of l, in the equations (43) and (44) of Chapter V, with some transformation these equations convert into the equations (23) or (24). In other words, the equation (30), as relation between the integration constants of a standing wave, is the necessary and sufficient condition that this standing wave be a free oscillation.

64. A single term of a free oscillation of a circuit, with the distance counted from one end of the circuit, that is, one point of zero power, thus is represented by equations (23) or (24), respectively.

Reversing the sign of l, that is, counting the distance in the opposite direction, and substituting $B = \pm 2 A \sqrt{\dfrac{L}{C}}$, these equations assume a more convenient form, thus:

for
$$e = 0 \text{ at } l = 0,$$

and
$$\left.\begin{aligned} e &= B\varepsilon^{-ut} \sin kl \sin (qt - \gamma) \\ i &= B\sqrt{\frac{C}{L}}\varepsilon^{-ut} \cos kl \cos (qt - \gamma), \end{aligned}\right\} \quad (31)$$

and for
$$i = 0 \text{ at } l = 0,$$

and
$$\left.\begin{aligned} e &= B\varepsilon^{-ut} \cos kl \cos (qt - \gamma) \\ i &= B\sqrt{\frac{C}{L}}\varepsilon^{-ut} \sin kl \sin (qt - \gamma). \end{aligned}\right\} \quad (32)$$

Introducing again the velocity of propagation as unit distance,
$$\left.\begin{aligned} \lambda &= \sigma l, \\ \sigma &= \sqrt{LC}, \end{aligned}\right\} \quad (33)$$

from equation (5) of Chapter IV and (33) we get:
$$\begin{aligned} kl &= \lambda \sqrt{q^2 + m^2} \\ &= q\lambda \sqrt{1 + \left(\frac{m}{q}\right)^2}; \end{aligned}$$

hence, if m is small compared with q,

$$kl = q\lambda, \tag{34}$$

and substituting (33) in (34) gives

$$k = \sigma q = q\sqrt{LC}, \tag{35}$$

and from (14) and (15), for a quarter-wave oscillation, we have

and
$$\left.\begin{aligned} k &= \frac{(2n+1)\pi}{2l_0} \\[2mm] q &= \frac{(2n+1)\pi}{2l_0\sqrt{LC}}; \end{aligned}\right\} \tag{36}$$

for a half-wave oscillation,

$$\left.\begin{aligned} k &= \frac{n\pi}{l_0}, \\[2mm] q &= \frac{n\pi}{l_0\sqrt{LC}}. \end{aligned}\right\} \tag{37}$$

Denoting the length of the circuit in a quarter-wave oscillation by

$$\lambda_1 = \sigma l_0, \tag{38}$$

and the length of the circuit in a half-wave oscillation by

$$\lambda_2 = \sigma l_0, \tag{39}$$

the wave length of the fundamental or lowest frequency of oscillation is

$$\lambda_0 = 4\lambda_1 = 2\lambda_2; \tag{40}$$

or the length of the fundamental wave, with the velocity of propagation as distance unit, in a quarter-wave oscillation is

and in a half-wave oscillation is
$$\left.\begin{aligned} \lambda_0 &= 4l_0\sqrt{LC}, \\[2mm] \lambda_0 &= 2l_0\sqrt{LC}. \end{aligned}\right\} \tag{41}$$

Substituting (41) into (36) and (37) for a quarter-wave oscillation gives

$$
\left.
\begin{aligned}
k &= (2\,n + 1)\,\frac{2\,\pi\,\sqrt{LC}}{\lambda_0} \\[2mm]
q &= (2\,n + 1)\,\frac{2\,\pi}{\lambda_0},
\end{aligned}
\right\}
\tag{42}
$$

and

and for a half-wave oscillation gives

$$
\left.
\begin{aligned}
k &= n\,\frac{2\,\pi\,\sqrt{LC}}{\lambda_0} \\[2mm]
q &= n\,\frac{2\,\pi}{\lambda_0}.
\end{aligned}
\right\}
\tag{43}
$$

and

Writing now

$$
\left.
\begin{aligned}
\theta &= \frac{2\,\pi}{\lambda_0}\,l, \\[2mm]
\tau &= \frac{2\,\pi}{\lambda_0}\,\lambda = \frac{2\,\pi\,\sqrt{LC}}{\lambda_0}\,l,
\end{aligned}
\right\}
\tag{44}
$$

that is, representing a complete cycle of the fundamental frequency, or complete wave in time, by $\theta = 2\,\pi$, and a complete wave in space by $\tau = 2\,\pi$, from (43) and (44) we have

$$
\left.
\begin{aligned}
kl &= n\tau \\
qt &= n\theta,
\end{aligned}
\right\}
\tag{45}
$$

and

where n may be any integer number with a half-wave oscillation, but only an odd number with a quarter-wave oscillation.

65. Substituting (45) into (31) and (32) gives as the complete expression of a free oscillation the following equation

  A. Quarter-wave oscillation.

   (a) $e = 0$ at $l = 0$ (or $\tau = 0$)

$$
\left.
\begin{aligned}
e &= \varepsilon^{-ut}\sum_0^\infty {}_n B_n \sin(2\,n + 1)\,\tau \sin[(2\,n + 1)\,\theta - \gamma_n] \\[2mm]
i &= \sqrt{\frac{C}{L}}\,\varepsilon^{-ut}\sum_0^\infty {}_n B_n \cos(2\,n+1)\,\tau \cos[(2\,n+1)\theta - \gamma_n];
\end{aligned}
\right\}
\tag{46}
$$

and

(b) $i = 0$ at $l = 0$ (or $\tau = 0$)

$$
\left.
\begin{aligned}
e &= \varepsilon^{-ut} \sum_{0}^{\infty}{}_n B_n \cos (2n+1)\,\tau \cos [(2n+1)\,\theta - \gamma_n] \\
\text{and} & \\
i &= \sqrt{\frac{C}{L}}\,\varepsilon^{-ut} \sum_{0}^{\infty}{}_n B_n \sin (2n+1)\,\tau \sin [(2n+1)\theta - \gamma_n].
\end{aligned}
\right\} \quad (47)
$$

B. Half-wave oscillation.

(a) $e = 0$ at $l = 0$ (or $\tau = 0$)

$$
\left.
\begin{aligned}
e &= \varepsilon^{-ut} \sum_{0}^{\infty}{}_n B_n \sin n\tau \sin (n\theta - \gamma_n) \\
\text{and} & \\
i &= \sqrt{\frac{C}{L}}\,\varepsilon^{-ut} \sum_{0}^{\infty}{}_n B_n \cos n\tau \cos (n\theta - \gamma_n);
\end{aligned}
\right\} \quad (48)
$$

(b) $i = 0$ at $l = 0$ (or $\tau = 0$)

$$
\left.
\begin{aligned}
e &= \varepsilon^{-ut} \sum_{0}^{\infty}{}_n B_n \cos n\tau \cos (n\theta - \gamma_n) \\
\text{and} & \\
i &= \sqrt{\frac{C}{L}}\,\varepsilon^{-ut} \sum_{0}^{\infty}{}_n B_n \sin n\tau \sin (n\theta - \gamma_n),
\end{aligned}
\right\} \quad (49)
$$

where

$$
\left.
\begin{aligned}
\theta &= \frac{2\pi}{\lambda_0}\,t, \\
\tau &= \frac{2\pi\sqrt{LC}}{\lambda_0}\,l,
\end{aligned}
\right\} \quad (44)
$$

$$
\left.
\begin{aligned}
\lambda_0 &= 4\,l_0\,\sqrt{LC} \text{ in a quarter-wave} \\
&= 2\,l_0\,\sqrt{LC} \text{ in a half-wave oscillation,}
\end{aligned}
\right\} \quad (41)
$$

and

$$
\varepsilon^{-ut} = \varepsilon^{-\frac{u\lambda_0\theta}{2\pi}}. \qquad (50)
$$

λ_0 is the wave length, and thus $\dfrac{1}{\lambda_0}$ the frequency, of the fundamental wave, with the velocity of propagation as distance unit.

It is interesting to note that the time decrement of the free oscillation, ε^{-ut}, is the same for all frequencies and wave lengths, and that the relative intensity of the different harmonic components of the oscillation, and thereby the wave shape of the oscillation, remains unchanged during the decay of the oscillation.

This result, analogous to that found in the chapter on traveling waves, obviously is based on the assumption that the constants of the circuit do not change with the frequency. This, however, is not perfectly true. At very high frequencies r increases, due to unequal current distribution in the conductor and the appearance of the radiation resistance, as discussed in Section III, L slightly decreases hereby, g increases by the energy losses resulting from brush discharges and from electrostatic radiation, etc., so that, in general, at very high frequency an increase of $\frac{r}{L}$ and $\frac{g}{C}$, and therewith of u, may be expected; that is, very high harmonics would die out with greater rapidity, which would result in smoot ing out the wave shape with increasing decay, making it more nearly approach the fundamental and its lower harmonics, as discussed in the Chapters on "Variation of Contents."

66. The equations of a free oscillation of a circuit, as quarter-wave or half-wave, (46) to (49), still contain the pairs of integration constants B_n and γ_n, representing, respectively, the intensity and the phase of the nth harmonic.

These pairs of integration constants are determined by the terminal conditions of time; that is, they depend upon the amount and the distribution of the stored energy of the circuit at the starting moment of the oscillation, or, in other words, on the distribution of current and e.m.f. at $t = 0$.

The e.m.f., e_0, and the current, i_0, at time $t = 0$, can be expressed as an infinite series of trigonometric functions of the distance l; that is, the distance angle τ, or a Fourier series of such character as also to fulfill the terminal conditions in space, as discussed above, that is, $e = 0$, and $i = 0$, respectively, at the ends of the circuit.

The voltage and current distribution in the circuit, at the starting moment of the oscillation, $t = 0$, or, $\theta = 0$, can be represented by the Fourier series, thus:

$$e_0 = \sum_0^\infty n \left(a_n \cos n\tau + a_n{}' \sin n\tau\right)$$

and

$$i_0 = \sum^\infty n \left(b_n \cos n\tau + b_n{}' \sin n\tau\right),$$

(51)

where

$$a_0 = \frac{1}{2\pi} \int_0^{2\pi} e_0 d\tau = \mathrm{avg}\,[e_0]_0^{2\pi},$$

$$a_n = \frac{1}{\pi} \int_0^{2\pi} e_0 \cos n\tau\, d\tau = 2\,\mathrm{avg}\,[e_0 \cos n\tau]_0^{2\pi},$$

$$a_n' = \frac{1}{\pi} \int_0^{2\pi} e_0 \sin n\tau\, d\tau = 2\,\mathrm{avg}\,[e_0 \sin n\tau]_0^{2\pi},$$

$$(52)$$

and analogously for b.

The expression $\mathrm{avg}\,[F]_{a_1}^{a_2}$ denotes the average value of the function F between the limits a_1 and a_2.

Since these integrals extend over the complete wave 2π, the wave thus has to be extended by utilizing the terminal conditions regarding τ, but the wave is symmetrical with regard to $l = 0$ and with regard to $l = l_0$, and this ature in the case of a quarter-wave oscillation excludes the existence of even values of n in equations (51) and (52).

67. Substituting in equations (46) to (49),

$$t = 0, \qquad \theta = 0,$$

and then equating with (51), gives, from (46),

$$e_0 = \sum_0^\infty {}_n B_n \sin(2n+1)\,\tau \sin \gamma_n = \sum_0^\infty {}_n [a_n \cos(2n+1)\tau + a_n' \sin(2n+1)\,\tau]$$

and

$$i_0 = \sqrt{\frac{C}{L}} \sum_0^\infty {}_n B_n \cos(2n+1)\,\tau \cos \gamma_n = \sum_0^\infty {}_n [b_n \cos(2n+1)\,\tau + b_n' \sin(2n+1)\,\tau];$$

hence,

$$a_n = 0, \qquad b_n' = 0,$$

$$B_n \sin \gamma_n = a_n' \quad \text{and} \quad \sqrt{\frac{C}{L}} B_n \cos \gamma_n = b_n.$$

Equation (46) gives the constants

$$a_n = 0; \quad b_n' = 0,$$

$$\left. \begin{aligned} B_n &= \sqrt{a_n'^2 + \frac{L}{C} b_n^2}, \\[2mm] \tan \gamma_n &= \frac{a_n'}{b_n} \sqrt{\frac{L}{C}}; \end{aligned} \right\} \tag{53}$$

in the same manner equation (243) gives the constants

$$a_n' = 0; \quad b_n = 0,$$

$$\left. \begin{aligned} B_n &= \sqrt{a_n^2 + \frac{L}{C} b_n'^2}, \\[2mm] \tan \gamma_n &= \frac{b_n'}{a_n} \sqrt{\frac{C}{L}}. \end{aligned} \right\} \tag{54}$$

Equation (48) gives the same values as (46), and (49) the same values as (47).

Examples.

68. As first example may be considered the discharge of a transmission line: A circuit of length l_0 is charged to a uniform voltage E, while there is no current in the circuit. This circuit then is grounded at one end, while the other end remains insulated.

Let the distance be counted from the grounded end, and the time from the moment of grounding, and introducing the denotations (39).

The terminal conditions then are:

(a) $\tau = 0 \quad e = 0,$

$\tau = \dfrac{\pi}{2} \quad i = 0.$

(b) at $\theta = 0$

$e = 0$ for $\tau = 0$; $e = E$ for $\tau \neq 0$,

$i = 0$ for $\tau \neq 0$; $i =$ indefinite for $\tau = 0$.

The distribution of e.m.f., e_0, and current, i_0, in the circuit, at the starting moment $\theta = 0$, can be expressed by the Fourier series (51), and from (52),

$$a_n' = 2 \operatorname{avg}[E \sin (2\pi + 1)\tau] = \frac{4E}{(2n+1)\pi} \tag{55}$$

and

$$b_n = 0,$$

and from (249),

$$\left. \begin{array}{c} B_n = \dfrac{4E}{(2n+1)\pi} \text{ and } \tan \gamma_n = \infty \, ; \\[3mm] \text{hence,} \qquad \gamma_n = \dfrac{\pi}{2}, \end{array} \right\} \tag{56}$$

and substituting (56) into (46),

$$\left. \begin{array}{c} e = \dfrac{4E}{\pi} \varepsilon^{-ut} \displaystyle\sum_{1}^{\infty} {}_n \dfrac{\sin (2n+1)\tau \cos (2n+1)\theta}{2n+1} \\[5mm] \text{and} \\[3mm] i = \dfrac{4E}{\pi} \sqrt{\dfrac{C}{L}} \varepsilon^{-ut} \displaystyle\sum_{1}^{\infty} {}_n \dfrac{\cos (2n+1)\tau \sin (2n+1)\theta}{2n+1} . \end{array} \right\} \tag{57}$$

From (44) it follows that

$$t = \frac{2 l_0 \sqrt{LC}}{\pi} \theta.$$

$\theta = 2\pi$ gives the period,

$$t_1 = 4 l_0 \sqrt{LC},$$

and the frequency,

$$f_1 = \frac{1}{4 l_0 \sqrt{LC}},$$

and $\tau = 2\pi$ gives the wave length,

$$l_w = 4 l_0,$$

of the fundamental wave, or oscillation of lowest frequency and greatest wave length.

Choosing the same line constants as in paragraph 16, namely: $l_0 = 120$ miles; $r = 0.41$ ohm per mile; $L = 1.95 \times 10^{-3}$ henry per mile; $g = .25 \times 10^{-6}$ mho per mile, and $C = 0.0162 \times 10^{-6}$ farad per mile, we have

$$u = 113,$$

$$ut = \frac{2\,l_0\,\sqrt{LC}}{\pi}\,u\theta = 0.0485\,\theta,$$

and the fundamental frequency of oscillation is

$$f_1 = 371 \text{ cycles per second.}$$

If now the e.m.f. to which the line is charged is

$$E = 40{,}000 \text{ volts,}$$

substituting these values in equations (57) **gives**

$$
\left.
\begin{aligned}
e &= 51{,}000\,\varepsilon^{-0.0485\,\theta}\left\{\sin \tau \cos \theta + \tfrac{1}{3}\sin 3\,\tau \cos 3\,\theta\right. \\
&\quad \left. + \tfrac{1}{5}\sin 5\,\tau \cos 5\,\theta + \ldots\right\}, \text{ in volts} \\[4pt]
\text{and} \\[2pt]
i &= 147\,\varepsilon^{-0.0485\,\theta}\left\{\cos \tau \sin \theta + \tfrac{1}{3}\cos 3\,\tau \sin 3\,\theta\right. \\
&\quad \left. + \tfrac{1}{5}\cos 5\,\tau \sin 5\,\theta + \ldots\right\}, \text{ in amp.}
\end{aligned}
\right\} \quad (58)
$$

The maximum value of e is

$$e = E = 40{,}000 \text{ volts,}$$

and the maximum current of i is

$$i = I = 115.5 \text{ amp.}$$

Since

$$
\left.
\begin{aligned}
\sum_{1}^{\infty} {}_n \frac{\sin (2\,n-1)\,a \cos (2\,n-1)\,b}{2\,n-1} &= 0, \text{ if } b-\frac{\pi}{2}<a<b, \\[8pt]
&\quad \text{or } b+\frac{\pi}{2}<a<b+\pi, \\[8pt]
&= +\frac{\pi}{2}, \text{ if } b<a<b+\frac{\pi}{2} \\[8pt]
&= -\frac{\pi}{2}, \text{if } b<a<b-\frac{\pi}{2},
\end{aligned}
\right\} \quad (59)
$$

and

applying (59) to (58) we have at any point τ of the line, at the time θ given by

$$0 < \theta < \tau: \qquad e = E\varepsilon^{-ut}; \qquad \dot{\imath} = 0.$$

$$\tau < \theta < \tau + \frac{\pi}{2}: \qquad e = 0; \qquad \dot{\imath} = I\varepsilon^{-ut}.$$

$$\tau + \frac{\pi}{2} < \theta < \tau + \pi: \qquad e = -E\varepsilon^{-ut}; \qquad \dot{\imath} = 0.$$

$$\tau + \pi < \theta < \tau + \frac{3\pi}{2}: \qquad e = 0; \qquad \dot{\imath} = -I\varepsilon^{-ut}.$$

$$\tau + \frac{3\pi}{2} < \theta < \tau + 2\pi; \qquad e = E\varepsilon^{-ut}; \qquad \dot{\imath} = 0, \text{ etc.}$$

At any moment of time θ one part of the line has voltage $e = E\varepsilon^{-ut}$ and zero current, and the other part of the line has current $\dot{\imath} = I\varepsilon^{-ut}$ and zero voltage, and the dividing line between the two sections of the line is at $\tau = \theta \pm \dfrac{n\pi}{2}$, hence moves along the line at the rate $\tau = \theta$.

69. As second example may be considered the discharge of a live line into a dead line: A circuit of length l_1, charged to a uniform voltage E, but carrying no current, is connected to a circuit of the same constants, but of length l_2, and having neither voltage nor current, otherwise both circuits are insulated.

Let the total length of the circuit be denoted by

$$2l = l_1 + l_2,$$

and let the time be counted from the moment where the circuits l_1 and l_2 are connected together, the distance from the beginning of the live circuit l_1, whose other end is connected to the dead circuit l_2.

Introduce again the denotations (44), and represent the total length of the line $2l = l_1 + l_2$ by $\tau = \pi$, then write

$$\tau_1 = \frac{l_1}{l_1 + l_2}\pi.$$

As the voltage is E from $\tau = 0$ to $\tau = \tau_1$, and 0 from $\tau = \tau_1$ to $\tau = \pi$, the mean value of voltage, or the voltage which will be left on the line after the transient phenomenon has passed, is

$$e_0 = \frac{\tau_1}{\pi}E,$$

and the terminal conditions of voltage and current are

$$\theta = 0$$
$$e = E - e_0 \text{ for } 0 < \tau < \tau_1,$$
$$e = - e_0 \quad \text{for } \tau_1 < \tau < \pi,$$
$$i = 0.$$

Proceeding then in the same manner as in paragraph 34, in the present case the equations (49) and (52) apply, and

$$a_n = \frac{2}{\pi} \left\{ \int_0^{\tau_1} (E - e_0) \cos n\theta \, d\theta - \int_{\tau_1}^{\pi} e_0 \cos n\theta \, d\theta \right\}$$

$$= \frac{2 E \sin n\tau_1}{n},$$

$$a_n' = b_n = b_n' = 0;$$

hence,

$$\gamma_n = 0$$

and

$$
\begin{aligned}
e &= \frac{2 E}{\pi} \left\{ \frac{\tau_1}{2} + \varepsilon^{-ut} \sum_1^{\infty} n \frac{\sin n\tau_1}{n} \cos n\tau \cos n\theta \right\}, \\
i &= \frac{2 E}{\pi} \sqrt{\frac{C}{L}} \varepsilon^{-ut} \sum_1^{\infty} n \frac{\sin n\tau_1}{n} \sin n\tau \sin n\theta.
\end{aligned}
\right\} \tag{60}
$$

Choosing the same line constants as in paragraph 35, and assuming

$$l_1 = 120 \text{ miles and } l_2 = 80 \text{ miles,}$$

we have

$$l = 100 \text{ miles and } \tau_1 = 0.6 \, \pi.$$

Let $E = 40{,}000$ volts, $ut = 0.0404 \, \theta$, and the fundamental frequency of oscillation, $f_1, = 445$ cycles per second; then

$$
\begin{aligned}
e = 24{,}000 + 25{,}500 \, \varepsilon^{-0.0404\,\theta} & \left\{ \sin 108° \cos \tau \cos \theta + \tfrac{1}{2} \sin 216° \right. \\
& \left. \cos 2 \tau \cos 2 \theta + \tfrac{1}{3} \sin 348° \cos 3 \tau \cos 3 \theta + \cdots \right\} \text{ volts}
\end{aligned}
$$

and

$$
\begin{aligned}
i = 73.5 \, \varepsilon^{-0.0404\,\theta} & \left\{ \sin 108° \sin \tau \sin \theta + \tfrac{1}{2} \sin 216° \sin 2 \tau \right. \\
& \left. \sin 2 \theta + \tfrac{1}{3} \sin 348° \sin 3 \tau \sin 3 \theta + \cdots \right\} \text{ amp.}
\end{aligned}
$$

$$\tag{61}$$

CHAPTER VIII.

TRANSITION POINTS AND THE COMPOUND CIRCUIT.

70. The discussions of standing waves and free oscillations in Chapters V and VII, and traveling waves in Chapter VI, apply directly only to simple circuits, that is, circuits comprising a conductor of uniformly distributed constants r, L, g, and C. Industrial electric circuits, however, never are simple circuits, but are always complex circuits comprising sections of different constants, — generator, transformer, transmission lines, and load, — and a simple circuit is realized only by a section of a circuit, as a transmission line or a high-potential transformer coil, which is cut off at both ends from the rest of the circuit, either by open-circuiting, $i = 0$, or by short-circuiting, $e = 0$. Approximately, the simple circuit is realized by a section of a complex or compound circuit, connecting to other sections of very different constants, so that the ends of the circuit can, approximately, be considered as reflection points. For instance, an underground cable of low L and high C, when connected to a large reactive coil of high L and low C, may approximately, at its ends be considered as having reflection points $i = 0$. A high-potential transformer coil of high L and low C, when connected to a cable of low L and high C, may at its ends be considered as having reflection points $e = 0$. In other words, in the first case the reactive coil may be considered as stopping the current, in the latter case the cable considered as short-circuiting the transformer. This approximation, however, while frequently relied upon in engineering practice, and often permissible for the circuit section in which the transient phenomenon originates, is not permissible in considering the effect of the phenomenon on the adjacent sections of the circuit. For instance, in the first case above mentioned, a transient phenomenon in an underground cable connected to a high reactance, the current and e.m.f. in the cable may approximately be represented by considering the reactive coil as a reflection point, that is, an open circuit, since only a small current

565

exists in the reactive coil. Such a small current in the reactive coil may, however, give a very high and destructive voltage in the reactive coil, due to its high L, and thus in the circuit beyond the reactive coil. In the investigation of the effect of a transient phenomenon originating in one section of a compound circuit, as an oscillating arc on an underground cable, on other sections of the circuit, as the generating station, even a very great change of circuit constants cannot be considered as a reflection point. Since this is the most important case met in industrial practice, as disturbances originating in one section of a compound circuit usually develop their destructive effects in other sections of the circuit, the investigation of the general problem of a compound circuit comprising sections of different constants thus becomes necessary. This requires the investigation of the changes occurring in an electric wave, and its equations, when passing over a transition point from one circuit or section of a circuit into another section of different constants.

71. The equations (53) to (60) of Chapter I, while most general, are less convenient for studying the transition of a wave from one circuit to another circuit of different constants, and since in industrial high-voltage circuits, at least for waves originating in the circuits, q and k are very large compared with s and h, as discussed in paragraph 16, s and h may be neglected compared with q and k. This gives, as discussed in paragraph 9,

$$\left.\begin{aligned} h &= \sigma s, \\ k &= \sigma q, \\ c_1 = c_2 &= \sqrt{\frac{L}{C}} = c, \\ c_1{}' = c_2{}' &= \frac{m}{q}\sqrt{\frac{L}{C}} = 0, \end{aligned}\right\} \tag{1}$$

where

$$\sigma = \sqrt{LC}, \tag{2}$$

and substituting

$$\lambda = \sigma l, \tag{3}$$

that is

$$\left.\begin{aligned} kl &= q\lambda, \\ hl &= s\lambda, \end{aligned}\right\} \tag{4}$$

gives

$$
\begin{aligned}
i = \varepsilon^{-ut} \big\{ &\varepsilon^{-s(\lambda-t)} [C_1 \cos q\,(\lambda - t) + C_1' \sin q\,(\lambda - t)] \\
&- \varepsilon^{+s(\lambda+t)} [C_2 \cos q\,(\lambda + t) + C_2' \sin q\,(\lambda + t)] \\
&+ \varepsilon^{+s(\lambda-t)} [C_3 \cos q\,(\lambda - t) + C_3' \sin q\,(\lambda - t)] \\
&- \varepsilon^{-s(\lambda+t)} [C_4 \cos q\,(\lambda + t) + C_4' \sin q\,(\lambda + t)] \big\}
\end{aligned}
\tag{5}
$$

and

$$
\begin{aligned}
e = \sqrt{\frac{L}{C}}\,\varepsilon^{-ut} \big\{ &\varepsilon^{-s(\lambda-t)} [C_1 \cos q\,(\lambda - t) + C_1' \sin q\,(\lambda - t)] \\
&+ \varepsilon^{+s(\lambda+t)} [C_2 \cos q\,(\lambda + t) + C_2' \sin q\,(\lambda + t)] \\
&+ \varepsilon^{+s(\lambda-t)} [C_3 \cos q\,(\lambda - t) + C_3' \sin q\,(\lambda - t)] \\
&+ \varepsilon^{-s(\lambda+t)} [C_4 \cos q\,(\lambda + t) + C_4' \sin q\,(\lambda + t)] \big\}.
\end{aligned}
\tag{6}
$$

Substituting now

$$
\left.
\begin{aligned}
C_1^2 + C_1'^2 &= A^2, \\
C_2^2 + C_2'^2 &= B^2, \\
C_3^2 + C_3'^2 &= C^2, \\
C_4^2 + C_4'^2 &= D^2,
\end{aligned}
\right\}
\tag{7}
$$

$$
\left.
\begin{aligned}
\frac{C_1'}{C_1} &= \tan \alpha, \\[4pt]
\frac{C_2'}{C_2} &= \tan \beta, \\[4pt]
\frac{C_3'}{C_3} &= \tan \gamma, \\[4pt]
\frac{C_4'}{C_4} &= \tan \delta,
\end{aligned}
\right\}
\tag{8}
$$

gives

$$
\begin{aligned}
i = \varepsilon^{-ut} \big\{ & A\varepsilon^{-s(\lambda-t)} \cos [q\,(\lambda-t) - \alpha] - B\varepsilon^{+s(\lambda+t)} \cos [q\,(\lambda+t) - \beta] \\
&+ C\varepsilon^{+s(\lambda-t)} \cos [q\,(\lambda-t) - \gamma] - D\varepsilon^{-s(\lambda+t)} \cos [q\,(\lambda+t) - \delta] \big\}
\end{aligned}
\tag{9}
$$

and

$$e = \sqrt{\frac{L}{C}}\,\varepsilon^{-ut}\big\{A\varepsilon^{-s(\lambda-t)}\cos[q(\lambda-t)-\alpha]+B\varepsilon^{+s(\lambda+t)}\cos[q(\lambda+t)-\beta]$$
$$+C\varepsilon^{+s(\lambda-t)}\cos[q(\lambda-t)-\gamma]+D\varepsilon^{-s(\lambda+t)}\cos[q(\lambda+t)-\delta]\big\},$$
$$(10)$$

72. In these equations (9) and (10) λ is the distance coördinate, using the velocity of propagation as unit distance, and at a transition point from one circuit to another, where the circuit constants change, the velocity of propagation also changes, and thus t for the same time constants s and q, h and k also change, and therewith kl, but transformed to the distance variable λ, $q\lambda$ remains the same; that is, by introducing the distance variable λ, the distance can be measured throughout the entire circuit, and across transition points, at which the circuit constants change, and the same equations (9) and (10) apply throughout the entire circuit. In this case, however, in any section of the circuit,

$$\left.\begin{array}{l}\lambda = \sigma_i l \\ = l\sqrt{L_i C_i},\end{array}\right\} \qquad (11)$$

where L_i and C_i are the inductance and the capacity, respectively, of the section i of the circuit, per unit length, for instance, per mile, in a line, per turn in a transformer coil, etc.

In a compound circuit the time variable t is the same throughout the entire circuit, or, in other words, the frequency of oscillation, as represented by q, and the rate of decay of the oscillation, as represented by the exponential function of time, must be the same throughout the entire circuit. Not so, however, with the distance variable l; the wave length of the oscillation and its rate of building up or down along the circuit need not be the same, and usually are not, but in some sections of the circuit the wave length may be far shorter, as in coiled circuits as transformers, due to the higher L, or in cables due to the higher C. To extend the same equations over the entire compound circuit, it therefore becomes necessary to substitute for the distance variable l another distance variable λ of such character that the wave length has the same value in all sections of the compound circuit. As the wave, length of the section i is $\dfrac{l}{\sqrt{L_i C_i}}$, this is done by changing the unit distance by the factor $\sigma_i = \sqrt{L_i C_i}$. The distance unit of the new distance variable λ then is the distance traversed by the wave in unit time, hence different in linear measure for the different sections of the circuit, but offers the advantage of carrying the distance measurements across the entire circuit and over transition points by the same distance variable λ.

This means that the length l_i of any section i of the compound circuit is expressed by the length $\lambda_i = \sigma_i l_i$.

The introduction of the distance variable λ also has the advantage that in the determination of the constants r, L, g, C of the different sections of the circuit different linear distance measurements l may be used. For instance, in the transmission line, the constants may be given per mile, that is, the mile used as unit length, while in the high-potential coil of a transformer the turn, or the coil, or the total transformer may be used as unit of length l, so that the actual linear length of conductor may be unknown. For instance, choosing the total length of conductor in the high-potential transformer as unit length, then the length of the transformer winding in the velocity measure λ is $\lambda_0 = \sqrt{L_0 C_0}$, where L_0 = total inductance, C_0 = total capacity of transformer.

The introduction of the distance variable λ thus permits the representation in the circuit of apparatus as reactive coils, etc., in which one of the constants is very small compared with the other and therefore is usually neglected and the apparatus considered as "massed inductance," etc., and allows the investigation of the effect of the distributed capacity of reactive coils and similar matters, by representing the reactive coil as a finite (frequently quite long) section λ_0 of the circuit.

73. Let λ_0, λ_1, λ_2, $\ldots$ λ_n be a number of transition points at which the circuit constants change and the quantities may be denoted by index 1 in the section from λ_0 to λ_1, by index **2** in the section from λ_1 to λ_2, etc.

At $\lambda = \lambda_1$ it then must be $i_1 = i_2$, $e_1 = e_2$; thus substituting $\lambda = \lambda_1$ into equations (9) and (10) gives

$$\varepsilon^{-(u_1-s_1)t}\{A_1\varepsilon^{-s_1\lambda_1}\cos[q_1(\lambda_1-t)-\alpha_1]-B_1\varepsilon^{+s_1\lambda_1}\cos[q_1(\lambda_1+t)-\beta_1]\}$$
$$+\varepsilon^{-(u_1+s_1)t}\{C_1\varepsilon^{+s_1\lambda_1}\cos[q_1(\lambda_1-t)-\gamma_1]-D_1\varepsilon^{-s_1\lambda_1}\cos[q_1(\lambda_1+t)-\delta_1]\}$$
$$=\varepsilon^{-(u_2-s_2)t}\{A_2\varepsilon^{-s_2\lambda_1}\cos[q_2(\lambda_1-t)-\alpha_2]-B_2\varepsilon^{+s_2\lambda_1}\cos[q_2(\lambda_1+t)-\beta_2]\}$$
$$+\varepsilon^{-(u_2+s_2)t}\{C_2\varepsilon^{+s_2\lambda_1}\cos[q_2(\lambda_1-t)-\gamma_2]-D_2\varepsilon^{-s_2\lambda_1}\cos[q_2(\lambda_1+t)-\delta_2]\}.$$
$$(12)$$

Herefrom it follows that

$$q_2 = q_1;$$ (13)

that is, the frequency must be the same throughout the entire circuit as is obvious, and

$$u_2 \pm s_2 = u_1 \pm s_1.$$ (14)

Since $u_2 \neq u_1$, only one of the two waves can exist, the $A\ B$, or the $C\ D$, and since these two waves differ from each other only by the sign of s, by assuming now that s may be either positive or negative we can select one of the two waves, for instance, the second wave, but use $A,\ B,\ \alpha,\ \beta$ as denotations of the integration constants.

74. The equations (9) and (10) now assume the form

and

$$\left.\begin{aligned}
i &= \varepsilon^{-ut}\{A\varepsilon^{+s(\lambda-t)}\cos[q(\lambda-t)-\alpha] \\
&\quad - B\varepsilon^{-s(\lambda+t)}\cos[q(\lambda+t)-\beta]\} \\
\\
e &= \sqrt{\frac{L}{C}}\varepsilon^{-ut}\{A\varepsilon^{+s(\lambda-t)}\cos[q(\lambda-t)-\alpha] \\
&\quad + B\varepsilon^{-s(\lambda+t)}\cos[q(\lambda+t)-\beta]\},
\end{aligned}\right\}$$ (15)

or

and

$$\left.\begin{aligned}
i &= \varepsilon^{-(u+s)t}\{A\varepsilon^{+s\lambda}\cos[q(\lambda-t)-\alpha] \\
&\quad - B\varepsilon^{-s\lambda}\cos[q(\lambda+t)-\beta]\} \\
\\
e &= \sqrt{\frac{L}{C}}\varepsilon^{-(u+s)t}\{A\varepsilon^{+s\lambda}\cos[q(\lambda-t)-\alpha] \\
&\quad + B\varepsilon^{-s\lambda}\cos[q(\lambda+t)-\beta]\};
\end{aligned}\right\}$$ (16)

or, using equations (5) and (6) instead of (9) and (10), the corresponding equations are of the form

and

$$\left.\begin{aligned}
i &= \varepsilon^{-ut}\{\varepsilon^{+s(\lambda-t)}[A\cos q(\lambda-t)+B\sin q(\lambda-t)] \\
&\quad - \varepsilon^{-s(\lambda+t)}[C\cos q(\lambda+t)+D\sin q(\lambda+t)]\} \\
\\
e &= \sqrt{\frac{L}{C}}\varepsilon^{-ut}\{\varepsilon^{+s(\lambda-t)}[A\cos q(\lambda-t)+B\sin q(\lambda-t)] \\
&\quad + \varepsilon^{-s(\lambda+t)}[C\cos q(\lambda+t)+D\sin q(\lambda+t)]\},
\end{aligned}\right\}$$ (17)

or

$$i = \varepsilon^{-(u+s)t}\left\{\varepsilon^{+s\lambda}\left[A \cos q\,(\lambda - t) + B \sin q\,(\lambda - t)\right]\right.$$
$$\left. - \varepsilon^{-s\lambda}\left[C \cos q\,(\lambda + t) + D \sin q\,(\lambda + t)\right]\right\}$$

and

$$e = \sqrt{\frac{L}{C}}\,\varepsilon^{-(u+s)t}\left\{\varepsilon^{+s\lambda}\left[A \cos q\,(\lambda - t) + B \sin q\,(\lambda - t)\right]\right.$$
$$\left. + \varepsilon^{-s\lambda}\left[C \cos q\,(\lambda + t) + D \sin q\,(\lambda + t)\right]\right\},$$

$$(18)$$

where s may be positive or negative.

From equation (12) it then follows that

$$u_1 + s_1 = u_2 + s_2 = u_3 + s_3 = \ldots = u_n + s_n = u_0, \qquad (19)$$

where u_1, u_2, u_3, etc., u_n are the time constants of the individual sections of the complex circuit, $\dfrac{1}{2}\left(\dfrac{r}{L} + \dfrac{g}{C}\right)$, and u_0 may be called the *resultant time decrement of the complex circuit*.

75. Equation (12), by canceling equal terms on both sides, then assumes the form

$$A_1\varepsilon^{+s_1\lambda_1}\cos\left[q\,(\lambda_1 - t) - \alpha_1\right] - B_1\varepsilon^{-s_1\lambda_1}\cos\left[q\,(\lambda_1 + t) - \beta_1\right] =$$
$$A_2\varepsilon^{+s_2\lambda_1}\cos\left[q\,(\lambda - t) - \alpha_2\right] - B_2\varepsilon^{-s_2\lambda_1}\cos\left[q\,(\lambda_1 + t) - \beta_2\right],$$

and, resolved for $\cos qt$ and $\sin qt$, this gives the identities

$$A_1\varepsilon^{+s_1\lambda_1}\cos\,(q\lambda_1 - \alpha_1) - B_1\varepsilon^{-s_1\lambda_1}\cos\,(q\lambda_1 - \beta_1) =$$
$$A_2\varepsilon^{+s_2\lambda_1}\cos\,(q\lambda_1 - \alpha_2) - B_2\varepsilon^{-s_2\lambda_1}\cos\,(q\lambda_1 - \beta_2),$$
$$A_1\varepsilon^{+s_1\lambda_1}\sin\,(q\lambda_1 - \alpha_1) + B_1\varepsilon^{-s_1\lambda_1}\sin\,(q\lambda_1 - \beta_1) =$$
$$A_2\varepsilon^{+s_2\lambda_1}\sin\,(q\lambda_1 - \alpha_2) + B_2\varepsilon^{-s_2\lambda_1}\sin\,(q\lambda_1 - \beta_2). \qquad (20)$$

These identities resulted by equating $i_1 = i_2$ from equation (15). In the same manner, by equating e_1 and e_2 from equation (15) there result the two further identities

$$\sqrt{\frac{L_1}{C_1}}\left\{A_1\varepsilon^{+s_1\lambda_1}\cos\,(q\lambda_1 - \alpha_1) + B_1\varepsilon^{-s_1\lambda_1}\cos\,(q\lambda_1 - \beta_1)\right\} =$$

$$\sqrt{\frac{L_2}{C_2}}\left\{A_2\varepsilon^{+s_2\lambda_1}\cos\,(q\lambda_1 - \alpha_2) + B_2\varepsilon^{-s_2\lambda_1}\cos\,(q\lambda_1 - \beta_2)\right\},$$

$$\sqrt{\frac{L_1}{C_1}}\left\{A_1\varepsilon^{+s_1\lambda_1}\sin\left(q\lambda_1-\alpha_1\right)-B_1\varepsilon^{-s_1\lambda_1}\sin\left(q\lambda_1-\beta_1\right)\right.=$$

$$\sqrt{\frac{L_2}{C_2}}\left\{\dot{A}_2\varepsilon^{+s_2\lambda_1}\sin\left(q\lambda_1-\alpha_2\right)-B_2\varepsilon^{-s_2\lambda_1}\sin\left(q\lambda_1-\beta_2\right)\right\}. \quad (21)$$

Equations (20) and (21) determine the constants of ɐny section of the circuit, A_2, B_2, α_2, β_2, from the constants of the next section of the circuit, A_1, B_1, α_1, β_1.

Let

$$\left.\begin{aligned}
A\varepsilon^{+s\lambda_1}\cos\left(q\lambda_1-\alpha\right) &= A'; \\
A\varepsilon^{+s\lambda_1}\sin\left(q\lambda_1-\alpha\right) &= A''; \\
B\varepsilon^{-s\lambda_1}\cos\left(q\lambda_1-\beta\right) &= B', \\
B\varepsilon^{-s\lambda_1}\sin\left(q\lambda_1-\beta\right) &= B'';
\end{aligned}\right\} \quad (22)$$

$$\left.\begin{aligned}
c_1 &= \sqrt{\frac{L_1}{C_1}}, \\
c_2 &= \sqrt{\frac{L_2}{C_2}}.
\end{aligned}\right\} \quad (23)$$

Then

$$\left.\begin{aligned}
2\,c_2 A_2' &= (c_1+c_2)\,A_1' + (c_1-c_2)\,B_1', \\
2\,c_2 B_2' &= (c_1+c_2)\,B_1' + (c_1-c_2)\,A_1', \\
2\,c_2 A_2'' &= (c_1+c_2)\,A_1'' - (c_1-c_2)\,B_1'', \\
2\,c_2 B_2'' &= (c_1+c_2)\,B_1'' - (c_1-c_2)\,A_1'',
\end{aligned}\right\} \quad (24)$$

and since **by** (22):

$$A'^2 + A''^2 = A^2\varepsilon^{+2s\lambda_1} \text{ etc.,} \quad (25)$$

substituting herein (24),

$$4\,c_2^2 A_2^2\varepsilon^{+2s_2\lambda_1} = (c_1+c_2)^2\,A_1^2\varepsilon^{+2s_1\lambda_1} + (c_1-c_2)^2\,B_1^2\varepsilon^{-2s_1\lambda_1}. \quad (26)$$
$$+ 2\left(A_1'B_1' - A_1''B_1''\right)\left(c_1^2-c_2^2\right)$$

and

$$\tan(q\lambda_1 - \alpha_2) = \frac{1 - \dfrac{c_1 - c_2}{c_1 + c_2}\dfrac{B_1}{A_1}\varepsilon^{-2s_1\lambda_1}\dfrac{\sin(q\lambda_1 - \beta_1)}{\sin(q\lambda_1 - \alpha_1)}}{1 + \dfrac{c_1 - c_2}{c_1 + c_2}\dfrac{B_1}{A_1}\varepsilon^{-2s_1\lambda_1}\dfrac{\cos(q\lambda_1 - \beta_1)}{\cos(q\lambda_1 - \alpha_1)}}\tan(q\lambda_1 - \alpha_1)$$

$$\tan(q\lambda_1 - \beta_2) = \frac{1 - \dfrac{c_1 - c_2}{c_1 + c_2}\dfrac{A_1}{B_1}\varepsilon^{+2s_1\lambda_1}\dfrac{\sin(q\lambda_1 - \alpha_1)}{\sin(q\lambda_1 - \beta_1)}}{1 + \dfrac{c_1 - c_2}{c_1 + c_2}\dfrac{A_1}{B_1}\varepsilon^{+2s_1\lambda_1}\dfrac{\cos(q\lambda_1 - \alpha_1)}{\cos(q\lambda_1 - \beta_1)}}\tan(q\lambda_1 - \beta_1).$$

$$(27)$$

In the same manner, equating, for $\lambda = \lambda_1$, in equations (18) the current i_1, corresponding to the section from λ_0 to λ_1, with the current i_2, corresponding to the section from λ_1 to λ_2, and also the e.m.fs., $e_2 = e_1$, gives the constants in equations (18) and (17), of one section, λ_1 to λ_2, expressed by those of the next adjoining section, λ_0 to λ_1, as

$$A_2 = \varepsilon^{-s_2\lambda_1}\left\{a_1\varepsilon^{+s_1\lambda_1}A_1 + b_1\varepsilon^{-s_1\lambda_1}(C_1\cos 2q\lambda_1 + D_1\sin 2q\lambda_1)\right\}$$
$$B_2 = \varepsilon^{-s_2\lambda_1}\left\{a_1\varepsilon^{+s_1\lambda_1}B_1 + b_1\varepsilon^{-s_1\lambda_1}(C_1\sin 2q\lambda_1 - D_1\cos 2q\lambda_1)\right\}$$
$$C_2 = \varepsilon^{+s_2\lambda_1}\left\{a_1\varepsilon^{-s_1\lambda_1}C_1 + b_1\varepsilon^{+s_1\lambda_1}(A_1\cos 2q\lambda_1 - B_1\sin 2q\lambda_1)\right\}$$
$$D_2 = \varepsilon^{+s_2\lambda_1}\left\{a_1\varepsilon^{-s_1\lambda_1}D_1 + b_1\varepsilon^{+s_1\lambda_1}(A_1\sin 2q\lambda_1 - B_1\cos 2q\lambda_1)\right\}$$

$$(28)$$

where

$$a_1 = \frac{c_1 + c_2}{2c_2},$$
$$b_1 = \frac{c_1 - c_2}{2c_2},$$

$$(29)$$

$$c_1 = \sqrt{\frac{L_1}{C_1}},$$
$$c_2 = \sqrt{\frac{L_2}{C_2}}.$$

$$(30)$$

76. The general equation of current and e.m.f. in a complex circuit thus also consists of two terms, the main wave A in equations (15), (16), and its reflected wave B.

The factor $\epsilon^{-(u+s)t} = \epsilon^{-u_0 t}$ in equations (16) and (18) represents the time decrement, or the decrease of the intensity of the wave with the time, and as such is the same throughout the entire circuit. In an isolated section, of time constant u, the time decrement, from Chapters V and VII, is, however, ϵ^{-ut}; that is, with the decrement ϵ^{-ut} the wave dies out in the isolated section at the rate at which its stored energy is dissipated by the power lost in resistance and conductance. In a section of the circuit connected to other sections the time decrement $\epsilon^{-u_0 t}$ does not correspond to the power dissipation in the section; that is, the wave does not die out in each section at the rate as given by the power consumed in this section, or, in other words, power transfer occurs from section to section during the oscillation of a compound circuit

If s is negative, u_0 is less than u, and the wave dies out in that particular section at a lesser rate than corresponds to the power consumed in the section, or, in other words, in this section of the compound circuit more power is consumed by r and g than is supplied by the decrease of the stored energy, and this section, therefore, must receive energy from adjoining sections. Inversely, if s is positive, $u_0 > u$, the wave dies out more rapidly in that section than its stored energy is consumed by r and g; that is, a part of the stored energy of this section is transferred to the adjoining sections, and only a part — occasionally a very small part — dissipated in the section, and this section acts as a store of energy for supplying the other sections of the system.

The constant s of the circuit, therefore, may be called *energy transfer constant*, and positive s means transfer of energy from the section to the rest of the circuit, and negative s means reception of energy from other sections. This explains the vanishing of s in a standing wave of a uniform circuit, due to the absence of energy transfer, and the presence of s in the equations of the traveling wave, due to the transfer of energy along the circuit, and in the general equations of alternating-current circuits.

It immediately follows herefrom that in a compound circuit some of the s of the different sections must always be positive, some negative.

In addition to the time decrement $\epsilon^{-(u+s)t} = \epsilon^{-u_0 t}$ the waves in equations (16) and (18) also contain the distance decrement $\epsilon^{+s\lambda}$ for the main wave, $\epsilon^{-s\lambda}$ for the reflected wave. Negative s therefore means a decrease of the main wave for increasing λ, or in the direction of propagation, and a decrease of the reflected wave for decreasing λ, that is, also in the direction of propagation; while positive s means increase of main wave as well as reflected wave in the direction of propagation along the circuit. In other words, if s is negative and the section consumes more power than is given by its stored energy, and therefore receives power from the adjoining sections, the electric wave decreases in the direction of its propagation, or builds down, showing the gradual dissipation of the power received from adjoining sections. Inversely, if s is positive and the section thus supplies power to adjoining sections, the electric wave increases in this section in the direction of its propagation, or builds up.

In other words, in a compound circuit, in sections of low power dissipation, the wave increases and transfers power to sections of high power dissipation, in which the wave decreases.

This can still better be seen from equations (15) and (17). Here the time decrement ϵ^{-ut} represents the dissipation of stored energy by the power consumed in the section by r and g. The time distance decrement, $\epsilon^{+s(\lambda - t)}$ for the main wave, $\epsilon^{-s(\lambda + t)}$ for the reflected wave, represents the decrement of the wave for constant $(\lambda - t)$ or $(\lambda + t)$ respectively; that is, shows the change of wave intensity during its propagation. Thus for instance, following a wave crest, the wave decreases for negative s and ·increases for positive s, in addition to the uniform decrease by the time constant ϵ^{-ut}; or, in other words, for positive s the wave gathers intensity during its progress, for negative s it loses intensity in addition to the loss of intensity by the time constant of this particular section of the circuit.

77. Introducing the resultant time decrement u_0 of the compound circuit, the equations of any section, (16) and (18), can also be expressed by the resultant time decrement of the entire compound circuit, u_0, and the energy transfer constant of the individual section; thus

$$s = u_0 - u, \tag{31}$$

$$i = \varepsilon^{-u_0 t} \{ A \varepsilon^{+s\lambda} \cos [q(\lambda - t) - \alpha] - B \varepsilon^{-s\lambda} \cos [q(\lambda + t) - \beta] \}$$

and

$$e = \sqrt{\frac{L}{C}} \varepsilon^{-u_0 t} \{ A \varepsilon^{+s\lambda} \cos [q(\lambda - t) - \alpha] + B \varepsilon^{-s\lambda} \cos [q(\lambda + t) - \beta] \}, \qquad (32)$$

or

$$i = \varepsilon^{-u_0 t} \{ \varepsilon^{+s\lambda} [A \cos q(\lambda - t) + B \sin q(\lambda - t)]$$
$$- \varepsilon^{-s\lambda} [C \cos q(\lambda + t) + D \sin q(\lambda + t)] \}$$

and

$$e = \sqrt{\frac{L}{C}} \varepsilon^{-u_0 t} \{ \varepsilon^{+s\lambda} [A \cos q(\lambda - t) + B \sin q(\lambda - t)]$$
$$+ \varepsilon^{-s\lambda} [C \cos q(\lambda + t) + D \sin q(\lambda + t)] \}. \qquad (33)$$

The constants A, B, C, D are the integration constants, and are such as given by the terminal conditions of the problem, as by the distribution of current and e.m.f. in the circuit at the starting moment, for $t = 0$, or at one particular point, as $\lambda = 0$.

78. The constants u_0 and q depend upon the circuit conditions. If the circuit is closed upon itself — as usually is the case with an electrical transmission or distribution circuit — and Λ is the total length of the closed circuit, the equations must give for $\lambda = \Lambda$ the same values as for $\lambda = 0$, and therefore q must be a complete cycle or a multiple thereof, $2\,n\pi$; that is,

$$q = \frac{2\,n\pi}{\Lambda}, \qquad (34)$$

and the least value of q, or the fundamental frequency of oscillation, is

$$q_0 = \frac{2\,\pi}{\Lambda} \qquad (35)$$

and

$$q = nq_0. \qquad (36)$$

If the compound circuit is open at both ends, or grounded at both ends, and thus performs a half-wave oscillation, and $\Lambda_1 =$ total length of the circuit,

$$q_0 = \frac{\pi}{\Lambda_1} \text{ and } q = nq_0, \qquad (37)$$

and if the circuit is open at one end, grounded at the other end, thus performing a quarter-wave oscillation, and $\Lambda_2 =$ total length of circuit, it is

$$q_0 = \frac{\pi}{2 \Lambda_2} \quad \text{and} \quad q = (2 n - 1) q_0, \qquad (38)$$

while, if the length of the compound circuit is very great compared with the frequency of the oscillation, q_0 may have any value; that is, if the wave length of the oscillation is very short compared with the length of the circuit, any wave length, and therefore any frequency, may occur. With uniform circuits, as transmission lines, this latter case, that is, the response of the line to any frequency, can occur only in the range of very high frequencies. Even in a transmission line of several hundred miles' length the lowest frequency of free oscillation is fairly high, and frequencies which are so high compared with the fundamental frequency of the circuit that, considered as higher harmonics thereof, they overlap (as discussed in the above), must be extremely high—of the magnitude of million cycles. In a compound circuit, however, the fundamental frequency may be very much lower, and below machine frequencies, as the velocity of propagation $\frac{1}{\sqrt{LC}}$ may be quite low in some sections of the circuit, as in the high-potential coils of large transformers, and the presence of iron increases the inconstancy of L for high frequencies, so that in such a compound circuit, even at fairly moderate frequencies, of the magnitude of 10,000 cycles, the circuit may respond to any frequency.

79. The constant u_0 is also determined by the circuit constants. Upon u_0 depends the energy transfer constant of the circuit section, and therewith the rate of building up in a section of low power consumption, or building down in a section of high power consumption. In a closed circuit, however, passing around the entire circuit, the same values of e and i must again be reached, and the rates of building up and building down of the wave in the different sections must therefore be such as to neutralize each other when carried through the entire circuit; that is, the total building up through the entire compound circuit must be zero. This gives an equation from which u_0 is determined.

In a complex circuit having n sections of different constants and therefore n transition points, at the distances

$$\lambda_1, \lambda_2 \ldots \lambda_n \qquad ((39)$$

where $\lambda_{n+1} = \lambda_1 + \Lambda$, and $\Lambda =$ the total length of the circuit, the equations of i and e of any section i are given by equations (33) containing the constants A_i, B_i, C_i, D_i.

The constants A, B, C, D of any section are determined by the constants of the preceding section by equations (28) to (30). The constants of the second section thus are determined by those of the first section, the constants of the third section by those of the second section, and thereby, by substituting for the latter, by the constants of the first section, and in this manner, by successive substitutions, the constants of any section i can be expressed by the constants of the first section as linear functions thereof.

Ultimately thereby the constants of section $(n + 1)$ are expressed as linear functions of the constants of the first section:

$$\left.\begin{aligned}
A_{n+1} &= a'A_1 + a''B_1 + a'''C_1 + a''''D_1, \\
B_{n+1} &= b'A_1 + b''B_1 + b'''C_1 + b''''D_1, \\
C_{n+1} &= c'A_1 + c''B_1 + c'''C_1 + c''''D_1, \\
D_{n+1} &= d'A_1 + d''B_1 + d'''C_1 + d''''D_1,
\end{aligned}\right\} \qquad ((40)$$

where a', a'', a''', a'''', b', b'', etc., are functions of s_i and λ_i.

The $(n + 1)$st section, however, is again the first section, and it is thereby, by equations (33) and (39),

$$\left.\begin{aligned}
A_{n+1} &= A_1\varepsilon^{-s_1\Lambda}, \\
B_{n+1} &= B_1\varepsilon^{-s_1\Lambda}, \\
C_{n+1} &= C_1\varepsilon^{+s_1\Lambda}, \\
D_{n+1} &= D_1\varepsilon^{+s_1\Lambda},
\end{aligned}\right\} \qquad ((41)$$

and substituting (41) into (40) gives four symmetrical linear equations in A_1, B_1, C_1, D_1, from which these four constants can be eliminated, as n symmetrical linear equations with

n variables are dependent equations, containing an identity, thus:

$$\left.\begin{array}{l} (a' - \varepsilon^{-s_1\Lambda_1})\, A_1 + a''B_1 + a'''C_1 + a''''D_1 = 0; \\[4pt] b'A_1 + (b'' - \varepsilon^{-s_1\Lambda})\, B_1 + a'''C_1 + b''''D_1 = 0; \\[4pt] c'A_1 + c''B_1 + (c''' - \varepsilon^{+s_1\Lambda})\, C_1 + c''''D_1 = 0; \\[4pt] d'A_1 + d''B_1 + d'''C_1 + (d'''' - \varepsilon^{+s_1\Lambda})\, D_1 = 0, \end{array}\right\} \qquad ((42)$$

and herefrom

$$\begin{vmatrix} (a' - \varepsilon^{-s_1\Lambda}) & a'' & a''' & a'''' \\[6pt] b' & (b'' - \varepsilon^{-s_1\Lambda}) & b''' & b'''' \\[6pt] c' & c'' & (c''' - \varepsilon^{+s_1\Lambda}) & c'''' \\[6pt] d' & d'' & d''' & (d'''' - \varepsilon^{+s_1\Lambda}) \end{vmatrix} = 0. \; ((43)$$

Substituting in this determinant equation for s_i the values from (19))

$$s_i = u_0 - u_i \qquad\qquad ((44)$$

gives an exponential equation in u_0, thus:

$$F(u_0, u_i, \lambda_i, c_i) = 0, \qquad\qquad ((45)$$

from which the value u_0, or the resultant time decrement of the circuit, is determined.

In general, this equation (45) can be solved only by approximation, except in special cases.

CHAPTER IX.

POWER AND ENERGY OF THE COMPOUND CIRCUIT.

80. The free oscillation of a compound circuit differs from that of the uniform circuit in that the former contains exponential functions of the distance λ which represent the shifting or transfer of power between the sections of the circuit.

Thus the general expression of one term or frequency of current and voltage in a section of a compound circuit is given by equations (33) of Chapter VIII;

$$
\begin{aligned}
i &= \varepsilon^{-u_0 t} \left\{ \varepsilon^{+s\lambda} [A \cos q \, (\lambda - t) + B \sin q \, (\lambda - t)] \right. \\
&\left. \quad - \varepsilon^{-s\lambda} [C \cos q \, (\lambda + t) + D \sin q \, (\lambda + t)] \right\}
\end{aligned}
$$

and

$$
\begin{aligned}
e &= \sqrt{\frac{L}{C}} \, \varepsilon^{-u_0 t} \left\{ \varepsilon^{+s\lambda} [A \cos q \, (\lambda - t) + B \sin q \, (\lambda - t)] \right. \\
&\left. \quad + \varepsilon^{-s\lambda} [C \cos q \, (\lambda + t) + D \sin q \, (\lambda + t)] \right\},
\end{aligned}
$$

where $q = n q_0$, $q_0 = \dfrac{2\,\pi}{\Lambda}$, $\Lambda =$ total length of circuit, expressed in the distance coördinate $\lambda = \sigma l$, l being the distance coördinate of the circuit section in any measure, as miles, turns, etc., and r, L, g, C the circuit constants per unit length of l,

$$
\sigma = \sqrt{LC},
$$

$$
c = \sqrt{\frac{L}{C}},
$$

$$
u = \frac{1}{2} \left(\frac{r}{L} + \frac{g}{C} \right) = \text{time constant of circuit section,}
$$

$u_0 = u + s =$ resultant time decrement of compound circuit,

$s = u_0 - u =$ energy transfer constant of circuit section.

The instantaneous value of power at any point λ of the circuit at any time t is

$$p = ei$$

$$= \sqrt{\frac{L}{C}} \varepsilon^{-2u_0 t} \{ \varepsilon^{+2s\lambda} [A \cos q (\lambda - t) + B \sin q (\lambda - t)]^2$$
$$- \varepsilon^{-2s\lambda} [C \cos q (\lambda + t) + D \sin q (\lambda + t)]^2 \}$$

$$= \frac{1}{2} \sqrt{\frac{L}{C}} \varepsilon^{-2u_0 t} \{ [\varepsilon^{+2s\lambda} (A^2 + B^2) - \varepsilon^{-2s\lambda} (C^2 + D^2)]$$
$$+ [\varepsilon^{+2s\lambda} (A^2 - B^2) \cos 2 q (\lambda - t) - \varepsilon^{-2s\lambda} (C^2 - D^2) \cos 2 q (\lambda + t)]$$
$$+ 2 [AB \varepsilon^{+2s\lambda} \sin 2 q (\lambda - t) - CD \varepsilon^{-2s\lambda} \sin 2 q (\lambda + t)] \}; \qquad (1)$$

that is, the instantaneous value of power consists of a constant term and terms of double frequency in $(\lambda - t)$ and $(\lambda + t)$ or in distance λ and time t.

Integrating (1) over a complete period in time gives the *effective* or *mean value of power at any point λ* as

$$P = \frac{1}{2} \sqrt{\frac{L}{C}} \varepsilon^{-2u_0 t} \{ \varepsilon^{+2s\lambda} (A^2 + B^2) - \varepsilon^{-2s\lambda} (C^2 + D^2) \}; \qquad (2)$$

that is, the effective power at any point of the circuit is the difference between the effective power of the main wave and that of the reflected wave, and also, the instantaneous power at any time and any point of the circuit is the difference between the instantaneous power of the main wave and that of the reflected wave.

The effective power at any point of the circuit gradually decreases in any section with the resultant time decrement of the total circuit, $\varepsilon^{-2u_0 t}$, and varies gradually or exponentially with the distance λ, the one wave increasing, the other decreasing, so that at one point of the circuit or circuit section the effective power is zero; which point of the circuit is a power node, or point across which no energy flows. It is given by

$$\varepsilon^{+2s\lambda} (A^2 + B^2) = \varepsilon^{-2s\lambda} (C^2 + D^2),$$

or

$$\left. \begin{array}{l} \varepsilon^{4s\lambda} = \dfrac{C^2 + D^2}{A^2 + B^2}, \\[2ex] \lambda = \dfrac{1}{4s} \log \dfrac{C^2 + D^2}{A^2 + B^2}. \end{array} \right\} \qquad (3)$$

The difference of power between two points of the circuit, λ_1 and λ_2, that is, the power which is supplied or received (depending upon its sign) by a section $\lambda' = \lambda_2 - \lambda_1$ of the circuit, is given by equation (2) as

$$P_0 = \frac{1}{2}\sqrt{\frac{L}{C}}\varepsilon^{-2u_0 t}\left\{(\varepsilon^{+2s\lambda_2} - \varepsilon^{+2s\lambda_1})(A^2 + B^2)\right.$$
$$\left. - (\varepsilon^{-2s\lambda_2} - \varepsilon^{-2s\lambda_1})(C^2 + D^2)\right\}. \quad (4)$$

If P_0 is > 0, this represents the power which is supplied by the section λ' to the adjoining section of the circuit; if $P_0 < 0$, this is the power received by the section from the rest of the complex circuit.

If $s\lambda_2$ and $s\lambda_1$ are small quantities, the exponential function can be resolved into an infinite series, and all but the first term dropped, as of higher orders, or negligible, and this gives the approximate value

$$\varepsilon^{\pm 2s\lambda_2} - \varepsilon^{\pm 2s\lambda_1} = \pm 2s(\lambda_2 - \lambda_1) = \pm 2s\lambda'; \quad (5)$$

hence,

$$P_0 = s\lambda'\sqrt{\frac{L}{C}}\varepsilon^{-2u_0 t}\{A^2 + B^2 + C^2 + D^2\}; \quad (6)$$

that is, the power transferred from a section of length λ' to the rest of the circuit, or received by the section from the rest of the circuit, is proportional to the length of the section, λ', to its transfer constant, s, and to the sum of the power of main wave and reflected wave.

81. The energy stored by the inductance L of a circuit element $d\lambda$, that is, in the magnetic field of the circuit, is

$$dw_1 = \frac{L'i^2}{2}d\lambda,$$

where $L' = $ inductance per unit length of circuit expressed by the distance coördinate λ.

Since $L = $ the inductance per unit length of circuit, of distance coördinate l, and $\lambda = \sigma l$,

$$L' = \frac{L}{\sigma} = \frac{L}{\sqrt{LC}} = \sqrt{\frac{L}{C}};$$

hence,

$$dw_1 = \frac{1}{2}\sqrt{\frac{L}{C}}\, i^2 d\lambda. \tag{7}$$

In general, the circuit constants r, L, g, C, per unit length, $l = 1$ give, per unit length, $\lambda = 1$, the circuit constants

$$\left. \begin{array}{l} \dfrac{r}{\sigma}\,;\ \dfrac{L}{\sigma}\,;\ \dfrac{g}{\sigma}\,;\ \dfrac{C}{\sigma}\,, \\[3mm] \text{or} \\[3mm] \dfrac{r}{\sqrt{LC}}\,;\ \dfrac{L}{\sqrt{LC}} = \sqrt{\dfrac{L}{C}}\,;\ \dfrac{g}{\sqrt{LC}}\,;\ \dfrac{C}{\sqrt{LC}} = \sqrt{\dfrac{C}{L}}\,. \end{array} \right\} \tag{8}$$

Substituting (290) in equation (309) gives

$$\begin{aligned} \frac{dw_1}{d\lambda} &= \frac{1}{2}\sqrt{\frac{L}{C}}\,\varepsilon^{-2u_0 t}\{\varepsilon^{+2s\lambda}\,[A\cos q\,(\lambda - t) + B\sin q\,(\lambda - t)]^2 \\ &\quad + \varepsilon^{-2s\lambda}\,[C\cos q\,(\lambda + t) + D\sin q\,(\lambda + t)]^2 \\ &\quad - 2\,[A\cos q\,(\lambda - t) + B\sin q\,(\lambda - t)] \\ &\quad\quad [C\cos q\,(\lambda + t) + D\sin q\,(\lambda + t)]\} \\[2mm] &= \frac{1}{4}\sqrt{\frac{L}{C}}\,\varepsilon^{-2u_0 t}\,\{[\varepsilon^{+2s\lambda}\,(A^2 + B^2) + \varepsilon^{-2s\lambda}\,(C^2 + D^2)] \\ &\quad + [\varepsilon^{+2s\lambda}\,(A^2 - B^2)\cos 2q\,(\lambda - t) \\ &\quad + \varepsilon^{-2s\lambda}\,(C^2 - D^2)\cos 2q\,(\lambda + t)] \\ &\quad + 2\,[AB\varepsilon^{+2s\lambda}\sin 2q\,(\lambda - t) + CD\varepsilon^{-2s\lambda}\sin 2q\,(\lambda + t)] \\ &\quad - 2\,[(AC - BD)\cos 2q\lambda + (AD + BC)\sin 2q\lambda] \\ &\quad - 2\,[(AC + BD)\cos 2qt + (AD - BC)\sin 2qt]\}. \end{aligned} \tag{9}$$

Integrating over a complete period in time gives the effective energy stored in the magnetic field at point λ as

$$\begin{aligned} \frac{dW_1}{d\lambda} &= \frac{1}{2\pi}\int_0^{2\pi}\frac{dw_1}{d\lambda}\,dt \\[2mm] &= \frac{1}{4}\sqrt{\frac{L}{C}}\,\varepsilon^{-2u_0 t}\,\{[\varepsilon^{+2s\lambda}\,(A^2 + B^2) + \varepsilon^{-2s\lambda}\,(C^2 + D^2)] \\ &\quad - 2\,[(AC - BD)\cos 2q\lambda + (AD + BC)\sin 2q\lambda]\}, \tag{10} \end{aligned}$$

and integrating over one complete period of distance λ, or one complete wave length, this gives

$$\frac{dW_1^0}{d\lambda} = \frac{1}{2\pi} \int_0^{2\pi} \frac{dW_1}{d\lambda} d\lambda = \frac{1}{4} \sqrt{\frac{L}{C}} \varepsilon^{-2u_0 t} \left\{ [\varepsilon^{+2s\lambda} (A^2 + B^2) \right.$$

$$\left. + \varepsilon^{-2s\lambda} (C^2 + D^2) \right\}. \tag{11}$$

The energy stored by the inductance L, or in the magnetic field of the conductor, thus consists of a constant part,

$$\frac{dw_0}{d\lambda} = \frac{1}{4} \sqrt{\frac{L}{C}} \varepsilon^{-2u_0 t} \left\{ \varepsilon^{+2s\lambda} (A^2 + B^2) + \varepsilon^{-2s\lambda} (C^2 + D^2), \tag{12} \right.$$

a part which is a function of $(\lambda - t)$ and $(\lambda + t)$,

$$\frac{dw'}{d\lambda} = \frac{1}{4} \sqrt{\frac{L}{C}} \varepsilon^{-2u_0 t} \left\{ [\varepsilon^{+2s\lambda} (A^2 - B^2) \cos 2q (\lambda - t) \right.$$

$$+ \varepsilon^{-2s\lambda} (C^2 - D^2) \cos 2q (\lambda + t)]$$

$$+ 2 [AB\varepsilon^{+2s\lambda} \sin 2q (\lambda - t)$$

$$+ CD\varepsilon^{-2s\lambda} \sin 2q (\lambda + t)] \right\}, \tag{13}$$

a part which is a function of the distance λ only but not of time t,

$$-\frac{dw''}{d\lambda} = \frac{1}{2} \sqrt{\frac{L}{C}} \varepsilon^{-2u_0 t} \left\{ (AC - BD) \cos 2q\lambda + (AD + BC) \sin 2q\lambda, \right.$$

$$\tag{14}$$

and a part which is a function of time t only but not of the distance λ,

$$-\frac{dw'''}{d\lambda} = \frac{1}{2} \sqrt{\frac{L}{C}} \varepsilon^{-2u_0 t} \left\{ (AC + BD) \cos 2qt + (AD - BC) \sin 2qt \right\},$$

$$\tag{15}$$

and the total energy of the electromagnetic field of circuit element $d\lambda$ at time t is

$$\frac{dw_1}{d\lambda} = \frac{dw_0}{d\lambda} + \frac{dw'}{d\lambda} - \frac{dw''}{d\lambda} - \frac{dw'''}{d\lambda}. \tag{16}$$

82. The energy stored in the electrostatic field of the conductor or by the capacity C is given by

$$dw_2 = \frac{C' e^2}{2} d\lambda;$$

or, substituting (8),

$$\frac{dw_2}{d\lambda} = \frac{1}{2}\sqrt{\frac{C}{L}}\, e^2, \tag{17}$$

and substituting in (17) the value of e from equation (33) of Chapter VIII gives the same expression as (9) except that the sign of the last two terms is reversed; that is, the total energy of the electrostatic field of circuit element $d\lambda$ at time t is

$$\frac{dw^2}{d\lambda} = \frac{dw_0}{d\lambda} + \frac{dw'}{d\lambda} + \frac{dw''}{d\lambda} + \frac{dw'''}{d\lambda} \tag{18}$$

and adding (16) and (18) gives the total stored energy of the electric field of the conductor,

$$\frac{dw}{d\lambda} = \frac{dw_1}{d\lambda} + \frac{dw_2}{d\lambda} = 2\frac{dw_0}{d\lambda} + 2\frac{dw'}{d\lambda}, \tag{19}$$

and integrated over a complete period of time this gives

$$\frac{dW}{d\lambda} = 2\frac{dw_0}{d\lambda} = \frac{1}{2}\sqrt{\frac{L}{C}}\,\varepsilon^{-2u_0 t}\left\{\varepsilon^{+2s\lambda}\,(A^2+B^2)+\varepsilon^{-2s\lambda}\,(C^2+D^2)\right\}. \tag{20}$$

The last two terms, $\dfrac{dw''}{d\lambda}$ and $\dfrac{dw'''}{d\lambda}$, thus represent the energy which is transferred, or pulsates, between the electromagnetic and the electrostatic field of the circuit; and the term $\dfrac{dw'}{dt}$ represents the alternating (or rather oscillating) component of stored energy.

83. The energy stored by the electric field in a circuit section λ', between λ_1 and λ_2, is given by integrating $\dfrac{dW}{d\lambda}$ between λ_2 and λ_1, as

$$W = \frac{1}{4s}\sqrt{\frac{L}{C}}\,\varepsilon^{-2u_0 t}\left\{(\varepsilon^{+2s\lambda_2} - \varepsilon^{+2s\lambda_1})\,(A^2 + B^2)\right.$$
$$\left. - (\varepsilon^{-2s\lambda_2} - \varepsilon^{-2s\lambda_1})\,(C^2 + D^2)\right\}; \tag{21}$$

or, substituting herein the approximation (307),

$$W = \frac{1}{2}\lambda'\sqrt{\frac{L}{C}}\,\varepsilon^{-2u_0 t}\left\{A^2 + B^2 + C^2 + D^2\right\}. \tag{22}$$

Differentiating (22) with respect to t gives the power supplied by the electric field of the circuit as

$$P = -\frac{dW}{dt} = u_0\lambda'\sqrt{\frac{L}{C}}\,\varepsilon^{-2u_0 t}\left\{A^2 + B^2 + C^2 + D^2\right\}, \quad (23)$$

or, more generally,

$$P = \frac{u_0}{2\,s}\sqrt{\frac{L}{C}}\,\varepsilon^{-2u_0 t}\left\{(\varepsilon^{+2s\lambda_2} - \varepsilon^{+2s\lambda_1})\,(A^2 + B^2)\right.$$
$$\left. - (\varepsilon^{-2s\lambda_2} - \varepsilon^{-2s\lambda_1})\,(C^2 + D^2)\right\}. \quad (24)$$

84. The power dissipated in the resistance $r'd\lambda = \dfrac{rd\lambda}{\sqrt{LC}}$ of a conductor element $d\lambda$ is

$$dp' = r'i^2 d\lambda \quad (25)$$
$$= \frac{2\,r}{L}\,dw_1;$$

hence, substituting herein equation (16) gives the power consumed by resistance of the circuit element $d\lambda$ as

$$\frac{dp'}{d\lambda} = \frac{2\,r}{L}\left\{\frac{dw_0}{d\lambda} + \frac{dw'}{d\lambda} - \frac{dw''}{d\lambda} - \frac{dw'''}{d\lambda}\right\}, \quad (26)$$

and the power consumed by the conductance $g'd\lambda = \dfrac{g}{\sqrt{LC}}\,d\lambda$ of a conductor element $d\lambda$ is

$$dp'' = g'e^2 d\lambda \quad (27)$$
$$= \frac{2\,g}{C}\,dw_2;$$

hence the power consumed by conductance of circuit element $d\lambda$ is

$$\frac{dp''}{d\lambda} = \frac{2\,g}{C}\left\{\frac{dw_0}{d\lambda} + \frac{dw'}{d\lambda} + \frac{dw''}{d\lambda} + \frac{dw'''}{d\lambda}\right\}, \quad (28)$$

and the *total power dissipated* in the circuit element $d\lambda$ is

$$\frac{dp_1}{d\lambda} = \frac{dp'}{d\lambda} + \frac{dp''}{d\lambda} = 4\,u\left(\frac{dw_0}{d\lambda} + \frac{dw'}{d\lambda}\right) - 4\,m\left(\frac{dw''}{d\lambda} + \frac{dw'''}{d\lambda}\right), \quad (29)$$

where, as before,

$$u = \frac{1}{2}\left(\frac{r}{L} + \frac{g}{C}\right)$$

and

$$m = \frac{1}{2}\left(\frac{r}{L} - \frac{g}{C}\right),$$

(30)

and integrating over a complete period

$$\frac{dP_1}{d\lambda} = 4\,u\,\frac{dw_0}{d\lambda} - 4\,m\,\frac{dw''}{d\lambda},$$ (31)

the power dissipated in the circuit thus contains a constant term,

$4\,u\,\dfrac{dw_0}{d\lambda}$, and a term which is a periodic function of the distance λ,

$4\,m\,\dfrac{dw''}{d\lambda}$, of double frequency.

Averaged over a half-wave of the circuit, or a multiple thereof, the second term disappears, and

$$\frac{dP_1^0}{d\lambda} = 4\,u\,\frac{dw_0}{d\lambda};$$

or, substituting (12),

$$\frac{dP_1^0}{d\lambda} = u\sqrt{\frac{L}{C}}\,\varepsilon^{-2u_0 t}\left\{\varepsilon^{+2s\lambda}\,(A^2 + B^2) + \varepsilon^{-2s\lambda}\,(C^2 + D^2)\right\},$$ (32)

thus the power dissipated in a section $\lambda' = \lambda_2 - \lambda_1$ of the circuit is, by integrating between limits λ_1 and λ_2,

$$P_1^0 = \frac{u}{2\,s}\sqrt{\frac{L}{C}}\,\varepsilon^{-2u_0 t}\left\{(\varepsilon^{+2s\lambda_2} - \varepsilon^{+2s\lambda_1})\,(A^2 + B^2)\right.$$
$$\left. + (\varepsilon^{-2s\lambda_2} - \varepsilon^{2s\lambda_1})\,(C^2 + D^2)\right\},$$ (33)

or, approximately,

$$P_1^0 = u\lambda'\sqrt{\frac{L}{C}}\,\varepsilon^{-2u_0 t}\left\{A^2 + B^2 + C^2 + D^2\right\}.$$ (34)

85. Writing, therefore,

$$H^2 = (A^2 + B^2 + C^2 + D^2)\sqrt{\frac{L}{C}},$$ (35)

the *energy stored in the electric field of the circuit section of length* λ' is

$$W = \frac{1}{2} \lambda' H^2 \varepsilon^{-2 u_0 t}; \tag{36}$$

the *power supply to the conductor by the decay of the electric field of the circuit* is

$$P = u_0 \lambda' H^2 \varepsilon^{-2 u_0 t}; \tag{37}$$

the *power dissipated in the circuit section* λ' by its effective resist ance and conductance is

$$P_1{}^0 = u \lambda' H^2 \varepsilon^{-2 u_0 t}, \tag{38}$$

and the *power transferred from the circuit section* λ' to the rest of the circuit is

$$P_0 = s \lambda' H^2 \varepsilon^{-2 u_0 t}; \tag{39}$$

that is, $\dfrac{u}{u_0}$ = ratio of power dissipated in the section to that supplied to the section by its stored energy of the electric field.

$\dfrac{s}{u_0}$ = fraction of power supplied to the section by its electric field, which is transferred from the section to adjoining sections (or, if $s < 0$, received from them).

$\dfrac{s}{u}$ = ratio of power transferred to other sections to power dissipated in the section.

$u_0 \div u \div s$ thus is the ratio of the power supplied to the section by its electric field, dissipated in the section, and transferred from the section to adjoining sections.

These relations obviously are approximate only, and applicable to the case where the wave length is short.

86. Equation (4), of the power transferred from a section to the adjoining section, can be arranged in the form

$$P_0 = \frac{1}{2} \sqrt{\frac{L}{C}} \varepsilon^{-2 u_0 t} \{ [\varepsilon^{+2 s \lambda_2} (A^2 + B^2) - \varepsilon^{-2 s \lambda_2} (C^2 + D^2)]$$
$$- [\varepsilon^{+2 s \lambda_1} (A^2 + B^2) - \varepsilon^{-2 s \lambda_1} (C^2 + D^2)] \}; \tag{40}$$

that is, it consists of two parts, thus:

$$P_0{}' = \frac{1}{2} \sqrt{\frac{L}{C}} \varepsilon^{-2 u_0 t} \{ \varepsilon^{+2 s \lambda_2} (A^2 + B^2) - \varepsilon^{-2 s \lambda_2} (C^2 + D^2) \}, \tag{41}$$

which is the power transferred from the section to the next following section, and

$$P_0'' = +\frac{1}{2}\sqrt{\frac{L}{C}}\,\varepsilon^{-2u_0\ell}\left\{\varepsilon^{+2s\lambda_1}\,(A^2+B^2)-\varepsilon^{-2s\lambda_1}\,(C^2+D^2)\right\}, \quad (42)$$

which is the power received from the preceding section, and the difference between the two values,

$$P_0 = P_0' - P_0'', \tag{43}$$

therefore, is the excess of the power given out over that received, or the resultant power supplied by the section to the rest of the circuit.

An approximate idea of the value of the power transfer constant can now be derived by assuming H^2 as constant throughout the entire compound circuit, which is approximately the case.

In this case, as the total power transferred between the sections must be zero, thus:

$$\Sigma P_0 = 0;$$

hence, substituting (341),

$$\Sigma s_i \lambda_i' = 0, \tag{44}$$

and, since

$$s_i = u_0 - u_i,$$
$$u_0 \Lambda = \Sigma w_i \lambda_i'; \tag{45}$$

that is, the resultant circuit decrement multiplied by the total length of the circuit equals the sum of the time constants of the sections multiplied with the respective length of the section, or, if $\zeta_1, \zeta_2 \ldots \zeta_1$ = length of the circuit section, as fraction of the total circuit length Λ,

$$u_0 = \Sigma \zeta_i u_i. \tag{46}$$

Whether this expression (46) is more general is still unknown.

87. As an example assume a transmission line having the following constants per wire: $r_1 = 52; L_1 = 0.21; g_1 = 40 \times 10^{-6}$, and $C_1 = 1.6 \times 10^{-6}$.

Further assume this line to be connected to step-up and step-down transformers having the following constants per transformer high-potential circuit: $r_2 = 5$, $L_2 = 3$; $g_2 = 0.1 \times 10^{-6}$, and $C_2 = 0.3 \times 10^{-6}$; then

$$\lambda_1' = \sigma_1 = \sqrt{L_1 C_1} = 0.58 \times 10^{-3}, \quad \lambda_2' = \sigma_2 = 0.95 \times 10^{-3},$$
$$u_1 = 136, \qquad\qquad u_2 = 1.$$

The circuit consists of four sections of the lengths

$$\lambda_1' = 0.58 \times 10^{-3}, \; \lambda_2' = 0.95 \times 10^{-3}, \; \lambda_3' = 0.58 \times 10^{-3}, \; \lambda_4' = 0.95 \times 10^{-3};$$

hence a total length

$$\Lambda = 3.06 \times 10^{-3},$$

and the resultant circuit decrement is

$$u_0 = 2\frac{\lambda_1'}{\Lambda}u_1 + 2\frac{\lambda_2'}{\Lambda}u_2 = 51.6 + 0.59 = 52.2;$$

hence,

$$s_1 = -83.8 \text{ and } s_2 = +51.2.$$

If now the current in the circuit is $i_0 = 100$ amperes, the e.m.f. $e_0 = 40,000$ volts, the total stored energy is

$$W = i_0^2 (L_1 + L_2) + e_0^2 (C_1 + C_2)$$
$$= 32,000 + 3000 = 35,000 \text{ joules,}$$

and from equation (36) then follows, for $t = 0$,

$$\tfrac{1}{2} \Lambda H^2 = 35,000,$$

$$H^2 = \frac{70,000}{\Lambda} = 22.8 \times 10^6,$$

which gives

$$u_0 = 52.2,$$
$$H^2 = 22.8 \times 10^6,$$
$$W = 35,000.$$

		Line.	Step-up Transformer.	Line.	Step-down Transformer.
Length of section,	$\lambda' =$	0.58×10^{-3}	0.95×10^{-3}	0.58×10^{-3}	0.95×10^{-3}
Time constant,	$u =$	136	1	136	1
Transfer constant,	$s =$	-83.8	$+51.2$	-83.8	$+51.2$
Energy of electric field,	$W =$	6.650	10.850	6.650	10.850 kilojoules
Power supplied by electric field,	$P =$	690	1132	690	1132 kilowatts
Power dissipated,	$P_0^0 =$	1800	22	1800	22 kilowatts
Power transferred,	$\overset{\cdot}{P}_0 =$	-1110	1110	-1110	1110 kilowatts

Thus, of the total power produced in the transformers by the decrease of their electric field, only 22 kw. are dissipated as heat in the transformer, and 1110 kw. transferred to the transmission line. While the power available by the decrease of the electric field of the transmission line is only 690 kw., the line dissipates energy at the rate of 1800 kw., receiving 1110 kw. from the transformers.

CHAPTER X.

REFLECTION AND REFRACTION AT TRANSITION POINT.

88. The general equation of the current and voltage in a section of a compound circuit, from equations (33) of Chapter VIII, is

$$\begin{aligned}
i &= \varepsilon^{-u_0 t} \{\varepsilon^{+s\lambda} [A \cos q \, (\lambda - t) + B \sin q \, (\lambda - t)] \\
&\quad - \varepsilon^{-s\lambda} [C \cos q \, (\lambda + t) + D \sin q \, (\lambda + t)]\} \\
e &= c\varepsilon^{-u_0 t} \{\varepsilon^{+s\lambda} [A \cos q \, (\lambda - t) + B \sin q \, (\lambda - t)] \\
&\quad + \varepsilon^{-s\lambda} [C \cos q \, (\lambda + t) + D \sin q \, (\lambda + t)]\},
\end{aligned} \tag{1}$$

where

$\lambda = \sigma l =$ distance variable with velocity as unit;

$\sigma = \sqrt{LC}$;

$c = \sqrt{\dfrac{L}{C}}$;

$u_0 = u + s =$ resultant time decrement;

$u = \dfrac{1}{2}\left(\dfrac{r}{L} + \dfrac{g}{C}\right) =$ time constant, and

$s =$ energy transfer constant of section.

At a transition point λ_1 between section 1 and section 2 the constants change by equations (28) and (29) of Chapter VIII

$$\begin{aligned}
A_2 &= \varepsilon^{-s_2\lambda_1}\{a_1\varepsilon^{+s_1\lambda_1}A_1 + b_1\varepsilon^{-s_1\lambda_1}(C_1 \cos 2\, q\lambda_1 + D_1 \sin 2\, q\lambda_1)\} \\
B_2 &= \varepsilon^{-s_2\lambda_1}\{a_1\varepsilon^{+s_1\lambda_1}B_1 + b_1\varepsilon^{-s_1\lambda_1}(C_1 \sin 2\, q\lambda_1 - D_1 \cos 2\, q\lambda_1)\} \\
C_2 &= \varepsilon^{+s_2\lambda_1}\{a_1\varepsilon^{-s_1\lambda_1}C_1 + b_1\varepsilon^{+s_1\lambda_1}(A_1 \cos 2\, q\lambda_1 + B_1 \sin 2\, a\lambda_1)\} \\
D_2 &= \varepsilon^{+s_2\lambda_1}\{a_1\varepsilon^{-s_1\lambda_1}D_1 + b_1\varepsilon^{+s_1\lambda_1}(A_1 \sin 2\, q\lambda_1 - B_1 \cos 2\, q\lambda_1)\},
\end{aligned} \tag{2}$$

where

$$a_1 = \frac{c_1 + c_2}{2\, c_2} \quad \text{and} \quad b_1 = \frac{c_1 - c_2}{2\, c_2}. \tag{3}$$

Choosing now the transition point as zero point of λ, so that $\lambda < 0$ is section 1, $\lambda > 0$ is section 2, equations (2) assume the form

$$\left.\begin{aligned}
A_2 &= a_1 A_1 + b_1 C_1, \\
B_2 &= a_1 B_1 - b_1 D_1, \\
C_2 &= a_1 C_1 + b_1 A_1, \\
D_2 &= a_1 D_1 - b_1 B_1.
\end{aligned}\right\} \quad (4)$$

From equations (4) and (3) it follows that

$$\left.\begin{aligned}
c_2 \left(A_2^2 - C_2^2\right) &= c_1 \left(A_1^2 - C_1^2\right) \\
\text{and} \\
c_2 \left(B_2^2 - D_2^2\right) &= c_1 \left(B_1^2 - D_1^2\right).
\end{aligned}\right\} \quad (5)$$

If now a wave in section 1, $A\,B$, travels towards transition point $\lambda = 0$, at this point a part is reflected, giving rise to the reflected wave $C\,D$ in section 1, while a part is transmitted and appears as main wave $A\,B$ in section 2. The wave $C\,D$ in section 2 thus would not exist, as it would be a wave coming towards $\lambda = 0$ from section 2, so not a part of the wave coming from section 1. In other words, we can consider the circuit as comprising two waves moving in opposite direction:

(1) A main wave $A_1 B_1$, giving a transmitted wave $A_2 B_2$ and reflected wave $C_1 D_1$.

(2) A main wave $C_2 D_2$, giving a transmitted wave $C_1' D_1'$ and reflected wave $A_2' B_2'$.

The waves moving towards the transition point are single main waves, $A_1 B_1$ and $C_2 D_2$, and the waves moving away from the transition point are combinations of waves reflected in the section and waves transmitted from the other section.

89. Considering first the main wave moving towards rising λ: in this $C_2 = 0 = D_2$, hence, from (4)

$$\left.\begin{aligned}
a_1 C_1 + b_1 A_1 &= 0 \\
\text{and} \\
a_1 D_1 - b_1 B_1 &= 0,
\end{aligned}\right\} \quad (6)$$

and herefrom

$$\left.\begin{aligned}
C_1 &= -\frac{b_1}{a_1}\, A_1 = +\frac{c_2 - c_1}{c_2 + c}\, A_1 \\
\text{and} \\
D_1 &= +\frac{b_1}{a_1}\, B_1 = +\frac{c_1 - c_2}{c_1 + c_2}\, B_1;
\end{aligned}\right\} \quad (7)$$

which substituted in (349) gives

$$A_2 = a_1 A_1 - \frac{b_1^2}{a_1} A_1 = \frac{a_1^2 - b_1^2}{a_1} A_1 = \frac{2\,c_1}{c_1 + c_2} A_1$$

and

$$B_2 = a_1 B_1 - \frac{b_1^2}{a_1} B_1 = \frac{a_1^2 - b_1^2}{a_1} B_1 = \frac{2\,c_1}{c_1 + c_2} B_1. \tag{8}$$

Then for the *main wave* in section 1,

$$i_1 = \varepsilon^{-u_0 t}\,\varepsilon^{+s_1\lambda}\left\{A_1 \cos q\,(\lambda - t) + B_1 \sin q\,(\lambda - t)\right\}$$

and

$$e_1 = c_1\,\varepsilon^{-u_0 t}\,\varepsilon^{+s_1\lambda}\left\{A_1 \cos q\,(\lambda - t) + B_1 \sin q\,(\lambda - t)\right\}. \tag{9}$$

When reaching a transition point $\lambda = 0$, the wave resolves into the *reflected wave*, turned back on section 1, thus:

$$i_1' = -\frac{c_2 - c_1}{c_2 + c_1}\varepsilon^{-u_0 t}\varepsilon^{-s_1\lambda}\left\{A_1 \cos q\,(\lambda+t) - B_1 \sin q\,(\lambda+t)\right\}$$

and

$$e_1' = +c_1\frac{c_2 - c_1}{c_2 + c_1}\varepsilon^{-u_0 t}\varepsilon^{-s_1\lambda}\left\{A_1 \cos q\,(\lambda+t) - B_1 \sin q\,(\lambda+t)\right\}. \tag{10}$$

The *transmitted wave*, which by passing over the transition point enters section 2, is given by

$$i_2 = \frac{2\,c_1}{c_1 + c_2}\varepsilon^{-u_0 t}\varepsilon^{+s_2\lambda}\left\{A_1 \cos q\,(\lambda - t) + B_1 \sin q\,(\lambda - t)\right\}$$

and

$$e_2 = c_2\frac{2\,c_1}{c_1 + c_2}\varepsilon^{-u_0 t}\varepsilon^{+s_2\lambda}\left\{A_1 \cos q\,(\lambda - t) + B_1 \sin q\,(\lambda - t)\right\}. \tag{11}$$

The reflection angle, $\tan(i_1') = -\dfrac{B_1}{A_1}$, is supplementary to the impact angle, $\tan(i_1) = +\dfrac{B_1}{A_1}$, and transmission angle, $\tan(i_2) = +\dfrac{B_1}{A_1}$.

Reversing the sign of λ in the equation (10) of the reflected wave, that is, counting the distance for the reflected wave also in the direction of its propagation, and so in opposite direction as in the main wave and the transmitted wave, equations (10) become

$$
\left.\begin{aligned}
i_1'' &= + \frac{c_2 - c_1}{c_2 + c_1} \varepsilon^{-u_0 t}\, \varepsilon^{+s_1 \lambda'} \left\{ A_1 \cos q\,(\lambda' - t) + B_1 \sin q\,(\lambda - t) \right\}, \\[2ex]
e_1'' &= + c_1 \frac{c_2 - c_1}{c_2 + c_1} \varepsilon^{-u_0 t}\, \varepsilon^{+s_1 \lambda'} \left\{ A_1 \cos q\,(\lambda' - t) + B_1 \sin q\,(\lambda' - t) \right\},
\end{aligned}\right\} \quad (12)
$$

and then

or

$$
\left.\begin{aligned}
i_2 + i_1'' &= i_1, \\[1.5ex]
i_1 + i_1' &= i_2, \\[1.5ex]
\frac{c_1}{c_2} e_2 + e_1'' &= e_1, \\[1.5ex]
e_1 + e_1'' &= e_2.
\end{aligned}\right\} \quad (13)
$$

(1) In a single electric wave, current and e.m.f. are in phase with each other. Phase displacements between current and e.m.f. thus can occur only in resultant waves, that is, in the combination of the main and the reflected wave, and then are a function of the distance λ, as the two waves travel in opposite direction.

(2) When reaching a transition point, a wave splits up into a reflected wave and a transmitted wave, the former returning in opposite direction over the same section, the latter entering the adjoining section of the circuit.

(3) Reflection and transmission occur without change of the phase angle; that is, the phase of the current and of the voltage in the reflected wave and in the transmitted wave, at the transition point, is the same as the phase of the main wave or incoming wave. Reflection and transmission with a change of phase angle can occur only by the combination of two waves traveling in opposite direction over a circuit; that is, in a resultant wave, but not in a single wave.

(4) The sum of the transmitted and the reflected current equals the main current, when considering these currents in their respective direction of propagation.

The sum of the voltage of the main wave and the reflected wave equals the voltage of the transmitted wave.

The sum of the voltage of the reflected wave and the voltage of the transmitted wave reduced to the first section by the ratio of voltage transformation $\frac{c_2}{c_1}$, equals the voltage of the main wave.

(5) Therefore a voltage transformation by the factor $\frac{c_2}{c_1}$

$= \sqrt{\frac{L_2}{C_2}\frac{C_1}{L_1}}$ occurs at the transition point; that is, the transmitted wave of voltage equals the difference between main wave and reflected wave multiplied by the transformation ratio $\frac{c_2}{c_1}$;

$e_2 = \frac{c_1}{c_2}(e_1 - e_1'')$. As result thereof, in passing from one section of a circuit to another section, the voltage of the wave may decrease or may increase. If $\frac{c_2}{c_1} > 1$, that is, when passing from a section of low inductance and high capacity into a section of high inductance and low capacity, as from a transmission line into a transformer or a reactive coil, the voltage of the wave is increased; if $\frac{c_2}{c_1} < 1$, that is, when passing from a section of high inductance and low capacity into a section of low inductance and high capacity, as from a transformer to a transmission line, the voltage of the wave is decreased.

This explains the frequent increase to destructive voltages, when entering a station from the transmission line or cable, of an impulse or a wave which in the transmission line is of relatively harmless voltage.

The ratio of the transmitted to the reflected wave is given by

$$\frac{i_2}{i_1''} = \frac{2c_1}{c_2 - c_1} = \frac{2\sqrt{L_1 C_2}}{\sqrt{L_2 C_1} - \sqrt{L_1 C_2}} = \frac{2}{\sqrt{\frac{L_2}{L_1}\frac{C_1}{C_2}} - 1}$$

and

$$\frac{e_2}{e_1''} = \frac{2c_2}{c_2 - c_1} = \frac{2\sqrt{L_2 C_1}}{\sqrt{L_2 C_1} - \sqrt{L_1 C_2}} = \frac{2}{1 - \sqrt{\frac{L_1}{L_2}\frac{C_2}{C_1}}}.$$

(14)

90. Example:

Transmission line	Transformer
$L_1 = 1.95 \times 10^{-3}$	$L_2 = 1$
$C_1 = 0.0162 \times 10^{-6}$	$C_2 = 0.4 \times 10^{-6}$
$c_1 = 346$	$c = 1580$
$\dfrac{i_2}{i_1''} = 0.56$	$\dfrac{e_2}{e_1''} = 2.56$

And in the opposite direction

$$\frac{i_2}{i_1''} = -2.56 \qquad\qquad \frac{e_2}{e_1''} = -0.56.$$

The ratio $\dfrac{e_2}{e_1''}$ becomes a maximum, $= \infty$, for $\dfrac{L_1}{C_1} = \dfrac{L_2}{C_2}$, but in this case $e_1'' = 0$; that is, no reflection occurs, and the reflected wave equals zero, the transmitted wave equals the incoming wave.

$$\frac{i_2}{i_1} = \frac{2\,c_1}{c_2 + c_1};$$

hence, becomes a maximum for $c_2 = 0$, or $c_1 = \infty$ and then $= 2$; in which case $e_2 = 0$.

$$\frac{e_2}{e_1} = \frac{2\,c_2}{c_2 + c_1};$$

(15)

hence, becomes a maximum for $c_1 = 0$, or $c_2 = \infty$ and then $= 2$; in which case $i_2 = 0$. From the above it is seen that the maximum value to which the voltage can build up at a single transition point is twice the voltage of the incoming wave, and this occurs at the open end of the circuit, or, approximately, at a point where the ratio of inductance to capacity very greatly increases.

$$\frac{i_1''}{i_1} = \frac{c_2 - c_1}{c_2 + c_1};$$

hence, becomes a maximum, and equal to 1, for $c_1 = 0$, or $c_2 = \infty$.

(16)

$$\frac{e_1''}{e_1} = \frac{c_2 - c_1}{c_2 + c_1}$$

has the same value as the current-ratio.

91. Consider now a wave traversing the circuit in opposite direction; that is, $C_2 D_2$ is the main wave, $A_2 B_2$ the reflected wave, $C_1 D_1$ the transmitted wave, and $A_1 = 0 = B_1$. In equation (4) this gives

$$A_2 = b_1 C_1;$$
$$B_2 = -b_1 D_1;$$
$$C_2 = a_1 C_1,$$

and
$$D_2 = a_1 D_1;$$

hence,

$$\left.\begin{aligned}
C_1 &= \frac{1}{a_1} C_2 = \frac{2 c_2}{c_1 + c_2} C_2; \\[4pt]
D_1 &= \frac{1}{a_1} D_2 = \frac{2 c_2}{c_1 + c_2} D_2; \\[4pt]
A_2 &= \frac{b_1}{a_1} C_2 = \frac{c_1 - c_2}{c_1 + c_2} C_2, \\[4pt]
B_2 &= -\frac{b_1}{a_1} D_2 = -\frac{c_1 - c_2}{c_1 + c_2} D_2 = \frac{c_2 - c_1}{c_2 + c_1} D_2;
\end{aligned}\right\} \quad (17)$$

and

that is, the same relations as expressed by equations (7) and (8) for the wave traveling in opposite direction.

The equations of the components of the wave then are:

Main wave:

$$\left.\begin{aligned}
\bar{i}_2 &= - \varepsilon^{-u_0 t} \varepsilon^{-s_2 \lambda} \left\{ C_2 \cos q \,(\lambda + t) + D_2 \sin q \,(\lambda + t) \right\} \\
\bar{e}_2 &= + c_2 \varepsilon^{-u_0 t} \varepsilon^{-s_2 \lambda} \left\{ C_2 \cos q \,(\lambda + t) + D_2 \sin q \,(\lambda + t) \right\}:
\end{aligned}\right\} \quad (18)$$

Transmitted wave:

$$\left.\begin{aligned}
\bar{i}_1 &= - \frac{2 c_2}{c_1 + c_2} \varepsilon^{-u_0 t} \varepsilon^{-s_1 \lambda} \left\{ C_2 \cos q \,(\lambda + t) + D_2 \sin q \,(\lambda + t) \right\} \\
\bar{e}_1 &= + c_1 \frac{2 c_2}{c_1 + c_2} \varepsilon^{-u_0 t} \varepsilon^{-s_1 \lambda} \left\{ C_2 \cos q \,(\lambda + t) + D_2 \sin q \,(\lambda + t) \right\};
\end{aligned}\right\} \quad (19)$$

Reflected wave:

$$\left.\begin{aligned}
\bar{i}_2' &= \frac{c_1 - c_2}{c_1 + c_2} \varepsilon^{-u_0 t} \varepsilon^{+s_1 \lambda} \left\{ C_2 \cos q \,(\lambda - t) - D_2 \sin q \,(\lambda - t) \right\} \\
\bar{e}_2' &= c_2 \frac{c_1 - c_2}{c_1 + c_2} \varepsilon^{-u_0 t} \varepsilon^{+s_1 \lambda} \left\{ C_2 \cos q \,(\lambda - t) - D_2 \sin q \,(\lambda - t) \right\};
\end{aligned}\right\} \quad (20)$$

or, in the direction of propagation, that is, reversing the sign of λ:

$$\bar{i}_2'' = - \frac{c_1-c_2}{c_1+c_2}\varepsilon^{-u_0 t}\varepsilon^{-s_1\lambda'}\left\{C_2\cos q\,(\lambda'+t)+D_2\sin q\,(\lambda'+t)\right\}$$

$$\bar{e}_2'' = c_2\frac{c_1-c_2}{c_1+c_2}\varepsilon^{-u_0 t}\varepsilon^{-s_1\lambda'}\left\{C_2\cos q\,(\lambda'+t)+D_2\sin q\,(\lambda'+t)\right\}.$$

$$(21)$$

92. The compound wave, that is, the resultant of waves passing the transition point in both directions, then is

$$i_1^{\,0} = \bar{i}_1 + i_1' + i_1$$
$$e_1^{\,0} = \bar{e}_1 + e_1' + e_1$$
$$i_2^{\,0} = i_2 + \bar{i}_2 + i_2'$$
$$e_2^{\,0} = e_2 + \bar{e}_2 + \bar{e}_2'$$

$$(22)$$

In the neighborhood of the transition point, that is, for values λ which are sufficiently small, so that $\varepsilon^{+s\lambda}$ and $\varepsilon^{-s\lambda}$ can be dropped as being approximately equal to 1, by substituting equations (9) to (11) and (18) to (21) into (22) we have

$$i_1^{\,0} = \varepsilon^{-u_0 t}\ \ [\{A_1\cos q\,(\lambda-t)+B_1\sin q\,(\lambda-t)\}$$
$$-\frac{c_2-c_1}{c_1+c_2}\{A_1\cos q\,(\lambda+t)-B_1\sin q\,(\lambda+t)\}$$
$$-\frac{2c_2}{c_1+c_2}\{C_2\cos q\,(\lambda+t)+D_2\sin q\,(\lambda+t)\}];$$

$$e_1^{\,0} = c_1\varepsilon^{-u_0 t}\ \ [\{A_1\cos q\,(\lambda-t)+B_1\sin q\,(\lambda-t)\}$$
$$+\frac{c_2-c_1}{c_2+c_1}\{A_1\cos q\,(\lambda+t)-B_1\sin q\,(\lambda+t)\}$$
$$+\frac{2c_2}{c_1+c_2}\{C_2\cos q\,(\lambda+t)+D_2\sin q\,(\lambda+t)\}];\quad (23)$$

$$i_2^{\,0} = -\varepsilon^{-u_0 t}\ \ [\{C_2\cos q\,(\lambda+t)+D_2\sin q\,(\lambda+t)\}$$
$$-\frac{c_1-c_2}{c_1+c_2}\{C_2\cos q\,(\lambda-t)-D_2\sin q\,(\lambda-t)\}$$
$$-\frac{2c_1}{c_1+c_2}\{A_1\cos q\,(\lambda-t)+B_1\sin q\,(\lambda-t)\}];$$

$$e_2^0 = c_2\varepsilon^{-u_0 t}\ \left[\{C_2 \cos q\,(\lambda + t) + D_2 \sin q\,(\lambda + t)\}\right.$$

$$+\frac{c_1 - c_2}{c_1 + c_2}\{C_2 \cos q\,(\lambda - t) - D_2 \sin q\,(\lambda - t)\}$$

$$\left.+\frac{2\,c_1}{c_1 + c_2}\{A_1 \cos q\,(\lambda - t) + B_1 \sin q\,(\lambda - t)\}\right].$$

In these equations the first term is the main wave, the second term its reflected wave, and the third term the wave transmitted from the adjoining section over the transition point.

Expanding and rearranging equations (23) gives

$$i_1^0 = \frac{2\,\varepsilon^{-u_0 t}}{c_1 + c_2}\left[\{(c_1 A_1 - c_2 C_2)\cos q\lambda + c_2\,(B_1 - D_2)\sin q\lambda\}\cos qt\right.$$
$$\left. - \{(c_1 B_1 + c_2 D_2)\cos q\lambda - c_2\,(A_1 + C_2)\sin q\lambda\}\sin qt\right];$$

$$e_1^0 = \frac{2\,c_1}{c_1 + c_2}\varepsilon^{-u_0 t}\left[\{c_2\,(A_1 + C_2)\cos q\lambda + (c_1 B_1 + c_2 D_2)\sin q\lambda\}\cos qt\right.$$
$$\left. - \{c_2\,(B_1 - D_2)\cos q\lambda - (c_1 A_1 - c_2 C_2)\sin q\lambda\}\sin qt\right];$$

$$(24)$$

$$i_2^0 = \frac{2\,\varepsilon^{-u_0 t}}{c_1 + c_2}\left[\{(c_1 A_1 - c_2 C_2)\cos q\lambda + c_1\,(B_1 - D_2)\sin q\lambda\}\cos qt\right.$$
$$\left. - \{(c_1 B_1 + c_2 D_2)\cos q\lambda - c_1\,(A_1 + C_2)\sin q\lambda\}\sin qt\right];$$

$$e_2^0 = \frac{2\,c_2}{c_1 + c_2}\varepsilon^{-u_0 t}\left[\{c_1(A_1 + C_1)\cos q\lambda + (c_1 B_1 + c_2 D_2)\sin q\lambda\}\cos qt\right.$$
$$\left. - \{c_1(B_1 - D_2)\cos q\lambda - (c_1 A_1 - c_2 C_2)\sin q\lambda\}\sin qt\right]$$

93. This gives the distance phase angle of the waves:

$$\left.\begin{aligned}\tan i_1^0 &= \frac{c_2\{B_1 - D_2)\cos qt + (A_1 + C_2)\sin qt\}}{(c_1 A_1 - c_2 C_2)\cos qt - (c_1 B_1 + c_2 D_2)\sin qt},\\[2mm]\tan i_2^0 &= \frac{c_1\{(B_1 - D_2)\cos qt + (A_1 + C_2)\sin qt\}}{(c_1 A_1 - c_2 C_2)\cos qt - (c_1 B_1 + c_2 D_2)\sin qt};\end{aligned}\right\}\quad(25)$$

hence,

$$\frac{\tan i_2^0}{\tan i_1^0} = \frac{c_1}{c_2} = \sqrt{\frac{L_1 C_2}{L_2 C_1}};\qquad(26)$$

$$\left.\begin{aligned}\tan e_1^0 &= \frac{(c_1 B_1 + c_2 D_2)\cos qt + (c_1 A_1 - c_2 C_2)\sin qt}{c_2\{(A_1 + C_2)\cos qt - (B_1 - D_2)\sin qt\}},\\[2mm]\tan e_2^0 &= \frac{(c_1 B_1 + c_2 D_2)\cos qt + (c_1 A_1 - c_2 C_2)\sin qt}{c_1\{(A_1 + C_2)\cos qt - c_1\,(B_1 - D_2)\sin qt\}};\end{aligned}\right\}\quad(27)$$

hence,

$$\frac{\tan e_2{}^0}{\tan e_1{}^0} = \frac{c_2}{c_1} = \sqrt{\frac{L_2 C_1}{L_1 C_2}} \; ; \tag{28}$$

that is, at a transition point the distance phase angle of the wave changes so that the ratio of the tangent functions of the phase angle is constant, and the ratio of the tangent functions of the phase angle of the voltages is proportional, of the currents inversely proportional to the circuit constants $c = \sqrt{\dfrac{L}{C}}$.

In other words, the transition of an electric wave or impulse from one section of a circuit to another takes place at a constant ratio of the tangent functions of the phase angle, which ratio is a constant of the circuit sections between which the transition occurs.

This law is analogous to the law of refraction in optics, except that in the electric wave it is the ratio of the tangent functions, while in optics it is the ratio of the sine functions, which is constant and a characteristic of the media between which the transition occurs.

Therefore this law may be called the *law of refraction of a wave* at the boundary between two circuits, or at a transition point.

The law of refraction of an electric wave at the boundary between two media, that is, at a transition point between two circuit sections, is given by the constancy of the ratio of the tangent functions of the incoming and refracted wave.

CHAPTER XI.

INDUCTIVE DISCHARGES.

94. The discharge of an inductance into a transmission line may be considered as an illustration of the phenomena in a compound circuit comprising sections of very different constants; that is, a combination of a circuit section of high inductance and small resistance and negligible capacity and conductance, as a generating station, with a circuit of distributed capacity and inductance, as a transmission line. The extreme case of such a discharge would occur if a short circuit at the busbars of a generating station opens while the transmission line is connected to the generating station.

Let r = the total resistance and L = the total inductance of the inductive section of the circuit; also let $g = 0, C = 0$, and L_0 = inductance, C_0 = capacity, r_0 = resistance, g_0 = conductance of the total transmission line connected to the inductive circuit.

In either of the two circuit sections the total length of the section is chosen as unit distance, and, translated to the velocity measure, the length of the transmission line is

$$\lambda_0 = \sigma = \sqrt{L_0 C_0},$$

and the length of the inductive circuit is

$$\lambda_1 = \sigma_1 = \sqrt{L_1 C_1} = 0; \qquad (1)$$

that is, the inductive section of zero capacity has zero length when denoted by the velocity measure λ, or is a "massed inductance."

It follows herefrom that throughout the entire inductive section $\lambda = 0$, and current i_1 therefore is constant throughout this section.

Choosing now the transition point between the inductance and the transmission line as zero of distance, $\lambda = 0$, the inductance is massed at point $\lambda = 0$, and the transmission line extends from $\lambda = 0$ to $\lambda = \lambda_0$.

Denoting the constants of the inductive section by index 1, those of the transmission line by index 2, the equations of the two circuit sections, from (33) of Chapter VIII, are

$$
\left.\begin{aligned}
i_1 &= \varepsilon^{-u_0 t}\left\{(A_1 - C_1)\cos qt - (B_1 + D_1)\sin qt\right\}, \\
e_1 &= c_1\,\varepsilon^{-u_0 t}\left\{(A_1 + C_1)\cos qt - (B_1 - D_1)\sin qt\right\};
\end{aligned}\right\} \quad (2)
$$

$$
\left.\begin{aligned}
i_2 &= \varepsilon^{-u_0 t}\left\{\varepsilon^{+s\lambda}\left[A_2\cos q\,(\lambda - t) + B_2\sin q\,(\lambda - t)\right]\right. \\
&\qquad\left. -\varepsilon^{-s\lambda}\left[C_2\cos q\,(\lambda + t) + D_2\sin q\,(\lambda + t)\right]\right\}, \\
e_2 &= c_2\,\varepsilon^{-u_0 t}\left\{\left[\varepsilon^{+s\lambda}\left[A_2\cos q\,(\lambda - t) + B_2\sin q\,(\lambda - t)\right]\right.\right. \\
&\qquad\left.\left. +\varepsilon^{-s\lambda}\left[C_2\cos q\,(\lambda + t) + D_2\sin q\,(\lambda + t)\right]\right\},\right.
\end{aligned}\right\} \quad (3)
$$

and the constants of the second section are related on those of the first section by the equations (28) to (30) of Chapter VIII:

$$
\left.\begin{aligned}
A_2 &= a_1 A_1 + b_1 C_1, & C_2 &= a_1 C_1 + b_1 A_1, \\
B_2 &= a_1 B_1 - b_1 D_1, & D_2 &= a_1 D_1 - b_1 B_1,
\end{aligned}\right\} \quad (4)
$$

where

$$
\left.\begin{aligned}
a_1 &= \frac{c_1 + c_2}{2\,c_2}, \\[2mm]
b_1 &= \frac{c_1 - c_2}{2\,c_2},
\end{aligned}\right\} \quad (5)
$$

and

$$
c = \sqrt{\frac{L}{C}}. \quad (6)
$$

95. In the inductive section having the constants L and r, that is, at the point $\lambda = 0$ of the circuit, current i_1 and voltage e_1 must be related by the equation of inductance,

$$
e_1 = ri_1 - L\frac{di_1}{dt}. \quad (7)
$$

Substituting (2) in (7), and expanding, gives

$$
\begin{aligned}
c_1\left\{(A_1 + C_1)\cos qt - (B_1 - D_1)\sin qt\right\} \\
= (r + u_0 L)\left\{(A_1 - C_1)\cos qt - (B_1 + D_1)\sin qt\right\} \\
+ qL\left\{(A_1 - C_1)\sin qt + (B_1 + D_1)\cos qt\right\},
\end{aligned}
$$

and herefrom the identities

$$c_1 (A_1 + C_1) = (r + u_0L) (A_1 - C_1) + qL (B_1 + D_1), \\ c_1 (B_1 - D_1) = (r + u_0L) (B_1 + D_1) - qL (A_1 - C_1). \quad (8)$$

Writing

and

$$A_1 - C_1 = M \\ B_1 + D_1 = N \quad (9)$$

gives

and

$$c_1 (A_1 + C_1) = (r + u_0L) M + qLN \\ c_1 (B_1 - D_1) = (r + u_0L) N - qLM, \quad (10)$$

which substituted in (4) of Chapter X gives

$$A_2 = \frac{1}{2\,c} \{(c + r + u_0L) M + qLN\} \cong \frac{1}{2} (M + pN),$$

$$B_2 = \frac{1}{2\,c} \{(c + r + u_0L) N - qLM\} \cong \frac{1}{2} (N - pM),$$

$$C_2 = \frac{1}{2\,c} \{qLN - (c - r - m_0L) M\} \cong \frac{1}{2} (pN - M),$$

$$D_2 = \frac{1}{2\,c} \{qLM + (c - r - m_0L) N\} \cong \frac{1}{2} (pM + N), \quad (11)$$

where in the second expression terms of secondary order have been dropped.

$$c = \sqrt{\frac{L_0}{C_0}}, \qquad p = \frac{qL}{c}.$$

Then substituting in (2) gives the equations of massed inductance:

$$i_1 = \varepsilon^{-u_0 t} \{M \cos qt - N \sin qt\}$$

$$e_1 = \varepsilon^{-u_0 t} \{[(r + u_0L) M + qLN] \cos qt - [(r + u_0L) N - qLM] \sin qt\}. \quad (12)$$

If at $t = 0$, $e_1 = 0$, that is, if at the beginning of the transient discharge the voltage at the inductance is zero, as for instance the inductance had been short-circuited, then, substituting in

(12), and denoting by i_0 the current at the moment $t = 0$, or at the moment of start, we have

$$t = 0, \; i_1 = i_0, \; e_1 = 0; \text{ hence,}$$

$$\left. \begin{array}{l} M = i_0, \\[2mm] N = - \dfrac{r + u_0 L}{qL} \, i_0, \end{array} \right\} \tag{13}$$

and

$$\left. \begin{array}{l} i_1 = i_0 \varepsilon^{-u_0 t} \left\{ \cos qt + \dfrac{r + u_0 L}{qL} \sin qt \right\}, \\[3mm] e_1 = \dfrac{(qL)^2 + (r + u_0 L)^2}{qL} \, i_0 \varepsilon^{-u_0 t} \sin qt. \end{array} \right\} \tag{14}$$

In this case

$$\left. \begin{array}{l} A_2 = \dfrac{i_0}{2} ; \\[3mm] B_2 = - \dfrac{i_0}{2 \, cqL} \left\{ (qL)^2 + (r + u_0 L)^2 + c \, (r + u_0 L) \right\} ; \\[3mm] C_2 = - \dfrac{i_0}{2} ; \\[3mm] D_2 = \dfrac{i_0}{2 \, cqL} \left\{ (qL)^2 + (r + u_0 L)^2 - c \, (r + u_0 L) \right\}. \end{array} \right\} \tag{15}$$

96. In the case that the transmission line is open at its end, at point $\lambda = \lambda_0$,

$$\lambda = \lambda_0,$$
$$i_2 = 0;$$

hence, this substituted in (3), expanded and rearranged as function of $\cos qt$ and $\sin qt$, gives the two identities

$$\left. \begin{array}{l} \varepsilon^{+s\lambda_0} (A_2 \cos q\lambda_0 + B_2 \sin q\lambda_0) = \varepsilon^{-s\lambda_0} (C_2 \cos q\lambda_0 + D_2 \sin q\lambda_0) \\ \text{and} \\ \varepsilon^{+s\lambda_0} (A_2 \sin q\lambda_0 - B_2 \cos q\lambda_0) = -\varepsilon^{-s\lambda_0} (C_2 \sin q\lambda_0 - D_2 \sin q\lambda_0). \end{array} \right\} \tag{16}$$

Squared and added these two equations (16) give

$$\varepsilon^{+2 s\lambda_0} (A_2^2 + B_2^2) = \varepsilon^{-2 s\lambda_0} (C_2^2 + D_2^2). \tag{17}$$

Divided by each other and expanded equations (16) give

$$(A_2 C_1 - B_2 D_2) \sin 2 \, q\lambda_0 = (A_2 D_2 + B_2 B_2) \cos 2 \, q\lambda_0. \quad (18)$$

Substituting (381) into equations (17) and (18) gives

$$\epsilon^{+2\,s\lambda_0}\left\{(qL)^2 + (c + r + u_0 L)^2\right\} = \epsilon^{-2\,s\lambda_0}\left\{(qL)^2 + (c - r - u_0 L)^2\right\} \quad (19)$$

$$\left\{(qL)^2 + (r + u_0 L)^2 - c^2\right\} \sin 2 \, q\lambda_0 = 2 \, cqL \cos 2 \, q\lambda_0. \quad (20)$$

Since $2 \, s\lambda_0$ is a small quantity, in equation (19) we can substitute

$$\epsilon^{\pm 2\,s\lambda_0} = 1 \pm 2 \, s\lambda_0;$$

hence, rearranging (19) and substituting

$$s = u_0 - u$$

gives

$$c \, (r + u_0 L) - (u - u_0) \, \lambda_0 \left\{(qL)^2 + (r + u_0 L)^2 + c^2\right\} = 0. \quad (21)$$

Since $(r + u_0 L)$ is a small quantity compared with $c^2 \, (qL)^2$, it can be neglected, and equation (20) and (21) assume the form

$$\left\{(qL)^2 - c^2\right\} \sin 2 \, q\lambda_0 = 2 \, cqL \cos 2 \, q\lambda_0 \quad (22)$$

$$c \, (r + u_0 L) - (u - u_0) \, \lambda_0 \left\{(qL)^2 + c^2\right\} = 0, \quad (23)$$

and, transformed, equation (22) assumes the form

or

or

$$\left.\begin{aligned}
\tan 2 \, q\lambda_0 &= \frac{2 \, cqL}{(qL)^2 - c^2} \, ; \\[2mm]
q &= -\frac{c}{L} \, \tan q\lambda_0, \\[2mm]
q &= +\frac{c}{L} \, \cot q\lambda_0,
\end{aligned}\right\} \quad (24)$$

hence $\tan 2 \, q\lambda_0$ is *positive* if $qL > c$, as is usually the case.

Expanded for u_0, equation (23) assumes the form

or

$$\left.\begin{aligned}
u_0 &= \frac{u\lambda_0 \left\{(qL)^2 + c^2\right\} - cr}{\lambda_0 \left\{(qL)^2 + c^2\right\} + cL}, \\[3mm]
1 - \frac{u_0}{u} &= \frac{c \left(L + \dfrac{r}{w}\right)}{cL + \lambda_0 \left\{(qL)^2 + c^2\right\}} \, .
\end{aligned}\right\} \quad (25)$$

$$s = - \, (u - u_0).$$

From equations (24) q is calculated by approximation, and then from (25) u_0 and s.

As seen, in all these expressions of q, u_0, s, etc., the integration constants M and N eliminate; that is, the frequency, time attenuation constant, power transfer, etc., depend on the circuit constants only, but not on the distribution of current and voltage in the circuit.

97. At any point λ of the circuit, the voltage is given by equation (3), which, transposed, gives

$$
\begin{aligned}
e = c\varepsilon^{-u_0 t}\big\{\varepsilon^{+s\lambda}&[(A_2 \cos q\lambda + B_2 \sin q\lambda) \cos qt \\
&+ (A_2 \sin q\lambda - B_2 \cos q\lambda) \sin qt] \\
+ \varepsilon^{-s\lambda}&[(C_2 \cos q\lambda + D_2 \sin q\lambda) \cos qt \\
&- (C_2 \sin q\lambda - D_2 \cos q\lambda) \sin qt]\big\},
\end{aligned}
$$

or approximately,

$$
\begin{aligned}
e = c\varepsilon^{-u_0 t}\big\{&[(A_2 + C_2) \cos q\lambda + (B_2 + D_2) \sin q\lambda] \cos qt \\
&+ [(A_2 - C_2) \sin q\lambda - (B_2 - D_2) \cos q\lambda] \sin qt\big\}.
\end{aligned}
$$

Similarly to equation (381),

$$
\left.
\begin{aligned}
A_2 + C_2 &= pN; \\
A_2 - C_2 &= M; \\
B_2 + D_2 &= N; \\
B_2 - D_2 &= -pM,
\end{aligned}
\right\} \tag{26}
$$

where

$$
p = \frac{qL}{c},
$$

then

$$
\left.
\begin{aligned}
e_2 &= \varepsilon^{-u_0 t}(qL \cos q\lambda + c \sin q\lambda)(N \cos qt + M \sin qt), \\
i_2 &= \varepsilon^{-u_0 t}\left(\cos q\lambda - \frac{qL}{c} \sin q\lambda\right)(M \cos qt - N \sin qt);
\end{aligned}
\right\} \tag{27}
$$

$$
\left.
\begin{aligned}
e_1 &= qL\varepsilon^{-u_0 t}(N \cos qt + M \sin qt), \\
i_1 &= \varepsilon^{-u_0 t}(M \cos qt - N \sin qt).
\end{aligned}
\right\} \tag{28}
$$

If $\qquad e_1 = 0$ for $t = 0$,

$\qquad\qquad N = 0;$

hence,

$$
\left.
\begin{aligned}
i_1 &= i_0 \varepsilon^{-u_0 t} \cos qt, \\
e_1 &= qL i_0 \varepsilon^{-u_0 t} \sin qt; \\
i_2 &= i_0 \varepsilon^{-u_0 t} \left(\cos q\lambda - \frac{qL}{c} \sin q\lambda\right) \cos qt, \\
e_2 &= i_0 \varepsilon^{-u_0 t} \left(qL \cos q\lambda + c \sin q\lambda\right) \sin qt.
\end{aligned}
\right\}
\tag{29}
$$

Writing

$$
\sqrt{\tfrac{1}{2}\left(M^2 + N^2\right)} = I_0,
\tag{30}
$$

the effective values of the quantities are

$$
\left.
\begin{aligned}
I_1 &= I_0 \varepsilon^{-u_0 t}, \\
E_1 &= qL I_0 \varepsilon^{-u_0 t}; \\
I_2 &= I_0 \varepsilon^{-u_0 t} \left(\cos q\lambda - \frac{qL}{c} \sin q\lambda\right), \\
E_2 &= I_0 \varepsilon^{-u_0 t} \left(qL \cos q\lambda + c \sin q\lambda\right).
\end{aligned}
\right\}
\tag{31}
$$

Herefrom it follows that

$\qquad I_2 = 0$ for $\lambda = \lambda_0$ by the equation

$$
\cos q\lambda_0 - q\frac{L}{c} \sin q\lambda_0 = 0,
$$

or

$$
q = \frac{c}{L} \cot q\lambda_0,
\tag{32}
$$

while

$$
q = -\frac{c}{L} \tan q\lambda_0
\tag{33}
$$

gives

$$
qL \cos q\lambda_0 + c \sin q\lambda_0 = 0;
$$

that is,

$$
E_2 = 0 \text{ at } \lambda = \lambda_0.
$$

At the open end of the line $\lambda = \lambda_0$ the voltage E_2 by substituting (32) into (31) is

$$E_2{}^0 = \frac{qL}{\cos q\lambda_0}\, I_0\varepsilon^{-u_0 t}. \tag{34}$$

At the grounded end of the line $\lambda = \lambda_0$ the current I_2, by substituting (403) into (401), is

$$I_2{}^0 = \frac{I_0\varepsilon^{-u_0 t}}{\cos q\lambda_0}. \tag{(35)}$$

An inductance discharging into the transmission line thus gives an oscillatory distribution of voltage and current along the line.

98. As example may be considered the three-phase high-potential circuit, comprising a generating system of $r = 2$ ohms and $L = 0.5$ henry per phase and connected to a long-distance transmission line of $r_0 = 0.4$ ohm, $L_0 = 0.002$ henry, $g_0 = 0.2 \times 10^{-6}$ mho, $C_0 = 0.016 \times 10^{-6}$ farad per mile of conductor or phase, and of $l_0 = 80$ miles length.

$$c = \sqrt{\frac{L_0}{C_0}} = 354, \qquad c^2 = 125{,}300;$$

$$\sigma_0 = \sqrt{L_0 C_0} = 5.66 \times 10^{-6};$$

$$\lambda_0 = l_0\sigma_0 = 0.453 \times 10^{-3};$$

$$u = \frac{1}{2}\left(\frac{r_0}{L_0} + \frac{g_0}{C_0}\right) = 106;$$

$$\frac{c}{L} = 708,$$

and herefrom, substituting in equations (34) and (35),

$$q = -\,708 \tan (0.0259\, q)° \quad \text{(zero voltage)}$$
$$= +\,708 \cot (0.0259\, q)° \quad \text{(zero current)},$$

$$1 - \frac{u_0}{u} = \frac{1}{0.618\, q^2\, 10^{-6} + 1.28}$$

$q\lambda_0 =$	100.35°	185.64°	273.83°	362.89°	452.32°	541.94°
$q =$	3875	7168	10,572	14,010	17,463	20,920
$1 - \dfrac{u_0}{u} =$	0.0946	0.0302	0.0142	0.00816	0.0047	0.0037
$u_0 =$	95.8	102.8	104.5	105.1	105.5	105.6
$-s =$	10.2	3.2	1.5	0.87	0.5	0.4

By equation (31) the effective values of the first six harmonics are given as

(1) Quarter-wave: 100.35°.

$$q_1 = 3875;$$
$$u_0 = 95.8;$$
$$I = i_0 \varepsilon^{-u_0 t} (\cos q\lambda - 5.48 \sin q\lambda),$$
$$E = 1939 \, i_0 \varepsilon^{-u_0 t} (\cos q\lambda + 0.182 \sin q\lambda).$$

(2) Half-wave: 185.64°.

$$q_2 = 7168;$$
$$u_0 = 102.8;$$
$$I = i_0 \varepsilon^{-u_0 t} (\cos q\lambda - 10.14 \sin q\lambda);$$
$$E = 3585 \, i_0 \varepsilon^{-u_0 t} (\cos q\lambda + 0.098 \sin q\lambda).$$

(3) Three-quarter wave: 273.83°.

$$q_3 = 10,572;$$
$$u_0 = 104.5;$$
$$I = i_0 \varepsilon^{-u_0 t} (\cos q\lambda - 14.90 \sin q\lambda);$$
$$E = 5287 \, i_0 \varepsilon^{-u_0 t} (\cos q\lambda + 0.067 \sin q\lambda).$$

(4) Full wave: 362.89°.

$$q_4 = 14,010;$$
$$u_0 = 105.1;$$
$$I = i_0 \varepsilon^{-u_0 t} (\cos q\lambda - 19.8 \sin q\lambda);$$
$$E = 7005 \, i_0 \varepsilon^{-u_0 t} (\cos q\lambda + 0.050 \sin q\lambda).$$

(5) Five-quarter wave: 452.32°.

$$q_5 = 17,463;$$
$$u_0 = 105.5;$$
$$I = i_0 \varepsilon^{-u_0 t} (\cos q\lambda - 24.65 \sin q\lambda);$$
$$E = 8732 \, i_0 \varepsilon^{-u_0 t} (\cos q\lambda + 0.040 \sin q\lambda).$$

(6) Three-half wave: $541.94°$.

$$q_6 = 20,920;$$

$$u_0 = 105.6;$$

$$I = i_0 \varepsilon^{-u_0 t} (\cos q\lambda - 29.6 \sin q\lambda);$$

$$E = 10,460 \, i_0 \varepsilon^{-u_0 t} (\cos q\lambda + 0.033 \sin q\lambda).$$

SECTION V

VARIATION OF CIRCUIT CONSTANTS

CHAPTER I.

1. In the preceding investigations on transients, the usual assumption is made, that the circuit constants: resistance r, inductance L, capacity C and shunted conductance g, are constant. While this is true, with sufficient approximation, for the usual machine frequencies and for moderately high frequencies, experience shows that it is not even approximately true for very high frequencies and for very sudden circuit changes, as steep wave front impulses, etc.

If r, L, C and g are assumed as constant, it follows that the attenuation is independent of the frequency, that is, waves of all frequencies decay at the same rate, and as the result, a complex wave or an impulse traversing a circuit dies out without changing its wave shape or the steepness of its wave front.

Experience, however, shows that steep wave fronts are dangerous only near their origin, and rapidly lose their destructiveness by the flattening of the wave front when running along the circuit. Experimentally, small inductances shunted by a spark gap, inserted in transmission lines for testing for high frequencies or steep wave fronts, show appreciable spark lengths, that is, high voltage gradients, only near the origin of the disturbance.

The rectangular wave of starting a transmission line by connecting it to a source of voltage, which is given by the theory under the assumption of constant r, L, C and g, is not shown by oscillograms of transmission lines.

If r and L are constant, the power factor of the line conductor, $\dfrac{r}{\sqrt{r^2 + (2\pi fL)^2}}$, should with increasing frequency continuously decrease, and reach extremely low values, at very high frequencies, so that at these, an oscillatory disturbance should be sustained over very many cycles, and show with increasing frequency an increasing liability to become a sustained or cumulative oscillation.   Experience, however, shows that high frequency oscillations die out much more rapidly than accounted for by the standard theory, and show at very high frequency practically no tendency to become cumulative.

Therefore, when dealing with transients containing very high frequencies or steep wave fronts, the previous theory, which is based on the assumption of constant r, L, C and g, correctly represents the transient only in its initial stage and near its origin, but less so its course after its initial stage and at some distance from the origin, especially with high frequency transients or steep wave fronts.

It therefore, is of importance to investigate the factors which cause a change of the line constants r, L, C and g, with increasing frequency or steepness of wave front, and the effect produced on the course of the transient as regard to duration and wave shape, by the variation of the line constants.

The two most important factors in the variation of the circuit constants r, L, C and g seem to be the *unequal current distribution in the conductor* and the *finite velocity of the electric field.*

UNEQUAL CURRENT DISTRIBUTION IN THE CONDUCTOR.

2. The magnetic field of the current surrounds this current and fills all the space outside thereof, up to the return current. Some of the magnetic field due to the current in the interior and in the center of a conductor carrying current thus is inside of the conductor, while all the magnetic field of the current in the outer layer of the conductor is outside of it.   Therefore, more magnetic field surrounds the current in the interior of the conductor than the current in its outer layer, and the inductance therefore increases from the outer layer of the conductor toward its interior, by the "internal magnetic field."   In the interior of the conductor, the reactance voltage thus is higher than on the outside.

At low frequency, with moderate size of conductor, this difference is inappreciable in its effect.   At higher frequencies, however, the higher reactance in the interior of the conductor, due to this internal magnetic field, causes the current density to decrease toward the interior of the conductor, and the current to lag, until finally the current flows practically only through a thin layer of the conductor surface.

As the result thereof, the effective resistance of the conductor is increased, due to the uneconomical use of the conductor material caused by the lower current density in the interior, and due to the phase displacement, which results in the sum of the currents in the successive layers of the conductor being larger than the resultant current. Due to this unequal current distribution, the internal reactance of the conductor is decreased, as less current penetrates to the interior of the conductor, and thus produces less magnetic field inside of the conductor.

The derivation of the equations of the effective resistance of unequal current distribution in the conductor, r_1, and of the internal reactance x_1 under these conditions is give in Section III, Chapter VII. It is interesting to note that effective resistance and internal reactance, with increasing frequency, approach the same limit, and become proportional to the square root of the frequency, the square root of the permeability, and the square root of the resistivity of the conductor material, while at low frequencies the resistance is independent of the frequency and directly proportional to the resistivity, and the internal reactance is independent of the resistivity and directly proportional to the frequency.

FINITE VELOCITY OF THE ELECTRIC FIELD.

3. The derivation of the equations of the effective resistance of magnetic radiation, and in general of the effects of the finite velocity of the electric field on the line constants, are given in Section III, Chapter VIII.

The magnetic radiation resistance is proportional to the square of the frequency (except at extremely high frequencies). It therefore is negligible at low and medium frequencies, but becomes the dominating factor at high frequencies. It is proportional to the distance of the return conductor, but entirely independent of size, shape, or material of the conductor, as is to be expected, since it represents the energy dissipated into space. Only at extremely high frequencies the rise of radiation resistance becomes less than proportional to the square of the frequency. It becomes practically independent of the distance of the return conductor, when the latter becomes of the magnitude of the quarter wave length.

The same applies to the capacity. Due to the finite velocity of propagation, the dielectric or electrostatic field lags behind the voltage which produces it, by the same angle by which the magnetic field lags behind the current, and the capacity current or charging current thus is not in quadrature with the voltage, or reactive, but displaced in phase by more than 90°, thus contains a—negative—energy component, which gives rise to a shunted conductance of dielectric radiation g. This gives rise to an energy dissipation by the conductor, at high frequencies, by dielectric radiation into space, of the same magnitude as the energy dissipation by magnetic radiation, above considered.

The term "shunted conductance" g has been introduced into the general equations of the electric circuit largely from theoretical reasons, as representing the power consumption proportional to the voltage. Most theoretical investigations of transmission circuits consider only r, L and C as the circuit constants, and omit g, since under average transmission line conditions, at low and moderate frequencies, g usually is negligible. In communication circuits, as telegraph and telephone, there is a "leakage," which would be represented by a shunted conductance, and in underground cables there is a considerable energy consumption by dielectric losses in the insulation, as the investigations of the last years have shown, which gives a shunted conductance. In overhead power lines, however, energy losses depending on the voltage—and leading to a term g—have been known only at such high voltages where corona appears.

It is interesting, therefore, to note that at high frequencies "shunted conductance" g may reach very formidable values even in transmission lines, due to electrostatic radiation.

In investigating the effect of the finite velocity of the electric field on the inductance L and the capacity C, it is seen that the equations of external inductance and of capacity are not affected, but remain the same as the usual values derived by neglecting the velocity of the electric field, except at extremely high frequencies, when the distance of the return conductor approaches quarter wave length.

EQUATIONS OF ELECTRICAL CONSTANTS, AND NUMERICAL VALUES.

4. In the following are given, compiled from Section III, Chapters VII to IX, the equations of the components of the electrical constants, as functions of the frequency, for conductors with return conductor, and also for conductors without return conductor (as approximated by lightning strokes or wireless antennæ):

Resistance: True ohmic resistance or effective resistance of unequal current distribution, and magnetic radiation resistance.

Reactance: Low frequency internal reactance or internal reactance of unequal current distribution, and external reactance.

Inductance: Low frequency internal inductance or internal inductance of unequal current distribution, and external inductance.

Shunted conductance and capacity are not so satisfactorily represented, and therefore, instead of representing energy storage and power dissipation depending on the voltage by a conductance g and a capacity C or susceptance b, in shunt with each other, it is more convenient to represent them by an effective resistance, the dielectric radiation resistance r_c, and a capacity reactance x_c, in series with each other.

Equations of Electrical Constants.

Let l = length of conductor, cm.

l_r = radius of conductor, cm.

l_1 = circumference of conductor, cm.

l_2 = shortest circumference of conductor, cm.

l' = distance of return conductor

f = frequency, cycles per second

λ = electrical conductivity, mhos per cm.[3]

μ = magnetic permeability

$S = 3 \times 10^{10}$ = velocity of radiation in empty space; it is, then, log denoting the natural logarithm.

Resistances:

True ohmic resistance (thermal):

$$r_0 = \frac{l}{\lambda \pi l_r^2} \text{ ohms.}$$

Effective resistance of unequal current distribution (thermal):

$$r_1 = \frac{\pi l}{l_1} \sqrt{\frac{0.4\ \mu f}{\lambda}}\ 10^{-4}\ \text{ohms.}$$

Effective magnetic radiation resistance:

(a) Return conductor at distance l':

$$r_3 = \frac{8\ \pi^2 f^2 l' l}{S}\ 10^{-9}\ \text{ohms;}$$

at extremely high frequencies:

$$r_4 = 4\ \pi f l\ \left\{ \frac{\pi}{2} - \text{col}\ \frac{2\ \pi f l'}{S} \right\}\ 10^{-9}\ \text{ohms.}$$

(b) Conductor without return conductor:

$$r_2 = 2\ \pi^2 f l\ 10^{-9}\ \text{ohms.}$$

Effective dielectric radiation (shunted) resistance:*

(a) Return conductor at distance l':

$$r_c = \frac{2\ l'S}{l}\ 10^{-9}\ \text{ohms,}$$

at extremely high frequencies:

$$r_c = \frac{S^2}{\pi f l}\ \left\{ \frac{\pi}{2} - \text{col}\ \frac{2\ \pi f l'}{S} \right\}\ 10^{-9}\ \text{ohms.}$$

(b) Conductor without return conductor:

$$r_c = \frac{S}{2\ fl}\ 10^{-9}\ \text{ohms.}$$

Reactances:

Low frequency internal reactance:

$$x_{10} = \pi f l \mu\ 10^{-9}\ \text{ohms.}$$

Internal reactance of unequal current distribution:

$$x_1 = \frac{\pi l}{l_1} \sqrt{\frac{0.4\ \mu f}{\lambda}}\ 10^{-4}\ \text{ohms} = r_1.$$

*As shunted resistance and reactance, r_c and x_c are inverse proportional to the length of the conductor l, that is, the longer the conductor, the more current is shunted across, and the lower therefore are r_c and x_c. For this reason, the shunt constants usually are given as conductance g and susceptance b. In the present case, however, r and x give simpler expressions.

External reactance:

(a) Return conductor at distance l':

$$x_0 = 4\,\pi fl \log\frac{l'}{l_r}\,10^{-9} = 4\,\pi fl \log\frac{2\,\pi l'}{l_2}\,10^{-9}\text{ ohms};$$

at extremely high frequencies:

$$x_4 = 4\,\pi fl\left\{\log\frac{S}{2\,\pi fl_r} - 0.5772 - \operatorname{sil}\frac{2\,\pi fl'}{S}\right\}10^{-9}\text{ ohms.}$$

(b) Conductor without return conductor:

$$x_2 = 4\,\pi fl\left\{\log\frac{S}{2\,\pi fl_r} - 0.5772\right\}10^{-9}\text{ ohms.}$$

Shunted capacity reactance:

(a) Return conductor at distance l':

$$x_c = \frac{S^2\log\dfrac{l'}{l_r}}{\pi fl}\,10^{-9} = \frac{S^2\log\dfrac{2\,\pi l'}{l_2}}{\pi fl}\,10^{-9}\text{ ohms};$$

at extremely high frequencies:

$$x_c = \frac{S^2}{\pi fl}\left\{\log\frac{S}{2\,\pi fl_r} - 0.5772 - \operatorname{sil}\frac{2\,\pi fl'}{S}\right\}10^{-9}\text{ ohms.}$$

(b) Conductor without return conductor:

$$x_c = \frac{S^2}{\pi fl}\left\{\log\frac{S}{2\,\pi fl_r} - 0.5772\right\}10^{-9}\text{ ohms.}$$

Inductances:

Low frequency internal inductance:

$$L_{10} = \frac{l\mu}{2}\,10^{-9}\text{ henrys.}$$

Internal inductance of unequal current distribution:

$$L_1 = \frac{l}{l_1}\sqrt{\frac{0.1\,\mu}{\lambda f}}\,10^{-4}\text{ henrys.}$$

External inductance:

(a) Return conductor at distance l':

$$L_0 = 2\,l\log\frac{l'}{l_r}\,10^{-9} = 2\,l\log\frac{2\,\pi l'}{l_2}\,10^{-9}\text{ henrys};$$

at extremely high frequencies:

$$L_4 = 2\,l\left\{\log\frac{S}{2\,\pi fl_r} - 0.5772 - \operatorname{sil}\frac{2\,\pi fl'}{S}\right\}10^{-9}\text{ henrys.}$$

(b) Conductor without return conductor:

$$L_2 = 2\, l \left\{ \log \frac{S}{2\ \pi f l_r} - 0.5772 \right\} 10^{-9} \text{ henrys.}$$

Capacity:

(a) Return conductor at distance l':

$$C_0 = \frac{10^9}{2\ S^2 \log \dfrac{l'}{l_r}} \text{ farads;}$$

at extremely high frequencies:

$$C_4 = \frac{10^9}{2\ S^2 \left\{ \log \dfrac{S}{2\ \pi f l_r} - 0.5772 - \text{sil} \dfrac{2\ \pi f l'}{S} \right\}} \text{ farads:}$$

(b) Conductor without return conductor:

$$C_2 = \frac{10^9}{2\ S^2 \left\{ \log \dfrac{S}{2\ \pi f l_r} - 0.5772 \right\}} \text{ farads.}$$

5. Herefrom then follow the Circuit Constants:

At Low Frequencies (Machine Frequencies up to 10^3 cycles, approx.):

$$r = r_0$$
$$x = x_1 + x_0$$
$$L = L_{10} + L_0$$
$$C = C_0$$
$$g = 0$$

At Medium Frequencies (10^3 to 10^5 cycles, approx.):

$$r = r_1$$
$$x = x_{10} + x_0$$
$$L = L_1 + L_0$$
$$C = C_0$$
$$g = 0$$

At High Frequencies (10^5 to 10^7 cycles, approx.):

$$r = r_1 + r_3 \quad \text{(with return conductor)}$$
$$= r_1 + r_2 \quad \text{(without return conductor)}$$
$$x = x_1 + x_0 \ \text{(with return conductor)}$$
$$= x_1 + x_2 \ \text{(without return conductor)}$$
$$L = L_1 + L_0 \ \text{(with return conductor)}$$
$$= L_1 + L_2 \ \text{(without return conductor)}$$
$$C = C_0 \qquad \text{(with return conductor)}$$
$$= C_2 \qquad \text{(without return conductor)}$$

(approximately, or represented by x_c and r_c).

g represented by r_c and x_c:

$$g = \frac{r_c}{r_c{}^2 + x_c{}^2}; \; b = \frac{x_c}{r_c{}^2 + x_c{}^2} = 2\pi f C;$$

at extremely high frequencies (above 10^7 cycles, approx.):

$$r = r_1 + r_4 \text{ (with return conductor)}$$
$$= r_1 + r_2 \text{ (without return conductor)}$$
$$x = x_1 + x_4 \text{ (with return conductor)}$$
$$= x_1 + x_2 \text{ (without return conductor)}$$

C and g represented by r_c and x_c, thus:

$$g = \frac{r_c}{r_c{}^2 + x_c{}^2}.$$

$$C = \frac{x_c}{2\pi f(r_c{}^2 + x_c{}^2)}.$$

From these follow the derived circuit constants:
Magnetic attenuation:

$$u_1 = \frac{r}{2L}$$

Dielectric attenuation:

Usually zero at low and medium frequencies,

$$u_2 = \frac{g}{2C} = \frac{\pi f r_c}{x_c}$$

at high and very high frequencies.

Attenuation constant:

$$u = u_1 + u_2 = \tfrac{1}{2}\left(\frac{r}{L} + \frac{g}{C}\right)$$

Series power factor:

$$\cos \omega = \frac{r}{\sqrt{r^2 + x^2}}.$$

Shunt power factor:

Zero at low and medium frequencies,

$$\cos \omega_c = \frac{r_c}{\sqrt{r_c{}^2 + x_c{}^2}}$$

at high and very high frequencies.

Duration of oscillation:*

$$t_0 = \frac{1}{u} \text{ seconds}$$

$$N_0 = \frac{f}{u} \text{ cycles.}$$

6. As seen, four successive stages may be distinguished in the expressions of the circuit constants as functions of the frequency.

1. Low frequencies, such as the machine frequencies of 25 and 60 cycles. The resistance is the true ohmic resistance, the internal reactance and inductance that corresponding to uniform current density throughout the conductor, with conductors of moderate size, and of non-magnetic material.

2. Medium frequencies, of the magnitude of a thousand to ten thousand cycles. Resistance and internal reactance or inductance are those due to unequal current distribution in the conductor, that is, the resistance is rapidly increasing, and the internal inductance decreasing. The conductance g is still negligible, radiation effects still absent, and all the energy loss that of thermal resistance.

3. High frequencies, of the magnitude of one hundred thousand to one million cycles. The radiation resistance is appreciable and becomes the dominating factor in the energy dissipation. The internal inductance has practically disappeared, due to the current penetrating only a thin surface layer. A considerable shunted conductance exists due to the dielectric radiation.

4. Extremely high frequencies, of the magnitude of many millions of cycles, when the quarter wave length has become of the same magnitude or less than the distance of the return conductor. Radiation effects entirely dominate, and the usual expressions of inductance and of capacity have ceased to apply.

This last case is of little industrial importance, as such extremely high frequencies propagate only over short distances.

---

* Under "Duration" of a transient is understood the time (or the number of cycles), which the transient would last, that is, the time in which it would expend its energy, if continuing at its initial intensity. With a simple exponential transient, this is the time during which it decreases to $\frac{1}{\epsilon}$ or 36.8 per cent. of its initial value. It decreases to one-tenth of its initial value in 2.3 times this time.

It would come into consideration only in calculating the flattening of the wave front of a rectangular impulse in the immediate neighborhood of its origin, and similar problems.

Thus far, a general investigation does not seem feasible. Substituting the equations of the circuit constants, as functions of the frequency, into the general equations of the electric circuit, leads to expressions too complex for general utility, and the investigation thus must largely be made by numerical calculations.

Only when the frequencies which are of importance in the problem lie fairly well in one of the four ranges above discussed —as is the case in the investigation of the flattening of a steep wave front in moderate distances from its origin—a more general theoretical investigation becomes possible at present.

CHAPTER II.

7. From the equations given in Chapter I, numerical values of the line constants are calculated and given in Table IV, for average transmission line conditions, that is, a copper wire No. 00, with 6 ft. = 182 cm. between the conductors, and an average height of 30 ft. = 910 cm. above ground, for the two conditions:

(a) A high frequency oscillation between two line conductors.

(b) A high frequency oscillation between one line conductor and the ground.

The table gives:

The thermal resistance r_1, the radiation resistance r_3, and the total resistance $r = r_1 + r_3$.

The internal reactance x_1, the external reactance x_3, and the total reactance $x = x_1 + x_3$.

The magnetic attenuation $u_1 = \dfrac{r}{2L}$, the dielectric attenuation $u_2 = \dfrac{g}{2C}$, and the total attenuation $u = u_1 + u_2$.

The table also gives the duration of a transient in micro-seconds t and in cycles N, that is, the time which a high frequency oscillation of the frequency f would last, if continuing with its initial intensity and the number of cycles which it would perform. It also gives the power factor, in per cent, of the series circuit, as determined by resistance and inductance, and of the shunt circuit, as determined by shunted conductance and capacity.

As seen, the attenuation constant u is constant up to nearly one thousand cycles. Thus in this range, all the frequencies die out at the same rate. From about one thousand cycles up to about 100,000 cycles, the attenuation constant gradually increases, and thus oscillations die out the more rapidly, the higher the frequency, as seen by the gradual decrease of the duration t. However, as the increase of the attenuation constant and thus the increase of the rapidity of the decay of the disturbance, in this range, is smaller than the increase of frequency, the number of cycles performed by the oscillation increases. Thus, at 25 or 60 cycles, the stored energy, which supplies the oscillating power,

626

TABLE IV.—ATTENUATION CONSTANTS OF COPPER WIRE No. 00 B. & S. G.

(Per kilometer of conductor.)

	25	60	500	750	1000	2000	5000	7500	10,000	15,000
$f =$	25	60	500	750	1000	2000	5000	7500	10,000	15,000
$lf =$	1.398	1.778	2.699	2.875	3	3.301	3.699	3.875	4	4.176
$r_1 =$	0.24	0.24	0.24	0.24	0.273	0.386	0.612	0.750	0.865	1.060
$x_1 =$	0.008	0.019	0.157	0.236	0.273	0.386	0.612	0.750	0.865	1.060
$r_3 =$							0.0012	0.0027	0.0048	0.0108
$r =$	0.24	0.24	0.24	0.24	0.273	0.386	0.613	0.753	0.870	1.071
$x_3 =$	0.187	0.45	3.75	5.625	7.5	15.0	37.5	56.25	75.00	112.5
$x =$	0.195	0.469	3.907	5.861	7.773	15.39	38.11	57.0	75.86	113.6
$L =$	1.245	1.245	1.245	1.245	1.24	1.23	1.22	1.215	1.21	1.205
$\mu_1 = \dfrac{r}{2L} =$	96	96	96	96	110	157	251	310	360	445
$\mu_2 = \dfrac{g}{2C} =$			0.005	0.011	0.020	0.080	0.5	1.125	2.0	4.5
$\mu = \mu_1 + \mu_2 =$	96	96	96	96	110	157	252	311	362	450
$\dfrac{x}{\mu} =$	1.982	1.982	1.982	1.982	2.041	2.196	2.401	2.493	2.559	2.653
$N =$	10,400	10,400	10,400	10,400	9,090	6,370	3,970	3,210	2,760	2,220
$\dfrac{t}{N} =$	0.26	0.62	5.2	7.8	9.09	12.74	19.85	24.1	27.6	33.3
$\cos \omega =$			6.16	4.10	3.51	2.52	1.61	1.32	1.15	0.94
$\cos \omega_c =$									0.01	0.01
$r_3 =$							0.012	0.027	0.048	0.108
$r =$	0.24	0.24	0.24	0.24	0.273	0.386	0.624	0.777	0.913	1.168
$x_3 =$	0.26	0.624	5.2	7.8	10.4	20.8	52.0	78.0	104.0	156.0
$x =$	0.268	0.643	5.357	8.036	10.67	21.16	52.61	78.75	104.9	157.1
$L =$	1.705	1.705	1.705	1.705	1.70	1.69	1.68	1.675	1.67	1.665
$\mu_1 = \dfrac{r}{2L} =$	70	70	70	70	80	114	186	232	273	350
$\mu_2 = \dfrac{g}{2C} =$			0.04	0.08	0.14	0.58	3.6	8.1	14.4	32.5
$\mu = \mu_1 + \mu_2 =$	70	70	70	70	80	115	190	240	287	383
$\dfrac{x}{\mu} =$	1.845	1.845	1.845	1.845	1.903	2.061	2.279	2.380	2.458	2.583
$N =$	14,300	14,300	14,300	14,300	12,500	8,700	5,260	4,170	3,480	2,610
$\dfrac{t}{N} =$	0.36	0.86	7.15	10.7	12.5	17.4	26.3	31.3	34.8	39.1
$\cos \omega =$			4.48	2.98	2.56	1.89	1.19	0.98	0.87	0.74
$\cos \omega_c =$						0.01	0.02	0.03	0.05	0.07

(a) Return conductor at 6 ft = 182 cm. distance.

(b) Conductor 30 ft. above ground as return; $l' = 1820$ cm.

TABLE IV.—*Continued.*

	$20{,}000$	$50{,}000$	$75{,}000$	10^5	2×10^5	5×10^5	7.5×10^5	10^6	2×10^6	5×10^6	Cycles
$f =$											
$1/t =$	4.301	4.699	4.875	5	5.301	5.699	5.875	6	6.301	6.699	
$r_1 =$	1.225	1.93	2.37	2.73	3.86	6.12	7.50	8.65	12.25	19.3	Ohms
$x_1 =$	1.225	1.93	2.37	2.73	3.86	6.12	7.50	8.65	12.25	19.3	Ohms
(a) Return conductor at 6 ft. = 182 cm. distance.											
$r_3 =$	0.0192	0.12	0.27	0.48	1.92	12	27	48	192	$1{,}200$	Ohms
$r =$	1.243	2.05	2.64	3.21	5.78	18.1	34.5	56.6	204	$1{,}219$	Ohms
$x_3 =$	150	375	562.5	750	$1{,}500$	$3{,}750$	$5{,}625$	$7{,}500$	$15{,}000$	$37{,}500$	Ohms.
$x =$	151.2	376.9	564.9	752.7	$1{,}504$	$3{,}756$	$5{,}632$	$7{,}509$	$15{,}012$	$37{,}520$	Ohms
$L =$	1.205	1.20	1.20	1.20	1.20	1.195	1.195	1.195	1.195	1.195	Henrys
$\mu_1 = \dfrac{r}{2L} =$	516	855	$1{,}100$	$1{,}340$	$2{,}410$	$7{,}570$	$14{,}450$	$23{,}700$	$85{,}600$	$510{,}000$	
$\mu_2 = \dfrac{\theta}{2C} =$	8.0	50	112.5	200	800	$5{,}000$	$11{,}200$	$20{,}000$	$80{,}000$	$500{,}000$	
$\mu = \mu_1 + \mu_2 =$	524	905	$1{,}212$	$1{,}540$	$3{,}210$	$12{,}570$	$25{,}650$	$43{,}700$	$165{,}600$	$1{,}010{,}000$	
$1/\mu =$	2.719	2.957	3.084	3.187	3.506	4.099	4.409	4.640	5.219	6.004	
$t =$	$1{,}910$	$1{,}100$	825	649	311	79.6	39.0	22.9	6.04	1.00	Micro-seconds
$N =$	38.2	55.0	61.9	64.9	62.2	39.8	29.2	22.9	12.1	5.0	Cycles
$\cos \omega =$	0.82	0.54	0.47	0.43	0.38	0.48	0.61	0.75	1.36	3.25	Per cent.
$\cos \omega_e =$	0.01	0.03	0.05	0.06	0.13	0.32	0.48	0.64	1.27	3.18	Per cent.
(b) Conductor 30 ft. above ground as return; $l = 1820$ cm.											
$r_3 =$	0.192	1.2	2.7	4.8	19.2	120	270	480	$1{,}920$	$12{,}000$	Ohms
$r =$	1.417	3.13	5.07	7.53	23.06	126.1	277.5	488.6	$1{,}932$	$12{,}020$	Ohms
$x_3 =$	208	520	780	$1{,}040$	$2{,}080$	$5{,}200$	$7{,}800$	$10{,}400$	$20{,}800$	$52{,}000$	Ohms
$x =$	209.2	522	782	$1{,}043$	$2{,}084$	$5{,}206$	$7{,}807$	$10{,}409$	$20{,}812$	$52{,}020$	Ohms
$L =$	1.665	1.66	1.66	1.66	1.66	1.655	1.655	1.655	1.655	1.655	Henrys
$\mu_1 = \dfrac{r}{2L} =$	424	942	$1{,}520$	$2{,}270$	$6{,}930$	$38{,}200$	$83{,}800$	$147{,}500$	$583{,}000$	$3{,}630{,}000$	
$\mu_2 = \dfrac{\theta}{2C} =$	57.6	360	810	$1{,}440$	$5{,}760$	$36{,}000$	$81{,}000$	$144{,}000$	$576{,}000$	$3{,}600{,}000$	
$\mu = \mu_1 + \mu_2 =$	482	$1{,}302$	$2{,}330$	$3{,}710$	$12{,}690$	$74{,}200$	$164{,}800$	$291{,}500$	$1{,}159{,}000$	$7{,}230{,}000$	
$1/\mu =$	2.683	3.115	3.367	3.569	4.103	4.870	5.217	5.465	6.064	6.859	
$t =$	$2{,}070$	768	429	269	78.8	13.5	6.07	3.43	0.86	0.14	Micro-seconds
$N =$	41.4	38.4	31.2	26.9	15.8	6.75	4.55	3.43	1.72	0.69	Cycles
$\cos \omega =$	0.68	0.60	0.65	0.71	1.10	2.42	3.55	4.70	9.24	22.5	Per cent.
$\cos \omega_e =$	$.09$	0.23	0.35	0.46	0.92	2.31	3.46	4.62	9.22	22.5	Per cent.

would be expended in less than one cycle, that is, a real oscillation would hardly materialize (except by other sources of energy, as the stored magnetic energy of a transformer connected to the line). At 1000 cycles, the oscillation would already last 9 to 12 cycles, and at still higher frequencies reach a maximum of 41.4

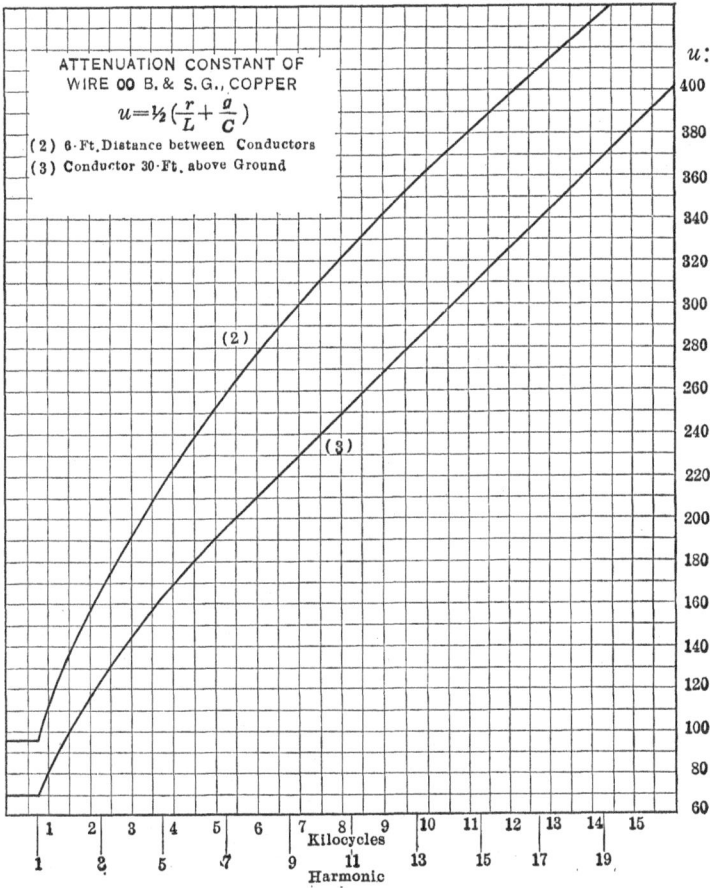

FIG. 103.

cycles at 20,000 cycles frequency, in the oscillation against ground; 64.9 cycles at 100,000 cycles, in the oscillation between line conductors. This represents already a fairly well sustained oscillation, in which the cumulative effect of successive cycles may be considerable. Above 100,000 cycles the attenuation constant begins to rise rapidly, and reaches enormous values, due to the rapidly increasing energy dissipation by radiation. As

ATTENUATION CONSTANT

$$u = \tfrac{1}{2}\left(\frac{r}{L} + \frac{g}{C}\right)$$

(1) No.00 B. & S.G. Copper 18″ Dist.
(2) ,, ,, ,, ,, ,, ,,   ,,   6′ ,,
(3) ,, ,, ,, ,, ,, ,,   ,,   30′ Above Ground
(4) ,, ,, ,, ,, ,, ,, Iron ,,   ,,
(5) ,, 4 ,, ,, ,, ,, Copper ,,   ,,

Fɪɢ. 104.

the result, the duration of the oscillation very rapidly decreases, and the number of cycles performed by the oscillation decreases, until, beyond a million cycles, the energy dissipation is so rapid that practically no oscillation can occur; the oscillation dying out in a cycle or less, thus being practically harmless.

The attenuation constant is plotted, up to 15,000 cycles, in Fig. 103, with the frequency as abscissæ, and is plotted in Fig. 104 in logarithmic scale, as (2) and (3).

Noteworthy is the great difference between the oscillation against ground, and the oscillation between line conductors; the oscillation against ground is more persistent at low frequencies, due to the greater amount of stored energy in the electric field of the conductor, which reaches all the distance to the ground. When reaching into very high frequencies, however, the energy dissipation by radiation becomes appreciable at lower frequencies in the oscillation against ground than in the oscillation between line conductors, and reaches much higher values, with the result that the decay of an oscillation between line and ground is much more rapid at high frequencies than the decay of an oscillation between line conductors. For instance, at 100,000 cycles, the latter performs 65 cycles before dying out, while the former has dissipated its energy in 27 cycles, that is, less than half the time.

8. To further investigate this, in Tables V and VI the numerical values of effective resistance, power factor, attenuation constant and duration of a transient oscillation, in cycles, are given for six typical conductors and circuits, for frequencies from 10 cycles to five million cycles, and plotted in Figs. 104, 105 and 106 in logarithmic scale.

1, 2 and 3 are lines of high power; copper conductor No. 00 B. & S. G., in 1 with 18 in. = 45.5 cm. between conductors, corresponding about to average distribution conductors; in 2 with 6 ft. = 182 cm., between conductors, corresponding to about average transmission line conductors with the oscillation between two lines, and in 3 with 60 ft. = 1830 cm. between conductor and return conductor, corresponding to an oscillation between line and ground, under average transmission line conditions, with the conductor 30 ft. above ground. 4, 5 and 6 give the same condition of an oscillation between line and ground, but in 4 an iron wire of the size of No. 00 B. & S. G., such as has been proposed for the station end of transmission lines, to oppose the approach of high frequency disturbances. In 5 a copper wire No. 4 B. & S. G., that is, a low power transmission line, is represented, and in 6 a stranded aluminum conductor of the same conductivity as copper wire No. 00 B. & S. G.

The equations of the constants for these six circuits are given in

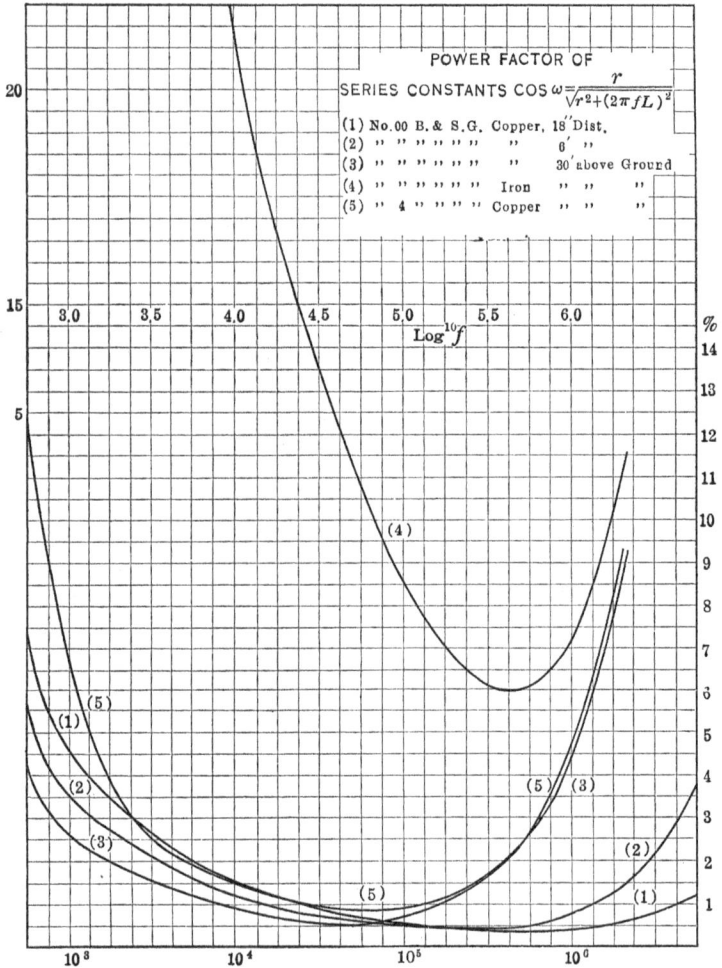

FIG. 105.

Table V. This table also gives the limiting frequencies, between which the various formulas apply with sufficient accuracy for practical purposes, and the lower limits, where the effects become appreciable, in the various conductors.

In Fig. 104 the attenuation constants are plotted, in Fig. 105 the power factors and in Fig. 106 the duration, in cycles.

As such a transient oscillation dies out exponentially, theoret-

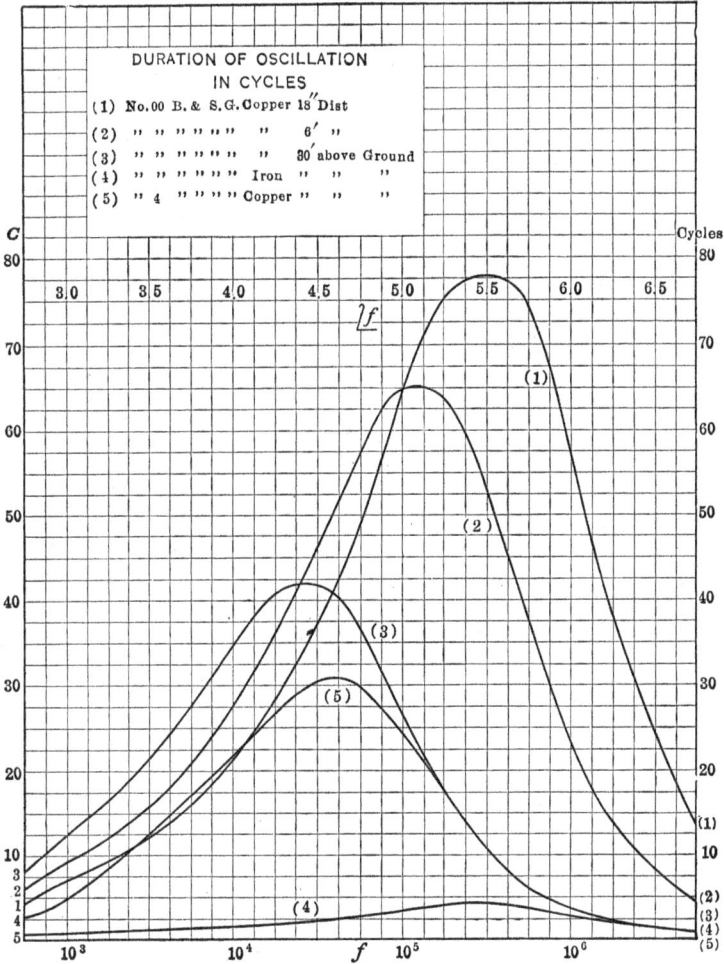

DURATION OF OSCILLATION
IN CYCLES
(1) No.00 B. & S.G. Copper 18" Dist
(2) " " " " " " " 6' "
(3) " " " " " " " " 30' above Ground
(4) " " " " " " " Iron " " "
(5) " 4 " " " " " Copper " " "

FIG. 106.

ically it has no definite duration, but lasts forever, though practically it may have ceased in a few micro-seconds. Thus as duration is defined the time, or the number of cycles, which the oscillation would last if maintaining its initial intensity. In reality, in this time the duration has decreased to $\frac{1}{\epsilon}$, or 37 per cent. of its initial value. Physically, at 37 per cent. of its initial value, or $0.37^2 = 0.135$ of its initial energy, it has become practically harmless, so that this measure of duration probably is the most representative.

From Tables V and VI it is seen, that there is no marked difference between the stranded aluminum conductor (6), and the solid copper wire of the same conductivity (3) and the values of (6) are not plotted in the figures, but may be represented by (3).

9. The attenuation constant μ, in Fig. 104, is plotted in logarithmic scale, with $\sqrt{f}$ as abscissæ. In such scale, a difference of one unit means ten times larger or smaller, and a straight line means proportionally to some power of the frequency. This figure well shows the three ranges: the initial horizontal range at low frequency, where the attenuation is constant; the approximately straight moderate slope of medium frequency, where the attenuation constant is proportional to the square root of the frequency, the unequal current distribution in the conductor predominating, and the steep slope at high frequencies, where the radiation resistance predominates, which is proportional to the square of the frequency.

It is interesting to note that at high frequencies the distance of the return conductor is the dominating factor, while the effect of conductor size and material vanishes; in copper wire No. 00 the rate of decay is practically the same as in copper wire No. 4, though the latter has more than three times the resistance; or in the iron wire, which has nearly six times the resistance and 200 times the permeability. The permeability of the iron wire has been assumed as $\mu = 200$, representing load conditions, where by the passage of the low frequency power current the iron is magnetically near saturation, and its permeability thus lowered. However, the decay of the oscillation between conductor and ground is 6 to 7 times more rapid than that between conductors 6 ft. apart, and that between conductors 18 in. apart about 3 times less.

This shows, that to produce quicker damping of high frequency waves, such as are instrumental in steep wave fronts, the most effective way is to separate the conductors as far as possible, perhaps even lead them to the station by separate single conductor

TABLE V.—LINE CONSTANTS.

f = frequency in cycles per second.
S = 3 × 10^10 cm. sec. = propagation velocity in empty space.
κ = 1 = specific capacity.
l_0 = length of conductor, in kilometers = 10^5 cm.

			Per kilometer of conductor:					
			(1)	(2)	(3)	(4)	(5)	(6)
Size of round conductor, B. & S. G			00	00	00	00	4	(00)
Radius of conductor, cm		$i_r =$	0.463	0.463	0.463	0.463	0.258	0.65
Distance of return conductor, cm		$l =$	45.5	182	1820	1820	1820	1820
Material of conductor			Copper	Copper	Copper	Iron	Copper	Aluminum
Conductivity		$\lambda = 10^5 \times$	6.2	6.2	6.2	1.1	6.2	3.8
Permeability		$\mu =$	1	1	1	200	1	1
Low frequency								
Ohmic resistance	$r_0 = \dfrac{10^7 l_0}{\pi l_r^2 \lambda}$ ohms =		0.240	0.240	0.240	1.35	0.770	0.240
Internal reactance	$x_0' = 0.1 \pi \mu f l_0 \, 10^{-3}$ ohms =	$f \times 10^{-3} \times$	0.314	0.314	0.314	62.83	0.314	0.314
Limit, cycles		$f =$	10^3	10^3	10^3	10	2 × 10^3	10^3
High frequency								
Thermal resistance / Internal reactance	$r_1 = x_1 = \dfrac{5 l_0}{l_r}\sqrt{\dfrac{0.4\,\mu f}{\lambda}}$ ohms =	$10^{-3} \times \sqrt{f} \times$	8.65	8.65	8.65	290	15.5	7.86
Lower limit, cycles		$f =$	2 × 10^5	10^5	2 × 10^4	2 × 10^5	2 × 10^4	2 × 10^4
Radiation resistance	$r_2 = \dfrac{0.8 \pi^2 f^2 l_0}{S} 10^{-3}$ ohms =	$f^2 \times 10^{-9} \times$	0.012	0.048	0.48	0.48	0.48	0.48
Upper limit, cycles		$f =$	10^8	3 × 10^7	3 × 10^6	3 × 10^6	3 × 10^6	3 × 10^6
External reactance	$X_0 = 0.4 \pi f l_0 \log \dfrac{l}{l_r} 10^{-3}$ ohms =	$f \times 10^{-3} \times$	5.76	7.5	10.4	10.4	11.1	9.95
Total resistance / Total reactance	$r = r_1 \,(\text{or } r_0) + r_2$ $x = x_1 \,(\text{or } x_0') + x_0$							

	Inductance..............	$L = \frac{x}{2\pi f}$ henrys							
	Magnetic attenuation.....	$\frac{r}{2L}$							
	Dielectric radiation resistance..................	$r_c = \frac{2Sl'}{\kappa l_0} 10^{-14}$ ohms =	....	0.0273	0.1092	1.092	1.092	1.092	1.092
	Capacity reactance.......	$x_c = \frac{S^2}{\pi f \kappa l_0} \log \frac{l'}{l_r} 10^{-14}$ ohms =	$\frac{10^6}{f} \times$	13.2	17.2	23.7	23.7	25.4	22.8
	Dielectric attenuation....	$\frac{g}{2C} = \pi f \frac{r_c}{x_c} = \frac{2\pi^2 f^2 l'}{S \log \frac{l'}{l_r}} =$	$f^2 \times 10^{-6} \times$	0.0065	0.020	0.144	0.144	0.135	0.151
	Attenuation constant.....	$u = \frac{1}{2}\left(\frac{r}{L} + \frac{g}{C}\right)$							

lines; but the use of high resistance conductors, or of magnetic material, as iron, offers little or practically no advantage in damping very high frequencies or flattening steep wave fronts.

At medium and low frequencies, however, the relation reverses, and the decay of the wave is the smaller, the greater the distance of the return conductor. The reason is, that in this range the effective resistance is still independent of the conductor distance, while the inductance increases with increasing distance. At medium and low frequencies, the iron conductor offers an enormously increased attenuation: from 10 to 20 times that of nonmagnetic conductors.

10. The power factor of the conductor is plotted in Fig. 105. As seen, it decreases, from unity at very low frequencies, to a minimum at medium high frequencies, and then increases again to very high values at very high frequencies. The minimum value is a fraction of one per cent. except with the iron conductor, where the minimum is very much higher. The power factor is of importance as it indicates the percentage of the oscillating energy, which is dissipated per wave of oscillation. This is represented still better by Fig. 106, the duration of the oscillation in cycles, that is, the number of cycles which an oscillation lasts before dissipating the stored energy which causes it.

At medium high frequencies, the oscillation is the more persistent the lower the ohmic resistance of the conductor and the further away the return conductor, while at very high frequencies the reverse is the case, and the oscillation is the more persistent, the shorter the distance of the return conductor, while the size and material of the conductor ceases to have any effect.

The maximum number of cycles is reached at medium high frequencies, in the range between 20,000 and 100,000 cycles—depending on conductor size and distance of return conductor. It thus is in this range of frequencies, where an oscillation caused by some disturbance lasts the greatest number of cycles, that the possibility, by some energy supply by means of an arc, etc., to form a stationary oscillation or even a cumulative oscillation, thus to become continuous or "undamped," is greatest.

It would thus appear, that this range of frequencies, of 20,000 to 100,000, represents what may be called the "danger frequencies" of transmission systems. It is interesting to note, that experimental investigations have shown, that the natural fre-

TABLE VI.—ATTENUATION CONSTANTS.

	f =	10	60	500	1000	2000	5000 c	10 kc	20	50	100	200	500	1000	2000	5000 kc
1. Copper. No. 00 B. & S. G. $l_r = 0.463$ cm. $l' = 45.5$ cm.	$r =$	0.24	0.24	0.24	0.273	0.386	0.612	0.866	1.230	1.96	2.85	4.34	9.12	20.65	60.25	319.3
	$\cos \omega =$	96.9	54.9	7.86	4.53	3.26	2.07	1.48	1.055	0.674	0.494	0.376	0.316	0.358	0.524	1.105%
	$u =$	124	124	124	142	204	327	466	666	1,081	1,615	2,620	6,595	17,740	58,800	336,500
	$N =$	0.08	0.48	4.03	7.04	9.80	15.3	21.5	30.2	46.2	64.5	76.4	76.0	56.4	34.0	14.9 c
2. Copper. $l' = 182$ cm.	$r =$	0.24	0.24	0.24	0.273	0.386	0.613	0.870	1.249	2.05	3.21	5.78	18.1	56.6	204.2	1,219
	$\cos \omega =$	95.1	45.4	6.16	3.51	2.52	1.61	1.15	0.822	0.544	0.427	0.384	0.483	0.753	1.36	3.25%
	$u =$	96	96	96	110	157	252	362	524	905	1,540	3,210	12,570	43,700	165,600	1,010,000
	$N =$	0.10	0.62	5.20	9.09	12.74	19.85	27.6	38.2	55.0	64.9	62.2	39.8	22.9	12.1	5.0 c
3. Copper. $l' = 1,820$ cm.	$r =$	0.24	0.24	0.24	0.273	0.386	0.624	0.913	1.417	3.13	7.53	23.06	126.1	488.6	1,932	12,020
	$\cos \omega =$	91.3	34.9	4.48	2.56	1.83	1.19	0.870	0.677	0.600	0.706	1.105	2.42	4.70	9.24	22.5%
	$u =$	70	70	70	80	115	190	287	482	1,302	3,710	12,690	74,200	291,500	1,159,000	7,230,000
	$N =$	0.14	0.86	7.15	12.5	17.4	26.3	34.8	41.4	38.4	26.9	15.8	6.75	3.43	1.72	0.69 c
4. Iron. $l' = 1,820$ cm.	$r =$	1.35	2.25	6.48	9.16	12.95	20.51	29.05	41.29	66.0	96.4	148.9	325	770	2,331	12,650
	$\cos \omega =$	(80)	61.7	48.5	42.4	35.8	27.1	21.5	16.0	11.3	8.5	6.7	6.0	7.2	11.0	23.3%
	$u =$	(25)	148	873	1,470	2,410	4,444	6,914	10,210	18,060	28,200	48,200	130,400	371,000	1,266,000	7,370,000
	$N =$	(0.40)	0.40	0.57	0.68	0.83	1.13	1.45	1.96	2.77	3.55	4.17	3.83	2.70	1.58	0.68 c
5. Copper. No. 4 B. & S. G. $l_r = 0.258$ cm. $l' = 1,820$ cm.	$r =$	0.77	0.77	0.77	0.77	0.772	1.112	1.598	2.382	4.66	9.70	26.1	131	495	1,940	12,030
	$\cos \omega =$	98.9	42.5	13.4	6.75	3.38	1.96	1.42	1.06	0.84	0.87	1.17	2.34	4.44	8.7	21.1%
	$u =$	212	212	212	212	213	311	458	721	1,647	4,080	12,750	70,550	274,500	1,087,000	6,775,000
	$N =$	0.047	0.28	2.36	4.72	9.38	16.05	21.8	27.8	30.4	24.5	15.7	7.1	3.65	1.84	0.75
6. 7 strand Al. equivalent to No. 00 Copper, $l_r = 0.65$ cm. $l' = 1,820$ cm.	$r =$	0.24	0.24	0.24	0.248	0.351	0.568	0.834	1.302	2.950	7.28	22.71	125.6	487.9	1,931	12,018
	$\cos \omega =$	91.9	36.0	4.64	2.42	1.72	1.12	0.83	0.65	0.59	0.73	1.14	2.5	4.9	9.8	23.7
	$u =$	73	73	73	76	109	181	275	467	1,298	3,780	13,130	77,000	303,000	1,207,000	7,580,000
	$N =$	0.14	0.82	6.85	13.2	18.3	27.6	36.4	42.8	38.5	26.5	15.2	6.5	3.3	1.66	0.65

quency of oscillation of the high voltage windings of large power transformers usually is within this range of danger frequencies. The possibility of the formation of destructive cumulative oscillations or stationary waves in the high voltage windings of large power transformers thus is greater than probably with any other class of circuits, so that such high potential transformer windings require specially high disruptive strength and protection. This accounts for the not infrequent disastrous experience with such transformers, before this matter was realized.

11. Figure 106 also shows, that the duration of an oscillation in iron wire, in cycles, is very low at all frequencies. Thus the formation of a stationary oscillation in an iron conductor is practically excluded, but such conductors would act as a dead resistance, damping any oscillation by rapid energy dissipation.

The duration of high frequency oscillation, in cycles, increases with increasing frequency, to a maximum at medium frequencies. This obviously does not mean that the time during which the oscillation lasts increases; the time, in micro-seconds, naturally decreases with increasing frequency, due to the increasing attenuation constant. Thus in conductor (2) for instance, the oscillation lasts 65 cycles at a frequency of 100,000 cycles, but only 9 cycles at 1000-cycle frequency. However, the 65 cycles are traversed in 650 micro-seconds, while the 9 cycles last 9000 microseconds, or 14 times as long. It is not the total time of oscillation, but the cumulative effect due to the numerous and only slowly decreasing successive waves, which increases as represented in Fig. 106.

An oscillation between copper wires No. 4 B. & S. G., 6 ft. apart, would give a duration curve, which at moderate frequencies follows (5) of Fig. 106, but at high frequencies follows (2). Thus the average duration and average rate of decay would be about the same as (3), an oscillation between copper wire No. 00 and ground. However, the oscillation would be more persistent in such a conducto at high, and less persistent at low frequencies. A complex wave, containing all the harmonics from low to very high ones, such as a steep wave front impulse or an approximately rectangular wave, as may be produced by a spark discharge, etc.; such a wave would have about the same average rate of decay in a copper wire No. 4 with return at 6 ft., as in a copper wire No. 00 with ground return. The wave front would flatten, and the wave round off, approach more and more sine shape, due to the more rapid disappearance of the higher frequencies, while at the same time decreasing in amplitude. The wave thus would pass through many intermediate shapes. But these intermediate shapes would be materially different with wire No. 4 and return at 6 ft., as with No. 00 and ground return; in the latter, the flattening of the steep wave front, and rounding of the wave, would be much more rapid at the beginning, due to the shorter duration of the transient, and while such wave would last about the same time, that is, pass over the lines to about the same distance, it would carry steep wave fronts to a much lesser distances, that is, its danger zone would be materially less than that of the wave in copper wire No. 4 with return at 6 ft.

It therefore, is of great interest to further investigate the effect of the changing attenuation constant on complex waves, and more particularly those with steep wave fronts, as the rectangular waves of starting or disconnecting lines, etc.

CHAPTER III.

ATTENUATION OF RECTANGULAR WAVES.

12. The destructiveness of high frequencies or step wave fronts in industrial circuits is rarely due to over-voltage between the circuit conductors or between conductor and ground, but is due to the piling up of the voltage locally, in inductive parts of the circuit, such as end turns of transformers or generators, current transformers, potential regulators, etc., or inside of inductive windings as the high potential coils of power transformers, by the formation of nodes and wave crests. Such effects may be produced by high frequency oscillations sustained over a number of cycles, as discussed in Chapter II, by oscillations lasting only a very few cycles or a fraction of a cycle, or due to non-oscillating transients, as single impulses, etc. As the high rate of change of voltage with the time, and the correspondingly high voltage gradients along the conductor are the source of danger, to calculate and compare oscillatory and non-oscillatory effects in this respect, it has become customary in the last years, to speak of an "equivalent frequency" of impulses, wave fronts or other non-oscillatory transients.

As "effective" or "equivalent" frequency of an impulse, wave front, etc., is understood the frequency of an oscillation, which has the same maximum amplitude, e or i, and the same maximum gradient $\dfrac{de}{dt}$ or $\dfrac{di}{dt}$. Thus assuming an impulse which reaches a maximum voltage $e = 60,000$, and has a maximum rate of increase of voltage of $\dfrac{de}{dt} = 10^{11}$, that is, a maximum voltage rise at the rate of 10^{11} volts per second, or 10,000 volts per micro-second. As the average voltage rise of a sine wave is $\dfrac{2}{\pi}$ times the maximum, the average rise of an oscillation of the same maximum gradient as the impulse, would be $\dfrac{2}{\pi}\dfrac{de}{dt} = \dfrac{20,000}{\pi}$ volts per micro-second and the total voltage rise of $e = 60,000$ thus would occur in

$$\frac{e}{\frac{2}{\pi} \frac{de}{dt}} = \frac{60,000 \, \pi}{20,000} = 9.4 \text{ micro-seconds.}$$ A complete cycle of this

oscillation thus would last $4 \times 9.4 = 37.6$ micro-seconds, and

the equivalent frequency of the impulse would be $f = \frac{.0^6}{37.6} =$

26,600 cycles or 26.6 kilo cycles. The equivalent frequency of a perfectly rectangular wave front, if such could exist, obviously would be infinity.

QUARTER WAVE CHARGING OR DISCHARGING OSCILLATION OF A LINE.

13. Considering first the theoretically rectangular wave of connecting a transmission line to a circuit, or disconnecting it from the circuit.

Suppose a transmission line, open at the distant end, is connected to a voltage E. At this moment, the voltage of the line is zero. It should be, in permanent conditions, E. Thus the circuit voltage consists of a permanent voltage E (which is the instantaneous value of the alternating supply voltage at this moment, $E_0 \sin \varphi$) and the transient voltage $- E$. We thus have a transient voltage, which is uniformly $= - E$ all along the line except at the switching point $l = 0$, where the transient voltage is zero.

Or, suppose a transmission line, open at the far end, is connected to a source of voltage, and at the moment where this voltage is E, the line short circuits at some point, by a spark discharge, flash-over, etc. Thus at this moment, the voltage $= 0$ at the point of short circuit, and is $= E$ everywhere between this point and the end of the line. Thus we get a line discharge leading to the same transient, a theoretically rectangular wave. In the part of the line between generator and short circuit, we have a different transient, a circuit of voltage $e = 0$ at one end, $e = E$ throughout the entire length at time $t = 0$, and $e = E$ continuously at the other end, where the generator maintains the voltage.

However, this again leads to the same transient, of a theoretically rectangular wave.

Assuming thus, as an instance, a transmission line of 100 km. length, of copper wire No. 00 B. & S. G., 30 ft. above ground, open circuited at the other end $l = 100$ km., and connected to a source of voltage E at the beginning, $l = 0$.

Then the beginning of the line, $l = 0$, is grounded, at the time $t = 0$, thus giving a quarter wave oscillation, with the terminal conditions:

Voltage along the line constant $= E$, at time $t = 0$, except at the beginning of the line, $l = 0$, where the voltage is 0.

Current along the line $= 0$ at $t = 0$, except at $l = 0$, where the current is indefinite.

14. The equation of the quarter wave oscillation of the line conductor against ground, then is (Chapter VII (57)):

$$e = \frac{4E}{\pi} \epsilon^{-ut} \sum_{0}^{\infty} {}_{n} \frac{\sin (2n + 1)\tau \cos (2n + 1)\vartheta}{n + 1} \tag{1}$$

where:

ϑ is the time angle of the fundamental wave of oscillation, of frequency:

$$f_0 = \frac{S}{4l_0} = \frac{3 \times 10^{10}}{4 \times 100 \times 10^5} = 750 \text{ cycles.} \tag{2}$$

τ is the distance angle, for $l_0 = 100$ km. $= 90° = \frac{\pi}{2}$, that is,

$$\left. \begin{aligned} \vartheta &= 2\pi f_0 t \\ \tau &= \frac{\pi l}{2l_0} \end{aligned} \right\} \tag{3}$$

Equation (1), however, assumes that u, and thus r, L, C and g are constant for all frequencies. As this is not the case, but u is a function of the frequency, and thus of n: u_n, ϵ^{-ut} can not be taken out of the summation sign. Equation (1) thus must be written:

$$e = \frac{4E}{\pi} \sum_{0}^{\infty} {}_{n} \epsilon^{-u_n t} \frac{\sin (2n + 1)\tau \cos (2\pi + 1)\vartheta}{2n + 1} \tag{4}$$

where u_n is the value of u for the frequency: $f = (2n + 1)f_0$.

From (4) follows, as the voltage gradient along the line:

$$\frac{de}{dt} = \frac{4E}{\pi} \sum_{0}^{\infty} n \, \epsilon^{-unt} \cos (2n + '1) \, \tau \cos (2n + 1) \, \vartheta \qquad (5)$$

$$= \frac{2E}{\pi} \left\{ \sum_{0}^{\infty} n \, \epsilon^{-\mu nt} \cos (2n + 1)(\tau + \vartheta) \right.$$

$$\left. + \sum_{0}^{\infty} n \, \epsilon^{-\mu nt} \cos (2n - 1) \, (\tau - \vartheta) \quad (6) \right.$$

The maximum voltage gradient occurs at the wave front, that is, for $\vartheta = \tau$. Substituting this, and substituting further, from (3):

$$d\tau = \frac{\pi}{2l_0} \, dl \qquad (7)$$

gives, as the maximum voltage gradient,

$$G = \frac{de}{dl} = \frac{E}{l_0} \left\{ \sum_{0}^{\infty} n \, \epsilon^{-unt} \cos (2n + 1) \, 2\tau + \sum_{0}^{\infty} u \, \epsilon^{-\mu nt} \right\} \qquad (8)$$

It is, however,

$$\sum_{0}^{\infty} n \cos (2n + 1) \, 2\tau = 0 \qquad (9)$$

for all values of τ except $\tau = 0$ and $\tau = \pi$, that is, the beginning of the line, $l = 0$.

Thus, approximately, as μ_n varies gradually:

$$\sum_{0}^{\infty} n \, \epsilon^{-unt} \cos (2\pi + 1) \, 2\tau = 0 \qquad (10)$$

except for values of $\tau = 0$ or very near thereto.

Substituting (10) into (8) gives

$$G = \frac{E}{l_0} \sum_{0}^{\infty} n \, \epsilon^{-unt} \qquad (11)$$

as the approximate expression of the maximum voltage gradient, that is, the steepness of the wave front, at time t, that is, at distance from the origin of the wave.

$$l = St = 3 \times 10^{10} \, t \qquad (12)$$

If ϑ differs materially from τ, the term with $(\tau - \vartheta)$ in equation (6) also vanishes, and $\frac{de}{dt} = 0$, that is, there is no voltage gradient except at and near the wave front.

From (11) are now calculated numerical values of the steepness of the wave front G, for various times t after its origin, and thus (by 12), various distances l from the origin. These numerical values are given in table VII, for $E = 60{,}000$ volts.

At a fundamental frequency of $f_0 = 750$ cycles, successive harmonics differ from each other by 1500 cycles, and for every value of t, values of ϵ^{-unt} thus have to be calculated for the frequencies:

$$n = \quad 0 \quad\quad 1 \quad\quad 2 \quad\quad 3 \quad\quad 4 \quad\quad 5 \quad\quad 6 \quad \text{etc.}$$
$$f = \quad 750 \quad 2250 \quad 3750 \quad 5250 \quad 6750 \quad 8250 \quad 9750 \text{ cycles, etc.,}$$

until the further terms add no further appreciable amount to $\Sigma\epsilon^{-unt}$. In calculating, it is found that for instance for $t = 5$ micro-seconds, or $l = 1.5$ km., this occurs at $f = 5 \times 10^6$ cycles, thus beyond the $\dfrac{5 \times 10^6}{750} = 6670$th harmonic. Thus 6670 terms of the series would have to be calculated to get this one point of the wave gradient: more terms for shorter, less terms for longer distances l of wave travel. This obviously is impossible, and some simpler approximation, of sufficient occuracy, thus is required.

This may be done as follows:

In the range from 5×10^5 to 10^6 cycles for instance, there are $\dfrac{10^6 - 5 \times 10^5}{1500} = 333$ harmonics. Instead of calculating u and ϵ^{-ut} for each of these harmonics, calculate ϵ^{-ut} for the average value of these 333 harmonics, and multiply by 333.

Thus, dividing the entire frequency range (beyond the lowest harmonics, which are calculated separately), into groups, and calculating one average value for each group, the calculation of $\Sigma\epsilon^{-ut}$ becomes feasible.

Since ϵ^{-ut} is calculated through $lg\epsilon^{-ut} = -utlg\epsilon$, as values of t, multiples of $\dfrac{1}{lg\epsilon}$ have been chosen, to still further simplify the calculation, in deriving a curve of gradients G.

15. Table VII gives the values of the maximum voltage gradient of the wave front, in volts per meter at 60,000 volts maximum initial line voltage, the equivalent frequency of the wave front, in kilocycles, and the length of the wave front, in meters, for various times of wave travel, from .03 micro-seconds up, and corresponding distances of wave travel, from 10 meters from the origin as rectangular wave, up to thousands of km., for copper wire No. 00 B. & S. G., 30 ft. = 910 cm. above ground, with the ground as return.

TABLE VII.—ATTENUATION OF WAVE FRONT OF QUARTER-WAVE OSCILLATION, OF 100 KM. LINE, 60,000 VOLTS.

Time t, Microseconds.	Distance l		f, Where Σ^{-u}. Vanishes.	Voltage Gradient.		Wave Front.	
	Km.	Miles.		$\frac{1}{E}\frac{de}{dt}$	G, Volts per Meter.	Equivalent Kilo-cycles.	Length, Meters.
From origin of wave.							
(3)	Copper wire No. 00 B. & S. G., 30 ft. = 910 cm. above ground.						
0	0	0	....	∞	60,000	∞	0
0.03	0.01	0.0062	....	5,800	5,500	8,800	6
0.575	0.1725	0.107	10×10^6	1,370	1,290	2,060	73
1.15	0.345	0.215	8×10^6	1,025	967	1,540	98
2.3	0.69	0.43	6×10^6	670	630	1,000	150
11.5	3.45	2.15	4×10^6	300	280	450	330
23	6.9	4.3	2×10^6	218	205	330	460
92	27.6	17.2	10^6	104	98	156	960
230	69	43	0.7×10^6	62	58	92	1,630
2,300	690	430	0.2×10^6	13	12	19	7,900
23,000	6,900	4,300	50,000	1.3	1.2	1.9	79,000
(1)	Copper wire No. 00 B. & S. G., return conductor at 18 in. = 45.5 cm. distance.						
2.3	0.69	0.43	20×10^6	3,450	3,240	5,160	29
23	6.9	4.3	10×10^6	1,020	960	1,530	98
230	69	43	4×10^6	250	234	370	400
(5)	Copper wire No. 4 B. & S. G., 30 ft. = 910 cm. above ground.						
23	6.9	4.3	2×10^6	227	213	340	440
230	69	43	0.7×10^6	64	60	94	1,570
(4)	Iron wire No. 00 B. & S. G., 30 ft. = 910 cm. above ground.						
23	6.9	4.3	2×10^6	143	135	215	700
230	69	43	0.3×10^6	11	10	16	9,400

For comparison are given some data of the same conductor, with the return conductor at 18 in. = 45.5 cm., and also for a copper wire No. 4, and an iron wire No. 00, with the ground as return.

These data are plotted in Fig. 107, showing the wave front, as it gradually flattens out in its travel over the line, from the very steep wave at 170 meters from the origin, to the wave with a front of 1630 meters, 69 km. away.

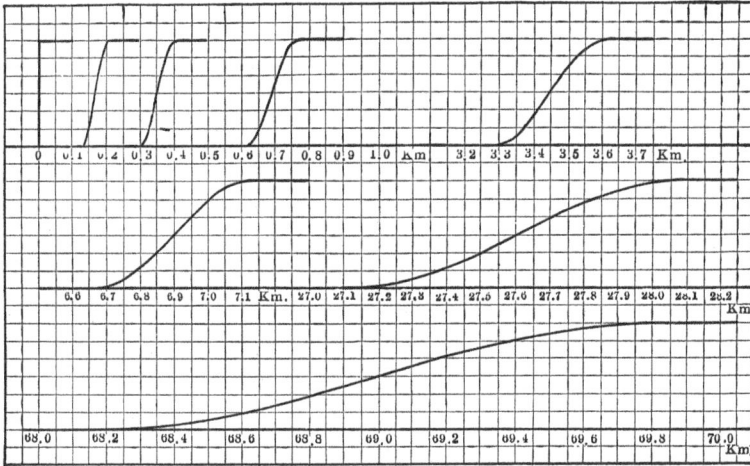

FIG. 107.

A comparison of the data of the four circuit conditions is given in Table VIII, and plotted in Fig. 108.

TABLE VIII.—ATTENUATION OF WAVE FRONT OF QUARTER-WAVE OSCILLA-
TION, OF 100 KM. LINE, 60,000 VOLTS.

	(3)	(1)	(5)	(4)
Conductor...............	(3)	(1)	(5)	(4)
Size No...................	00	00	4	00
Material.................	Copper	Copper	Copper	Iron
Distance of return conductor, cm....................	1820	45.5	1820	1820
After 23 microseconds, 6.9 km.:				
Gradient, volts per meter.....	205	960	213	135
Length of wave front, meters..	460	98	440	700
Equivalent kilocycles.........	330	1530	340	215
After 230 microseconds, 69 km.:				
Gradient, volts per meter.....	58	234	60	10
Length of wave front, meters..	1630	400	1570	9400
Equivalent kilocycles.........	92	370	94	16

It is interesting to note, that there is practically no difference in the flattening of the wave front on a low resistance conductor, No. 00, and a high resistance conductor, No. 4. There is, however, an enormous difference due to the effect of the closeness of the return conductor: with the return conductor at 18 in. distance, the wave front is still materially steeper at 6.9 km. distance, than it is in the conductor with ground return at 0.69 km. distance. That is, the flattening of the wave front in the conductor with ground return, is more than ten times as rapid, than in the same conductor with the return conductor closely adjacent. Or in other words, the danger zone of steep wave front, extends

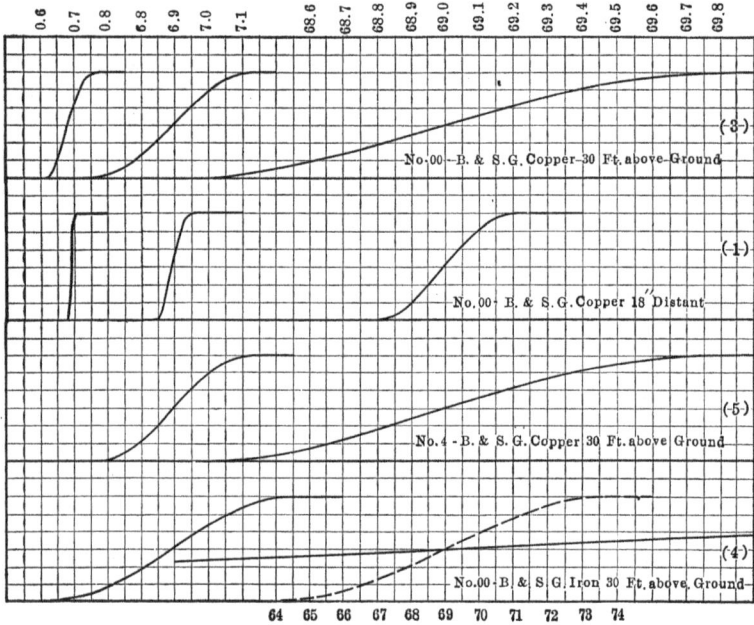

Fig. 108.

in a conductor with the return conductor closely adjacent, to more than ten times the distance than in the conductor with ground return.

This means, where it is desired to transmit a high-frequency impulse or steep wave front to the greatest possible distance, it is essential to arrange conductor and return conductor as closely adjacent as possible. But where it is essential to limit the harmful effect of very high frequency or steep wave front as much as possible to the immediate neighborhood of its origin, the return conductor should be separated as far as possible.

The data on iron wire are very disappointing: there is an enormous increase in the flattening of the wave front at great dis-

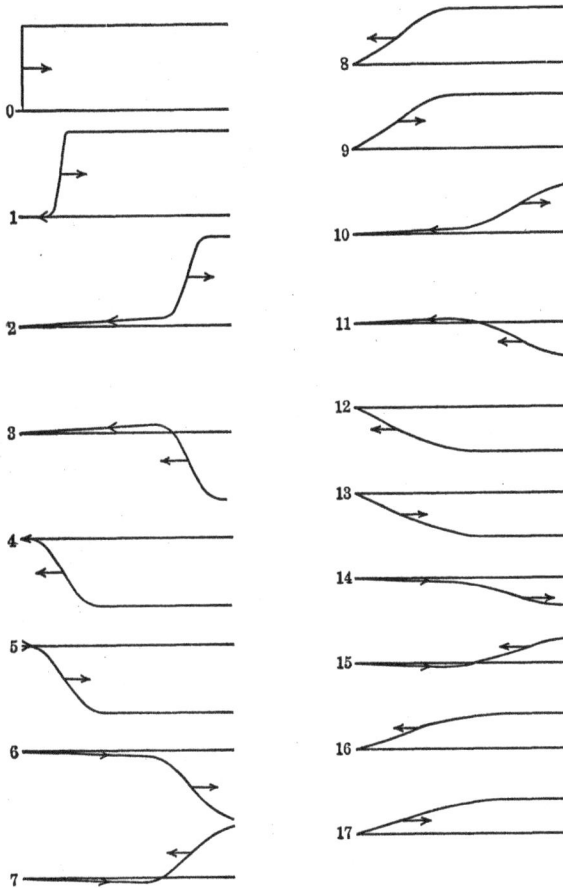

tances, by the use of iron as conductor material, so much so that the wave front of the iron conductor at 69 km. distance had to be shown (dotted) at one-tenth the scale as for the other conductors, in Fig. 108. But at moderate distances, 6.9 km. from the origin, the flattening of the wave front in the iron conductor is only little greater than in a copper conductor of the same size: 215 kilocycles against 330 kilocycles. At short distances, the difference almost entirely ceases, and within 1 km. from the origin, the wave front in an iron conductor is nearly as ·steep as in a copper conductor of the same size. Thus a short length of iron wire between station and line would exert practically no protection against very high-frequency oscillations or steep wave fronts, such as may be produced by lightning strokes in the neighborhood of the station.

This was to be expected from the shape of the curve of the attenuation constant, shown in Fig. 104.

16. From the data in Table VII then are constructed and shown in Fig. 109, the successive curves of voltage distribution in the line, as it is discharging (or charging), by the originally rectangular wave running over the line, reflecting at the end of the line and running back, then reflecting again at the beginning of the line and once more traversing it, etc., until gradually the transient energy is dissipated and the line voltage reaches its average, zero in discharge, or the supply voltage in charge. The direction of the wave travel in the successive positions is shown by the arrows in the center of the wave front; the existence and direction of current flow in the line by the arrows in the (nearly) horizontal part of the diagram. As seen, after four to five reflections, the voltage distribution in the line is practically sinoidal.

However, in these diagrams, Figs. 107 to 109, the wave front has been constructed from the maximum voltage gradient derived by the calculation, assuming as approximation the shape of the wave front as sinoid. This usually is a sufficient approximation, since the important feature is the maximum gradient, that is, the steepest part of the wave front, which was given by the calculation; but it is not strictly correct, and the wave front differs from sine shape. It thus is of interest to investigate the exact shape of the wave front, in its successful stages of flattening.

RECTANGULAR TRAVELING WAVE.

17. For this purpose may be investigated as further instance the gradual destruction of wave shape and decay of a 60,000-cycle rectangular traveling wave, during its passage over a transmission line, changing from the original rectangular wave shape produced at its origin by lightning stroke, spark discharge, etc., into practically a sine wave.

For this purpose, for the elementary symmetrical traveling wave, the equation is

$$e = \frac{4\,E}{\pi} \sum_0^\infty n\,\epsilon^{-u_n t} \sin (2\,n + 1)\varphi \qquad (13)$$

where $\varphi = \vartheta = \tau$ is the running time coordinate, in angular expression.

Values of e are calculated, for various times t, from 1 microsecond to 360 microseconds, for all the angles φ, where e has not yet become constant and equal to E.

TABLE IX.—ATTENUATION OF 60,000-CYCLE RECTANGULAR WAVE, IN LINE OF NO. 00 B. & S. G., COPPER, 30 FT. ABOVE GROUND.

		0	1	2	5	10	20	40	100	360	
Time, t =		0	1	2	5	10	20	40	100	360	ms.
Wave travel, l =		0	0.3	0.6	1.5	3	6	12	30	108	km.
φ = 0 degrees		0.785	0	0	0	0	0	0	0	0	
φ = 1 degree		.....	0.250	0.170	0.106	0.074	0.052	0.347	0.022	0.010	
φ = 5 degrees		.....	0.785	0.645	0.475						
φ = 10 degrees		.....	.....	0.785	.....	0.606	0.468	0.582	0.217	0.100	
φ = 20 degrees		.....	.....	.....	0.783	0.770	0.707	0.709	0.405	0.193	
φ = 30 degrees		.....	.....	.....	.....	0.780	0.762	0.740	0.547	0.282	
φ = 40 degrees		.....	.....	.....	.....	.....	0.775	0.758	0.639	0.360	
φ = 50 degrees		.....	.....	.....	.....	.....	.....	0.765	0.691	0.426	
φ = 60 degrees		.....	.....	.....	.....	.....	.....	.....	0.722	0.476	
φ = 70 degrees		.....	.....	.....	.....	.....	.....	.....	0.730	0.513	
φ = 80 degrees		.....	.....	.....	.....	.....	.....	.....	.....	0.535	
φ = 90 degrees		0.785	0.785	0.785	0.783	0.780	0.775	0.765	0.730	0.540	
Wave Front	Degrees........	0	10	16	26	44	70	90	140	180	
	Meters........	0	140	220	340	620	980	1250	1940	2500	
	Kilo-cycles. Max.	∞	1110	750	470	325	230	165	103	63	
	Kilo-cycles. Avg.	∞	1080	670	420	268	155	120	77	60	
Highest apprec. harmonic.		∞	61	45	33	15	9	7	3	1	
Decrease of wave max....		0	.....	.....	.....	0.6	1.3	2.6	7.0	31.2	%
Maximum gradient, volts per meter............		∞	690	470	295	205	140	100	65	40	

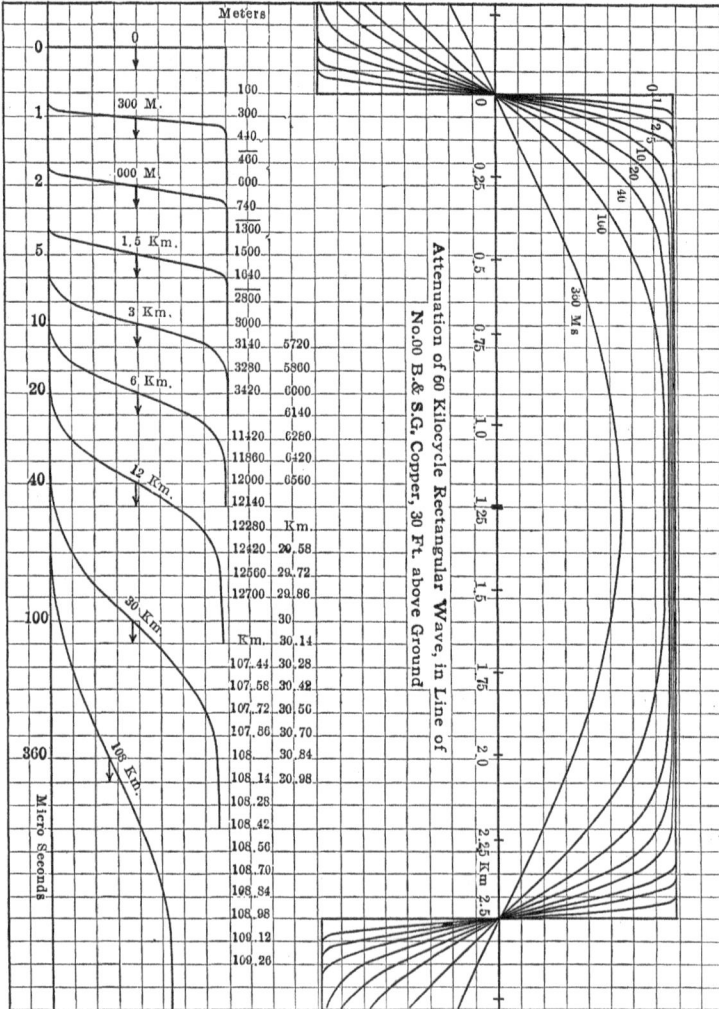

Fig. 110.

In Table IX are given numerical values, and plotted in Fig. 110, of

$$e \frac{\pi}{4 E} = \sum_0^\infty {}'n \epsilon^{-u_n t} \sin (2 n + 1)\varphi,$$

from 0 to 90°, and from 1 to 360 microseconds, and derived therefrom the values of the length of wave front, in degrees and in meters, and the equivalent frequency, in kilocycles. Two values are given, the maximum kilocycles, derived from the steepest part of the wave front, and the average kilocycles, derived from the total wave front. The difference indicates the deviation of the shape of the wave front from a sinoid.

18. As seen, due to the high fundamental frequency, 60,000 cycles, the number of significant harmonics is very greatly reduced, until frequencies are reached, where the attenuation is so enormous, that the destruction of the wave by its energy dissipation occurs in a few meters.

Thus already after 1 microsecond or 300-meter wave travel, the calculation needs to extend only to the 61st harmonic; after 2 microseconds, to the 45th harmonic, etc., while after 100 microseconds, or 30-km. travel, only the third harmonic is still appreciable, and after 360 microseconds, or 108-km. travel, even this has disappeared, and the wave is essentially a sine wave.

In 30-km. travel, the wave maximum has decreased 7 per cent; in 108 km., it has decreased 31.2 per cent.

It is important to note, however, that waves of relatively high frequency, within the range of the danger frequencies between 20 and 100 kilocycles, can travel considerable distances, if no other causes of rapid attenuation are at work, but those considered here, and still retain a large part of their energy. Thus the 60,000-cycle wave, after traversing 100 km. of line, still retains about 70 per cent of its amplitude, that is, about one-half of its energy. Thus the danger from resonance of power transformer windings with such frequencies is not local but rather extends over a large part of the system.

The lower part of Fig. 110 shows the shape of the wave front at various distances from the origin; the upper part shows the gradual change of the wave from rectangular to flat top to sine wave.

Plotting for the 750-cycle quarter-wave oscillation and the 60,000-cycle traveling wave, the logarithm of the length of the wave front, and of the equivalent frequency of the wave front, against the logarithm of the distance of wave travel, gives practically straight lines (except for very great distances of wave travel, where the lower harmonics predominate), and from the slope of these lines follows that:

The length of wave front is approximately proportional, and the equivalent frequency of the wave front approximately inverse proportional to the square root of the distance of wave travel:

$$l_f = c_1 \sqrt{l} \tag{14}$$

$$f_f = \frac{c_2}{\sqrt{l}} \tag{15}$$

This would give a wave front constant c_1 and an equivalent frequency constant c_2 of the circuit.

CHAPTER IV.

FLATTENING OF STEEP WAVE FRONTS.

19. A rectangular wave is represented by the equation

$$e = \frac{4E}{\pi} \sum_{0}^{\infty} n \, \epsilon^{-ut} \frac{\sin (2n + 1)\tau \cos (2n + 1)\vartheta}{2n + 1} \qquad (1)$$

where

$$\vartheta = 2 \pi f_0 t = \text{time angle}, \qquad (2)$$

$$\tau = 2\pi \frac{l}{l_0} = \text{distance angle}, \qquad (3)$$

$$f_0 = \text{fundamental frequency},$$

$$l_0 = \frac{S}{f_0} = \text{wave length}, \qquad (4)$$

$$S = 3 \times 10^{10} = \text{velocity of propagation},$$

$$E = \text{maximum voltage of wave}.$$

The voltage gradient is ￬

$$\frac{de}{dl} = \frac{de}{d\tau} \frac{d\tau}{dl}$$

$$= \frac{4E}{\pi} \frac{d\tau}{dl} \sum_{0}^{\infty} n \, \epsilon^{-ut} \cos (2n + 1)\tau \cos (2n + 1)\vartheta. \qquad (5)$$

Substituting for $\frac{d\tau}{dl}$ from (3), and resolving the cos-product, gives

$$\frac{de}{dl} = \frac{4E}{l_0} \left\{ \sum_{0}^{\infty} n \, \epsilon^{-ut} \cos (2n+1)(\tau - \vartheta) + \sum_{0}^{\infty} n \, \epsilon^{-ut} \cos(2n+1)(\tau + \vartheta) \right\} \qquad (6)$$

$$\sum_{0}^{\infty} n \, \epsilon^{-ut} \cos (2n + 1) (\tau + \vartheta) \text{ approaches zero for } \tau + \vartheta \gneq 0, \qquad (7)$$

thus,

$$\frac{de}{dl} = \frac{4E}{l_0} \sum_{0}^{\infty} n \, \epsilon^{-ut} \cos (2n + 1)(\tau - \vartheta). \qquad (8)$$

655

The maximum voltage gradient occurs for $\vartheta = \tau$, and thus is

$$G = \frac{de}{dl} = \frac{4 E}{l_0} \sum_0^\infty n \, \epsilon^{-ut} \tag{9}$$

This can be written in the form

$$G = \frac{2 E}{f_0 l_0} \sum_0^\infty n \, 2 f_0 \, \epsilon^{-ut} \tag{10}$$

It is, however, by (4),

$$f_0 l_0 = S \tag{11}$$

thus,

$$G = \frac{2 E}{S} \sum_0^\infty n \, 2 f_0 \, \epsilon^{-ut} \tag{12}$$

The values of ϵ^{-ut} are taken for all values of frequency, differing from each other by $2 f_0$, and, at and near the wave front, the value of $\sum_0^\infty n \, 2 f_0 \, \epsilon^{-ut}$ thus approaches the value of $\int_0^\infty \epsilon^{-ut} df$.

Substituting this into (10), gives as the equation of the maximum steepness of wave front

$$G = \frac{de}{dl} = \frac{2 E}{S} \int_0^\infty \epsilon^{-ut} \, df. \tag{12}$$

As seen, in this expression (12), wave length and frequency have disappeared. Equation (12) thus applies to any wave, whether of finite length or not. That is, it represents broadly (though only approximately) the maximum gradient of any steep wave front or impulse, at the time t after its origin as rectangular wave front or impulse.

As u is not a simple function of f, the integration of $\int_0^\infty \epsilon^{-ut} \, df$ in its general form meets with difficulties, and the integral may be evaluated thusly:

Equation (12) can be written in the form

$$G = \frac{2 E}{S} \int_{-\infty}^{+\infty} f \epsilon^{-ut} \, d \log f. \tag{13}$$

Plotting then $f \epsilon^{-ut}$ as ordinates, with $\log f$ as abscissæ, gives a curve, and the area of this curve gives the integral.

20. Instead of using $\log f$ as abscissæ, it is more convenient to use the common logarithm f, and divide the measured area by $\epsilon = .0.4343$.

Numerically, the integration is done by calculating $f\epsilon^{-ut}$ for (approximately) constant intervals of f, adding all the values,

TABLE X.—ATTENUATION OF WAVE FRONT OF RECTANGULAR IMPULSE.
Copper Wire No. 00 B. & S. G., 30 Ft. above Ground, Ground Return.

f	u	fe^{-ut} $t=10^{-3}$	10^{-4}	10^{-5}	10^{-6}	10^{-7}
1	70	1	1			
2	70	2	2			
5	70	5	5			
10	70	9	10			
20	70	19	20	} 88		
50	70	47	50			
10^2	70	93	99	100		
2×10^2	70	186	198	200		
5×10^2	70	468	496	500		
10^3	80	923	992	1,000	} 8,888	
2×10^3	115	1,760	1,970	2,000		
5×10^3	190	4,150	4,920	5,000		
10^4	287	7,500	2,720	9,999	10,000	
2×10^4	482	10,200	19,000	19,900	20,000	} 88,888
5×10^4	1,300	10,800	44,000	49,400	50,000	
10^5	3,710	2,450	69,000	96,400	99,500	100,000
2×10^5	12,690	1	56,100	176,000	197,000	200,000
5×10^5	74,200	0	240	190,000	466,000	498,000
10^6	291,500	......	0	54,300	748,000	970,000
2×10^6	1,160,000	......	....	18	625,000	1,780,000
5×10^6	7,230,000	......	....	0	3,660	1,860,000
10^7	28,800,000	......	....	....	1	562,000
2×10^7	115,000,000	......	....	....	0	200
$\Sigma =$		38,660	206,800	605,000	2,230,000	6,060,000
$\div 3 \times 0.4343 =$		29,700	158,700	464,000	1,710,000	4,650,000
$G = \dfrac{2E}{S} \times i =$		0.119	0.635	1.86	6.84	18.6
$[l =$		300	30	3	0.3	0.3 km.]

and multiplying the sum by the (average) difference between successive f.

As an instance is given, in Table X, the calculation for circuit (3) of the preceding, that is, copper wire No. 00 B. & S. G., 30 ft. above ground, with the ground as return conductor, for the times $t = 0.1$, 1, 10, 100 and 1000 microseconds, corresponding to the distance of travel of the wave front of 0.03, 0.3, 3, 30 and 300 km. As frequency intervals are selected: 1, 2, 5, 10, 20, 50, 100, 200, etc., giving the logarithms: 0, 0.3, 0.6, 1, 1.3, etc.

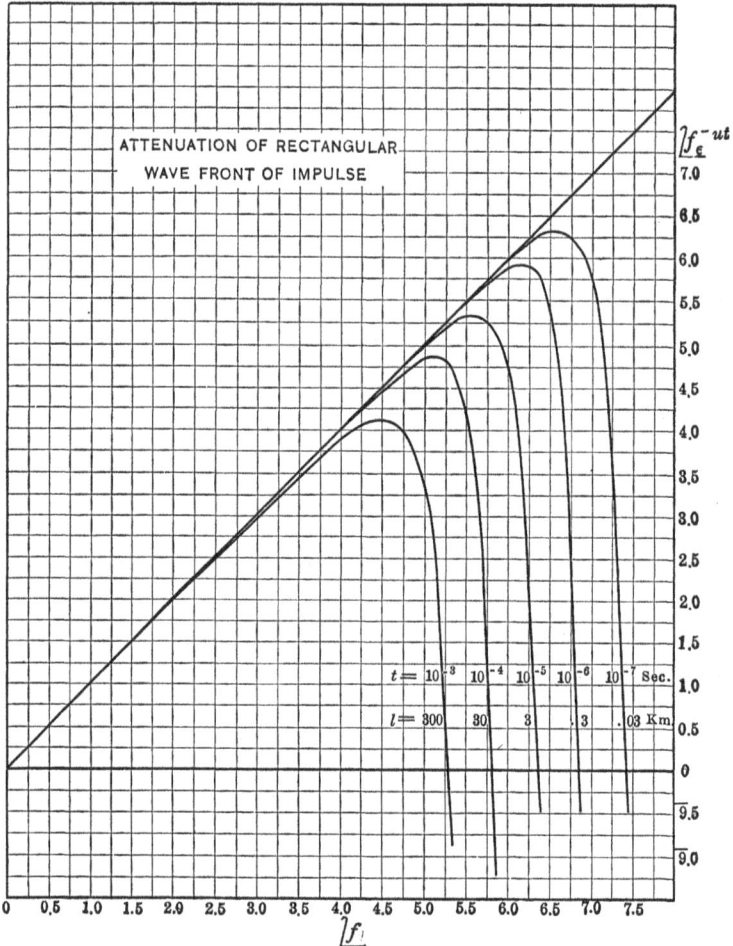

FIG. 111.

These curves are plotted, in logarithmic scale, in Fig. 111.

As seen, all these curves rise as straight lines under 45 degrees, and then very abruptly drop to negligible values.

In Table X, the values of $f\epsilon^{-ut}$ are added, then divided by 3, since there are three intervals for each unit of $]f$, and multiplied by $]\epsilon$, to reduce to natural logarithms   Multiplying by $\dfrac{E}{S}$ then gives the gradient G.

For medium and high frequencies, the attenuation constant u is given by the preceding equations as

$$u = \frac{r_1 + r_3}{2 (L_1 + L_0)} + \frac{g}{2 C}. \tag{14}$$

Neglecting the internal inductance L_1, as small compared with the external inductance, this gives

$$u = \frac{\dfrac{\pi}{l_1} \sqrt{2.5 \dfrac{\mu}{\lambda} \, 10^4}}{\log \dfrac{l'}{l_r}} \sqrt{f} + \frac{4 \, \pi^2 l'}{S \log \dfrac{l'}{l_r}} f^2 \Bigg\}$$
$$= m_1 \sqrt{f} + m_2 f^2 \tag{15}$$

where

$$m_1 = \frac{\pi \sqrt{2.5 \dfrac{\mu}{\lambda}} \, 10^4}{l_1 \log \dfrac{l'}{l_r}} \tag{16}$$

$$m_2 = \frac{4 \, \pi^2 l'}{S \log \dfrac{l'}{l_r}} \tag{17}$$

For the conductor (31) in Table X, it is
$$\begin{aligned} m_1 &= 2.6 \\ m_2 &= 0.29 \times 10^{-6} \end{aligned} \Bigg\} \tag{18}$$

This expression (15) of u holds for all frequencies except very low frequencies—below 1000 cycles—and extremely high frequencies—many millions of cycles. The latter are of little importance, as they are wiped out in the immediate neighborhood of the origin of the rectangular impulse. At the low frequencies, the attenuation is so small, within the distances which come into consideration in the wave travel, and these low frequencies give such a small part of the wave front gradient, that the error made by the use of (15) is negligible. For instance, in the case of Table X, even at $t = 100$ microseconds, or 10 km. wave travel, the error made in the voltage gradient by altogether neglecting the attenuation of the frequencies up to 1000 cycles, would be only 0.01 per cent.

Thus the equation (15) of the attenuation constant can safely be used for all practical purposes.

As the first term in equation (15) is proportional to $\sqrt{f}$, the second term to f^2, the second term is negligible at low and medium frequencies, while the first term is negligible at high frequencies.

Both terms are equal at

$$f = \left(\frac{m_1}{m_2}\right)^{2/3} \tag{19}$$

That is, in the above instance, at

$$f = 43,000 \text{ cycles.}$$

21. Thus, for high frequencies, that is, within moderate distances from the origin of the rectangular impulse—up to some kilometers—the first term can be neglected and the attenuation constant expressed by

$$u = m_2 f^2 \tag{20}$$

The integral in equation (12) then becomes

$$F = \int_0^\infty \epsilon^{-m_2 t\, f^2}\, df. \tag{21}$$

Substituting,

$$x^2 = m_2 t f^2$$

gives

$$df = \frac{dx}{\sqrt{m_2 t}}$$

and

$$F = \frac{1}{\sqrt{m_2 t}} \int_0^\infty \epsilon^{-x^2} dx.$$

It is, however,

$$\int_0^\infty \epsilon^{-x^2}\, dx = \frac{1}{2}\sqrt{\pi} \tag{22}$$

thus,

$$F = \frac{1}{2}\sqrt{\frac{\pi}{m_2 t}} \tag{23}$$

and

$$G = \frac{E\sqrt{\pi}}{S\sqrt{m_2 t}} \tag{24}$$

or, since

$$l = St:$$ (25)

$$G = \frac{E\sqrt{\pi}}{\sqrt{m_2 \, Sl}}$$ (26)

Substituting (17) into (26), gives, in cm. and volts per cm.

$$G = \frac{E}{2} \sqrt{\frac{\log \frac{l'}{l_r}}{\pi l' l}}$$

$$= 0.282 \, E \sqrt{\frac{1}{l'l} \log \frac{l'}{l_r}}$$ (27)

Thus, approximately:

The maximum gradient, or steepness of the wave front of a rectangular impulse, in the neighborhood and at moderate distances from its origin, decreases inverse proportional with the square root of the distance or time of wave travel.

It decreases with increasing distance l' of the return conductor, nearly inverse proportional to the square root of l'.

22. For the six types of circuits considered in the previous instances, it is:

$$G = \quad ?\overline{G\sqrt{l'}} =$$

(1) Copper wire 00 B. & S. G., 18 in. = 45.5 cm. from return conductor: $\dfrac{1}{\sqrt{l}} \times 1700 \quad 3.230$

(2) Copper wire 00 B. & S. G., 6 ft. = 182 cm. from return conductor: $\qquad\quad 968 \quad 2.986$
(3) Copper wire 00 B. & S. G., 60 ft. = 1820 cm. from return conductor: $\qquad\quad 360 \quad 2.556$
(4) Iron   wire 00 B. & S. G., 60 ft. = 1820 cm. from return conductor: $\qquad\quad 360 \quad 2.556$
(5) Copper wire 4 B. & S. G., 60 ft. = 1820 cm. from return conductor: $\qquad\quad 373 \quad 2.572$
(6) Aluminum, stranded, same conductivity and arrangement as (3): $\qquad\quad 353 \quad 2.548$

where G is given in volts per meter, at $E = 60,000$ volts, and l in kilometers:

$$\left. \begin{aligned} G &= 89 \times 10^{-3} \, E \sqrt{\frac{1}{l'l} \log \frac{l'}{l_r}} \\[2mm] &= 135 \times 10^{-3} \, E \sqrt{\frac{1}{l'l} \rceil \frac{l'}{l_r}} \\[2mm] &= 8100 \sqrt{\frac{1}{l'l} \rceil \frac{l'}{l_r}} \end{aligned} \right\}$$ (28)

From Tables X, IX, and VII are collected the values of wave gradients G, and given in Table XI, together with their logarithms and the $\rceil G_0$, calculated from equation (28), for comparison of the different methods of calculation. Table XI then gives the difference Δ, and its value in per cent.

The values of G are plotted in Fig. 112. The drawn line gives the values calculated by equation (28); the three-cornered stars the values from Table X, the crosses the values from Table IX, and the circles the values from Table VII.

TABLE XI.—CALCULATION OF WAVE FRONT.

Copper Wire No. 00 B. & S. G., 30 Ft. above Ground.

Dist. l km.	Gradient, Volts per meter at $E = 60,000$ V.				
	G	ηG	$\eta G_o{}^*$	Δ	$= \%$
	Impulse, $fo = 0$ (Table X).				
0.03	1860	3.269	3.317	+0.048	+11.7
0.3	684	2.835	2.817	−0.018	− 4.3
3	186	2.269	2.317	+0.048	+11.7
30	63.5	1.803	1.817	+0.014	+ 3.3
300	11.9	1.075	1.317	(+0.242)	
	Traveling Wave, $fo = 60,000$ (Table IX).				
0.3	690	2.839	2.817	−0.022	−5.2
0.6	470	2.672	2.667	−0.005	−1.1
1.5	295	2.470	2.468	−0.002	−0.5
3	205	2.312	2.317	+0.005	+1.1
6	140	2.146	2.167	+0.021	+5.0
12	100	2.000	2.016	+0.016	+3.7
30	65	1.813	1.817	+0.004	+0.9
108	40	1.602	1.539	(−0.063)	
	Quarter Wave, $fo = 750$ (Table VII).				
0.01	5500	3.740	3.556	−0.184	
0.1725	1290	3.111	2.938	−0.173	
0.345	967	2.985	2.787	−0.198	
0.69	630	2.799	2.637	+0.162	
3.45	280	2.447	2.287	−0.160	
6.9	205**	2.312	2.137	−0.175	
27.6	98	1.991	1.836	−0.155	
69	58	1.763	1.637	(−0.125)	
690	12	1.079	1.137	(+0.058)	
6900	1.2	0.079	0.637	(+0.558)	
	Same, 18 in. = 45.5 cm. between conductors.				
0.69	3240	3.510	3.311	−0.199	
6.9	960	2.982	2.811	−0.171	
69.0	234	2.369	2.311	(−0.058)	

* Calculated by equation (28).

** 131 to 293

23. As seen from Table XI and Fig. 112, the agreement of the equations (27) and (28) is satisfactory with the values of the wave front gradient taken from Tables X and IX.

The values of G from Table X differ erratically. This table was calculated by graphical integration, and it is probable that the intervals have been chosen too large, in that range where the curve drops very abruptly, as seen in Fig. 111.

The agreement with the values from Table IX is very close. In this table, representing the course of 60.000-cycle rectangular

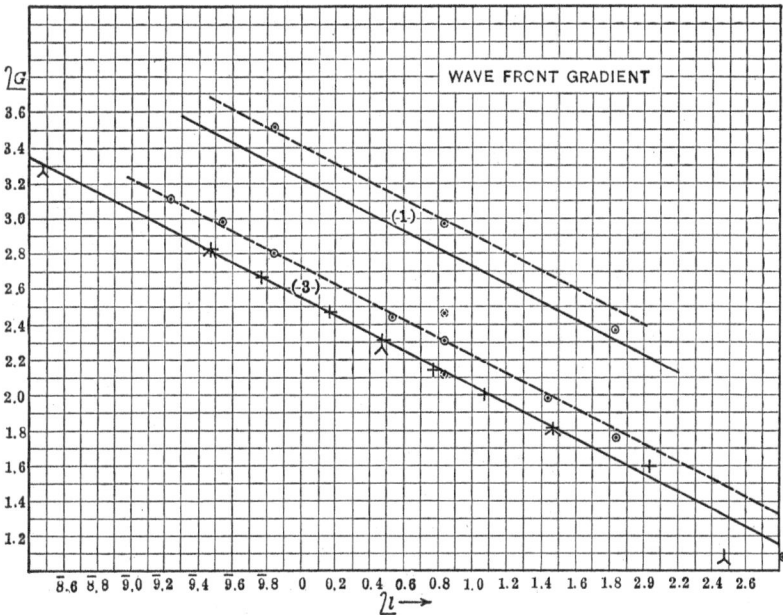

Fig. 112.

traveling wave, the individual harmonics have been calculated, as they were relatively few, due to the high frequency of the fundamental. This agreement is important as it justifies equation (27). In deriving (27), we have substituted integration for summation, that is, have replaced the discontinuous values of the individual harmonics by a continuous curve. Any error resulting from this should be greatest where the number of discontinuous harmonics is the least. Table XI and Fig. 112, however, show that the agreement of the gradient of the 60,000-cycle traveling wave with equation (27), is good even at $l = 30$ km., where only two significant harmonics are left, the fundamental and the third harmonic.

Thus the method of deriving an equation for G by integration is justified.

Unsatisfactory, however, is the agreement of equation (27) with the values of the gradient of the quarter-wave oscillation, taken from Table VII. These values lay on a straight line in Fig. II, given as dotted line, showing proportionality with the square root of the distance, but there is a constant error, and the gradients given in Table VII are about 50 per cent greater than those given by equation (27).

The cause of the discrepancy probably is in the method used in calculating the gradients of the quarter-wave oscillation given in Table VII. Due to the impossibility of calculating individually thousands of harmonics, the number of harmonics in successive intervals of frequency, has been multiplied with the average attenuation ϵ^{-ut} in this frequency range, and as the average has been used the mean value of the ϵ^{-ut} for the two extremes of the frequency range. However, ϵ^{-ut} drops exponentially and with great rapidity, so that the true average value is much lower than the average of maximum and minimum. Thus for instance in the range from 5×10^5 to 10×10^5 cycles, containing 334 harmonics, ϵ^{-ut} for $f = 5 \times 10^5$ [at $t = 23 \times 10^{-6}$ seconds], is 1.22×10^{-3} and for $f = 10 \times 10^5$, it is 181×10^{-3}, giving an average of 91.1×10^{-3}. However, for $f = 7.5 \times 10^5$, ϵ^{-ut} is 22.5×10^{-3}, thus less than a quarter of the average.

To check this, one value, at $t = 23 \times 10^{-6}$ or $l = 6.9$ km., has been re-calculated, by using not the average, but the maximum and the minimum of ϵ^{-ut} in each interval, and the two gradients derived therefrom: $G = 293$ and 131, are marked with dotted circles in Fig. 112. As seen, the use of the minimum value of ϵ^{-ut} agrees nearer with equation (27), as was to be expected.

It appears probable that the equation (27) gives more reliable values of wave front gradient, within the range of its applicability, than the method of calculation used in Table VII, and as it is much simpler, it is preferable.

As seen from Table XI and Fig. 112, the parabolic law of wave front flattening, given by equation (27) and (28), holds good up to about 30 km. distance of wave travel, and with fair approximation even to 100 km. In the range beyond this—which is of lesser importance, as the flattening of the wave front has greatly reduced its danger at these distances—the values of the gradient G decrease with increasing rapidity, with distance and time, due to the medium high harmonics showing in the attenuation.

24. At great distances from the origin of the rectangular wave, when the very high harmonics have practically died out and the wave attenuation is determined by the medium frequency harmonics, the second term of u in equation (15) becomes negligible, and u can be approximated by

$$u = m_1\sqrt{f} \tag{37}$$

The integral in equation (12) then becomes

$$F = \int_0^\infty \epsilon^{-m_1 t \sqrt{f}} df \tag{38}$$

Substituting,

$$m_1 t \sqrt{f} = x$$

gives

$$df = \frac{2\ x\,dx}{m_1{}^2 t^2}$$

and

$$F = \frac{2}{m_1{}^2 t^2} \int_0^\infty \epsilon\,x^{-x} dx$$

It is, however,

$$\int_0^\infty x\epsilon^{-x}\,dx = 1 \tag{39}$$

thus,

$$F = \frac{2}{m_1{}^2 t^2} \tag{40}$$

and

$$G = \frac{4\ E}{S m_1{}^2 t^2} \tag{41}$$

or,

$$G = \frac{4\ SE}{m_1{}^2 l^2} \tag{42}$$

Substituting (16) gives

$$G = \frac{1.6 \, S\lambda l_1{}^2 \left(\log \frac{l'}{l_r}\right)^2}{\pi^2 \mu l^2} E \qquad (43)$$

That is, at great distances (or considerable time) from the origin, the flattening of the wave front approaches inverse proportionality with the square of the distance (or time) of the wave travel.

However, this range is of little importance.

APPENDIX

VELOCITY FUNCTIONS OF THE ELECTRIC FIELD

1. In the study of the propagation of the electric field through space (wireless telegraphy and telephony, lightning discharges and other very high-frequency phenomena), a number of new functions appear (Section III, Chapter VIII).

By the following equations these functions are defined, and related to the "Sine-Integral" Si x, the "Cosine-Integral" Ci x, and the "Exponential Integral," Ei x, of which tables were calculated by J. W. L. Glaisher (Philosophical Transactions of the Royal Society of London, 1870, Vol. 160):

$$[\underline{\,n} \text{ here denotes } 1 \times 2 \times 3 \times \ldots \times n]$$

$$\text{col } x = \int_x^\infty \frac{\sin u}{u}\, du$$

$$= \int_x^\infty \left\{ u - \frac{u^3}{\underline{|3}} + \frac{u^5}{\underline{|5}} - \frac{u^7}{\underline{|7}} + - \ldots \right\} \frac{du}{u}$$

$$= \left/ u - \frac{1}{3}\frac{u^3}{\underline{|3}} + \frac{1}{5}\frac{u^5}{\underline{|5}} - \frac{1}{7}\frac{u^7}{\underline{|7}} + - \ldots \right/_x^\infty$$

$$= \frac{\pi}{2} - \left\{ x - \frac{1}{3}\frac{x^3}{\underline{|3}} + \frac{1}{5}\frac{x^5}{\underline{|5}} - \frac{1}{7}\frac{x^7}{\underline{|7}} + - \ldots \right\}$$

$$= \cos x \left\{ \frac{1}{x} - \frac{\underline{|2}}{x^3} + \frac{\underline{|4}}{x^5} - \frac{\underline{|6}}{x^7} + - \ldots \right\}$$

$$+ \sin x \left\{ \frac{1}{x^2} - \frac{\underline{|3}}{x^4} + \frac{\underline{|5}}{x^6} - \frac{\underline{|7}}{x^8} + - \ldots \right\}$$

$$= \frac{\pi}{2} - \text{Si } x$$

$$= \frac{\pi}{2} - \int_0^x \frac{\sin u}{u}\, du.$$

Herefrom follows:

$$\int_x^\infty \frac{\sin au}{u}\, du = \text{col } ax$$

$$\mathrm{sil}\ x = \int_x^\infty \frac{\cos u}{u}\,du$$

$$= \int_x^\infty \left\{1 - \frac{u^2}{\underline{|2}} + \frac{u^4}{\underline{|4}} - \frac{u^6}{\underline{|6}} + -\dots\right\}\frac{du}{u}$$

$$= \Big/ \log u - \frac{1}{2}\frac{u^2}{\underline{|2}} + \frac{1}{4}\frac{u^4}{\underline{|4}} - \frac{1}{6}\frac{u^6}{\underline{|6}} + -\dots \Big/_x^\infty$$

$$= -\left\{\gamma + \frac{1}{4}\log x^4\right\} + \left\{\frac{1}{2}\frac{x^2}{\underline{|2}} - \frac{1}{4}\frac{x^4}{\underline{|4}} + \frac{1}{6}\frac{x^6}{\underline{|6}} - \frac{1}{8}\frac{x^8}{\underline{|8}} + -\dots\right\}$$

$$= \cos x \left\{\frac{1}{x^2} - \frac{\underline{|3}}{x^4} + \frac{\underline{|5}}{x^6} - \frac{\underline{|7}}{x^8} + -\dots\right\}$$

$$\hspace{4cm} - \sin x\left\{\frac{1}{x} - \frac{\underline{|2}}{x^3} + \frac{\underline{|4}}{x^5} - \frac{\underline{|3}}{x^7} + -\dots\right\}$$

$$= -\mathrm{Ci}\,(x)$$

$$= -\int_\infty^x \frac{\cos u}{u}\,du;$$

where

$$\gamma = 0.5772156\ \dots$$

herefrom follows:

$$\int_x^\infty \frac{\cos au}{u}\,du = \mathrm{sil}\,ax$$

$$\mathrm{Expl}\ x = \int_x^q \frac{\varepsilon^u}{u}\,du = \int_{-x}^\infty \frac{\varepsilon^{-u}}{u}\,du$$

$$= \int_x^q \left\{1 + u + \frac{u^2}{\underline{|2}} + \frac{u^3}{\underline{|3}} + \frac{u^4}{\underline{|4}} + \dots\right\}\frac{du}{u}$$

$$= \Big/ \log u + u + \frac{1}{2}\frac{u^2}{\underline{|2}} + \frac{1}{3}\frac{u^3}{\underline{|3}} + \frac{1}{4}\frac{u^4}{\underline{|4}} + \dots \Big/_x^q$$

$$= -\left\{\gamma + \frac{1}{4}\log x^4 + x + \frac{1}{2}\frac{x^2}{\underline{|2}} + \frac{1}{3}\frac{x^3}{\underline{|3}} + \frac{1}{4}\frac{x^4}{\underline{|4}} + \dots\right\}$$

$$= \varepsilon^x\left\{\frac{1}{x} + \frac{1}{x^2} + \frac{\underline{|2}}{x^3} + \frac{\underline{|3}}{x^4} + \frac{\underline{|4}}{x^5} + \dots\right\}$$

$$= -\mathrm{Ei}\,x$$

$$= -\int_\infty^{-x} \frac{\varepsilon^{-u}}{u}\,du;$$

where

q is given by: expl $q = 0$ as: $q = 0.37249680\ldots$

herefrom follows:

$$\int_x^q \frac{\epsilon^{au}}{u}\,du = expl\ ax$$

$$\text{Expl}\,(-x) = \int_x^\infty \frac{\epsilon^{-u}}{u}\,du$$

$$= \int_x^\infty \left\{ 1 - u + \frac{u^2}{\underline{|2}} - \frac{u^3}{\underline{|3}} + \frac{u^4}{\underline{|4}} - + \ldots \right\} \frac{du}{u}$$

$$= \left/ \log u - u + \frac{1}{2}\frac{u^2}{\underline{|2}} - \frac{1}{3}\frac{u^3}{\underline{|3}} + \frac{1}{4}\frac{u^4}{\underline{|4}} - + \ldots \right/_x^\infty$$

$$= -\left\{ \gamma + \frac{1}{4}\log x^4 \right\} + \left\{ x - \frac{1}{2}\frac{x^2}{\underline{|2}} + \frac{1}{3}\frac{x^3}{\underline{|3}} - \frac{1}{4}\frac{x^4}{\underline{|4}} + - \ldots \right\}$$

$$\dot= \epsilon^{-x}\left\{ \frac{1}{x} - \frac{1}{x^2} + \frac{\underline{|2}}{x^3} - \frac{\underline{|3}}{x^4} + \frac{\underline{|4}}{x^5} - + \ldots \right\}$$

$$= -\text{Ei}\,(-x)$$

$$= -\int_\infty^x \frac{\epsilon^{-u}}{u}\,du.$$

herefrom follows:

$$\int_x^\infty \frac{\epsilon^{-au}}{u}\,du = expl\,(-ax)$$

Tables of these four functions, reduced from the Glaisher tables, are given in the following for 6 decimals, and their first part plotted in Fig. 113.

2. As seen from the preceding equations, these functions have the following properties:

col $0 = \dfrac{\pi}{2}$ col ∞ $= 0$

sil $0 = +\infty$ sil ∞ $= 0$

expl $0 = +\infty$ expl $(+\infty) = -\infty$

expl $q = 0$ expl $(-\infty) = 0$

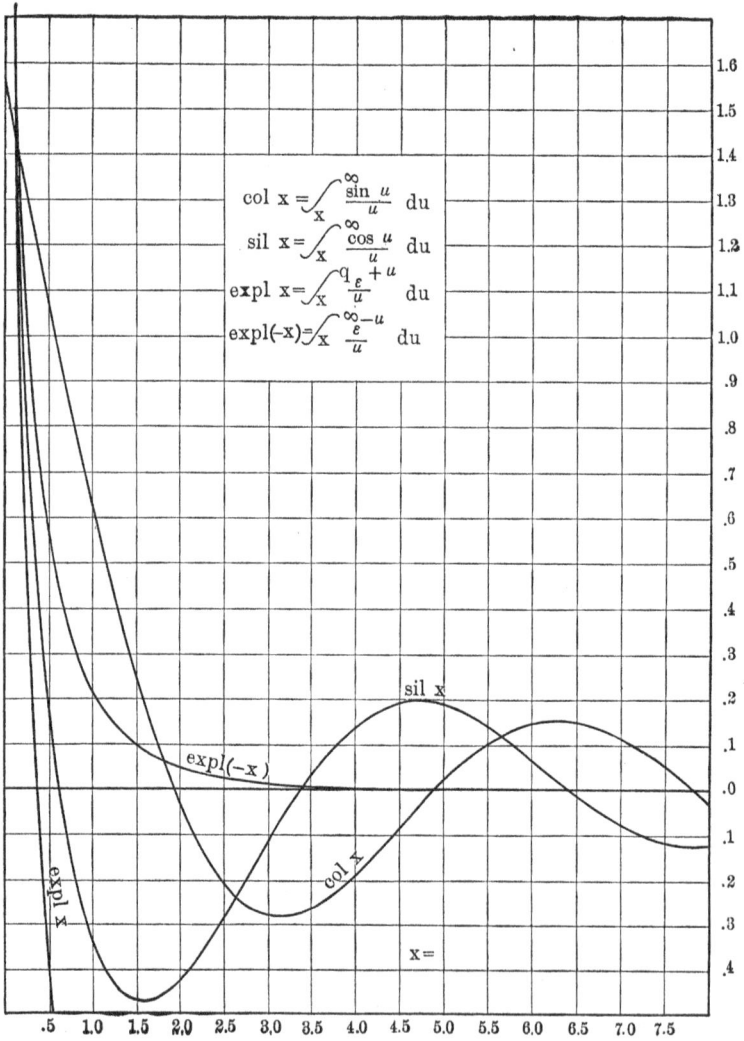

col $x = \int_x^\infty \dfrac{\sin u}{u}\ du$

sil $x = \int_x^\infty \dfrac{\cos u}{u}\ du$

expl $x = \int_x^q \dfrac{\epsilon^{+u}}{u}\ du$

expl$(-x) = \int_x^\infty \dfrac{\epsilon^{-u}}{u}\ du$

Fig. 113.

col x has maxima at the even, sil x at the odd quadrants, and these maxima are alternately positive and negative; that is,

$$\left.\begin{array}{l} \operatorname{col} \dfrac{\pi}{2} 2s \qquad = \text{max.} \\[3mm] \operatorname{sil} \dfrac{\pi}{2}(2s-1) = \text{max.} \end{array}\right\} s = \text{integer number.}$$

For large values of s, the numerical values of these maxima approach the values:

$$\operatorname{col} \dfrac{\pi}{2}\, 2s \qquad = (-1)^s \dfrac{1}{\pi s};$$

$$\operatorname{sil} \dfrac{\pi}{2}(2s-1) = (-1)^s \dfrac{2}{\pi(2s-1)},$$

for: $s > 40$, the approximation is correct to the 6th decimal.

From the series expressions of these functions follows:

$$\operatorname{col}(-x) \quad = \pi - \operatorname{col} x$$

$$\operatorname{sil}(-x) \quad = \operatorname{sil} x$$

$$\operatorname{expl}(jx) \quad = \operatorname{sil} x + j \operatorname{col} x - j\dfrac{\pi}{2}$$

$$\operatorname{expl}(-jx) = \operatorname{sil} x - j \operatorname{col} x + j\dfrac{\pi}{2}$$

$$\operatorname{col}(\pm jx) \quad = \pm \dfrac{j}{2}\{\operatorname{expl} x - \operatorname{expl}(-x)\} + j\pi$$

$$\operatorname{sil}(\pm jx) \quad = \dfrac{1}{2}\{\operatorname{expl} x + \operatorname{expl}(-x)\}$$

For small values of x, the approximations hold:

$$\operatorname{sil} x = \log \dfrac{1}{x} - 0.5772156 + \dfrac{x^2}{4}$$

$$= \log \dfrac{0.5615}{x} + \dfrac{x^2}{4}$$

$$\operatorname{col} x = \dfrac{\pi}{2} - x$$

These approximations are accurate within one per cent for values of x up to 0.67.

For very small values of x, the approximations hold:

$$\text{sil } x = \log \frac{1}{x} - 0.5772156$$

$$= \log \frac{0.5615}{x}$$

$$\text{col } x = \frac{\pi}{2}$$

These are accurate within one per cent for values of x up to 0.016.

For moderately large values of x the approximations hold:

$$\text{sil } x = -\frac{\sin x}{x} + \frac{\cos x}{x^2}$$

$$\text{col } x = \frac{\cos x}{x} + \frac{\sin x}{x^2}$$

These are accurate within one per cent for values of x down to 8.

For large values of x, the approximations hold:

$$\text{sil } x = -\frac{\sin x}{x}$$

$$\text{col } x = \frac{\cos x}{x}$$

or:

$$\text{sil } x = -\frac{\sin v}{x}$$

$$\text{col } x = \frac{\cos v}{x}$$

where:

$$x = 2\pi u + v$$

and $v < 2\pi$, that is, u is the largest multiple of 2π, contained in x.

These approximations are accurate within one per cent for values of x down to 12.

In some problems on the velocity of propagation of the electric field through space, such as the mutual inductance of two finite conductors at considerable distance from each other, or the capacity of a sphere in space, two further functions appear, coll x and sill x, which by partial integration can be reduced to the functions col x and sil x.

It is:

$$\text{coll } x = \int_x^\infty \frac{\cos u}{u^2} \, du$$

$$= -\int_x^\infty \cos u \, d\left(\frac{1}{u}\right)$$

$$= \frac{\cos x}{x} - \text{col } x$$

$$\text{sill } x = \int_x^\infty \frac{\sin u}{u^2} \, du$$

$$= -\int_x^\infty \sin u \, d\left(\frac{1}{u}\right)$$

$$= \frac{\sin x}{x} + \text{sil } x$$

Herefrom follows:

$$\int_x^\infty \frac{\cos au}{u^2} \, du = a \text{ coll } ax$$

$$= \frac{\cos ax}{x} - a \text{ sill } ax$$

$$\int_x^\infty \frac{\sin au}{u^2} \, du = a \text{ sill } ax$$

$$= \frac{\sin ax}{x} + a \text{ sil } ax$$

It is:

$$\text{coll } o = \infty \qquad\qquad \text{coll } \infty = o$$
$$\text{sill } o = \infty \qquad\qquad \text{sill } \infty = o$$

With increasing values of x, coll x and sill x decrease far more rapidly than col x and sil x.

For moderately large values of x, the approximations hold:

$$\text{coll } x = -\frac{\sin x}{x^2} - \frac{2 \cos x}{x^3}$$

$$\text{sill } x = \frac{\cos x}{x^2} + \frac{2 \sin x}{x^3}$$

These approximations are accurate within one per cent for values of x up to respectively

For large values of x, the approximations hold:

$$\text{coll } x = -\frac{\sin x}{x^2}$$

$$\text{sill } x = \frac{\cos x}{x^2}$$

These are accurate within one per cent for values of x up to respectively 12.

For low and medium values of x, coll x and sill x are preferably reduced to col x and sil x by above given equations.

TABLE I
Col x and sil x from 0.00 to 1.00

x	col x	$\varDelta$	sil x	$\varDelta$
.00	+1.570 796	. . . .	∞	. . . .
.01	1.560 796	10 000	+4.027 980	∞
.02	1.550 797	9 999	3.334 907	693 073
.03	1.540 798	9 999	2.929 567	405 340
.04	1.530 800	9 998	2.642 060	287 507
.05	1.520 803	9 997	2.419 142	222 918
.06	1.510 808	9 995	2.237 095	182 047
.07	1.500 815	9 993	2.083 269	153 826
.08	1.490 825	9 990	1.950 113	133 156
.09	1.480 837	9 988	1.832 754	117 359
.10	1.470 852	9 985	1.727 868	104 886
.11	1.460 870	9 982	1.633 083	94 785
.12	1.450 892	9 978	1.546 646	86 437
.13	1.440 918	9 974	1.467 227	79 419
.14	1.430 949	9 969	1.393 793	73 434
.15	1.420 984	9 965	1.325 524	68 269
.16	1.411 024	9 960	1.261 759	63 765
.17	1.401 069	9 955	1.201 957	59 802
.18	1.391 120	9 949	1.145 672	56 285
.19	1.381 177	9 943	1.092 527	53 145
.20	1.371 240	9 937	1.042 206	50 321
.21	1.361 310	9 930	.994 437	47 769
.22	1.351 387	9 923	.948 988	45 449
.23	1.341 471	9 916	.905 656	43 332
.24	1.331 563	9 908	.864 266	41 390
.25	1.321 663	9 900	.824 663	39 603
.26	1.311 771	9 892	.786 710	37 953
.27	1.301 888	9 883	.750 287	36 423
.28	1.292 013	9 875	.715 286	35 001
.29	1.282 148	9 865	.681 610	33 676
.30	1.272 292	9 856	.649 173	32 437
.31	1.262 447	9 845	.617 896	31 277
.32	1.252 611	9 836	.587 710	30 186
.33	1.242 786	9 825	.558 549	29 161
.34	1.232 972	9 814	.530 355	28 194
.35	1.223 169	9 803	.503 076	27 279
.36	1.213 378	9 791	.476 661	26 414
.37	1.203 599	9 779	.451 067	25 594
.38	1.193 832	9 767	.426 252	24 815
.39	1.184 077	9 755	.402 178	24 074
.40	1.174 335	9 742	.378 809	23 369
.41	1.164 606	9 729	.356 114	22 695
.42	1.154 891	9 715	.334 062	22 052
.43	1.145 189	9 702	.312 625	21 437
.44	1.135 501	9 688	.291 776	20 849
.45	1.125 828	9 673	.271 492	20 284
.46	1.116 170	9 658	.251 749	19 743
.47	1.106 526	9 644	.232 526	19 223
.48	1.096 898	9 628	.213 803	18 723
.49	1.087 285	9 613	.195 562	18 241
.50	+1.077 689	9 596	+0.177 784	17 778

APPENDIX

TABLE I—*Continued*

x	col x	Δ	sil x	Δ
.50	+1.077 689		+0.177 784	
		9 596		17 778
.51	1.068 108		.160 453	
		9 581		17 331
.52	1.058 545		.143 554	
		9 563		16 899
.53	1.048 998		.127 071	
		9 547		16 483
.54	1.039 448		.110 990	
		9 530		16 081
.55	1.029 956		.095 300	
		9 512		15 690
.56	1.020 461		.079 986	
		9 495		15 314
.57	1.010 985		.065 037	
		9 476		14 949
.58	1.001 527		.050 441	
		9 458		14 596
.59	.992 088		.036 190	
		9 439		14 251
.60	.982 667		.022 271	
		9 421		13 919
.61	.973 266		+ .008 675	
		9 401		13 516
.62	.963 885		− .004 606	
		9 381		13 281
.63	.954 523		− .017 582	
		9 362		12 976
.64	.945 182		− .030 260	
		9 341		12 678
.65	.935 861		− .042 650	
		9 321		12 390
.66	.926 561		− .054 758	
		9 300		12 108
.67	.917 282		− .066 591	
		9 279		11 833
.68	.908 024		− .078 158	
		9 258		11 567
.69	.898 788		− .089 463	
		9 236		11 305
.70	.889 574		− .100 515	
		9 214		11 052
.71	.880 382		− .111 318	
		9 192		10 803
.72	.871 213		− .121 879	
		9 169		10 561
.73	.862 066		− .132 203	
		9 147		10 324
.74	.852 942		− .142 296	
		9 124		10 093
.75	.843 842		− .152 164	
		9 100		9 868
.76	.834 765		− .161 810	
		9 077		9 646
.77	.825 713		− .171 240	
		9 052		9 430
.78	.816 684		− .180 458	
		9 029		9 218
.79	.807 680		− .189 470	
		9 004		9 012
.80	.798 700		− .198 279	
		8 980		8 809
.81	.789 746		− .206 889	
		8 954		8 601
.82	.780 817		− .215 305	
		8 929		8 416
.83	.771 913		− .223 530	
		8 904		8 225
.84	.763 035		− .231 568	
		8 878		8 038
.85	.754 184		− .239 423	
		8 851		7 855
.86	.745 358		− .247 098	
		8 825		7 675
.87	.736 560		− .254 597	
		8 798		7 499
.88	.727 788		− .261 923	
		8 772		7 326
.89	.719 043		− .269 079	
		8 745		7 156
.90	.710 325		− .276 068	
		8 718		6 989
.91	.701 636		− .282 893	
		8 689		6 825
.92	.692 974		− .289 558	
		8 662		6 665
.93	.684 340		− .296 064	
		8 634		6 506
.94	.675 735		− .302 415	
		8 605		6 351
.95	.667 158		− .308 614	
		8 577		6 199
.96	.658 610		− .314 662	
		8 548		6 048
.97	.650 092		− .320 563	
		8 518		5 901
.98	.641 602		− .326 319	
		8 490		5 756
.99	.633 143		− .331 931	
		8 459		5 612
1.00	+0.624 713		−0.337 404	
		8 430		5 473

TABLE II

Expl x and expl $(-x)$ from 0.00 to 1.00

x	expl x	Δ	expl $(-x)$	Δ
.00	$+\infty$	. . . .	∞	. . . .
		∞		∞
.01	$+4.017\ 929$	703 223	$+4.037\ 930$	684 222
.02	3.314 707	415 591	3.354 708	395 589
.03	2.899 116	297 859	2.959 119	277 855
.04	2.601 257	233 372	2.681 264	213 366
.05	2.367 885	192 602	2.467 898	172 591
.06	2.175 283	164 483	2.295 307	144 469
.07	2.010 800	143 916	2.150 838	123 897
.08	1.886 884	128 220	2.026 941	108 196
.09	1.738 664	115 851	1.918 745	95 821
.10	1.622 813	105 854	1.822 924	85 817
.11	1.516 959	97 609	1.737 107	77 565
.12	1.419 350	90 695	1.659 542	70 643
.13	1.328 655	84 814	1.588 899	64 753
.14	1.243·841	79 754	1.524 146	59 684
.15	1.164 086	75 355	1.464 462	55 275
.16	1.088 731	71 497	1.409 187	51 406
.17	1.017 234	68 087	1.357 781	47 985
.18	.949 148	65 052	1.309 796	44 938
.19	.884 096	62 335	1.264 858	42 207
.20	.821 761	59 889	1.222 651	39 749
.21	.761 872	57 676	1.182 902	37 522
.22	.704 195	55 666	1.145 380	35 497
.23	.648 529	53 832	1.109 883	33 648
.24	.594 697	52 153	1.076 235	31 952
.25	.542 543	50 611	1.044 283	30 394
.26	.491 932	49 191	1.013 889	28 956
.27	.442 741	47 878	.984 933	27 625
.28	.394 863	46 662	.957 308	26 390
.29	.348 202	45 533	.930 918	25 241
.30	.302 669	44 482	.905 677	24 171
.31	.258 186	43 503	.881 506	23 171
.32	.214 683	42 588	.858 335	22 234
.33	.172 095	41 732	.836 101	21 355
.34	.130 363	40 929	.814 746	20 531
.35	.089 434	40 176	.794 215	19 753
.36	.049 258	39 468	.774 462	19 021
.37	$+$ .009 790	38 801	.755 441	18 329
.38	$-$ .029 011	38 173	.737 112	17 675
.39	$-$ .067 185	37 581	.719 437	17 057
.40	$-$ .104 765	37 021	.702 380	16 470
.41	$-$ .141 786	36 492	.685 910	15 913
.42	$-$ .178 278	35 991	.669 997	15 384
.43	$-$ .214 270	35 517	.654 613	14 880
.44	$-$ .249 787	35 068	.639 733	14 402
.45	$-$ .284 855	34 642	.625 331	13 944
.46	$-$ .319 497	34 238	.611 387	13 510
.47	$-$ .353 735	33 854	.597 877	13 093
.48	$-$ .387 589	33 489	.584 784	12 695
.49	$-$ .421 078	33 142	.572 089	12 315
.50	$-0.454\ 220$		$+0.559\ 774$	

TABLE II—*Continued*

x	expl x	Δ	expl $(-x)$	Δ
.50	$-0.454\ 220$	33 142	$+0.559\ 774$	12 315
.51	$-\ .487\ 032$	32 812	$.547\ 822$	11 952
.52	$-\ .519\ 531$	32 498	$.536\ 220$	11 602
.53	$-\ .551\ 730$	32 200	$.524\ 952$	11 268
.54	$-\ .583\ 646$	31 915	$.514\ 004$	10 948
.55	$-\ .615\ 291$	31 645	$.503\ 364$	10 640
.56	$-\ .646\ 677$	31 387	$.493\ 020$	10 344
.57	$-\ .677\ 819$	31 141	$.482\ 960$	10 060
.58	$-\ .708\ 726$	30 907	$.473\ 173$	9 787
.59	$-\ .739\ 410$	30 684	$.463\ 650$	9 523
.60	$-\ .769\ 881$	30 472	$.454\ 380$	9 270
.61	$-\ .800\ 150$	30 269	$.445\ 353$	9 027
.62	$-\ .830\ 226$	30 076	$.436\ 562$	8 791
.63	$-\ .860\ 119$	29 892	$.427\ 997$	8 565
.64	$-\ .889\ 836$	29 717	$.419\ 652$	8 345
.65	$-\ .919\ 386$	29 550	$.411\ 517$	8 135
.66	$-\ .948\ 778$	29 392	$.403\ 586$	7 931
.67	$-\ .978\ 019$	29 241	$.395\ 853$	7 733
.68	$-1.007\ 116$	29 079	$.388\ 309$	7 544
.69	$-1.036\ 077$	28 960	$.380\ 950$	7 359
.70	$-1.064\ 907$	28 831	$.373\ 769$	7 181
.71	$-1.093\ 615$	28 707	$.366\ 760$	7 009
.72	$-1.122\ 205$	28 590	$.359\ 918$	6 842
.73	$-1.150\ 684$	28 479	$.353\ 237$	6 681
.74	$-1.179\ 058$	28 374	$.346\ 713$	6 524
.75	$-1.207\ 333$	28 275	$.340\ 341$	6 3 2
.76	$-1.235\ 513$	28 181	$.334\ 115$	6 226
.77	$-1.263\ 605$	28 092	$.328\ 032$	6 083
.78	$-1.291\ 613$	28 008	$.322\ 088$	5 944
.79	$-1.319\ 542$	27 929	$.316\ 277$	5 811
.80	$-1.347\ 397$	27 855	$.310\ 597$	5 680
.81	$-1.375\ 182$	27 785	$.305\ 043$	5 554
.82	$-1.402\ 902$	27 720	$.299\ 611$	5 432
.83	$-1.430\ 561$	27 659	$.294\ 299$	5 312
.84	$-1.458\ 164$	27 603	$.289\ 103$	5 196
.85	$-1.485\ 714$	27 550	$.284\ 019$	5 084
.86	$-1.513\ 216$	27 502	$.279\ 045$	4 974
.87	$-1.540\ 673$	27 457	$.274\ 177$	4 868
.88	$-1.568\ 089$	27 416	$.269\ 413$	4 764
.89	$-1.595\ 467$	27 379	$.264\ 749$	4 664
.90	$-1.622\ 812$	27 345	$.260\ 184$	4 565
.91	$-1.650\ 126$	27 314	$.255\ 714$	4 470
.92	$-1.677\ 413$	27 287	$.251\ 336$	4 378
.93	$-1.704\ 677$	27 264	$.247\ 050$	4 286
.94	$-1.731\ 920$	27 243	$.242\ 851$	4 199
.95	$-1.759\ 146$	27 226	$.238\ 738$	4 113
.96	$-1.786\ 357$	27 211	$.234\ 708$	4 030
.97	$-1.813\ 557$	27 200	$.230\ 760$	3 948
.98	$-1.840\ 749$	27 192	$.226\ 891$	3 869
.99	$-1.867\ 935$	27 186	$.223\ 100$	3 791
1.00	$-1.895\ 118$	27 183	$+0.219\ 384$	3 716

TABLE III

Col x and sil x from 0.0 to 5.0

x	col x	Δ	sil x	Δ
.0	$+1.570\ 796$	. . . .	∞	. . . .
.1	$1.470\ 852$		$+1.727\ 868$	
.2	$1.371\ 240$		$1.042\ 206$	
.3	$1.272\ 292$		$.649\ 173$	
.4	$1.174\ 335$		$.378\ 809$	
.5	$1.077\ 689$	. . . .	$.177\ 784$	. . . .
.6	$.982\ 667$		$+\ .022\ 271$	
.7	$.889\ 574$		$-\ .100\ 515$	
.8	$.798\ 700$		$-\ .198\ 279$	
.9	$.710\ 325$		$-\ .276\ 068$	
1.0	$.624\ 713$		$-\ .337\ 404$	. . . .
		$+82\ 602$		$+47\ 469$
1.1	$.542\ 111$	$79\ 362$	$-\ .384\ 873$	$35\ 586$
1.2	$.462\ 749$	$75\ 911$	$-\ .420\ 459$	$25\ 279$
1.3	$.387\ 838$	$72\ 269$	$-\ .445\ 739$	$16\ 268$
1.4	$.314\ 570$	$68\ 457$	$-\ .462\ 007$	$8\ 349$
1.5	$.246\ 113$		$-\ .470\ 356$	
		$64\ 497$		$+\ 1\ 377$
1.6	$.181\ 616$	$60\ 412$	$-\ .471\ 733$	$-\ 4\ 765$
1.7	$.121\ 204$	$56\ 225$	$-\ .466\ 968$	$-10\ 157$
1.8	$.064\ 979$	$51\ 958$	$-\ .456\ 811$	$-14\ 871$
1.9	$+\ .013\ 021$	$47\ 638$	$-\ .441\ 940$	$-18\ 959$
2.0	$-\ .034\ 617$		$-\ .422\ 981$	
		$43\ 285$		$-22\ 469$
2.1	$-\ .077\ 902$	$38\ 927$	$-\ .400\ 512$	$-25\ 437$
2.2	$-\ .116\ 839$	$34\ 582$	$-\ .375\ 075$	$-27\ 899$
2.3	$-\ .151\ 411$	$30\ 278$	$-\ .347\ 176$	$-29\ 884$
2.4	$-\ .181\ 689$	$26\ 035$	$-\ .317\ 292$	$-31\ 421$
2.5	$-\ .207\ 724$		$-\ .285\ 871$	
		$21\ 874$		$-32\ 534$
2.6	$-\ .229\ 598$	$17\ 818$	$-\ .253\ 337$	$-33\ 252$
2.7	$-\ .247\ 416$	$13\ 884$	$-\ .220\ 085$	$-33\ 597$
2.8	$-\ .261\ 300$	$10\ 094$	$-\ .186\ 488$	$-33\ 593$
2.9	$-\ .271\ 394$	$6\ 462$	$-\ .152\ 895$	$-33\ 265$
3.0	$-\ .277\ 856$		$-\ .119\ 630$	
		$+\ 3\ 007$		$-32\ 638$
3.1	$-\ .280\ 863$	$-\ \ \ 258$	$-\ .086\ 992$	$-31\ 735$
3.2	$-\ .280\ 605$	$-\ 3\ 319$	$-\ .055\ 257$	$-30\ 579$
3.3	$-\ .277\ 284$	$-\ 6\ 166$	$-\ .024\ 678$	$-29\ 196$
3.4	$-\ .271\ 118$	$-\ 8\ 789$	$+\ .004\ 518$	$-27\ 610$
3.5	$-\ .262\ 329$		$.032\ 129$	
		$-11\ 177$		$-25\ 846$
3.6	$-\ .251\ 152$	$-13\ 323$	$.057\ 974$	$-23\ 927$
3.7	$-\ .237\ 825$	$-15\ 231$	$.081\ 901$	$-21\ 877$
3.8	$-\ .222\ 594$	$-16\ 889$	$.103\ 778$	$-19\ 721$
3.9	$-\ .205\ 705$	$-18\ 298$	$.123\ 499$	$-17\ 483$
4.0	$-\ .187\ 407$		$.140\ 982$	
		$-19\ 460$		$-15\ 183$
4.1	$-\ .167\ 947$	$-20\ 375$	$.156\ 165$	$-12\ 848$
4.2	$-\ .147\ 572$	$-21\ 048$	$.169\ 013$	$-10\ 497$
4.3	$-\ .126\ 524$	$-21\ 486$	$.179\ 510$	$-\ 8\ 150$
4.4	$-\ .105\ 038$	$-21\ 694$	$.187\ 660$	$-\ 5\ 831$
4.5	$-\ .083\ 344$		$.193\ 491$	
		$-21\ 680$		$-\ 3\ 556$
4.6	$-\ .061\ 664$	$-21\ 455$	$.197\ 047$	$-\ 1\ 344$
4.7	$-\ .040\ 209$	$-21\ 030$	$.198\ 391$	$+\ \ \ 787$
4.8	$-\ .019\ 179$	$-20\ 416$	$.197\ 604$	$2\ 824$
4.9	$+\ .001\ 237$	$-19\ 628$	$.194\ 780$	$+\ 4\ 750$
5.0	$+0.020\ 865$		$+0.190\ 030$	

TABLE IV
expl x and expl $(-x)$ from 0.0 to 5.0

x	expl x	Δ	expl $(-x)$	Δ
.0	$+$ ∞	. . . .	∞	. . . .
.1	$+$ 1.622 813		$+$1.822 924	
.2	$+$ .821 761		1.222 651	
.3	$+$ .302 669		.905 677	
.4	$-$ .104 765		.702 380	
.5	$-$ .454 220	. . . .	.559 774	. . . .
.6	$-$ .769 881		.454 380	
.7	$-$ 1.064 907		.373 769	
.8	$-$ 1.347 397		.310 597	
.9	$-$ 1.622 812		.260 184	
1.0	$-$ 1.895 118	. . . .	.219 384	. . . .
1.1	$-$ 2.167 378	272 260	.185 991	33 393
1.2	$-$ 2.442 092	274 714	.158 408	27 583
1.3	$-$ 2.721 399	279 306	.135 451	22 957
1.4	$-$ 3.007 207	285 809	.116 219	19 232
1.5	$-$ 3.301 285	294 078	.100 020	16 199
1.6	$-$ 3.605 320	304 034	.086 308	13 712
1.7	$-$ 3.920 963	315 643	.074 655	11 653
1.8	$-$ 4.249 868	328 904	.064 713	9 942
1.9	$-$ 4.593 714	343 846	.056 204	8 509
2.0	$-$ 4.954 234	360 521	.048 901	7 303
2.1	$-$ 5.333 235	379 001	.042 614	6 287
2.2	$-$ 5.732 615	399 379	.037 191	5 423
2.3	$-$ 6.154 381	421 766	.032 502	4 689
2.4	$-$ 6.600 670	446 289	.028 440	4 062
2.5	$-$ 7.073 766	473 096	.024 915	3 525
2.6	$-$ 7.576 115	502 349	.021 850	3 065
2.7	$-$ 8.110 347	534 233	.019 182	2 668
2.8	$-$ 8.679 298	568 950	.016 855	2 327
2.9	$-$ 9.286 024	606 726	.014 824	2 031
3.0	$-$ 9.933 833	647 808	.013 048	1 776
3.1	$-$10.626 300	692 468	.011 494	1 554
3.2	$-$11.367 303	741 002	.010 133	1 361
3.3	$-$12 161 041	793 739	.008 939	1 194
3.4	$-$13.012 075	851 034	.007 891	1 048
3.5	$-$13 925 354	913 279	.006 970	921
3.6	$-$14.906 254	980 900	.006 160	810
3.7	$-$15.960 619	1.054 365	.005 448	712
3.8	$-$17.094 802	1.134 183	.004 820	628
3.9	$-$18.315 714	1.220 912	.004 267	553
4.0	$-$19.630 874	1.315 160	.003 779	488
4.1	$-$21.048 467	1.417 592	.003 349	430
4.2	$-$22.577 401	1.528 934	.002 969	380
4.3	$-$24.227 380	1.649 979	.002 633	336
4.4	$-$26.008 973	1.781 593	.002 336	297
4.5	$-$27.933 697	1.924 723	.002 073	263
4.6	$-$30.014 099	2.080 403	.001 841	232
4.7	$-$32.263 860	2.249 760	.001 635	206
4.8	$-$34.697 890	2.434 030	.001 453	182
4.9	$-$37.332 451	2.634 561	.001 291	162
5.0	$-$40.185 275	2.852 825	$+$0.001 148	143

TABLE V

Col x, sil x, expl x, and expl $(-x)$ from 0 to 15

x	col x	sil x	expl x	expl $(-x)$	x	$x°$
0	$+1.570796$	∞	∞	∞	0	0
1	$+ .624713$	$- .337404$	$- \quad 1.895118$	$+ .219384$	1	57.2958
2	$- .034617$	$- .422981$	$- \quad 4.954234$	$.048901$	2	114.5916
3	$- .277856$	$- .119630$	$- \quad 9.933833$	$.013048$	3	171.8874
4	$- .187407$	$+ .140982$	$- \quad 19.630874$	$.003779$	4	229.1832
5	$+ .020865$	$+ .190093$	$- \quad 40.185275$	$.001148$	5	286.479
6	$+ .146109$	$+ .068057$	$- \quad 85.990$	$.000360082$	6	343.775
7	$+ .116200$	$- .076695$	$- \quad 191.505$	$.000115482$	7	401.071
8	$- .003391$	$- .122434$	$- \quad 440.380$	$.000\ 37666$	8	458.366
9	$- .094244$	$- .055348$	$- \quad 1037.878$	$.000012447$	9	515.662
10	$- .087551$	$+ .045456$	$- \quad 2492.229$	$.000004157$	10	572.958
11	$- .007511$	$+ .089563$	$- \quad 6071.406$	$.000001400$	11	630.254
12	$+ .065825$	$+ .049780$	$- 14959$	$.000000475$	12	687.550
13	$+ .071435$	$- .026764$	$- 37198$	$.000000162$	13	744.846
14	$+ .014585$	$- .069396$	$- 93193$	$.000000056$	14	802.142
15	$- .047398$	$- .046279$	-234956	$+ .000000019$	15	859.438

TABLE **VI**

col x and sil x

x	co. x	sil x	x	col x	sil x
0	$+1.570\ 796$	∞	150	$+.004\ 6\mathrm{?}0$	$+.004\ 79\mathrm{?}$
5	$+\ .020\ 865$	$+.190\ 093$	160	$-\ .006\ 089$	$-.001\ 409$
10	$-\ .087\ 551$	$+.045\ 456$	170	$+.005\ 529$	$-.002\ 006$
15	$-\ .047\ 398$	$-.046\ 279$	180	$-.003\ 349$	$+.004\ 432$
20	$+\ .022\ 555$	$-.044\ 420$	190	$+.000\ 377$	$-.005\ 249$
25	$+\ .039\ 314$	$+.006\ 849$	200	$+.002\ 414$	$+.004\ 378$
30	$+\ .004\ 040$	$+.033\ 032$	300	$-.000\ 085$	$+.003\ 332$
35	$-\ .026\ 126$	$+.011\ 480$	400	$-.001\ 319$	$+.002\ 124$
40	$-\ .016\ 189$	$-.019\ 020$	500	$-.001\ 770$	$+.000\ 932$
45	$+\ .012\ 081$	$-.018\ 632$	600	$-.001\ 665$	$-.000\ 076$
			700	$-.001\ 198$	$-.000\ 779$
50	$+\ .019\ 179$	$+.005\ 628$	800	$-.000\ 559$	$-.001\ 118$
			900	$+.000\ 075$	$-.001\ 109$
55	$+\ .000\ 072$	$+.018\ 173$			
60	$-\ .015\ 949$	$+.004\ 813$	1 000	$+.000\ 563$	$-.000\ 826$
65	$-\ .006\ 675$	$-.012\ 847$			
70	$+\ .009\ 201$	$-.010\ 922$	2 000	$-.000\ 183$	$-.000\ 465$
75	$+\ .012\ 217$	$+.005\ 332$	3 000	$-.000\ 325$	$-.000\ 073$
80	$-\ .001\ 535$	$+.012\ 402$	4 000	$-.000\ 182$	$+.000\ 171$
85	$-\ .011\ 602$	$+.001\ 935$	5 000	$+.000\ 031$	$+.000\ 198$
90	$-\ .004\ 867$	$-.009\ 986$	6 000	$+.000\ 151$	$+.000\ 071$
95	$+\ .007\ 760$	$-.007\ 110$	7 000	$+.000\ 123$	$-.000\ 072$
			8 000	$+.000\ 008$	$-.000\ 125$
100	$+\ .008\ 571$	$+.005\ 149$	9 000	$-.000\ 088$	$-.000\ 068$
110	$-\ .009\ 084$	$+.000\ 320$	10 000	$-.000\ 095$	$+.000\ 031$
120	$+\ .007\ 824$	$-.004\ 781$			
130	$-\ .002\ 880$	$+.007\ 132$	11 000	$-.000\ 026$	$+.000\ 087$
140	$-\ .001\ 363$	$-.007\ 011$	100 000	$-.000\ 010$	$-.000\ 000$
			1 000 000	$+.000\ 001$	$+.000\ 000$
150	$+\ .004\ 630$	$+.004\ 796$			

TABLE VII

Maxima and Minima of $\operatorname{col}\frac{\pi}{2}x$ and $\operatorname{sil}\frac{\pi}{2}x$

x	$\operatorname{col}\dfrac{\pi}{2}x$	$\dfrac{2}{\pi x}-\operatorname{col}\dfrac{\pi}{2}x$	x	$\operatorname{sil}\dfrac{\pi}{2}x$	$\dfrac{2}{\pi x}-\operatorname{sil}\dfrac{\pi}{2}x$
0	+1.570 796		1	− .472 001	
2	− .281 141		3	+ .198 408	
4	+ .152 645		5	− .123 772	
6	− .103 966	$x(-1)^{\frac{x}{2}}$	7	+ .089 564	$x(-1)^{\frac{x+1}{2}}$
8	+ .078 635		9	− .070 065	
10	− .063 168	494	11	+ .057 501	374
12	+ .052 762	290	13	− .048 742	229
14	− .045 289	184	15	+ .042 292	149
16	+ .039 665	123	17	− .037 345	103
18	− .035 281	94	19	+ .033 432	74
20	+ .031 767	64	21	− .030 260	55
22	− .028 889	48	23	+ .027 637	42
24	+ .026 489	37	25	− .025 432	33
26	− .024 456	29	27	+ .023 552	26
28	+ .022 713	23	29	− .021 931	21
30	− .021 202	19	31	+ .020 519	17
32	+ .019 879	15	33	− .019 277	14
34	− .018 711	13	35	+ .018 177	12
36	+ .017 673	11	37	− .017 196	10
38	− .016 744	10	39	+ .016 315	9
40	+ .015 907	9	41	− .015 520	8
42	− .015 151	8	43	+ .014 799	7
44	+ .014 462	7	45	− .014 141	6
46	− .013 834	6	47	+ .013 540	5
48	+ .013 258	5	49	− .012 988	4
50	− .012 728	4	51	+ .012 480	3
52	+ .012 239	3	53	− .012 008	3
54	− .011 786	3	55	+ .011 572	3
56	+ .011 365	3	57	− .011 166	3
58	− .010 974	2	59	+ .010 788	2
60	+ .010 608	2	61	− .010 434	2
62	− .010 266	2	63	+ .010 103	2
64	+ .009 945	2	65	− .009 792	2
66	− .009 644	2	67	+ .009 500	2
68	+ .009 360	2	69	− .009 225	2
70	− .009 093	2	71	+ .008 965	1
72	+ .008 841	1	73	− .008 719	1
74	− .008 602	1	75	+ .008 487	1
76	+ .008 375	1	77	− .008 267	1
78	− .008 161	1	79	+ .008 057	1
80	+ .007 957	1	81	− .007 858	0

$x>80$

$$\operatorname{col}\frac{\pi}{2}x=(-1)^{\frac{x}{2}}\frac{2}{\pi x}$$

$x>79$

$$\operatorname{sil}\frac{\pi}{2}x=(-1)^{\frac{x+1}{2}}\frac{2}{\pi x}$$

INDEX

A

B

C

www.ingramcontent.com/pod-product-compliance
Lightning Source LLC
Chambersburg PA
CBHW031407180326
41458CB00043B/6648/J